Semiconductor Photovoltaic Cells

Chunfu Zhang · Jincheng Zhang · Xiaohua Ma ·
Qian Feng

Semiconductor Photovoltaic Cells

西安电子科技大学出版社
http://www.xduph.com

Springer

Chunfu Zhang
Xidian University
Xi'an, Shaanxi, China

Jincheng Zhang
Xidian University
Xi'an, Shaanxi, China

Xiaohua Ma
Xidian University
Xi'an, Shaanxi, China

Qian Feng
Xidian University
Xi'an, Shaanxi, China

ISBN 978-981-15-9479-3 ISBN 978-981-15-9480-9 (eBook)
https://doi.org/10.1007/978-981-15-9480-9

Jointly published with Xidian University Press
The print edition is not for sale in China (Mainland). Customers from China (Mainland) please order the print book from: Xidian University Press.

This Springer imprint is published by the registered company Springer Nature Singapore Pte Ltd.
The registered company address is: 152 Beach Road, #21-01/04 Gateway East, Singapore 189721, Singapore

Foreword

With the continuous development of society, people consume more and more energy. Among various forms of energy, electrical energy is undoubtedly the most widely used energy, and it is used in almost all aspects of human life. As people continue to pay attention to environmental protection, how to use clean and pollution-free methods to obtain electrical energy has become an important issue in today's society.

Unlike coal, oil, natural gas, and other limited and non-renewable resources, solar energy is inexhaustible, clean and pollution-free, and is an ideal source of energy. With the advent of solar photovoltaic devices, humans have obtained a method of directly generating high-quality energy from solar energy. In recent years, semiconductor solar photovoltaic devices have been used in many aspects of our lives from solar power stations to street lamp photovoltaic lighting systems. In the future, semiconductor solar photovoltaic devices will play an increasingly important role.

There are many types of semiconductor solar photovoltaic devices. This book introduces the most common and mature semiconductor solar photovoltaic devices in detail, including crystalline silicon solar cells, thin-film solar cells based on amorphous silicon, III-V single-junction and multi-junction solar cells, $Cu(InGa)Se_2$ solar cells, CdTe solar cells, dye-sensitized cells, and organic solar cells. These semiconductor solar photovoltaic devices are representative. Mastering the structure, principle, and manufacturing process of these devices, it is not a problem to understand other types of semiconductor solar photovoltaic devices.

In the process of writing this book, the author not only attached importance to the mutual connection between the chapters, but also properly maintained the independence of each chapter. Chapters 1 and 2 are introductions to the basic theory of this book. Readers should read these two chapters before reading the following chapters. Chapters 3 and 4 introduce various types of semiconductor solar photovoltaic devices. These chapters can be selected and read as needed without affecting the understanding of the chapter content. Chapter 5 further discusses the theoretical efficiency of semiconductor solar photovoltaic devices and ways to achieve maximum efficiency on the basis of the previous chapters, so that readers have a complete concept for the further development of semiconductor solar photovoltaic devices.

This book can be used by senior undergraduates and graduate students of semiconductor majors. It also has important reference value for researchers and engineers engaged in the research and production of semiconductor solar photovoltaic devices. Scientists and technicians in the fields of materials, energy, and information, and other related people interested in semiconductor photovoltaic devices can also obtain relevant knowledge of solar photovoltaic cells from this book.

During the preparation of this book, Wide-Band Gap Semiconductor Technology National Key Lab of Xidian University gave important support, and I sincerely thank you here. In addition, Dr. He Xi, Dr. Dazheng chen and the graduate students such as Aixue Su, Xueyi Li, and so on have also done a lot of work for this book, and I thank them here.

Due to the limited level of authors, there are inevitably some shortcomings in the book. I sincerely hope that the majority of readers will criticize and correct it.

Chunfu Zhang
Xidian Univerisity, Xi'an City, China

Jincheng Zhang
Xidian Univerisity, Xi'an City, China

Xiaohua Ma
Xidian Univerisity, Xi'an City, China

Qian Feng
Xidian Univerisity, Xi'an City, China

Brief Introduction

With the increasing demand for energy and the continuous emphasis on the living environment, people are gradually turning their attention to solar energy with extremely rich reserves that are clean and pollution-free. Therefore, semiconductor photovoltaic devices are increasingly valued by people. This book describes the basic working principle of semiconductor photovoltaic devices, discusses mainstream silicon-based III-V group compounds and traditional thin-film solar cells, and introduces the recently rapidly developing dye-sensitized solar cells and organic solar cells. Finally, some new ideas and new methods for exceeding the existing limits and obtaining high-efficiency solar cells are summarized.

This book can be used as a reference book for senior undergraduates and graduate students of semiconductor majors, and it also has important reference value for researchers and engineers engaged in semiconductor photovoltaic solar devices. Scientists and technicians in the fields of materials, energy, and information, as well as other relevant persons interested in semiconductor photovoltaic devices, can also learn about solar photovoltaic cells from this book.

Contents

Chapter 1
Introduction

With the rapid development of economy, the demand for energy is continuously increasing. How to realize the sustainable supply of energy has become the focus of social development. If the energy supply is insufficient, the economic development will gradually slow down. Accordingly, the issue concerning energy has aroused rising attention from the public, and the environmental contamination attributed to unsustainable resource exploitation has evidently lowered the quality of human life. For this reason, people are required to seek new, clean, and renewable green energy (e.g., wind energy, tidal energy, hydraulic energy, biological energy, and solar energy). As a matter of fact, the mentioned renewable energy sources substantially originate from solar energy. Solar energy can be exploited in a wide variety of manners, which covers exploiting the thermal effect of light to convert solar radiation into heat energy, as well as drawing upon the photogenic effect to directly transform solar radiation into electrical energy. For all the mentioned ways of energy exploitation, photoelectric exploitation of solar energy has become the research hotspot over the past few years [1]. In the present chapter, the development of solar cells, the characteristics exhibited by solar cells, and the applications of solar cells are elucidated in accordance with the principle of photovoltaic effect.

1.1 Energy Consumption and Solar Energy

After the industrial revolution was initiated, the world economy has been leaping forward. Economic growth displays tight relationships with the development and exploitation of energy resources. Accordingly, to develop the economy stably and sustainably, the exploitation of energy resources should be expounded, and a foundation should be laid for addressing subsequent or existing energy resource issues

C. Zhang et al., *Semiconductor Photovoltaic Cells*,
https://doi.org/10.1007/978-981-15-9480-9_1

that may or may not have occurred. To ensure the unconstrained economic development, the novel approaches to obtain and utilize energy resources are also urgently required.

1. **World energy status and characteristics**

Existing world energy status and characteristics include the following:

(1) **The booming of the economy and the growing population has led to an ever-increasing consumption of primary energy globally**

As the world economy expands ongoingly, global energy consumption continues to grow. In 1990, the GDP of all countries globally reached $26.5 trillion (as suggested by the constant price in 1995), and reached $34.3 trillion in 2000, a data even exceeding $72.6 trillion in 2012. In particular, China's economy has witnessed more rapid development. In 2012, its GDP reached over $8.3 trillion, ranking second worldwide [2]. In the wake of the economic growth, energy resources have been progressively demanded. As revealed from the statistics of the International Energy Agency (IEA), Organization for Economic Cooperation and Development (OECD), and World Bank, the average energy consumption was up-regulated from 19.4 to 21.3 MWH with the world population has risen from 5.26 billion in 1990 to 6.68 billion in 2008; as a result, the energy consumption increased from 102.3 MWH in 1990 to 142.3 MWH in 2008 at an increased ratio of 39% [3].

(2) **Energy consumption in different countries shows different growth patterns, and the growth rates of developing countries are significantly higher than those of developed countries**

Over the past two decades, the overall energy consumption has risen in the United States, the European Union, the Middle East, China, Latin America, Africa, and India, whereas the growth rates of North America and Europe, with the relatively developed economy, technology, and society, have been relatively slow. From 1990 to 2008, U.S. energy consumption showed an increase of only 20%, while EU growth had been even lower at only 7%. In contrast to the mentioned developed regions, energy consumption in developing regions has suggested a rapid increase. From 1990 to 2008, the Middle East, China, India, Africa, and Latin America had achieved increased energy consumption by 170%, 146%, 91%, 70%, and 66%, respectively. The proportion of energy consumption in developed countries to the global overall consumption has suggested a year-on-year decrease. The major reasons include the following: first, the economic development of developed countries has ushered into the post-industrialization stage, the economy is shifting toward the industrial structure with low energy consumption and high output, and the manufacturing industry with high energy consumption is progressively migrating to the developing countries; second, developed countries stress-energy conservation and make efforts to elevate energy efficiency.

(3) **Global energy consumption structure tends to exhibit high quality, whereas regional differences are still great**

Since the industrial revolution of the 1870s, the consumption of fossil fuel has witnessed a dramatic increase. In the preliminary stage, coal acted as the major source. Since the beginning of the twentieth century, especially after the Second World War, the production and consumption of oil and natural gases have been rising ongoingly. In the 1960s, oil surpassed coal for the first time and took up the dominant source of primary energy. Though the world has experienced two oil crises in the 1970s, the consumption of oil has not decreased on any occasion. Ever since then, the proportion of oil and coal has seen a slow decline, while the proportion of natural gas has risen. Meantime, other forms of new energy (e.g., nuclear energy, wind energy, water power, and geothermal energy) have been progressively developed and exploited; as a result, the existing energy structure has been formed, as dominated by fossil fuels with renewable and new energy coexisting. By the end of 2008, fossil energy remained major energy worldwide. Among the global energy consumption, oil, coal, and natural gas take up 33.5%, 26.8%, and 20.9%, respectively. Though non-fossil and renewable energy are booming, they still maintain a low proportion, with nuclear energy, water resources, and other renewable energy taking up 5.8%, 2.2%, and 10.6%, respectively.

As impacted by the abundant quantity and low cost of exploitation for oil and gas resources in the Middle East, the energy consumed there is primarily oil and natural gas, and its proportion, ranking first worldwide, is evidently higher than the world average. In the Asia Pacific region, countries (e.g., China and India) have sufficient reserves of coal, and coal accounts for a high proportion of energy consumption. Thus, the ratio of oil and natural gases is low in the energy structure of the Asia Pacific region.

(4) **Energy resources are relatively abundant worldwide, whereas energy trade and transportation are subject to growing pressure**

As revealed from the BP world energy statistics in 2013, by the end of 2012, the residual proven oil reserves worldwide were 166.89 billion barrels, 48.4, 13.2, 19.7, 8.4, 7.8, and 2.5% of which were in the Middle East, North America, Central, and South America, Europe and Eurasia, Africa, as well as the Asia Pacific region, respectively. The comparison of the oil production and consumption in different regions suggests that the Middle East should export large quantities of oil, and the oil production is greater than the consumption in Africa and Central and South America; note that a huge gap is identified between the consumption and production in the Asia Pacific, North America, and Europe.

Moreover, a huge imbalance remains in the distribution of coal resources. By the end of 2012, the global remaining recoverable reserves of coal reached 860.938 billion tons. Europe and Eurasia, North America, and the Asia Pacific are considered as the major distribution areas of coal, taking up nearly 94.8% of the global total quantity. Besides, the remaining recoverable reserves of natural gas have been 187.3

trillion cubic meters thus far. The Middle East and Europe exhibited the most abundant natural gas resources taking up approximately 74.2% of the global total amount, while other regions achieved a lower share. As energy resources are being consumed on earth, energy trade between regions and countries will be further promoted, and energy transportation will be increasingly demanded. Accordingly, issues (e.g., energy storage and transportation facilities and energy supply security) will be of rising significance.

(5) **Renewable energy, especially solar energy reserves, is abundant, whereas development and exploitation are relatively insufficient**

Renewable resources refer to energy originating from nature (e.g., solar energy, wind power, tidal energy, and geothermal energy), which are inexhaustible energies. It pertains to a type of energy relative to the non-renewable energy that tends to be exhausted. Solar energy alone can satisfy the energy needs 2850 times globally. Besides, wind energy, hydropower, biomass energy, and geothermal energy can meet 200 times, 3 times, 20 times, and 5 times the global energy needs, respectively. However, the actual exploitation of renewable energy is far below its potential to be developed. To be specific, in 2008, 19% of the global energy demand originated from renewable energy sources, of which 13% was conventional bioenergy mostly adopted for thermal energy (e.g., firewood), 3.2% was from water power, whereas only 2.7% was from emerging renewable energy sources (less than 20 MW of water power, modern biomass energy, wind energy, solar energy, and geothermal energy) [4].

2. **China's energy consumption status**

China is known as a country with considerable energy production and consumption, with abundant fossil energy resources. In 2006, the coal reserves were 1,034.5 billion tons in amount, of which the available reserves took up 13% of the whole world, ranking third worldwide. It is noteworthy that per capita energy resources are relatively low with coal and hydropower resources taking up only 50% of the world average level, and oil and natural gases reach about 1/15 of the world average level. The imbalance of energy resources is difficult to develop, and the proven reserves of high-quality energy (e.g., oil and natural gases) seriously lack. Given the technological backwardness and the low exploitation rate of energy resources, energy consumption in China is much faster than in other countries to satisfy the demands of high-speed economic growth; it is therefore revealed that energy exhaustion will occur earlier and is more serious than expected. The expected exploitable time of major conventional energy reserves of the world and China are listed in Table 1.1. As a result, in the context of the growing demand for energy resources, a way to break through the energy crisis should be urgently developed.

China is considered as one of the few countries taking coal as the major energy source. Approximately 65% of disposable energy is produced and consumed by coal. The wide use of coal resources has induced serious -environment issues; 66% of the atmospheric particulate matter content in Chinese cities and 22% of the urban air

Table 1.1 The expected exploitable time of major conventional energy reserves of the World and China in the future (year)

Types of energy	Solar	Coal	Oil	Natural gas	Uranium
World	Infinity	230	50	60	70
China	Infinity	100	40	40	50

sulfur dioxide content have not complied with National Air Quality Secondary Standard. Such coal-based energy structures and single energy consumption models have long caused severe environmental contamination. As the economy is booming, and energy demand is growing sustainably, greenhouse gas emissions from fossil fuel combustion have placed rising pressure on the environment. Under the serious environmental crisis caused by fossil energy consumption, the energy structure should be urgently adjusted [5]. Note that over the past few years, China has been covered with a large area of haze, and PM 2.5 content remains noticeably high in numerous regions; thus, the energy consumption should be regulated.

3. **Development and exploitation of solar energy resources**

The rising energy shortage and environmental contamination force us to seek new clean energy. New energy should meet the two conditions: First, it is abundant and will not be exhausted; second, it is safe and clean, not adversely affecting human health and environmental safety. At present, new energy primarily consists of solar energy, biological energy, geothermal energy, tidal energy, wind energy, nuclear energy, and hydrogen energy, among which the most ideal one tends to be solar energy. On the whole, solar Energy refers to the radiation energy of sunlight. Living creatures on earth rely primarily on the heat and light provided by the sun to survive. Since ancient times, humans know how to dry things in the sunshine as a way to preserve food (e.g., salt making and salted fish). However, it was not until the decrease in fossil fuels did human beings begin to be aware of further developing solar energy [6].

The exploitation of solar energy can fall into the following three types:

(1) Indirect use of solar energy: fossil energy and biomass energy (light energy converted into chemical energy);
(2) Direct use of solar energy: collectors (flat-plate collectors and concentrating collectors) (light energy converted into heat energy);
(3) Solar cells: generally employed in power stations, satellites, spacecraft, lighters, watches, etc. Recently, they have also become one of the vital sources of power energy in China (light energy converted into electrical energy).

The reason why we choose solar energy as the major energy resource in the future is primarily that the emerging clean and renewable solar energy exhibits unique advantages:

(1) Abundance: The energy emitted by the sun per second is equated with 1.6×10^{23} KW, of which only a tiny fraction (nearly one in 2.2 billion) reaches the Earth. Even so, the energy of the solar radiation reaching the earth's surface

per second is as high as 8×10^{13} KW, equaling 6×10^9 t SCE. Given such calculation, the total amount of solar energy reaching the ground in one year is converted into SCE, i.e., nearly 1.892×10^{16} t. Berko, a German solar expert, stated that by simply exploiting the solar power in parts of Africa, the electricity needs can be fulfilled worldwide. Furthermore, solar radiation is inexhaustible and can be continuously supplied to the earth.

(2) Universality: The distribution of solar energy is not identical to other energy resources that tend to be region-restricted. Numerous countries worldwide should spend huge transmission equipment or transportation costs for the imbalance of energy distribution. However, solar energy that can be smoothly exploited is conducive to alleviating the contradiction between energy supply and demand, releasing the pressure of energy transportation, and tackling down the energy supply issues of the remote area and island area under inconvenient transportation.

(3) Non-pollution: Human beings have been increasingly aware that environmental protection is vital to the sustainable development of the whole planet, suggesting that environment and development cannot be separated from each other. Among considerable environmental issues, the pollution caused by fossil fuel combustion is considered the most serious. However, the use of solar energy, producing no waste residue, waste gas, water discharging, noise, and harmful substances, is particularly valuable under the increasing environmental contamination in modern times.

(4) Economy: The costs of a power station are primarily composed of construction costs, fuel costs, operation management maintenance, and environmental protection investment. By comparison, solar energy that is pollution-free, inexhaustible, and ubiquitous will require significantly fewer costs for electricity generation in the long run. A consensus is reached in prediction and researches by experts that solar energy will act as the cleanest and cheapest energy for humans in the twenty-first century.

Indeed, solar energy exploitation also has its shortcomings (e.g., low energy density, difficulty to collect, instability, and varying with climate and weather); currently, the cost of power generation is relatively high. On the whole, however, solar energy exhibits superiority over other energy sources, and it has an extremely broad prospect [7].

China with its vast territory mostly situated in a low and middle latitude where the solar altitude angle is relatively large has been sufficiently radiated from the sun. It therefore becomes extremely abundant in solar energy, as presented in Fig. 1.1.

The annual total solar radiation in China reaches nearly 3300–8300 MJ/m^2. Given the difference in solar radiation and the characteristics of solar energy exploitation, the whole country can be split into four regional types.

(1) The most abundant areas: the Qinghai-Tibet Plateau and its marginal areas. The region's most significant features include high terrain, thin air, low moisture and dust content, high transparency, as well as significantly higher radiation levels than those of plain areas of the same latitude. Moreover, the mentioned areas

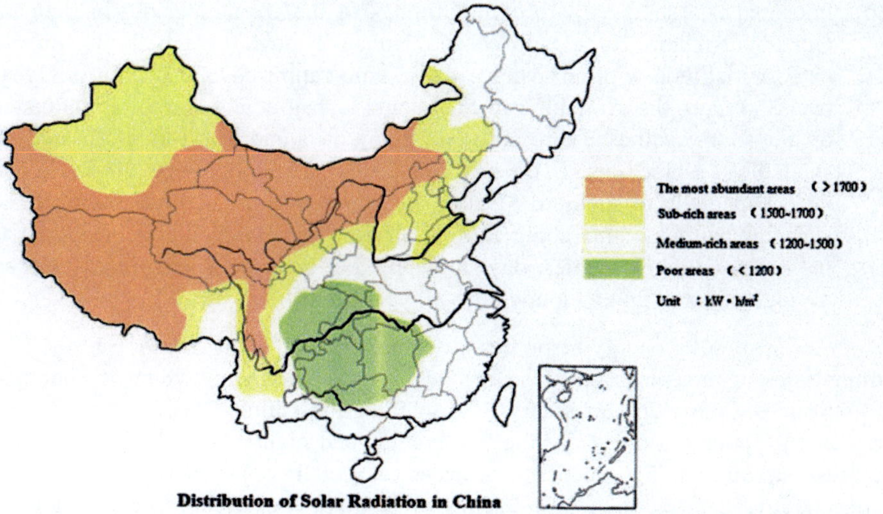

Fig. 1.1 Distribution of solar radiation in China (Red: The most abundant areas; Yellow: Sub-rich areas; White: Medium-rich areas; Green: Poor areas)

have less rainfall that mostly happens at night. The relatively longer sunshine hours generally range 2800–3200 h. On the whole, this place is extremely conducive to the exploitation of solar energy.

(2) Sub-rich areas: Xinjiang, most of Ningxia, Gansu, western Inner Mongolia, as well as Northern Shaanxi and Shanxi. The mentioned areas are mostly located in the arid area of Northwest China, exhibiting less cloudiness, more solar radiation, and the longest sunshine hours in the nation. On the whole, the annual sunshine time in the mentioned areas is beyond 3200 h, and the northern part of Xinjiang can last up to 3,500 h, which extremely facilitates the exploitation of solar energy. Even though the low-temperature wintertime is relatively long with a time of average temperature reaching over 0 °C from 100 to 150 days, the relatively high temperature in noon caused by the large diurnal temperature difference can lay favorable conditions for solar energy exploitation. More sandstorm days and higher wind speeds act as adverse factors in the region.

(3) Medium-rich areas: the northern parts of the Qinling-Huaihe River, the southern parts of Fujian and Guangdong, Taiwan, and most of Hainan. The advantages of exploiting solar in the northern region are that there are more sunny days, fewer clouds, and longer sunshine time generally beyond 2400 h a year. However, the wintertime with low temperatures and weak solar radiation generally takes up 3–5 months of the year, thereby creating the most difficulties to exploit solar in the area. Moreover, the northern region is generally windy, especially in spring and autumn, more significantly impacting the thermal efficiency of solar energy devices.

(4) Poor areas: the southern parts of the Qinling-Huaihe River, especially in Sichuan, Guizhou, southern Hubei, and Western Hunan. Though these areas are situated in a low-latitude with an average daily temperature beyond 0°C in one year over 350 days, the available sunshine hours is insufficient due to the influence of monsoon weather. There are many rainy days and heavy rainfall, and the available sunshine hours in the mentioned areas are no more than 2000 h in one year. Especially in southern Sichuan and northeastern Guizhou, the available sunshine hours are only about 1100 h. Sichuan and Guizhou are considered as the areas with the cloudiest days in China. For instance, Chengdu city has an average of about 244.6 cloudy days all year round.

In terms of solar energy resources in China, the total solar radiation in the first three types of areas is close to or reaches over 5500 MJ/m^2, which is abundant in resources, taking up about 2/3 of the total area in China. Even in the southern region with poor resources, for the low latitude and high solar radiation intensity, a great exploitation significance is revealed, especially solar energy devices with concentrating devices, so their effectiveness can be fully exerted. From a worldwide perspective, the use of solar energy is better in Europe, Japan, and other nations. The latitude of most parts of Europe is higher than that of China, and the latitude position of Japan is similar to that of North China; however, the climate of the mentioned areas is remarkably affected by the oceans, so these areas exhibit solar energy far less than that of China. The latitude of the United States is comparable to that of China, and the solar radiation is abundant as well, especially in the desert areas of the south and west. However, the unique solar energy resources (e.g., the Qinghai-Tibet region in China) are rare worldwide. The advantages of resources lay solid preconditions for the exploitation of solar energy in China [7].

1.2 Photovoltaic Effect and Its Application

In 1839, French scientist A.E. Becquerel initially reported the "Photovoltaic Effect". In brief, the so-called photovoltaic effect is that the distribution of charges in an object will vary when the object is illuminated, so an effect of electromotive force and current is exerted. First, it refers to an energy converting process, converting photons (light waves) into electrons, as well as transforming photon energy into electron energy. Second, it acts as the process to form voltage. Such effect can occur in gases, liquids, and solids; in solids, especially semiconductors, however, note that light energy is efficiently converted into electricity. Thus, the photovoltaic effect in semiconductors has aroused increasing attention.

Given the wave-particle duality of light, in the study on solar cells, light waves can be treated as particles, and the unit energy, proportional to the vibration frequency, is generally referred to as photons. The energy of the photons can be given by $E = h\upsilon$, where E denotes photon energy, υ is photon oscillation, and $\omega = 2\pi\upsilon$ represents the angular frequency. To ensure the effective absorption of photons, the energy of photons should satisfy

$$h\nu \geq E_g = h\nu_0 \tag{1.1}$$

where E_g denotes the value of the bandgap. Photon energy $h\nu_0$, equal to E_g, represents the least energy that may cause electron-hole pairs in semiconductors, which is termed as an intrinsic absorption edge. The corresponding maximum wavelength is expressed as

$$\lambda_0 = c\frac{2\pi}{\omega_0} = \frac{c}{\nu_0} \tag{1.2}$$

where c denotes the speed of light, about 3×10^8 cm/s in vacuum. Besides, the relationships between the maximum absorption wavelength and the bandgap E_g are expressed as

$$\lambda_0 = \frac{1.24}{E_g(\text{eV})}\,(\mu\text{m}) \tag{1.3}$$

If the energy of light is less than $h\nu_0$ or the wavelength is greater than λ_0, the absorption coefficient declines evidently. The relationship of $E = h\nu$ is critical to enhancing the characteristics and efficiency of solar cells.

Under the light incident perpendicularly to the PN junction surface, if the junction depth is sufficiently shallow to meet the $d < 1/\alpha$ condition (α denotes the absorption coefficient of the material, $\alpha = 4\pi\nu\kappa/c$ expresses wavelength, and κ represents the extinction coefficient), light can reach the barrier region or deeper. The photon with energy $h\nu \geq E_g = h\nu_0$ in the incident light will generate electron-hole pairs on both sides of the junction for intrinsic absorption. Under photoexcitation, the concentration of majority carriers is nearly unchanged, while the minority carrier concentration will vary noticeably.

As presented in Fig. 1.2, for the robust built-in electric field in the barrier region (from N region to p region), the photogenerated minority carriers on both sides of the junction migrate in opposite directions. The electrons in the p region pass through the PN junction into the N region, and the holes in the N region are also injected into the p region. The potential at the p region rises, while that at the N region falls. For this reason, the photovoltaic potential is generated at both ends of the PN junction, i.e., the photovoltaic effect of the PN junction. With the photogenerated carrier moving in the opposite directions, the photocurrent I_{ph} from N region to p region is generated in the PN junction. The photogenerated carriers drift across the barrier region and accumulate to develop an electric field opposite to the direction of the thermal equilibrium electric field, while generating a forward junction current opposite to the photocurrent; then, the junction electric field can be remedied. When the photocurrent equals the forward junction current, a stable potential difference is built between the two ends of the PN junction, i.e., the photovoltage. The photo-electromotive force generated by illumination at both ends of the PN junction is

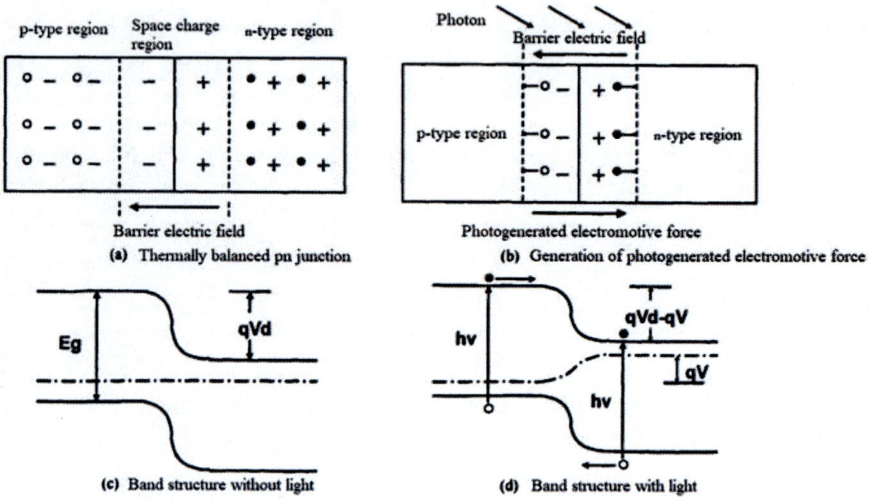

Fig. 1.2 Energy band diagram of photovoltaic effect

equated with a forward bias voltage V upon the junction, thereby reducing the barrier to $qV_D - qV$ and generating forward current I_F. In the case of the PN junction open circuit, under the equal photocurrent and forward current, a stable potential difference Voc is generated at both ends of the PN junction, acting as the open-circuit voltage of the solar cell. When connected to an external circuit, a stable flow of current will exist while being illuminated continuously, and the PN junction can act as the power supply.

As a matter of fact, photogenerated carriers do not overall contribute to the photocurrent. The photogenerated electron-hole pairs absorbed in the space charge region effectively absorb light, contributing to the photocurrent. Moreover, the diffusion distance of the holes in the N region is assumed as L_p in the lifetime τ_p, and the diffusion distance of electrons in the p region is L_n in the lifetime τ_n; thus, it yields $L_n + L_p = L$. The photogenerated carriers in the average diffusion distance L near the junction can be considered contributing to the photocurrent. For simplification, the space charge region is also covered in the length of L. Moreover, the electron-hole pairs with a position of more than L from the junction are all recombined during the diffusion process, which does not facilitate the photoelectric effect of the PN junction [8].

The exploitation of the photovoltaic effect is primarily achieved with the PN junction of the semiconductor, which converts light radiation energy into electric energy. The PN junction devices reach the highest conversion efficiency under illumination, so such photovoltaic devices are generally termed as solar cells, also known as photovoltaic cells or solar cells.

1.3 Development of Solar Cells

In 1954, Chapin, Fuller, and Pearson of Bell Laboratories in America developed the initial silicon solar cell with an efficiency of 6%. After being gradually improved, the solar cell achieved an efficiency of 10% and was equipped on Pioneer 1 satellite of the United States in 1958 for eight years. Before the 1970s, photovoltaic power was primarily employed in outer space. Thus far, photovoltaic power is still recognized as the major power source for spacecraft launched by countries, which has substantially boosted the advancement of the space industry. The energy crisis that happened in the 1970s evidently stimulated the demand for new energy and boosted the booming of solar cells. Since that, material, structure and manufacturing technology of solar cells have been greatly optimized, and the corresponding cost is continuously reduced. With the development of photovoltaic technology, the terrestrial photovoltaic applications became more and more, gradually extending this technology from space military field to ground civil field. However, due to the high price, the application market has long been restricted and obscured. It was not until 1997 did this stagnation begin to be broken. Because some developed countries rolled out the "Million Solar Roof Project", the growth rate of solar cell production reached 42% in 1997, greatly faster than average value of 12% in the previous years [9].

China's solar photovoltaic power industry was initiated in the 1970s and has ushered into a steady development stage in the mid-1990s. The output of solar cells and their components have suggested a steady year-on-year growth. After efforts of nearly four decades, the solar photovoltaic industry has been leaping forward in China. Moreover, a solid novel production chain has been established, and China's solar photovoltaic power generation has advanced as stimulated by national projects (e.g., "Brightness Program" and "Township Electrification Program" projects and other national projects) as well as the world PV market. At the end of August in 2013, the overall installed capacity of solar energy in China reached 8.98 GW [10]. A considerable number of enterprises were engaged in solar cell production, and the industrial chain composed of raw material production, and photovoltaic system construction had progressively become more stable and mature on a day-to-day basis. Note that the production of polycrystalline silicon materials has been significantly progressed, and the limitation in raw material production has been tackled down. Lastly, the large-scale development of photovoltaic power generation in China has been underpinned. The past decade can be considered witnessing the fastest growing period of China's solar photovoltaic industry.

Thus far, solar cells have developed into the second generation. In general, the first-generation solar cells are made from monocrystalline or polycrystalline silicon with the efficiency of industrial products of about 13–15%. At the current stage, solar cells that can be industrially produced and are profitable pertain to the first-generation cells. However, manufacturing technology and other factors raise high production costs. The second-generation solar cells are thin-film solar cells, costing less than that of the first generation, significantly expanding the manufacturing area.

However, the efficiency is relatively lower than that of the first generation. The third-generation solar cells should exhibit the thin film, high efficiency, abundant raw materials, and no toxicity. The available approaches to achieve the third-generation solar cells consist of tandem solar cells, multiband photovoltaic cells, hot-carrier cells, collision ionization solar cells, etc. In the chapters below, a range of solar cells is presented.

1.4 Characteristics of Solar Cells

Conventional solar cells are composed of two layers of semiconductors, one is positive (p-type) and the other is negative (n-type). The mentioned two types are close to forming a PN junction. When the semiconductor is illuminated, the incident photons with energy $h\nu$ exceeding the bandgap of the semiconductor are absorbed by the material; as a result, the electrons jump from the valence band into the conduction band to form a conductive current. Each negatively charged electron exhibits a corresponding positively charged hole. Electrons and holes near the junction are swept in opposite directions by the electric field. Besides, as impacted by the motive force of an electrode, the electrons are driven to the external circuit. In the external circuit, the mentioned electrons lose energy, and they return to the valance band of the material via the second electrode of the circuit. Conventional solar cells work abiding by the mentioned process, so some parameters or quality factors (e.g., open-circuit voltage, short-circuit current, filling factor, and photoelectric conversion efficiency) are required to characterize the performance of different solar cells. Moreover, identical terms (e.g., series resistance, photoelectric current, and quantum efficiency) are required to describe the specific working principle of solar cells.

1.4.1 Photocurrent and Quantum Efficiency

As mentioned above, when a beam of light exhibiting energy over the bandgap of the semiconductor material is incident on the PN junction, a stable potential difference, i.e., photogenerated voltage, will be generated at both ends of the PN junction. Moreover, the photovoltage refers to the voltage the solar cell may output. On open-circuit condition, the photovoltage pertains to open-circuit voltage. If the external circuit is short-circuited, the current of the external circuit will pertain to the short-circuit current, i.e., the photocurrent under ideal conditions. Short-circuit current corresponds to the intersection of the volt-ampere characteristic curve and ordinate coordinates, which is commonly expressed as I_{SC} (or short-circuit current density J_{SC}, generally, I represents current and J denotes current density). The short-circuit current I_{SC} is associated with the area of solar cells in positive correlation: larger I_{SC} for a large area. On the whole, the I_{SC} of silicon solar cells in the size of 1 cm^2 reaches 16–30 mA.

The current in solar cells is of three types (i.e., photocurrent I_{ph}, the PN junction forward current I_F under the photovoltage V, as well as the current flowing through the external circuit). Both I_{ph} and I_F will flow through the PN junction, but their flow directions are opposite.

As revealed from the rectification formula of PN junction, the forward current through the junction under the forward bias V is written as

$$I_F = I_0\left(e^{\frac{qV}{kT}} - 1\right) \tag{1.4}$$

where V denotes the photovoltaic voltage, i.e., the output voltage of solar cells; I_0 is the reverse saturated current; k represents the Boltzmann constant; T is the absolute temperature; q is the charge of electron for a solar cell in operation.

Under the solar cell irradiated with light exhibiting a certain intensity, the general intensity of light decreases exponentially with the depth of light penetration for the absorption. For this reason, the generation rate of photogenerated carriers decreases with the depth of light as well, i.e., the generation rate G expresses a function of the depth x. To gain direct insights into the photocurrent, the photogenerated current is estimated in this study. The average generation rate of non-equilibrium carriers in the diffusion length L is expressed as $\overline{G}$. This study assumes that the holes in the diffusion length L_p and the electrons in the diffusion length L_n can overall diffuse to the other side of the PN junction. Thus, the photocurrent is written as

$$I_{ph} = q\overline{G}AL \tag{1.5}$$

where A denotes the area of the PN junction. The photocurrent I_{ph} flows via the junction from the n-type region to the p-type region, which is opposite to I_F.

When the solar cell is linked to the load resistance, an ideal circuit of the solar cell will be built (as presented in Fig. 1.3a). The current via the load resistance is illustrated as follows:

$$I = I_{ph} - I_F = I_{ph} - I_0\left(e^{\frac{qV}{kT}} - 1\right) \tag{1.6}$$

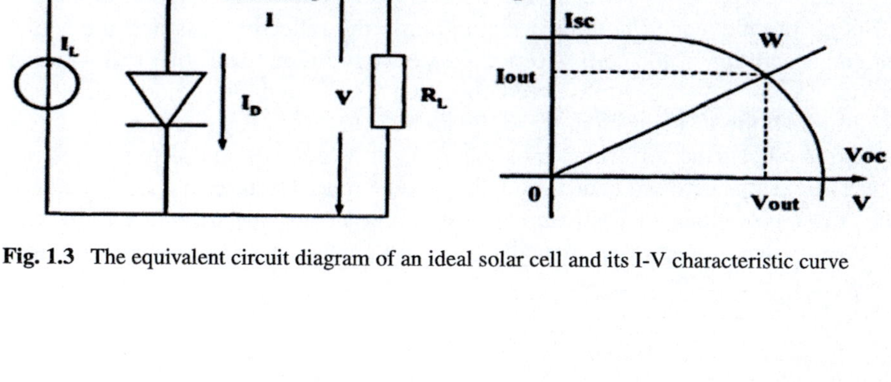

Fig. 1.3 The equivalent circuit diagram of an ideal solar cell and its I-V characteristic curve

Given the definition of short-circuit current, when $V = 0$, the corresponding current refers to the short-circuit current I_{SC}. As revealed in the above formula, I_{SC} is ideally equated with I_{ph}. However, a slight difference is identified between the two values due to a result of the non-ideal factors in actual circuits, as presented in the following chapters.

Quantum efficiency (QE) or collection efficiency is adopted to characterize the relationships between photocurrent and incident light. QE expresses the contribution of photons of different energies to short-circuit current I_{SC}. In terms of the whole incident spectrum, QE denotes the spectral response of the short-circuit. QE expresses a function of energy with two expressions. The first is external quantum efficiency (EQE), defining the probability that an incident photon of each wavelength λ can create an electron in the external circuit for the entire incident solar spectrum; it is expressed as

$$\mathrm{EQE}(\lambda) = \frac{I_{SC}(\lambda)}{qAQ(\lambda)} \tag{1.7}$$

where $Q(\lambda)$ denotes the spectral density of the incident photon; A is the area of the cells. It reflects the ratio of the density of photogenerated carriers to the density of incident photons that contributes to the short-circuit current.

The other expression of QE is the internal quantum efficiency (IQE), which is defined as the probability that an incident photon of wavelength λ absorbed by the cell can create an electron in the external circuit. IQE indicates the ratio of the number of photocarriers contributing to the short-circuit current to the number of photons absorbed by the cell.

$$\mathrm{IQE}(\lambda) = \frac{I_{SC}(\lambda)}{qA(1 - T(\lambda))(1 - R(\lambda))Q(\lambda)} \tag{1.8}$$

where $R(\lambda)$ denotes the reflectivity on the surface of the device; $T(\lambda)$ is the transmittance of light transmitted through the device.

As revealed from the comparison of the two definitions of quantum efficiency, the denominator of external quantum efficiency does not consider the loss factors of incident light reflection, material absorption, solar cell thickness, etc. Thus, EQE is usually less than 1. However, the denominator of internal quantum efficiency takes all the mentioned non-ideal factors (e.g., the reflection loss and the actual photon absorption of the cell). Accordingly, in terms of an ideal solar cell with the carrier lifetime $\tau \to \infty$, surface recombination $S \to 0$, and the sufficient thickness to absorb all the incident light, the IQE can be equal to 1. The quantum efficiency spectrum reflects the performance of solar cells from another perspective, revealing the relationships between material quality, geometrical structure, and technology of the device performance. The external quantum efficiency spectrum of the cell can be directly determined experimentally, while all the factors (e.g., reflection, optical thickness, and grid structure of cells) should be considered for the internal quantum efficiency [11].

1.4.2 Open-Circuit Voltage

According to the mentioned analysis, under an open circuit, the stable potential difference V_{OC} is generated across the PN junction when the photocurrent equals forward current, which indicates the open-circuit voltage of solar cells. In other words, the total output current is 0 pertaining to the open-circuit voltage state. Equation (1.6) reveals that the open-circuit voltage is ideal.

$$V_{OC} = \frac{kT}{q} \cdot \ln\left(\frac{I_{ph}}{I_0} + 1\right) \tag{1.9}$$

Ignoring the "+1" item, it yields

$$V_{OC} = \frac{kT}{q} \cdot \ln\left(\frac{I_{ph}}{I_0}\right) \tag{1.10}$$

It is suggested that the photovoltaic voltage rises with a decrease in reverse saturation current I_0 [12].

Based on the I-V characteristics of solar cells, under certain temperature and light irradiation, the terminal voltage of solar cells under an open circuit corresponds to the intersection of the I-V curve and abscissa coordinates, i.e., the open-circuit voltage V_{OC} [13]. A general solar cell can be approximated as an ideal solar cell, i.e., the cell exhibits zero series resistance and infinite shunt resistance. The open-circuit voltage V_{OC} of solar cells is independent of the solar cell area. On the whole, the open-circuit voltage of single-crystalline silicon cells is about 450–600 mV [9].

In actual solar cells, when the current flows via the load resistance, the loss is inevitable for the contacts on the surface and back of the solar cell and the resistivity of the material, which is expressed by a series resistance R_s in the equivalent circuit. Meantime, for leakage of the solar cell edge, metal bridge leakage takes place at the microcracks, scratches, and others. Thus, the current passing through the load is partially short-circuited. This effect can be expressed as an equivalent parallel resistance R_{sh}. The equivalent circuit at this time is illustrated in Fig. 1.4, and the output current I of the solar cell is expressed as

$$I = I_{ph} - I_0\left(e^{\frac{q(V+IR_s)}{nkT}} - 1\right) - \frac{V + IR_s}{R_{sh}} \tag{1.11}$$

In open-circuit cases, the load impedance is infinite, the output current I is zero, and the voltage at this time is referred to as open-circuit voltage. Set $I = 0$ in Eq. (1.11) above, then it yields

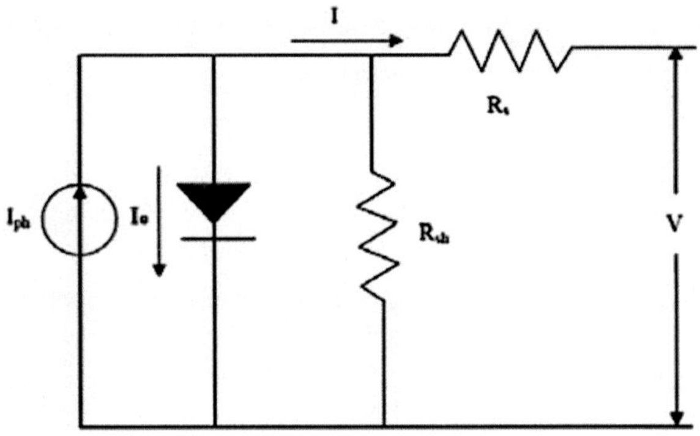

Fig. 1.4 The equivalent circuit diagram of a non-ideal diode

$$I_{\text{ph}} = I_0\left(e^{\frac{qV_{\text{OC}}}{nkT}} - 1\right) + \frac{V_{\text{OC}}}{R_{\text{sh}}} \tag{1.12}$$

$$V_{\text{OC}} = \frac{nkT}{q}\ln\left(\frac{I_{\text{ph}} - V_{\text{OC}}/R_{\text{sh}}}{I_0} + 1\right) \tag{1.13}$$

The above formula indicates that open-circuit voltage V_{OC} is not affected by series resistance R_{s}, whereas it shows relationships to shunt resistance R_{sh}. The ideal resistance of R_{sh} is infinite. It is suggested that with the decrease in R_{sh}, the V_{OC} will decrease. For the non-ideal resistance situation, the value of I_{SC} and the value of I_{ph} deviate from the ideal value.

1.4.3 Photoelectric Conversion Efficiency

The volt-ampere characteristics of ideal solar cells are presented in Fig. 1.3b. The load line in the figure is straight, of which the slope is determined by the size of the resistance. The intersection point W of the load line and volt-ampere characteristic curve acts as the working point. The power P_{out} obtained from the solar cell by the load resistance R_{L} is defined as

$$P_{\text{out}} = I_{\text{out}}V_{\text{out}} \tag{1.14}$$

which is the rectangular area in Fig. 1.3b. It is suggested that the working point W moves along the curve with the variation of the load resistance. There is a working point m, maximizing the product of the corresponding working voltage and the operating current (i.e., the maximum output):

$$P_{\text{m}} = I_{\text{m}}V_{\text{m}} \tag{1.15}$$

The corresponding load resistance is termed as the optimal load. Thus, the conversion efficiency of solar cells is defined as the ratio of output power to input power under the optimal load:

$$\eta = \frac{P_m}{AP_{in}} \times 100\% \tag{1.16}$$

where A denotes the active area of solar cells; P_{in} is the power of incident light per unit area. Moreover, the filling factor FF is defined as

$$FF = \frac{V_m I_m}{V_{OC} I_{SC}} \tag{1.17}$$

FF refers to one of the important indicators to measure the output characteristics of solar cells. For the characteristic curve under a determined open-circuit voltage and short-circuit current, the closer the filling factor is to 1, the squarer the curve will be, and then the higher the output power will be. Furthermore, the power conversion efficiency of solar cells can be expressed as

$$\eta = \frac{FF \times V_{OC} \times I_{SC}}{AP_{in}} \times 100\% \tag{1.18}$$

In actual solar cells considering the metal grid lines, the equation is modified to

$$\eta = \frac{V_m I_m}{A_t P_{in}} \times 100\% \tag{1.19}$$

where A_t denotes the total area of solar cells, which cover the metal grid lines area (also known as the whole area). Sometimes, active area A is adopted instead of A_t, i.e., the area occupied by the metal grid lines is removed from the total area; thus, the calculated efficiency is generally higher [9]. Photoelectric conversion efficiency critically determines the quality and technical level of solar cells, which displays relationships to the structural characteristics, material characteristics, working temperature, environmental temperature changes, etc. At constant temperature, the conversion efficiency increases with the increase in light intensity. For a given power output, the conversion efficiency of the solar cell determines the number of panels required; thus, it is extremely vital for solar cells to achieve the highest conversion efficiency. The principle can be complied with to enhance the conversion efficiency: the light intensity can be enhanced with a concentrator, as an attempt to down-regulate the number of the cells and lower the cost of power generation [14].

1.4.4 Series Resistance

Series resistance primarily originates from the body resistance of the solar cells, the contact resistance of the front electrode metal grid line, the resistance corresponding to the transverse current between the gate lines, the contact resistance of the back electrode, as well as the resistance of the metal. Under an open circuit with zero current, the open-circuit voltage is not affected by the series resistance. When the current is not zero, a voltage drop IR_s is achieved between the output terminals, so the series resistance noticeably impacts the fill factor. The larger the series resistance, the more significantly the short-circuit current will decrease [15]. It is difficult to determine the exact value of series resistance from the measured I-V curves and how to determine its exact value is one important work in theoretical and experimental studies. The series resistance significantly impacts its efficiency, so the series resistance of solar cells should be determined with accurate and simple methods to optimize the process technology, minimize series resistance, and enhance the efficiency of solar cells. In general, theoretically, the I-V characteristic of solar cells is a transcendental equation, so no measured I-V curve can be developed to approximate the series resistance. In practice, the method to measure series resistance can be derived from the slope of the I-V curve at the open-circuit voltage, which is a nonlinear interval. The light-dark curve comparison method and curve comparison method under different light intensities are available to extract series resistance, however, the methods are less practical for the high costs of the equipment and the sophisticated measuring process. The solution for the series resistance of solar cells can be yielded by numerical analysis of the I-V transcendental equation. For general solar cells, the series resistance, small in value, can only impact the efficiency by reducing the filling factor; as a result, the open-circuit voltage and the short-circuit current are not additionally impacted, and the photogenerated current is equated with the short-circuit current. It is therefore revealed that the higher the photogenerated current, the smaller the reverse saturated dark current and the series resistance will be, and the higher the efficiency will be.

1.4.5 Non-ideal Factors in Diode

The unidirectional conductivity refers to the major characteristic exhibited by the ideal diode (i.e., forward conductivity displaying low resistance and reverse non-conductivity equated with open circuit cases). The existence of reverse leakage current primarily refers to the non-ideal characteristic for diodes. Some impurities and defects are common in solar cells, originating from the silicon itself and the manufacturing process. Both of them can act as the recombination center, capturing or recombining holes and electrons. The compounding process is always accompanied by the directional movement of the carriers, and there is bound to be a small current. The mentioned currents contribute to the value of the dark current obtained by the test. The current contributed by the thin layer is termed as the thin layer leakage current, and the part contributed by the body region refers to the body leakage current; they overall lead to non-ideal diode characteristics.

It is noteworthy that I and J are adopted to express the current and current density, respectively. The mentioned two symbols will often appear in this book, and their differences should be highlighted.

1.5 Application of Solar Cells

Solar cells have broad applications in numerous fields [11]. Photovoltaic power generation can be split into the independent system, grid-connected system, and wind-solar-assisted system, as presented in Fig. 1.5. Each power generation system exhibits critical applications.

The earliest application of solar cells was as the power of a satellite. In 1958, the high-efficiency monocrystalline silicon solar cells were initially adopted in artificial

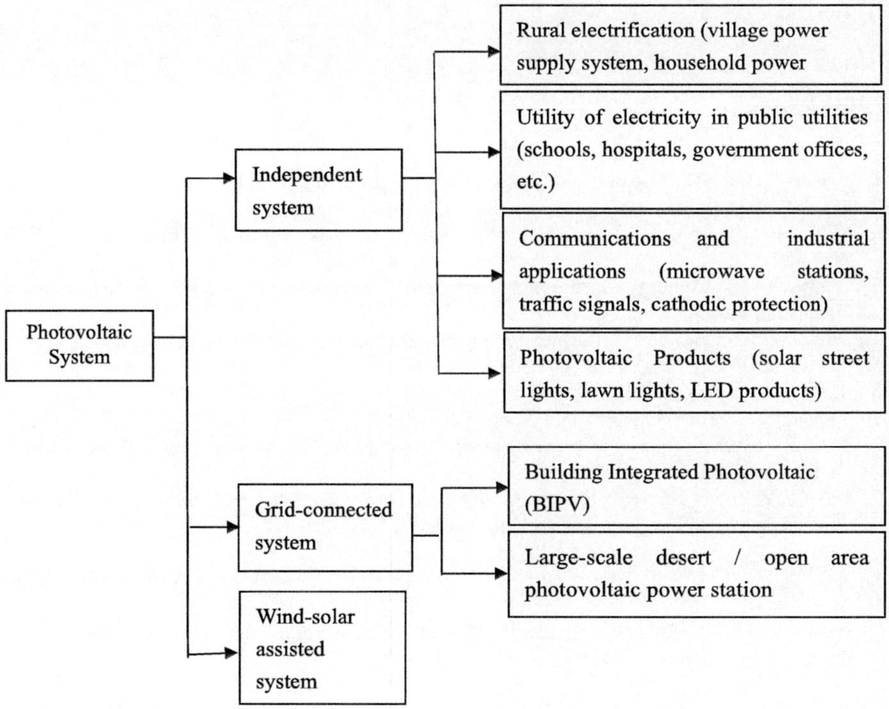

Fig. 1.5 Photovoltaic power generation system

satellites. In 1971, China successfully applied it to the Dong Fang Hong II satellite. Now, increasing solar cells have been adopted for civil use. In the Yangtze River Delta and Guangdong coastal pilot demonstration areas, large-scale grid-connected power generation has been achieved with ideal results, which lifts the tension of local power supply while protecting the environment. As residential areas and residential cities have been constructed, centralized grid-connected systems are progressively popularized. In the Western remote areas, solar energy is expected to solve the problem of long-term power supply. Solar panels on roofs and large buildings can both save electricity and protect the environment. This work has already been launched in developed countries (e.g., Germany), and China is also preparing for its implementation. Signal lights powered by solar energy have been extensively employed in some cities, and the application is expected to spread around the country very soon. Solar cells are capable of solving power shortages in areas far from a power grid. On August 5, 2007, the first solar photovoltaic power generation system "iPV Tracker" was formally linked to the grid in the beach volleyball stadium of the Beijing Olympic Games, in which the concept of humanistic Olympics, technology Olympics, and green Olympics were manifested. During the 2008 Olympic Games, the solar lamps used in the Bird's Nest are both energy-saving and beautiful. Furthermore, the great potential of solar energy is also well verified as solar cell phones, solar watches, solar cars, solar calculators, photovoltaic communications, and others have been extensively adopted.

References

1. Z. Cheng et al., The expatiates of the solar energy photovoltaic cell. Inf. Record. Mater. **8**(2), 41–47 (2007)
2. Wikipedia: Gross demistic product. https://en.wanweibaike.com/wiki-Gross%20domestic%20product
3. Wikipedia: World energy consumption. https://en.wanweibaike.com/wiki-Global%20energy%20consumption
4. Wikipedia: Renewable energy. https://en.wanweibaike.com/wiki-Renewable%20energy
5. 和讯网:中国面临的能源危机。
6. 凤凰网:太阳能资源的开发及利用。
7. Y. Zhao et al., Econ. Geogr. **18**(1), 56–611 (1998)
8. Y. Li, Research on the key technology of silicon solar cell, The Master Degree Dissertation of Heilongjiang university, 2009
9. S. Yu, J. Zhong et al., *Solar Energy Photovoltaic Device Technology* (University of Electronic Science and Technology Press, 2011) (in Chinese)
10. OFweek太阳能光伏网:我国太阳能累计装机容量8.98GW。
11. S. Xiong, M. Zhu et al., *Solar Cell Foundation and Application* (Science Press, Beijing, 2009) (in Chinese)
12. J. Si, Conversion efficiency calculation of the third generation solar cell, The Master degree dissertation of Hefei University of Technology, 2007

13. M. Pagliaro et al., *Flexible Solar Cells* (Wiley-VCH, 2008)
14. X. Yao, Research on key technologies of efficient solar cells, The Master degree dissertation of North China University, 2009
15. J. Gu et al., J. South-Cent. Univ. Natl. **28**(1), 57–61 (2009)

Chapter 2
Solar Cell Foundation

Before expounding semiconductor solar cells, several basic concepts of semiconductor physics should be presented, which are elucidated in most semiconductor physics textbooks. In the present chapter, these basic concepts are briefed. Since solar cells refer to devices converting solar energy into electrical energy, the solar spectrum should be clarified. Moreover, solar cell operation complies with the photovoltaic effect, which is inseparable from the existence of semiconductor materials and semiconductor junctions. Accordingly, the present chapter first briefs the concept of the solar spectrum and air mass and then elucidates the semiconductor materials (e.g., the basic properties of the semiconductor materials, the generation, recombination, and transmission of carriers in the semiconductor, as well as the relevant content of the semiconductor junction). Lastly, the highest efficiency of solar cells is discussed.

2.1 Photon and Solar Spectrum

2.1.1 Blackbody Radiation

As impacted by the thermal motion of atoms and molecules, all bodies generate electromagnetic radiation outward at the temperature above absolute zero. To be specific, electromagnetic radiation is generated by a hot body with wavelength or spectral distribution determined by the temperature of the body. A complete absorber is termed as a "blackbody". The spectral distribution of the radiation radiated by the black body abides by Planck's law. The optical radiation distribution $M(\lambda, T)$ is expressed as

$$M(\lambda, T) = \varepsilon(\lambda, T)\frac{2\pi hc^2}{\lambda^5}\frac{1}{\exp(hc/k\pi T) - 1} \tag{2.1}$$

C. Zhang et al., *Semiconductor Photovoltaic Cells*,
https://doi.org/10.1007/978-981-15-9480-9_2

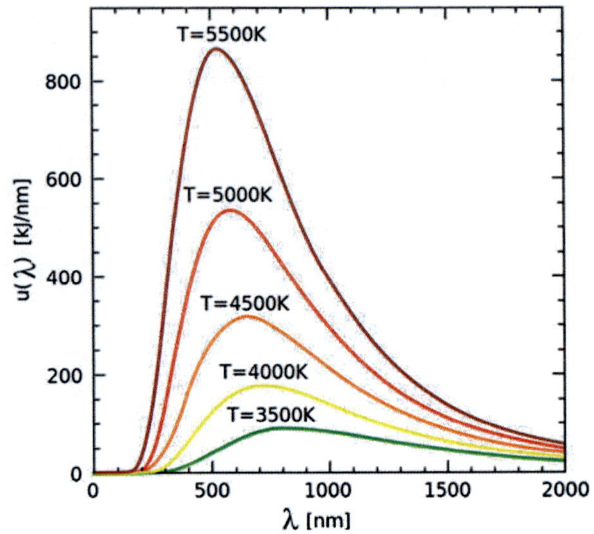

Fig. 2.1 Planck blackbody radiation distribution at different blackbody temperatures

where λ denotes the wavelength; k is the Boltzmann constant; h is Planck's constant; and c is the speed of light in a vacuum. The spectral emissivity $\varepsilon(\lambda, T)$ is equated with 1 only in the ideal blackbody. The energy distribution of the black body at varying temperatures is illustrated in Fig. 2.1. Two important characteristics are revealed: first, with the rise in the temperature of the heat source, the energy transmitted outward in all bands increases; second, the energy output in the short band increases rapidly [1].

A blackbody refers to an object that fully absorbs the frequencies of various electromagnetic radiations impinging on it. It reflects no light, so it appears to be black. A blackbody is considered an ideal model and the optimal radiator. Blackbody absorption and radiation are recognized as two aspects of coexistence. It is experimentally proved that the stronger the absorption capacity, the stronger the radiation ability will be. Under blackbody radiation, the color of the light varies with temperature, and the blackbody tends to vary as red-orange-yellow-yellow-white-white-blue-white. The color of the light emitted by a light source appears to be identical to that of the light emitted by the black body at a certain temperature. This temperature of the blackbody is termed as the color temperature of the light source. The higher the temperature of the black body, the darker the blue components and the less the red components will be in the spectrum. For instance, an incandescent lamp is colored as warm white, with a color temperature of 2700 K, while a daylight fluorescent lamp exhibits a color temperature of 6000 K.

Experiments show that for objects of different temperatures, the energy released by electrons transitioning from a high-energy state to a low-energy state is transmitted outward in the form of an electromagnetic wave. The electromagnetic radiation with frequency distribution varying according to temperature is termed thermal radiation. Blackbody radiation can be considered thermal radiation, i.e., a phenomenon of radiating electromagnetic waves outward for its own temperature. As revealed from the measured results, the solar spectrum is similar to the blackbody radiation spectrum at a temperature of 5758 K, so the sun is commonly considered a blackbody. In this regard, a range of characteristics of blackbody radiation can be directly exploited to express solar radiation [2].

The distribution of the radiant energy of an absolutely black body by wavelength displays exclusive association with temperature, instead of the nature of the object, while there is no such ideal blackbody in practice. For any wavelength, emissivity is defined as the ratio of the radiant energy of the real object in a small wavelength interval to the radiant energy of the blackbody at an identical temperature. Note that an emissivity is a positive number ranging from 0 and 1, and the general emissivity is determined by the material properties, environmental factors, and observation conditions. If emissivity is independent of the wavelength, the object can be termed as a gray body; otherwise, it is termed as a selective radiator [3, 4].

2.1.2 Solar Spectrum and Air Mass

The color of sunlight appears to be white. However, if a beam of sunlight passes through a glass prism, red, orange, yellow, green, cyan, blue, purple, and other colored bands of light will appear on the white curtain. Physically, such a colored band is termed as the solar spectrum. Figure 2.2 indicates that the visible spectrum only takes up a small fraction of the solar spectrum. The wavelength of the entire solar spectrum is significantly broad, ranging from a few angstroms (Å) to tens of meters (1 Å = 0.1 nm). Infrared rays, microwaves, radio waves, and others exhibit a wavelength longer than that of visible light. Ultraviolet rays, X-rays, and others exhibit a wavelength shorter than that of visible light. The solar spectrum refers to an absorption spectrum that distributes numerous dark lines on the background of a continuous spectrum. The reason for the solar spectrum is that the white light from the sun passes through the solar atmosphere, which is significantly cooler than the sun. In such a solar atmosphere, gases of considerable elements evaporate from the sun. When the sun passes through them, the identical element identification line of light is absorbed by the mentioned gases; thus, the absorption spectrum is formed when the sun reaches the earth. Over 99% of the solar radiation spectrum of the upper earth's atmosphere is at a wavelength of 0.15–4.0 μm. Nearly 50% of the solar radiation energy pertains to the visible region (wavelength 0.4–0.76 μm), 7% is in the ultraviolet spectral region (wavelength < 0.4 μm), 43% is in the infrared spectral region (wavelength > 0.76 μm), and the maximal energy is at a wavelength of 0.475 μm. Since the wavelength of solar radiation is significantly smaller than the wavelength of the ground and atmospheric radiations (about 3–120 μm), solar radiation is commonly termed as short-wave radiation and the ground and atmospheric radiation is referred to as long-wave radiation. Variations in solar activity and distance between the sun and the earth can cause solar radiation energy to vary in the upper atmosphere of the earth [5].

Fig. 2.2 Solar spectrum

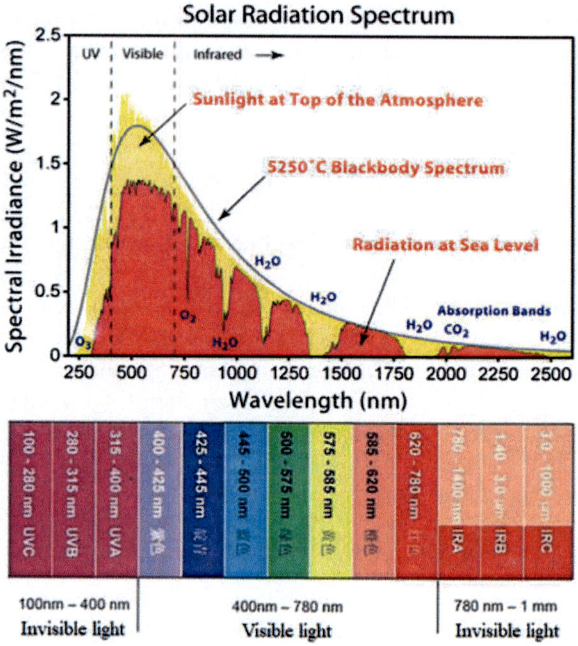

When the energy radiated from the surface of the sun reaches the atmosphere of the earth through the universe of approximately 100,005,000 km, the density of solar radiation energy on the surface perpendicular to the sunlight reaches about 1.395 KW/m^2, which is termed as the solar constant. The solar constant represents the solar radiation energy density per unit area on the plane perpendicular to the sunlight in the upper atmosphere when the earth is at an average distance from the sun. The value calculated by the International Earth Observation Year of 1964 has been generally adopted, i.e., the value of the solar constant is 1.382 KW/m^2.

The effects of atmospheric absorption on solar radiation anywhere on the ground can be ruled out. The actually measured sunlight is associated with the time and location of the test and with the local meteorological conditions. To express the impact of atmospheric absorption on solar radiation energy and its spectral distribution, the concept of air mass (AM) is developed. If the height of solar radiation passing through the atmosphere perpendicular to sea level at the solar zenith is adopted as the mass of the atmosphere, the air mass at any position of the sun will be defined as the ratio of the distance the sun passes through the atmosphere from the sea level to that through the atmosphere at the zenith. Thus, the air mass is commonly referred to as the "one air mass", and the air mass is a dimensionless quantity. Figure 2.3 gives a schematic diagram of the air mass, where *A* is a point on the earth's sea level. When the sun is at the zenith position *S*, the path of solar radiation passing through the atmosphere to point *A* refers to *OA*. When the sun is at any point *S'*, the path of solar radiation passing through the atmosphere is *O'A*. The air mass is defined as

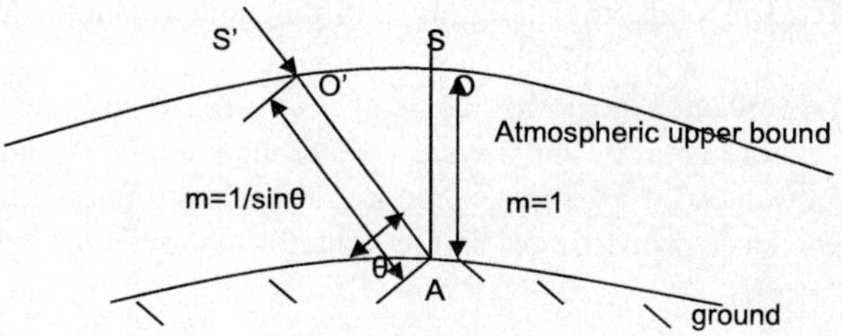

Fig. 2.3 Schematic diagram of air mass

$$AM = \frac{O'A}{OA} = \frac{1}{\sin\theta} \quad (2.2)$$

where θ is the angle between the incident ray of the sun and the horizontal plane, termed as the solar elevation angle. Given the difference in atmospheric pressure in various regions, i.e., reflecting the difference in the distance of sunlight via the atmosphere, it also reveals the difference in the mass of air contained in the atmospheric column per unit area. If point A is not at sea level, the air mass should be corrected as

$$AM = \frac{P}{P_0} \cdot \frac{1}{\sin\theta} \quad (2.3)$$

where P denotes the local atmospheric pressure; $P_0 = 101.3$ kPa, which is the standard atmospheric pressure.

It is suggested that the air mass at the sea level for the sun at the top is 1, which is termed as the AM1 condition. The case where the outer space does not pass through the atmosphere is termed as the AM0 condition, demonstrating that the air mass is zero. The solar constant S_0 is the solar radiation flux under AM0. AM0 spectrum has been primarily adopted to assess the performance of photovoltaic cells and components in space.

With the decrease in the height of the sun, the path of light through the atmosphere becomes longer, and the air mass reaches over 1. The irradiance of light reaching the ground is reduced with the increase in atmospheric absorption. Since the AM1 conditions on the ground significantly differ from the actual conditions of the human living area, the AM1.5 condition closer to the reality of human life is commonly taken as the standard for evaluating solar cells and components. Then, the solar elevation angle is about 48.2° [6].

2.2 Principles of Solar Cell Devices

The core portion of the solar cell devices consists of a semiconductor. Accordingly, the basic properties of semiconductor materials, the characteristics of carriers in semiconductor materials, and the semiconductor junctions formed by semiconductor material contact are critical to expound the working principle of solar cells and enhance the performance of solar cells. In numerous books concerning semiconductor materials and physics, the fundamental characteristics of semiconductors are introduced. Based on the mentioned existing foundations, the present section briefs the basic properties of semiconductor materials for solar cells.

2.2.1 Basic Properties of Semiconductor Materials

A semiconductor refers to a material with electrical conductivity between a metal and an insulator. Semiconductors can basically fall into two types, i.e., elemental semiconductor materials and compound semiconductor materials located in Group IV of the periodic table. Most compound semiconductor materials are formed by combining Group III and Group V elements. A semiconductor composed of one element is termed as an elemental semiconductor (e.g., Si and Ge). Silicon is a commonly employed semiconductor material in integrated circuits; it has increasingly extensive applications. Two-element compound semiconductors (e.g., GaAs or GaP) are formed by combining Group III and Group V elements. GaAs refers to one of the most extensively applied compound semiconductors. Its high optical properties support its broad applications in optical devices and in special applications where high-speed devices are required. It is also likely to manufacture a three-element compound semiconductor (e.g., $Al_xGa_{1-x}As$) in which the subscript x is a component of a low atomic number element. Even more complicated semiconductors can be formed, making the selection of material properties flexible.

Amorphous, polycrystalline, and single crystals are three basic types of solids. Each type of feature is dependent on the size of the ordered region in the material. An ordered region refers to a spatial category where atoms or molecules exhibit a regular or periodic geometric arrangement. Amorphous materials are ordered only on the scale of a few atoms or molecules. Polycrystalline materials are ordered on the scale of numerous atoms or molecules. The mentioned ordered regions are termed as single crystal regions and exhibit different sizes and orientations from each other. The single-crystal regions are referred to as crystal grains separated from each other by grain boundaries. Single-crystal materials have a high geometrical periodicity over the entire range. The merit of single crystal materials lies in that their electrical properties are generally better than those of amorphous materials, and the presence of grain boundaries can cause electrical properties to degrade.

1. **Crystal and band structure**

Semiconductor materials applied in solar cells are classified into single crystals (e.g., polycrystals), amorphous materials with atomic disorder, newly developed low-dimensional materials (e.g., nanocrystals, quantum dots, and superlattice quantum wells), and organic materials. To be specific, the crystal material is currently the largest cell application, and the nature of the single crystal material can also be referenced for discussing other materials. Crystals are formed by periodically repeating the arrangement of atoms in a three-dimensional space. The primitive cell is considered the smallest repeating unit of the crystal, and different types of crystals are formed in accordance with the structure of the cell. Taking the primitive cell of a cubic crystal as an example, it can be adopted to form three primitive cells: simple cube (a), face-centered cube (b), and body-centered cube (c) (Fig. 2.4). The ribbed length of the primitive cell is the lattice constant a.

For crystallography, Ge, Si, C exhibit a diamond structure as their crystal structure, which is a compound lattice where two face-centered cubes are combined in a cubic system. The two face-centered cubic lattices are embedded in the diagonal direction of the body [111] by 1/4 length (Fig. 2.5). A tetrahedron is the smallest repeating unit of a diamond structure, i.e., a physics primitive cell. A tetrahedron covers two atoms: four apex atoms contribute 1/4 atom and one central atom to the primitive cell, respectively.

Most of the III–V compounds and some II–VI compounds (e.g., CdTe) exhibit the cubic crystal sphalerite structure. The sphalerite structure is identical to the crystal lattices of the diamond structure, while the diamond structure covers the same type of atoms, and the sphalerite structure consists of two different atoms. Figure 2.6 illustrates the GaAs atomic arrangement structure, i.e., a face-centered cubic lattice of Ga atoms and a face-centered cubic lattice of atoms embedded in the diagonal direction of the body, taking up 1/4 length of the diagonal of the moving body. The

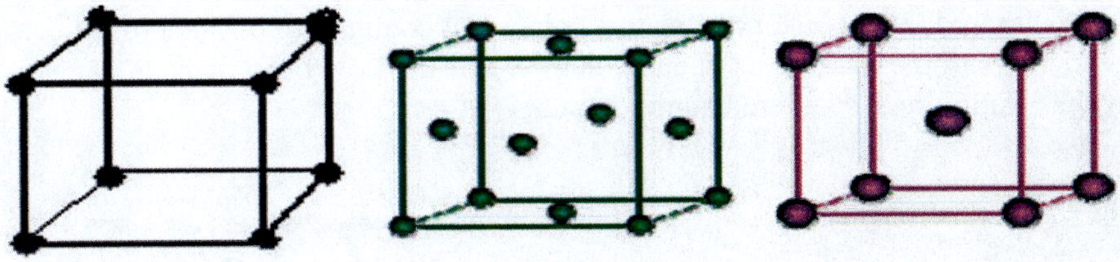

(a) simple cube; (b) face-centered cube; (c) body-centered cube

Fig. 2.4 Cubic crystal cell

Fig. 2.5 Diamond structure in a cubic system

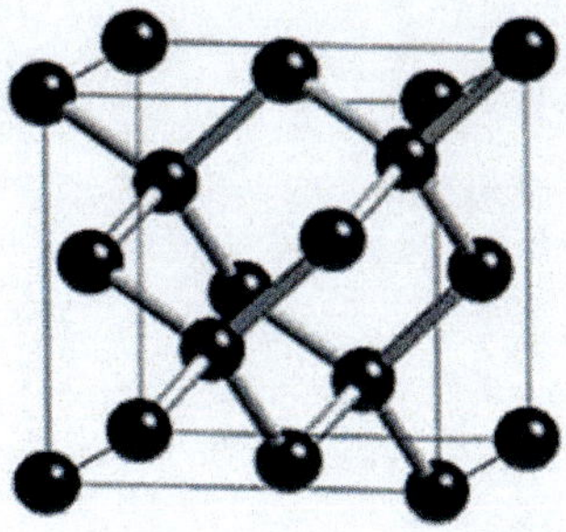

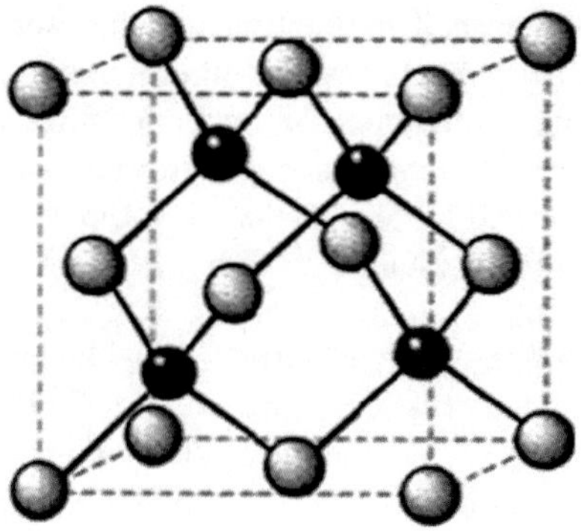

Fig. 2.6 Sphalerite structure in a cubic system

sphalerite structure also covers a tetrahedral physical primitive cell, whereas the central atom and the apical atom of the tetrahedron are different.

Other II–VI photovoltaic materials (e.g., CdS, ZnS, CdSe, and ZnSe) and common III–V GaN system materials belong to the simple cubic wurtzite structure (Fig. 2.7). Each atom in the structure is bonded to the nearest four atoms while forming a tetrahedron. Though the sphalerite structure and the wurtzite structure are both tetrahedral stacks composed of two different atoms, they exhibit different stacked configurations. Two tetrahedrons are combined by a covalent bond. If the projections of the other three keys on the vertical shared bond plane are coincident, they are termed as the coincidence configuration, and those that do not coincide are termed as staggered configurations. The projection of the six bonds of the wurtzite structure on the vertical shared bond plane pertains to a coincident configuration. The projection of the six keys of the sphalerite structure on the vertical shared bond plane is staggered by 60°, exhibiting a staggered configuration.

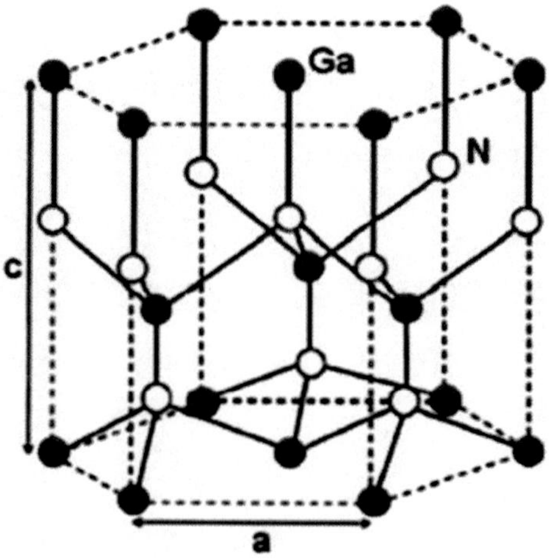

Fig. 2.7 Wurtz structure

The physical properties of semiconductor materials display close associations with the motion states of electrons and holes, and the description and insights into their motion states comply with the theory of energy bands. When a single atom consists of a crystal, the energy level of the atom will be not fixed. If two independent atoms far apart tend to approach each other, the electrons in the respective atom are affected by the potential field of their own atoms as well as the potential field of the other atom. The result suggests that according to the degeneracy of the electron energy level, the original single energy level will be split into m similar energy levels (m denotes the degeneracy of the energy level). If N atoms form a crystal, the energy level of each atom will fall into m similar energy levels, and the mN energy levels will form an energy band exhibiting similar energy. The total number of the mentioned splitting energy levels is so large that the energy levels in such a band can be considered continuous. Then, the shared electrons do not move in one energy level but between the energy bands of a crystal, which is termed as the permitted band. No electronic movement is identified between the permitted band, which is termed as the forbidden band. Figure 2.8 represents a schematic diagram of splitting into energy bands when the atomic energy level constitutes a crystal.

The inner shell of the atom exhibits low electron energy levels, a low degree of degeneracy, and a low degree of commonality. Thus, the energy level is less split, and the band is narrow. The outer shell exhibits extremely high electron energy (especially valence electrons), high degree of degeneracy as well as high degree of commonality, so the energy level is more split, and the band is wide.

Commonly, the lower energy bands are filled with electrons and termed as full bands. The highest energy band in the band diagram is commonly empty or half empty. This band is termed as the conduction band (with the conduction band bottom energy expressed as E_c). For the full band under the conduction band, its electrons may be transited to the conduction band, which is termed as the valence band (with the valence band top energy expressed as E_v). The area between them is termed as the forbidden band (with the forbidden bandwidth as E_g). The electrons can move

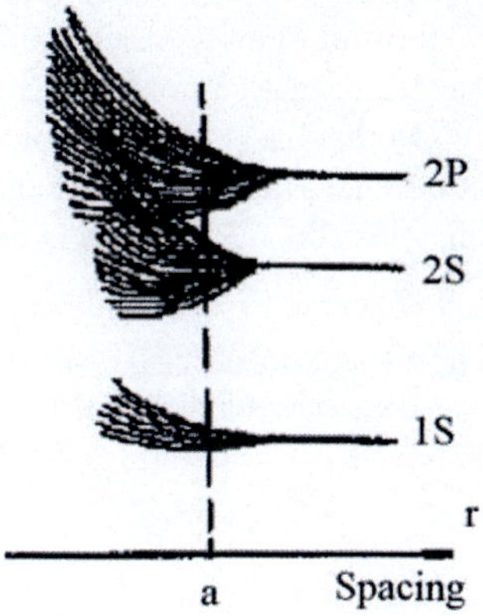

Fig. 2.8 Schematic diagram of splitting into energy bands when atomic energy levels form a crystal

in different energy bands, or transit between different energy bands, whereas they cannot move in the area between the energy bands.

For an energy band filled with electrons, though the in-band electrons can move under the external field, there are always pairs of electrons with the identical velocity and opposite direction, and thus the average current is zero. Thus, the full band that is fully taken up by the electrons is not electrically conductive. At absolute zero, both the insulator and the semiconductor are not electrically conductive. The valence band of the metal is half-filled, and the electron motion under the external field does not exhibit symmetry; as a result, high electrical conductivity is achieved. Though both the insulator and the semiconductor are not electrically conductive at absolute zero, their electrical conductivity remains different, which is attributed to the difference in the bandgap width E_g of the insulator and the semiconductor. The size of the E_g impacts the external field (e.g., the thermal field, the electric field, the light field, and the electromagnetic field) to excite electrons from the valence band to the conduction band. The insulator exhibits a large forbidden bandwidth E_g, commonly above 5.0 eV, thus, it is not electrically conductive. E_g of the semiconductor material is small in the range of 0.5–3.0 eV. The valence band electrons are more easily excited to the conduction band. Then, the valence band and the conduction band are not full. Though the conductivity is low, it can continuously exhibit conductivity and form a semiconductor with lower conductivity. The energy band diagrams of different materials are given in Fig. 2.9.

2. **Electronic states in semiconductors**

On the whole, the state of carriers in a semiconductor is studied with quantum statistical methods, as primarily represented by the state density and the Fermi statistical distribution in the energy level. Taking electrons as an example, when quantum statistics is adopted to process the state and distribution of electrons in a semiconductor, electrons are considered independent bodies, and the interaction between electrons is weak; electrons in the identical system are identical and indistinguishable (identical concepts in quantum mechanics), and the exchange of any two electrons does not cause a novel microscopic state; the number of electrons in the same energy level is no more than 2; since the spin quantum number of electrons is 1/2, a respective quantum state can only accommodate at most one electron.

On that basis, the distribution of electrons complies with the Fermi–Dirac distribution, i.e., the probability that the electron energy level of energy E is occupied by one electron $f(E)$, which is expressed as

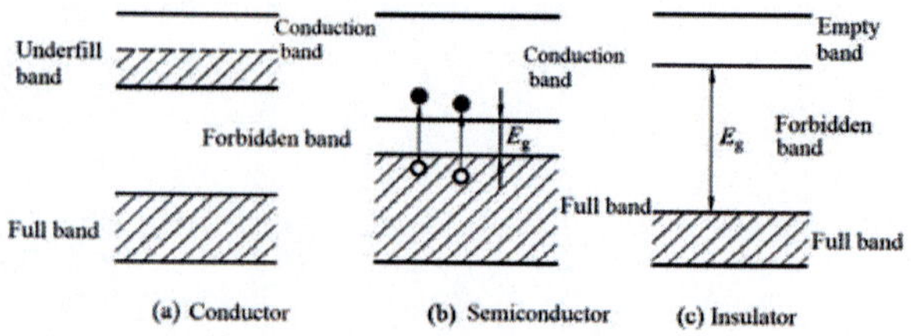

Fig. 2.9 Schematic diagram of energy bands for different materials

$$f(E) = \frac{1}{e^{\frac{E-E_F}{kT}} + 1} \tag{2.4}$$

where $f(E)$ denotes the Fermi distribution function; k is the Boltzmann constant; T is the thermodynamic temperature; and E_F is the Fermi level. When the energy level equals the Fermi level, the Fermi distribution function is expressed as

$$f(E_F) = \frac{1}{e^{\frac{E_F-E_F}{kT}} + 1} = \frac{1}{2} \tag{2.5}$$

In other words, the energy level of 1/2 of the electron occupancy rate refers to the Fermi level.

Figure 2.10 shows the variation of the Fermi distribution function $f(E)$ with energy. At $T = 0$ K,

$$\text{If } E < E_F, \text{ then } f(E) = 1$$
$$\text{If } E > E_F, \text{ then } f(E) = 0$$

It is suggested that at absolute temperature zero, the probability that the energy level smaller than E_F is taken up by electrons is 100%, and there is no empty energy level. While the energy level larger than E_F is occupied by electrons with zero probability, all energy levels are empty.

At $T > 0$ K, the probability that the energy level smaller than E_F is taken up by electrons progressively decreases as the energy level increases, and the probability that the energy level larger than E_F is occupied by electrons progressively increases with the decrease in the energy level. In other words, the electrons at the energy level near E_F, exhibiting the energy less than E_F, after absorbing energy, jump to an energy level greater than E_F, leaving a vacancy in the original place. Obviously, the transition of electrons from a low energy level to a high energy level is equivalent to the transition of holes from a high energy level to a low energy level. The higher the energy level taken up by electrons, the lower the energy level occupied by holes will be, and the higher the energy of the system will be. Thus, the probability of distribution of holes is $[1 - f(E)]$ with respect to the probability of distribution of electrons.

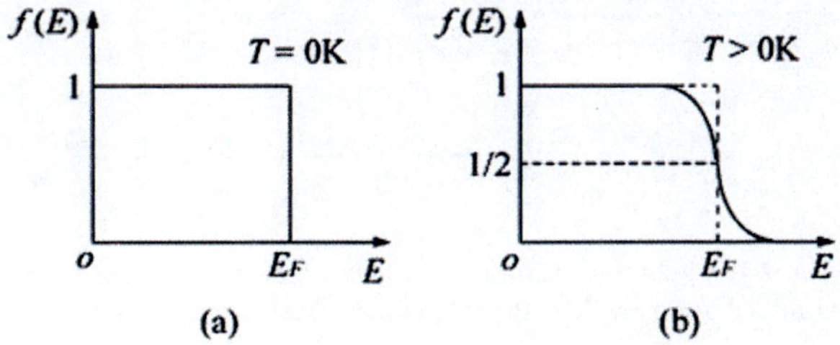

Fig. 2.10 Fermi distribution function *f*(*E*) varies with energy

When $(E - E_F) \gg kT$, $e^{\frac{E-E_F}{kT}} \gg 1$, the Fermi distribution function can be simplified to

$$f(E) \approx e^{\frac{E-E_F}{kT}} \tag{2.6}$$

Then, the Fermi distribution function is consistent with the classical Boltzmann distribution.

Silicon is adopted as an example to study the conductive properties of semiconductors. As shown in Fig. 2.11, when $T = 0$ K, there are eight valence electrons around each silicon atom, and the mentioned valence electrons are in the lowest energy state and are combined by covalent bonds. As the temperature rises from 0 K, electrons on some valence bands may be sufficiently heated to break the covalent bond and jump into the conduction band. Semiconductors are electrically neutral, indicating that once a negatively charged electron is removed from the original covalent bond position, it falls into a positively charged "empty state" at the same position in the valence band. With the continuous rise of temperature, more covalent bonds are broken, increasing electrons' jump into the conduction band, and the valence band also produces more positively charged "empty states."

On the whole, the motion of electrons in a crystal lattice differs from that in free space. The applied force, as well as the internal forces generated by positively charged ions and negatively charged electrons in the crystal, has effects on the motion of the electrons in the crystal lattice, which is written as

$$F_{\text{total}} = F_{\text{ext}} + F_{\text{int}} = ma \tag{2.7}$$

where F_{total}, F_{ext}, and F_{int} denote the total force, external force, and internal force of the particles in the crystal, respectively. The parameter a is the acceleration, and m represents the static mass of the particles.

Since it is difficult to consider the internal forces of the particles in sequence, the equation is written as

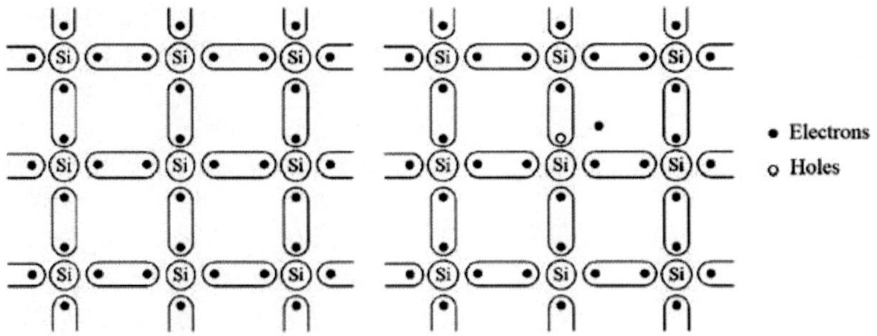

Fig. 2.11 Schematic diagram of the two-dimensional structure of the crystalline silicon covalent bond and the vacancy diagram of the electronic transition

$$F_{ext} = m * a \tag{2.8}$$

The acceleration a is directly associated with the external force. The parameter $m*$ is termed as the effective mass, summarizing the quality of the particles and the effect of the internal forces.

Effective mass is a parameter that combines quantum mechanical results with classical mechanical force equations. For the most part, the electrons at the bottom of the conduction band can be considered classical particles with motion abiding by Newtonian mechanical specifications; the internal forces and quantum mechanical properties are summarized as effective masses. If an electric field is introduced to the electrons at the bottom of the permitted band, the acceleration is written as

$$a = \frac{-eE}{m_n^*} \tag{2.9}$$

where m_n^* denotes the effective mass of the electron. The effective mass m_n^* of the electron near the bottom of the conduction band is a constant.

When a valence electron jumps into the conduction band, it leaves an "empty state" with a positive charge. When $T > 0$ K, all valence electrons may achieve thermal energy. If a valence electron receives thermal energy, it may leap into empty states. The movement of valence electrons in an empty state can be equated with the movement of those positively charged empty states themselves. There is now a second, equally important charge current in the crystal that can form currently. This type of charge carrier is termed as a hole. In the first chapter, the term "hole" has been adopted many times. The Hole can be considered a classic particle which obeys Newton's law of mechanics.

Also, given the electrons at the top of the permitted band, if an electric field is introduced, and the Newtonian mechanics equation is adopted, then

$$F = m * a = -eE \tag{2.10}$$

However, $m*$ is now negative, so it yields

$$a = \frac{-eE}{-|m*|} = \frac{+eE}{|m*|} \tag{2.11}$$

The direction of motion of the electrons near the top of the permitted band is identical to that of the applied electric field.

The density of the mentioned particles in the valence band is identical to the empty state of the electrons, which have a positive charge and a positive effective mass m_p^*. This new particle is a hole. It is precisely because the holes exhibit a positive effective mass and a positive charge such that their direction of motion is identical to the direction of the applied electric field [7].

3. **Doping**

The conduction of conductors (e.g., metal materials) is attributed to the movement of electrons. However, in solar photovoltaic materials (semiconductor materials), besides electrons, a positively charged hole can also conduct electricity, and the conductivity of the material also depends on the concentration, distribution, and mobility of electrons and holes. The mentioned conductive electrons and holes are termed as carriers, and their concentration is a basic parameter of the semiconductor material, which critically impacts the electrical properties of the semiconductor material.

On the whole, semiconductor materials exploit high-purity materials, and then different types and concentrations of impurities are artificially introduced to precisely control the concentration of electrons or holes. In high purity semiconductor materials without impurity, the concentrations of electrons and holes are equal, which is termed as an intrinsic semiconductor. If an impurity element is doped in an ultra-high-purity semiconductor material, the electron concentration reaches over the hole concentration, which is termed as n-type semiconductor. Then, the electron is termed as the majority carrier, and the hole is termed as the minority carrier. On the other hand, if an impurity element is doped in an ultra-high-purity semiconductor material, the hole concentration reaches over the electron concentration, which is termed as a p-type semiconductor; the hole here is termed as the majority carrier, and the electron is termed as the minority carrier. Accordingly, the mentioned impurities are referred to as n-type dopants (donor impurities) or p-type dopants (acceptor impurities).

For general conductive material, its conductivity σ can be expressed as follows:

$$\sigma = ne\mu \tag{2.12}$$

where n denotes the carrier concentration (number/cm^3); e is the charge amount of the electron; and μ is the mobility of the carrier (the moving speed of carriers per unit electric field intensity cm^2/(V·S)). The carriers are electrons here. For a semiconductor material, there are two types of carriers for the conductivity of electrons and holes, so it yields

$$\sigma = nq\mu_e + pq\mu_p \tag{2.13}$$

where n denotes the electron concentration; p is the hole concentration; q is the charge of the electron; and μ_e and μ_p are the mobility of electrons and holes, respectively. If the electron concentration n is significantly greater than the hole concentration p, the conductivity of the material is $\sigma \approx nq\mu_e$. In contrast, the conductivity of the material is $\sigma \approx pq\mu_p$.

For doped semiconductors, whether it is n-type or p-type, it is overall electrically neutral, the number of internal positive charges and negative charges is equal, and the externality does not exhibit electrical properties. This is because the single crystal semiconductor and the doping impurities are electrically neutral. In the process of

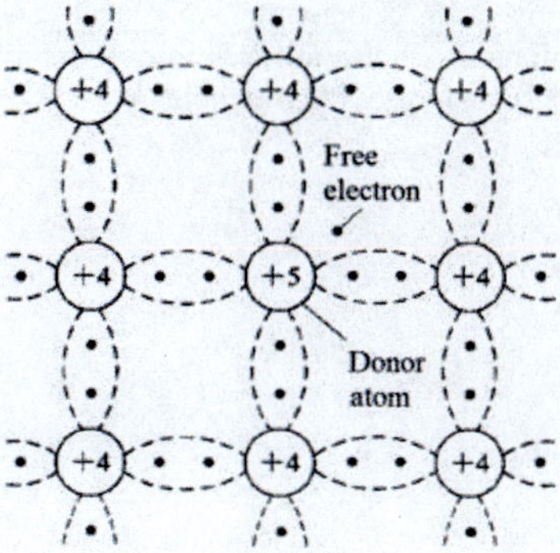

Fig. 2.12 Schematic diagram of phosphorus -doped into silicon to provide free electrons

doping, the charge is lost, and the charge is obtained from the outside, whereas the number of valence electrons doped into the impurity atom is one or less than the valence electron of the base material. As a result, considerable movable electrons or holes appear in the semiconductor, and the balance of positive and negative charges in the whole semiconductor is not broken.

It is now assumed that a group V element (e.g., phosphorus) is doped into the silicon as a substitute impurity. The group V element has five valence electrons, four of which are combined with a silicon atom to form a covalent bond, and the other fifth is loosely bound to the phosphorus atom. Figure 2.12 gives a schematic diagram of this phenomenon. The fifth valence electron is termed as the donor electron.

The phosphorus atom is positively charged after losing the donor electron. At very low temperatures, the donor electrons are trapped on the phosphorus atoms. However, the energy required to excite valence electrons into the conduction band is significantly smaller than the energy required to excite electrons bound by covalent bonds. If the donor electron acquires a small amount of energy (e.g., thermal energy), it can excite the conduction band, thereby leaving a positively charged phosphorus atom. The electrons in the conduction band can now move in the entire crystal to form a current, while the positively charged phosphorus atoms are immobilized. Since this type of impurity atom provides the electrons for the conduction band, they can be termed as the donor impurity atoms. As the donor impurity atoms increase the conduction band electrons, but do not generate valence band holes, the semiconductor material here is termed as an n-type semiconductor.

It is now assumed that a group III element (e.g., boron) is doped into silicon as a substitute impurity for silicon. Group III elements have three valence electrons and are combined with silicon to form covalent bonds; one covalent bond position is empty (Fig. 2.13). When the vacancies introduced by the boron atoms are filled, the other valence electron positions will be empty. The mentioned empty electronic locations can be considered holes in semiconductor materials.

The holes can move in the entire crystal to form a current, whereas the negatively charged boron atoms are immobilized. Group III element atoms acquire electrons from the valence band, which are termed as acceptor impurity atoms. The acceptor impurity atoms can generate holes in the valence band, whereas electrons in the

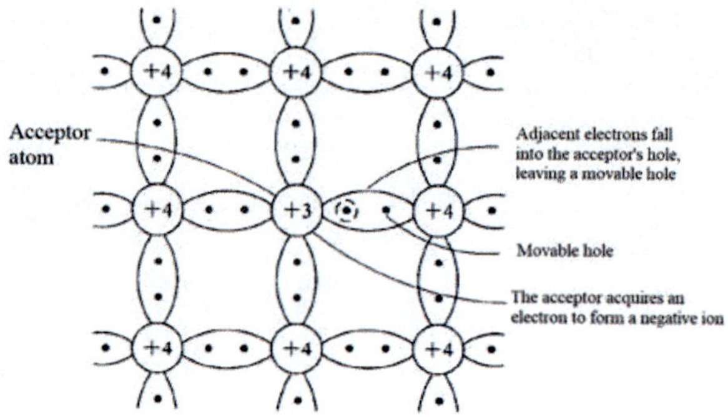

Fig. 2.13 Schematic diagram of boron-doped into silicon to provide free holes

conduction band are not generated; the semiconductor material here is termed as a p-type material.

A pure single crystal semiconductor is termed as an intrinsic semiconductor. When a quantitative impurity atom is doped, it becomes an extrinsic semiconductor. An extrinsic semiconductor has several dominant electrons or several dominant holes.

The case of compound semiconductors (e.g., GaAs) is more complicated. Group II elements (e.g., Be, Zn, and Cd) can enter the crystal lattices as substitute impurities, instead of the group III element Ga, to become acceptor impurities. Likewise, Group VI elements (e.g., Se and Te) can also enter the crystal lattices as substitute impurities, instead of the Group V element As, to be donor impurities. Group IV elements (e.g., Si and Ge) can also be impurity atoms in GaAs. If a Si atom replaces a Ga atom, the silicon impurity will become a donor. If a Si atom replaces an As atom, the silicon impurity will act as an acceptor, similar to Ge atoms as impurities. This impurity is termed as an amphoteric impurity. In the experiment of GaAs, it is found that Ge is primarily represented as acceptor impurities, while Si primarily exhibits donor impurities.

A compensation semiconductor refers to a semiconductor containing both donor and acceptor impurity atoms in an identical region. A compensation semiconductor can be formed by diffusing the acceptor impurity into the n-type material or diffusing the donor impurity into the p-type material. When $N_d > N_a$, an n-type compensation semiconductor is formed; when $N_d < N_a$, a p-type compensation semiconductor is formed; when $N_d = N_a$, a fully compensated semiconductor is generated, exhibiting the characteristics of an intrinsic semiconductor.

When a very large dopant concentration is added into the semiconductor, the dopant can no longer be considered only a perturbation to the system. Then, their impact on the band structure should be considered. This so-called heavy doping effect indicates that they can reduce the forbidden bandwidth E_g; as a result, the intrinsic carrier concentration increases. This narrowing of the forbidden bandwidth is adverse for the performance of the solar cell. The solar cell is generally designed to avoid this effect, though it may play a role in the heavily doped region near the solar cell electrode [8].

4. Semiconductor characteristics of equilibrium and non-equilibrium states

The nature of the semiconductor material strongly relies on its carrier concentration. In the case of a certain doping concentration, the carrier concentration is primarily determined by the temperature. In the case of intrinsic semiconductors at absolute zero, electrons are bound to the valence band; thus, the semiconductor material has no free electrons and holes, as well as no carriers. With the increase in the temperature, the electrons absorb energy from the lattice of thermal vibration and then transit from a low-energy state to a high-energy state (e.g., a transition from a valence band to a conduction band), forming free conduction band electrons and valence band holes, which is termed as intrinsic excitation. For impurity semiconductors, there is the ionization of impurities besides the intrinsic excitation. At very low temperatures, impurity electrons are also bound to the impurity level. With the rise in the temperature, the electrons absorb energy and then transit from a low-energy state to a high-energy state (e.g., the transition from the donor level to the conduction band to produce free conduction band electrons, or from the valence band transition to the acceptor level to produce free valence band holes). Accordingly, with the increase in temperature, carriers are continuously generated.

The equilibrium state or the thermal equilibrium state refers to a state in which the semiconductor is not externally impacted (e.g., voltage, electric field, magnetic field, and temperature gradient). In this state, all properties of the material are independent of time. In the absence of external light, electricity, magnetism, and others, at a certain temperature, carriers that transit from a low-energy state to a high-energy state also produce motion in the opposite direction, i.e., transition from a high-energy state to a low-energy state will release a certain amount of energy, termed as the recombination of carriers. Thus, at a certain temperature, while the carriers are continuously generated, the carrier recombination continues, and the final carrier concentration will reach a certain stable value, and the semiconductor pertains to a state of thermal equilibrium. In the state of thermal equilibrium, electrons are continuously excited from the valence band to the conduction band to produce electron–hole pairs. Meantime, they are continuously recombined, so the total carrier concentration is kept constant. For intrinsic semiconductors, the electron and hole concentrations are equal. For n-type semiconductors, the electron concentration reaches over the hole concentration, the electrons are majority carriers, and the holes act as minority carriers. For p-type semiconductors, the holes act as majority carriers, and the electrons are minority carriers.

The distribution of conduction band electrons is attributed to the density of the allowed quantum states in the conduction band and the probability of a quantum state being taken up by electrons, which is expressed as

$$n(E) = g_{\rm C}(E) f_{\rm F}(E) \tag{2.14}$$

where $f_{\rm F}(E)$ denotes the Fermi–Dirac probability distribution function; $g_{\rm C}(E)$ is the quantum state density in the conduction band. By integrating over the entire conduction band energy range, the total electron concentration per unit volume in the conduction band is calculated as

$$n_0 = N_C \exp\left[\frac{-(E_C - E_F)}{kT}\right] \tag{2.15}$$

Likewise, the distribution of holes in the valence band results from the valence band, making the density of a quantum state a probability that a quantum state is not occupied by electrons, which is expressed as

$$p(E) = g_V(E)[1 - f_F(E)] \tag{2.16}$$

By integrating over the entire valence band energy range, the total hole concentration per unit volume in the valence band is calculated by

$$p_0 = N_V \exp\left[\frac{-(E_F - E_V)}{kT}\right] \tag{2.17}$$

In the intrinsic semiconductor, the electron concentration in the conduction band is equated with the hole concentration in the valence band. The electron concentration and the hole concentration in the intrinsic semiconductor are, respectively, expressed as n_i, p_i, as commonly referred to as intrinsic electron concentration and intrinsic hole concentration. Since $n_i = p_i$, the intrinsic carrier concentration is simply expressed by n_i, suggesting the intrinsic electron concentration or the intrinsic hole concentration. The Fermi level of the intrinsic semiconductor is termed as the intrinsic Fermi level, or $E_F = E_{Fi}$. Subsequently, it yields

$$n_0 = n_i = N_C \exp\left[\frac{-(E_C - E_{Fi})}{kT}\right] \tag{2.18}$$

with

$$p_0 = p_i = n_i = N_V \exp\left[\frac{-(E_{Fi} - E_V)}{kT}\right] \tag{2.19}$$

Multiply the two equations, then it yields

$$n_i^2 = N_C N_V \exp\left[\frac{-(E_C - E_{Fi})}{kT}\right] \cdot \exp\left[\frac{-(E_{Fi} - E_V)}{kT}\right] \tag{2.20}$$

or

$$n_i^2 = N_C N_V \exp\left[\frac{-(E_C - E_V)}{kT}\right] = N_C N_V \exp\left[\frac{-E_g}{kT}\right] \tag{2.21}$$

where E_g denotes the forbidden bandwidth. For a given semiconductor material, when the temperature is constant, n_i is constant regardless of the Fermi level.

Since the electron and hole concentrations are equal, it yields

$$N_V \exp\left[\frac{-(E_{Fi} - E_V)}{kT}\right] = N_C \exp\left[\frac{-(E_C - E_{Fi})}{kT}\right] \tag{2.22}$$

It is suggested that

$$E_{Fi} = \frac{1}{2}(E_C + E_V) + \frac{3}{4}kT \ln\left(\frac{m_p^*}{m_n^*}\right) \tag{2.23}$$

The first item of the above formula refers to the exact intermediate energy value between E_C and E_V, i.e., the center of the forbidden band. If the effective mass of electrons and holes is $m_p^* = m_n^*$, the intrinsic Fermi level accurately falls in the center of the forbidden band. If $m_p^* < m_n^*$, the position of the intrinsic Fermi level will slightly overtop the center of the forbidden band. If $m_p^* > m_n^*$, the position of the intrinsic Fermi level will be slightly lower than the center of the forbidden band.

By introducing donor or acceptor atoms to a semiconductor, the distribution of electrons and holes in the material will vary. Since the Fermi level is associated with the distribution function, it also variations as the impurity atoms are doped. If the Fermi level deviates from the center of the forbidden band, the concentration of electrons in the conduction band and the concentration of holes in the valence band will vary.

After the deformation is derived, the thermal equilibrium electron concentration is also written as

$$n_0 = n_i \exp\left[\frac{E_F - E_{Fi}}{kT}\right] \tag{2.24}$$

with

$$p_0 = n_i \exp\left[\frac{-(E_F - E_{Fi})}{kT}\right] \tag{2.25}$$

When the donor or acceptor impurity is introduced, the Fermi level varies, and the above two equations reveal that as the Fermi level deviates from the intrinsic Fermi level, n_0 and p_0 also deviate from n_i. If $E_F > E_{Fi}$, there are $n_0 > n_i$ and $p_0 < n_i$. The n-type semiconductor is characterized by $E_F > E_{Fi}$, so $n_0 > p_0$. Likewise, in a p-type semiconductor, $E_F < E_{Fi}$, so $p_0 > n_i$, $n_0 < n_i$, and then $p_0 > n_0$.

In terms of intrinsic or extrinsic semiconductors in thermal equilibrium, it yields

$$n_i^2 = N_C N_V \exp\left[\frac{-(E_C - E_V)}{kT}\right] = N_C N_V \exp\left[\frac{-E_g}{kT}\right] \tag{2.26}$$

which is

$$n_0 p_0 = n_i^2 \tag{2.27}$$

For a given semiconductor material at a certain temperature, where the product of n_0 and p_0 is a constant on the whole, though this equation seems simple, it is a basic formula for thermally balanced state semiconductors.

Any deviation from the thermal equilibrium state in the semiconductor can cause the concentration of electrons and holes to vary. For instance, a sudden increase in temperature elevates the rate at which heat generates electrons and holes, causing their concentration to vary over time until a novel equilibrium value is reached. An additional stimulus (e.g., light) also produces electrons and holes, resulting in a non-equilibrium state. For an n-type semiconductor, holes are minority carriers. In the presence of non-equilibrium carriers, the holes are termed as non-equilibrium minority carriers. For a p-type semiconductor, the electrons in a non-equilibrium carrier are termed as non-equilibrium minority carriers. On the whole, the non-equilibrium carrier concentration is very low compared with the doping concentration (i.e., the majority carrier concentration) and has little effect on the majority carrier concentration. However, it is equivalent to the minority carrier concentration in the semiconductor, which seriously affects minority carrier concentration and relevant properties. Accordingly, in non-equilibrium carriers, the impact of non-equilibrium minority carriers on semiconductor performance is crucial.

The solar photovoltaic effect is the application of typical semiconductor materials in a non-equilibrium state. High-energy photons are assumed to be incident on the semiconductor, thereby causing electrons in the valence band to be excited into the conduction band. Then, an electron is generated in the conduction band, and a hole is generated in the valence band; as a result, an electron–hole pair is generated. The extra electron and hole become excess electrons and holes. The external action produces a certain ratio of excess electrons and holes. Set g_n' as the generation rate of excess electrons, and g_p' as the generation rate of excess holes. For direct band generation, excess electrons and holes appear in pairs, so it yields.

$$g_n' = g_p' \tag{2.28}$$

When non-equilibrium electrons and holes are generated, the concentration of electrons in the conduction band and the concentration of holes in the valence band are higher than those during their thermal equilibrium, which can be written as

$$n = n_0 + \delta n \tag{2.29}$$

with

$$p = p_0 + \delta p \tag{2.30}$$

where n_0 and p_0 denote the thermal equilibrium concentrations; δ_n and δ_p are the excess electron and hole concentrations. The equilibrium state is disturbed by external forces; thus, the semiconductor is no longer in thermal equilibrium. In the non-equilibrium state, $np \neq n_0 p_0 = n_i^2$.

Moreover, quasi-Fermi levels of electron concentration and hole concentration for non-equilibrium states can be defined. If δn and δp are the excess electron and hole concentrations, respectively, it yields

$$n_0 + \delta n = n_i \exp\left[\frac{E_{Fn} - E_{Fi}}{kT}\right] \tag{2.31}$$

with

$$p_0 + \delta p = n_i \exp\left[\frac{E_{Fi} - E_{Fp}}{kT}\right] \tag{2.32}$$

where E_{Fn} and E_{Fp} denote quasi-Fermi levels of electron concentration and hole concentration, respectively. The overall electron concentration and total hole concentration are functions of the Fermi level.

2.2.2 Carrier Generation, Recombination, and Transport

1. **Light absorption and photoelectric generation of carriers**

When the sunlight hits the solid surface, the interaction between photons and electrons in the solid complies with three basic processes, i.e., light absorption, spontaneous emission, and stimulated emission. Semiconductor materials typically absorb light energy strongly. Besides, the effective process of dominating solar cells and photodetectors is light absorption. There are various light absorption processes in semiconductors (e.g., intrinsic absorption between energy bands, exciton absorption, absorption between sub-bands, free carrier absorption from transitions of carriers in the same band, as well as lattice absorption associated with the transition between lattice vibrational energy levels). The absorption process reflects the mechanism of different electron or phonon transitions. The study on different absorption processes will effectively indicate crystal band structure and phonon spectrum. Intrinsic absorption between bands can cause electrons to transit from the valence band to the conduction band. To achieve the occurrence of intrinsic absorption, the photon energy should be greater than the forbidden bandwidth E_g. The electron transition associated with light absorption is ascertained by the band structure, as well as the conservation of energy and momentum.

At absolute zero, the valence band of an ideal semiconductor is filled with electrons, so the electrons in the valence band without external action cannot be excited to higher energy levels. Only when the photon absorbs sufficient energy can the electron be excited and jump over the forbidden band into the conduction band, which can leave a hole in the valence band and form the electron–hole pair. This absorption process, formed by the Band to band transition, is termed as intrinsic absorption.

Experiments reveal that only those photons with energy $h\nu$ equal to or greater than the forbidden bandwidth E_g can trigger the intrinsic absorption, which is expressed as

$$h\nu \geq h\nu_0 = E_g \tag{2.33}$$

or

$$\frac{hc}{\lambda} \geq \frac{hc}{\lambda_0} = E_g \tag{2.34}$$

where ν_0 and λ_0 denote the photon frequencies and wavelengths capable of only generating the intrinsic absorption, termed as the intrinsic absorption edge of semiconductors, which is introduced in the first chapter.

The formula of intrinsic absorption long-wavelength edge is obtained as

$$\lambda_0 = \frac{1.24}{E_g(eV)}(\mu m) \tag{2.35}$$

This formula has been mentioned in chapter 1. According to the different forbidden bandwidths of semiconductor materials, the corresponding intrinsic absorption long-wavelength edge can be calculated. For instance, the forbidden bandwidth Eg of silicon is 1.12 eV, so silicon's intrinsic absorption long-wavelength edge is 1.11 μm. GaAs exhibits a bandgap E_g of 1.43 eV, so λ_0 is 0.867 μm.

In the intrinsic absorption of electron–hole pairs, the conservation of energy should be observed, and the conservation of momentum should be complied with. If the minimum value of the conduction band bottom of the semiconductor material and the maximum value of the valence band top have identical wave loss k, the momentum of electrons in the valence band does not change when they transition to the conduction band, which is termed as the direct transition. Such semiconductors are termed as direct bandgap semiconductors (e.g., GaAs). If the bottom of the conduction band and the top of the valence band have different k, when the electrons in the valence band transit to the conduction band, the momentum will change. Besides the absorption of photon energy, electrons should also interact with the crystal lattice to emit or absorb phonons to achieve momentum conservation. This type of semiconductor is termed as the indirect bandgap semiconductor (e.g., Si and Ge). Thus, indirect transitions depend not only on the interaction of electrons and photons, but also on the interaction of electrons and lattices; as a result, the absorption coefficient is greatly reduced. Overall, the absorption coefficient of an indirect bandgap semiconductor is 2–3 orders of magnitude lower than that of a direct bandgap semiconductor, requiring thicker materials to absorb the energy of the same spectrum. For the indirect band semiconductor silicon, the thickness of several hundred microns or more is required to fully absorb the energy of the light waves in the sunlight larger than its forbidden band width. However, for a direct bandgap GaAs semiconductor, only a few microns of thickness should fully absorb the energy of the light in the sunlight that reaches over its forbidden bandwidth. Experiments have shown that wavelengths shorter than intrinsic absorption edge λ_0 can often be absorbed in the semiconductor. In fact, besides intrinsic absorption, there exist other light absorption processes (e.g., exciton absorption, impurity absorption, and free carrier absorption), which are termed as non-intrinsic absorption.

The critical absorption of light in the semiconductor is intrinsic absorption. The intrinsic absorption coefficient is several tens to tens of thousands more than other non-intrinsic absorption coefficients. Accordingly, only the intrinsic absorption can be considered under general illumination conditions.

When the photon energy reaches over or equals to the bandgap width, the intrinsic absorption can be generated, whereas it does not mean that light entering the semiconductor can be immediately absorbed. The absorption of light in the semiconductor is illustrated in Fig. 2.14. When the light of intensity I_0 accesses the interior of the semiconductor from the vertical surface, the light intensity entering the semiconductor is $I_0\ (1-R)$ after deducting the reflection. The light intensity I_x at the inside distance of the semiconductor surface x complies with the absorption law:

$$I_x = I_0(1 - R)e^{-\alpha x} \tag{2.36}$$

where α denotes absorption coefficient; R is reflectivity. Light intensity attenuation is a result that the solid absorbs a certain amount of energy and excites electrons from a lower energy state to a higher energy state.

2. **Carrier recombination**

For the random nature of the thermal process, electrons are constantly excited by heat to jump from the valence band to the conduction band. Meantime, electrons in the conduction band move randomly in the crystal; when approaching the hole, they may fall into an empty state in the valence band. This recombination process removes both electrons and holes. Since the net carrier concentration at thermal

Fig. 2.14 Light absorption in semiconductors (w: Total thickness of semiconductor)

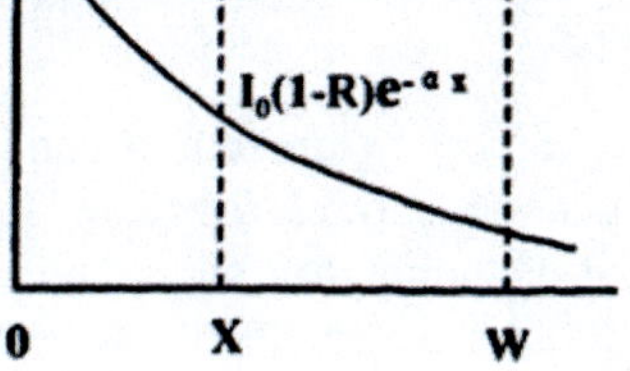

equilibrium is independent of time, electron and hole generation rates should equal their recombination rate.

Set G_{n0} and G_{p0} as the generation rates of electrons and holes, respectively. For direct bandgap generation, electrons and holes come in pairs, so it yields

$$G_{\mathrm{n0}} = G_{\mathrm{p0}} \tag{2.37}$$

Set R_{n0} and R_{p0} as the recombination rate of electrons and holes, respectively. For direct bandgap recombination, electrons and holes disappear in pairs, so it yields

$$R_{\mathrm{n0}} = R_{\mathrm{p0}} \tag{2.38}$$

For the thermal equilibrium state, the electron and hole concentrations are independent of time, so the probability of generation and recombination are equal. Thus, it yields

$$G_{\mathrm{n0}} = G_{\mathrm{p0}} = R_{\mathrm{n0}} = R_{\mathrm{p0}} \tag{2.39}$$

When the external effect of the non-equilibrium carrier is removed, the injected non-equilibrium carrier cannot exist all the time, and it will progressively disappear, i.e., the electrons excited to the conduction band will return to the valence band, and the electrons and holes will disappear in pairs again. Lastly, carrier concentration returns to the thermal equilibrium value. This process is termed as the recombination of non-equilibrium carriers.

When the external effect of non-equilibrium carriers is removed, the injected non-equilibrium carriers do not disappear immediately, but comply with a process, i.e., they have a certain survival time in the conduction band and the valence band. The average survival time of a non-equilibrium carrier termed as the life of a non-equilibrium carrier is expressed as τ. Compared with the non-equilibrium majority carrier, the impact of a non-equilibrium minority carrier is in the dominant and decisive position, so the non-equilibrium carrier life is commonly termed as non-equilibrium minority carrier life. Note that $1/\tau$ represents the probability of recombination of a non-equilibrium carrier per unit time. On the whole, the number of electron–hole pairs in which the net recombination per unit time and unit volume disappear is referred to as the recombination rate of the non-equilibrium carrier. The recombination rate of excess electrons is R'_{n}. and the recombination rate of excess holes is expressed as R'_{p}. Excess electrons and holes are compounded in pairs, so the recombination rate should be the same, which is written as

$$R'_{\mathrm{n}} = R'_{\mathrm{p}} \tag{2.40}$$

Since the direct inter-band recombination is a spontaneous behavior, the recombination rate of electrons and holes is constant relative time. Moreover, the probability of recombination should be proportional to the concentration of electrons and holes. If there are no electrons or no holes, recombination cannot be produced.

When the compounding process occurs, electrons in the conduction band jump to the bottom of the conduction band, and the energy is transferred to the crystal lattice as thermal energy. Subsequently, the electron transition from the bottom of the conduction band to the valence band recombines with the hole. This recombination is termed as direct recombination. If there are defect levels in the forbidden band, including those levels caused by defects in the body and surface states, electrons in the valence band will be excited to the defect level, and electrons in the defect level may be excited to the conduction band. In the presence of recombination, the electrons from the bottom of the conduction band will first jump to the defect level and then to the valence band and hole recombination, which is termed as indirect recombination; the defect is also termed as recombination center.

When the non-equilibrium carrier is recombined, the excess energy will be released by the transition from the high energy level to the low energy level. According to the way of energy release, recombination can be divided into the following three forms:

(1) When carrier recombination occurs, photons are emitted to generate a luminescence phenomenon, which is termed as radiation recombination or luminescence recombination;
(2) When carrier recombination occurs, phonons are emitted to transfer energy to the lattice and generate heat energy, which is termed as non-radiation recombination;
(3) When carrier recombination occurs, energy is transferred to other carriers to increase their energy, which is termed as Auger recombination.

It can be seen that, as impacted by external conditions, non-equilibrium carriers are generated and different forms of recombination appear. If the external action always exists, and the non-equilibrium carriers are constantly generated and recombined, the resulting non-equilibrium carriers and the composite unbalanced carriers will reach a novel balance. If the external interaction disappears, the generated unbalanced carrier will rapidly disappear for recombination and return to the original equilibrium state.

3. **Carrier transport**

Carrier transport refers to the transfer of charge, energy, heat, and other processes through the movement of carriers. There are two basic transport mechanisms in semiconductor crystals: drift motion, i.e., carrier motion caused by an electric field; diffusion motion, i.e., carrier flow caused by a concentration gradient.

If there are empty energy states in the conduction band and the valence band, the net acceleration and net displacement of electrons and holes in the semiconductor will be generated by the applied electric field force. The carrier motion as impacted by electric field force is termed as drift motion. The net drift of the carrier charge forms a drift current. As impacted by the electric field, the drift of free holes in the direction of the electric field, or the drift of electrons in the reverse electric field, will form a current. The carrier is expedited by the continuous acquisition of energy from the electric field, so its drift speed displays associations with the electric field.

On the other hand, carriers are scattered by the distorted potential deviating from the periodic field in the crystal field and lose their original direction of motion or energy. After re-acceleration, re-scattering and re-acceleration are conducted in sequence. Lastly, the scattering effect of deviating from the periodic potential makes the drift velocity of carriers not increase indefinitely. In terms of a constant electric field, the drift velocity v_D is proportional to the electric field intensity F, $v_d = \mu F$. The proportional coefficient μ is termed as mobility, defined as the drift velocity of the carrier under per unit electric field. In principle, mobility is a function of the electric field; in a weak field, however, mobility is independent of an electric field and can be considered a constant. Solar cells typically operate in low electric field conditions. The drift current density Jn of the electron concentration n is expressed as

$$J_{\mathrm{n}} = -qnv_{\mathrm{d}} = qn\mu_{\mathrm{n}}F \tag{2.41}$$

The drift current density Jp of the hole concentration p is defined as

$$J_{\mathrm{p}} = qpv_{\mathrm{d}} = qp\mu_{\mathrm{p}}F \tag{2.42}$$

The conductivity of n-type and p-type semiconductors is, respectively, expressed as

$$\sigma_{\mathrm{n}} = nq\mu_{\mathrm{n}}, \sigma_{\mathrm{p}} = pq\mu_{\mathrm{p}} \tag{2.43}$$

A situation where both electrons and holes contribute to conductance is

$$\sigma = \mathrm{q}(n\mu_{\mathrm{n}} + p\mu_{\mathrm{p}}) \tag{2.44}$$

The mentioned two formulas have also been touched before. Mobility is one of the main macroscopic parameters of semiconductor materials, whose unit is cm^2/(V·S). It complies with the scattering process encountered by carrier motion in solids. It relates to the elastic or inelastic scattering of carriers caused by lattice defects, impurities, and lattice vibrations in crystals. The parameters describing this scattering process are τ and the probability of scattering $\frac{1}{\tau}$. τ can be understood as the average time interval between carriers in the two scatterings, and its size directly reflects the migration ability of carriers in the crystal. The critical scattering mechanisms in solar cells are lattice scattering and ionization impurity scattering. Mobility is controlled by lattice scattering at low doping levels, whereas by ionized impurity scattering at high doping levels.

Diffusion motion occurs when the concentration of particles in solids (e.g., atoms, molecules, electrons, and holes) is not evenly distributed in space. The diffusion movement of carriers from high concentration to low concentration (just in the opposite direction of the gradient) is a critical mode of carrier transport. For instance, if a beam of light is incident on semiconductor material, the absorption of light by the semiconductor is attenuated along the incident direction. In the range of absorption depth from the surface, considerable electrons and holes will be excited to form an uneven distribution of photogenic carrier concentration from the surface to the body. In this scenario, though the semiconductor is at the same temperature, the carrier distribution refers to a function of the spatial position. Without light, the n-type semiconductor electron concentration space is evenly distributed as n0. After being illuminated, the electrons have a concentration distribution in the x-direction of the illumination as n(x). Furthermore, the concentration variation of the photogenerated electrons in the x-direction is defined as Δn(x) = n(x)-0, and the diffusion movement of electron diffusion flux density is expressed as

$$J_{n\ \mathrm{diffusion}} = \mathrm{q}D_{\mathrm{n}}\frac{\mathrm{d}\Delta n}{\mathrm{d}x} \tag{2.45}$$

The diffusion current density is opposite to the concentration gradient. However, the electron has a negative charge, so the diffusion current density of the electron $J_{\mathrm{n\ diffusion}}$ does not exhibit a negative sign. Likewise, the hole diffusion current density

$$J_{p\ \mathrm{diffusion}} = -\mathrm{q}D_{\mathrm{P}}\frac{\mathrm{d}\Delta p}{\mathrm{d}x} \tag{2.46}$$

where the proportional coefficients D_n and D_p are, respectively, the diffusion coefficients of electrons and holes, and the unit is cm^2/s.

Thus far, four independent currents have been known in semiconductors (i.e., electron drift current and diffusion current, hole drift current and diffusion current). The total current density is the sum of the four. For the one-dimensional case, it yields

$$J = en\mu_{\mathrm{n}}E_x + ep\mu_{\mathrm{p}}E_x + eD_{\mathrm{n}}\frac{\mathrm{d}n}{\mathrm{d}x} - eD_{\mathrm{p}}\frac{\mathrm{d}p}{\mathrm{d}x} \tag{2.47}$$

Promoted to 3D, it yields

$$J = en\mu_{\mathrm{n}}E + ep\mu_{\mathrm{p}}E + eD_{\mathrm{n}}\nabla n - eD_{\mathrm{p}}\nabla p \tag{2.48}$$

Electron mobility expresses the motion of electrons in a semiconductor under electric field forces. The diffusion coefficient of electrons expresses the motion of electrons in a semiconductor under a concentration gradient. Correlations are identified between electron mobility and diffusion coefficient. Likewise, the mobility and diffusion coefficients of holes are not independent of each other.

In the state of thermal equilibrium, there is no net hole flow or net electron flow, i.e., drift and diffusion currents must be strictly balanced. In non-degenerate materials, Einstein relations are defined as

$$\frac{D}{\mu} = \frac{kT}{q} \tag{2.49}$$

It allows the diffusion coefficient to be directly calculated from mobility [8].

When a beam of light with an energy greater than the gap width of the semiconductor material is incident vertically to the surface of the PN junction, the incident light will excite the electron–hole pairs in the junction region. The photogenerated electrons and hole in the space charge region are split by the junction electric field, and the photogenerated carriers in the diffusion length range near the junction are diffused into the space charge region, which is also separated by the electric field. As suggested in the first chapter, the electrons in the p region drift to the N region as impacted by the electric field, and the holes in the N region drift to the p region; as a result, the photogenerated current from the N region to the p region is formed. The photogenerated carriers drift and accumulate to form an electric field opposite to the direction of the thermal equilibrium junction electric field and produce a forward junction current opposite to the direction of the photogenerated current, which remedies the junction electric field. When the photogenerated current is equated with the forward junction current, a stable electric potential difference is established across the PN junction. Thus, the above description of the photoelectric process reveals that when light is irradiated to the solar cell, the diffusion motion and drift motion of carriers exist simultaneously.

2.2.3 Semiconductor Junction

1. **Semiconductor junction and generation of photovoltaic effect**

The PN junction is mentioned in the first chapter when the photovoltaic effect is introduced, which is elucidated here.

It is generally known that high-purity semiconductor materials have great resistance, and can form n-type or p-type semiconductors if a certain amount of impurities are mixed into the semiconductors. Rigorous control of the amount of impurity introduced can make the semiconductor resistivity meet the needs of the device. Silicon is a semiconductor material that has four valence electrons. As mentioned earlier, if an element having five valence electrons is incorporated as an impurity into silicon, it becomes an n-type semiconductor. Since the doped impurity has one more valence electron than silicon, this electron can act as a transport current in silicon; such type of impurity is termed as a donor impurity. The electron motion forms an electric current, which leaves the ionized donor impurities. For a large number of the mentioned electrons, they are termed as majority carriers. In n-type semiconductors, there are considerable negatively charged electrons and an equal number of positively charged ionized donor ions, thus maintaining electrical neutrality. Compared with the number of electrons, there are a few holes in n-type semiconductors, i.e., minority carriers. When a material with three valence electrons is mixed into silicon as an impurity, it turns out to be a p-type semiconductor. Since the impurity is one electron less than silicon, equivalent to a hole, the impurity is termed as the acceptor impurity. In p-type semiconductor materials, holes are majority carriers. In comparison with the number of holes, the electron is a minority carrier. In p-type semiconductors, there are considerable positively charged holes and an equivalent number of negatively charged ionized acceptor ions, so they remain electrically neutral.

An n-type semiconductor and a p-type semiconductor are in very close contact to form the PN junction. After their contact, most carrier electrons in the N region diffuse to the p region, and most carrier holes in the p region diffuse to the N region as impacted by the concentration difference between electrons and holes at the interface. After diffusion, a positively charged donor ion is left on the n side of the interface for building a positively charged region. Likewise, a negatively charged acceptor ion is left on the p side of the interface to form a negatively charged region. Thus, on both sides of the interface between the n-type region and the p-type region, a very thin region with a positive charge on one side and a negative charge on the other side is formed. The electron diffusion current consists of the diffusion electron flow due to the concentration difference, which is termed as a space charge region or PN junction. Diffusion electron flow formed by concentration difference constitutes electron diffusion current. The diffusion current includes electron diffusion current and hole diffusion current. In the PN junction, there is an electric field from the N region to the p region. As it is generated by the internal charge of the PN junction, it is termed as the built-in electric field or the self-built electric field. As impacted by the built-in electric field, the drift motion of carriers will be generated in the space charge area; as a result, the electrons pull from the p region to the N region, and the holes pull from the N region to the p region, and the direction is opposite to the direction of diffusion motion. Then, the diffusion motion predominates, the positive and negative charges on both sides of the space charge region progressively increase, the space charge region progressively widens, and the built-in electric field is enhanced progressively. With the enhancement of the built-in electric field, the drift motion is progressively enhanced, the diffusion motion begins to weaken, and lastly the diffusion motion and the drift motion tend to balance, the diffusion motion no longer develops, the thickness of the space charge region no longer increases, and the built-in electric field no longer increases. Then, the number of carriers that diffuse and drift is equal and the direction of motion is opposite to achieve dynamic equilibrium. In the state of dynamic equilibrium, the potential on both sides of the built-in electric field is different. The N region is higher than the p region, and there is a potential difference termed as a PN junction barrier, also termed as a built-in potential difference or a contact potential difference, which is represented by the symbol V_D. It can be seen from the flow of electrons from the N region to the p region that the potential difference of the p region for the N region is a negative value. Since the p region has a potential $-V_D$ with respect to the N region, all the electrons in the p-region have additional potential energy, whose value is.

$$\text{Electric potential energy} = \text{charge} \times \text{potential} = (-q) \times (-V_D) = q \times V_D$$

qV_D is commonly the height of the barrier, denoted by φ_B. The barrier height depends on the doping concentration of the n and p regions, and the higher the doping concentration, the higher the barrier height.

The so-called non-equilibrium PN junction refers to the PN junction under an external bias voltage. Then, the PN junction is unbalanced, and its potential barrier differs from that at the equilibrium. The diffusion current will not be able to offset the drift current, and there is a current passing through the PN junction. Since the barrier of the PN junction is a high resistance layer, when the voltage "V" is applied across the PN junction, this voltage will concentrate on the barrier region.

Consider the PN junction is applied with a forward voltage, i.e., the p-type side is linked to the positive electrode, and the n-type side is linked to the negative electrode. This voltage is opposite to the direction of the original contact potential difference V_D in the balanced PN junction. Accordingly, after the forward voltage is adopted, the electric field in the potential barrier decreases, the barrier width is narrowed, and the barrier height turns out to be $q(V_D + V)$, which decreases qV compared with the original barrier height. When the barrier height is lowered, the diffusion effect of carriers in the space charge region will be greater than the drift effect. The electrons in the N region are continuously diffused to the p region, and the holes in the p region are continuously diffused to the N region. Such injected carriers are termed as non-equilibrium minority carriers. The electrons injected into the p region first accumulate at the barrier boundary, and diffuse into the depth direction of the p region for the concentration gradient. In the diffusion process, the electrons are recombined with the holes, and the electron current is progressively converted into the hole current. After the distance of several electron diffusion lengths, the electrons are basically recombined with the hole, and the electron current injected into the p region in the N region is fully converted into the hole current in the p region. Likewise, the holes injected into the N region were firstly accumulated at the potential barrier boundary. Because of the concentration gradient, it diffuses in the depth of the N region. In the diffusion process, it is continuously recombined with electrons, and the hole current was progressively converted into the electron current. After the distance of several hole diffusion lengths, the hole is basically recombined with the electron, and the hole current injected into the N region in the p region is fully converted into the electron current in the N region. The region on both sides of the barrier is termed as the diffusion length of the diffusion region. The p zone is the electron diffusion zone, and the n zone is the hole diffusion zone. On the whole, though the total electron–hole current flowing through any section of the PN junction should be the same, the proportion of electron current and hole current in the total current in different regions is different. Overall, the hole current approaches the end near the p-type side, and the electron current approaches the end near the n-type side. The two parts of the current are alternated by the composite in the middle of the PN junction, and the total current remains unchanged. Obviously, as the forward voltage increases, the barrier height decreases further, and the electrons moving over the barrier from the N region to the p region and the holes moving from the p region to the N region increase rapidly, so the forward current of the PN junction increases rapidly. This current is termed as the forward current of the PN junction. Consider the case where the PN junction is introduced with a reverse voltage, i.e., the p-type side is linked to the negative electrode, and the n-type side is linked to the positive electrode. This voltage is in the same direction as the original built-in potential V_D

in a balanced PN junction. Thus, after the reverse voltage is applied, the electric field in the barrier will increase, the barrier area will become wider, and the barrier height will increase compared with the equilibrium, changing from the original qV to $q(V_D + V)$.

Due to the robust electric field in the potential barrier of the reverse PN junction, the minority carriers at the edge of the barrier are all pulled away by the electric field, and the minority carrier concentration in the diffusion region on both sides of the barrier is lower than the equilibrium concentration. Then, what happens in the diffusion region is the inverse process of recombination generation? The holes generated in the n-type diffusion region are diffused to the potential barrier region, pulled into the p-type region as impacted by a robust electric field, and merged with the holes generated in the diffusion region to move toward the negative electrode. The motion of electrons is fully similar. The electrons generated in the p-type diffusion region diffuse to the potential barrier region, and then, as impacted by the electric field, drift across the barrier region and join up with the electrons generated in the n-type diffusion region to move toward the positive electrode. To form the reverse current from the n-type region to the p-type region, electron current and hole current replace each other in the diffusion region of the potential barrier region. Since the electrons extracted from the n-type region and the p-type region are minority carriers, the minority carrier concentration at the two boundaries of the barrier region does not need significantly reverse voltage to be dropped to zero, and the voltage is zero. In other words, the minority carrier concentration gradient does not vary with the reverse voltage, and the magnitude of the reverse current depends on the minority carrier concentration gradient in the diffusion region. Since the gradient is not large, the reverse current of the PN junction is a small value and does not vary with the applied reverse voltage. This is the reverse saturation current of the PN junction. When the voltage V is applied, the current density J_{i} flowing through the PN junction can be expressed as follows:

$$J_{\mathrm{i}} = J_0(\mathrm{e}^{\frac{qV}{kT}} - 1) \tag{2.50}$$

where J_0 is the reverse saturation current density.

When the PN junction is introduced with reverse voltage, the space charge region is widened, and the electric field is enhanced. When the reverse voltage rises to a certain extent, the reverse current dramatically increases. If the external circuit cannot limit the reverse current, the reverse current will be sufficiently large to burn the PN junction. The voltage at which the reverse current suddenly increases is termed as the breakdown voltage. There are two basic breakdown mechanisms, namely tunnel breakdown (i.e., Zener breakdown) and avalanche breakdown. The former has a breakdown voltage of less than 0.6 V as well as a negative temperature coefficient. Besides, the latter has a breakdown voltage of greater than 0.6 V as well as a positive temperature coefficient. The positive and negative charges in the space charge region constitute a capacitive device when the PN junction is introduced with reverse voltage. Its capacitance varies with the applied voltage.

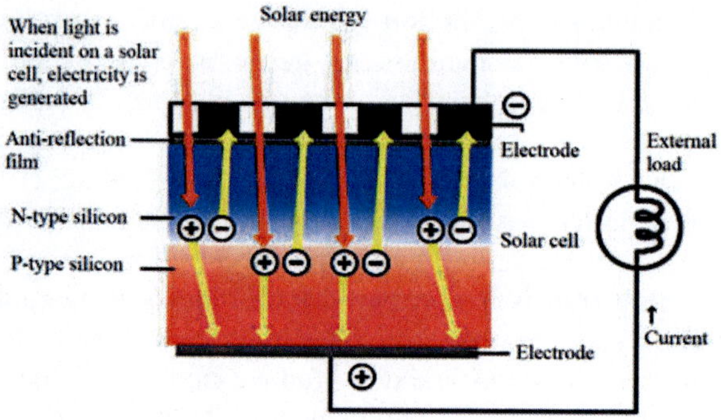

Fig. 2.15 Generation of photovoltaic effects in semiconductor junctions

The PN junction under illumination is a non-equilibrium PN junction as well. The injected non-equilibrium carrier is generated by electron–hole pairs after photoexcitation. If the energy of the photon is equal to or greater than E_g, photons are irradiated into the semiconductor to excite the electron from the valence band to the conduction band, thereby leaving a hole in the valence band and forming an electron–hole pair. The excited electrons spontaneously tend to re-jump back into the valence band and recombine with the hole, releasing the absorbed energy and restoring the equilibrium position. Accordingly, the electrons and holes should be separated before they recombine, so they will not recombine and achieve the purpose of converting light into electricity. Such separation can be achieved by the space charge region of the PN junction (Fig. 2.15).

The electron–hole pairs in the space charge region can be separated quickly to produce photocurrent as impacted by the built-in electric field. If the generated electron–hole pairs outside the space charge region have a long enough lifetime, then the light-generating minority carriers in the p-region and N-region will be diffused to the potential barrier area of the PN junction, and separated by the built-in electric field to generate the photocurrent. Photogenerated non-equilibrium minority carrier electrons are swept from the p region into the N region. Photogenic non-equilibrium minority carrier holes are swept from the N region into the p region. Thus, effective photoelectric conversion can be achieved [9].

2. **Types of semiconductor junctions**

The types of semiconductor junctions can be divided into the abrupt junction, graded junction, and hyper-abrupt junction according to the doping mode.

The main characteristic of the abrupt junction is that the impurity concentration in each doping region is uniformly distributed, and the concentration of impurities at the interface has a sudden jump. The characteristic of a linear graded junction is that the doping concentration on both sides of the metallurgical junction can be approximated by linear distribution. The characteristic of the hyper-abrupt junction is that the doping concentration starts to decrease from a metallurgical junction, which is a special PN junction.

The doping in the actual PN junction is often not uniform. In some practical electrical applications, specific non-uniform doping is often used to achieve the required capacitance characteristics of the PN junction.

Moreover, according to the similarities and differences of materials, the semiconductor junction can be divided into the homogeneous junction, heterogeneous junction, and metal–semiconductor junction.

A junction made of the same semiconductor material is termed as a homogenous junction. A junction composed of two different semiconductor materials is termed as a semiconductor heterogeneous junction. There are four basic types of heterojunctions. The junction of doped-type variation is termed inverse heterostructure junction. The inversion heterojunction can be made into an nP junction or Np junction, where capital letters indicate a wider bandgap material. A heterojunction with the same doping type is termed as a homogeneous heterojunction, which can be made into nN and pP homogeneous heterojunctions.

In fact, the PN junction is not a simple physical combination of p-type semiconductor photoelectric material and n-type semiconductor photoelectric material, but is formed by alloy method, diffusion method, ion implantation method, and film growth method. The simplest method is the diffusion method, which forms the PN junction by diffusing impurities and forming a layer of material on the substrate that is opposite to the conductive type of the substrate. The basic structure of solar cells can be divided into two types according to a different substrate and diffused impurities. One is that the substrate material is a p-type semiconductor photoelectric material, which diffuses impurities to provide electrons and forms an n-type material on the surface of the p-type substrate material to prepare the PN junction, in which the n-type material is the receiving surface. On the contrary, the substrate material is n-type semiconductor photoelectric material, and the diffused impurities provide holes. A p-type material is formed on the surface of the n-type substrate material, and the p-type material is the receiving surface.

The metallurgical contact formed by the metal–semiconductor is termed as metal–semiconductor junction (M–S junction) or metal–semiconductor contact.

A metal–semiconductor junction can be achieved by pressing the whisker-shaped metal stylus on a semiconductor crystal or by vapor-depositing a large-area metal film on a semiconductor surface under a high vacuum. The former is termed as point contact, and the latter is termed as surface contact correspondingly. The two critical effects of metal–semiconductor contacts are the rectification effect and the ohm effect. The former is termed as the rectification contact and rectifying junction. The latter is termed as ohmic contact and unrectified junction. Non-rectifying characteristics have a low ohmic voltage drop and no rectifying effect regardless of the polarity of the applied voltage. Such contact is an integral part of the development and production of almost all semiconductor devices, requiring ohmic contact to be connected to other devices or circuit elements.

3. **Semiconductor–semiconductor junction and metal–semiconductor junction**

The semiconductor materials can form homogenous and heterogenous junctions given their similarities and differences. A homogenous junction requires two different types of doping on a semiconductor material to form a PN junction. Moreover, the heterogeneous junction can be made into homogeneous and heterogenous hetero-junctions. The formation of semiconductor–semiconductor junctions results from the uneven distribution of carriers, thereby causing majority carriers to spread to the minority carrier region; as a result, immovable impurity ions are left in the original position. The mentioned impurity ions in turn form a built-in electric field in the junction region to prevent further diffusion of carriers to reach the final equilibrium state. The energy band structure is the energy band bending, and the Fermi energy levels are at the same level.

As a conductor, a metal commonly has no forbidden bandwidth, and free electrons can move freely in the conduction band, exhibiting a robust conductive ability. In metals, electrons also obey the Fermi distribution. Like semiconductor materials, at absolute zero, electrons fill energy levels below the Fermi level (E_{Fm}), and the energy level above the Fermi level is fully empty. With the rise in the temperature, electrons can absorb energy and jump from low to high levels, while levels are primarily below the Fermi level. Besides, only a few electrons near the Fermi level can jump above the Fermi level, while a very small number of high-energy electrons absorb enough energy to jump outside the metal. E_0 denotes the energy of a stationary electron outside the metal surface in a vacuum, so the minimal energy required for an electron to transition from the metal to the outside body is expressed as

$$W_{\mathrm{m}} = E_0 - E_{\mathrm{Fm}} \tag{2.51}$$

$\mathrm{W_m}$ is termed as the work function of the metal.

Likewise, for semiconductor materials, a certain amount of energy is required to transfer an electron from the conduction or valence band to the body externally. Similar to metals, if E_0 represents the energy of stationary electrons outside the semiconductor surface in a vacuum, the work function of the semiconductor refers to the difference between E_0 and Fermi energy levels (E_{Fs}); it yields

$$W_{\mathrm{s}} = E_0 - E_{\mathrm{Fs}} \tag{2.52}$$

When the metal contacts with n-type semiconductor materials, they exhibit an identical vacuum electron energy level. If the work function of the metal before contact reaches over the work function of the semiconductor, the Fermi energy level of the metal is lower than the Fermi energy level of the semiconductor; besides, the difference of the Fermi energy level of the two is equated with the difference of the work function, i.e., $E_{\mathrm{Fs}} - E_{\mathrm{Fm}} = W_{\mathrm{m}} - W_{\mathrm{s}}$. After the contact, though the electron concentration of the metal is higher than that of the semiconductor, the Fermi energy level of the metal is lower than that of the semiconductor; as a result, electrons in the semiconductor flow to the metal, and the electron concentration on the metal surface increases. Thus, the metal surface is negatively charged and the semiconductor surface is positively charged. Moreover, the semiconductor and the metal have an identical amount of positive and negative charges, and the whole metal–semiconductor system remains neutral, which only enhances the potential of the semiconductor and down-regulates the potential of the metal.

After the electrons flow from the semiconductor to the metal, the near surface of the n-type semiconductor leaves a positively charged donor ion, and the electrons flowing to the metal are concentrated on the metal side of the metal–semiconductor interface layer due to the attraction of the mentioned positively charged ions. The mentioned electrons together with the donor ion formed a certain thickness of the built-in electric field and the space charge area. The direction of the built-in electric field from the n-type semiconductor to metal, primarily falling on the near-surface layer of the semiconductor. Similar to the semiconductor PN junction, the built-in electric field creates a potential barrier, termed as the surface barrier of the metal–semiconductor contact, i.e., the electron blocking layer, which bends the energy band of the space charge region. In addition, the electrons are subjected to forces opposed to the direction of diffusion due to the built-in electric field, causing them to flow from the metal to the n-type semiconductor. When equilibrium is reached, the number of electrons flowing from the n-type semiconductor to the metal is equal to that from the metal to the semiconductor, the net current in the space charge area is zero, the Fermi energy levels of the metal and the semiconductor are the same, and the difference between the potentials on both sides of the barrier is termed as metal–semiconductor contact potential difference. The contact potential difference of the metal–semiconductor equals the difference between the Fermi level of the metal and the semiconductor contact, or the difference between the work functions, namely

$$V_{\mathrm{ms}} = \frac{1}{q}(W_{\mathrm{m}} - W_{\mathrm{s}}) = \frac{1}{q}(E_{\mathrm{Fs}} - E_{\mathrm{Fm}}) \tag{2.53}$$

If the work function of the metal before contact is less than the work function of the semiconductor, i.e., the Fermi level of the metal is higher than the Fermi level of the semiconductor, as revealed from identical analysis, before the metal and the semiconductor are in contact, a thin high-density hole layer is formed on the metal side approaching the interface, and a certain thickness of the electron accumulation region is formed on the semiconductor side; as a result, a space charge region is formed exhibiting high electronic conductivity. It is termed as an electronic high-conductance area, i.e., an anti-blocking area.

Likewise, for the contact between the metal and the p-type semiconductor, there is also a space charge area near the interface, forming a hole barrier area (blocking area) and a hole high-conductance area (anti-blocking area).

If a voltage is applied between the metal and the n-type semiconductor, it will affect the effect of the built-in electric field and the surface barrier, showing the rectifying effect of the contact between the metal and the semiconductor. When a metal is linked to a positive electrode and a semiconductor to a negative electrode, the applied electric field is directed from the metal to the semiconductor, as opposed to the built-in electric field. The applied electric field will offset a part of the built-in electric field, resulting in a lower electron barrier and a thinner electron blocking layer, increasing the flow of electrons from the n-type semiconductor to the metal and increasing the current. In contrast, when the metal is linked to the negative electrode and the semiconductor to the positive electrode, the applied electric field is directed from the semiconductor to the metal, in line with the built-in electric field; as a result, the electron barrier increases, and the electron blocking layer is thickened, the flow of electrons from the n-type semiconductor to the metal is reduced, and the current to almost zero is down-regulated. Such characteristic is identical to the current–voltage characteristics of the PN junction, and exerts a rectification effect [10].

Metal-to-semiconductor contact can be divided into rectified contact and non-rectified contact. When the work function of the metal W_{m} > the work function of the n-type semiconductor W_{n} or when W_{m} < the work function of the p-type semiconductor W_{p}, a rectifying contact is formed between the metal and this n-type or p-type semiconductor, respectively. When the work function of the metal W_{m} < the work function of n-type semiconductor W_{n}, or when W_{m} > the work function of the p-type semiconductor W_{p}, the non-rectifying contact, i.e., an ohmic contact, is formed between the metal and the n-type or p-type semiconductor, respectively.

An ohmic contact refers to a contact that does not produce significant additional impedance and does not significantly alter the concentration of equilibrium carriers in the semiconductor. Electrically speaking, when there is a current flowing through the ohmic contact, the voltage drop across it is significantly smaller than the voltage drop across the sample or device itself. This contact does not affect the I-V characteristics of the device. Ohmic contacts also show critical applications in practice. Overall, semiconducting devices exploit metal electrodes to input or output current, which requires a high ohmic contact between the metal and the semiconductor. Ohmic contact is one of the key problems in the design and manufacture of UHF and high-power devices. In the ohmic contact, for the small additional resistance, there are symmetrical current–voltage characteristics in the forward and reverse directions (Fig. 2.16).

When the metal and the semiconductor are in contact, if the doping concentration of the semiconductor is significantly high, the width of the barrier region becomes very thin, and electrons easily tunnel via the barrier for the tunneling effect to generate a considerable tunnel current, even exceeding the emission current of thermal electrons to become the major component of the current. When tunnel current is dominant, its contact resistance is small and can be used as an ohmic contact. Accordingly, when the semiconductor is heavily doped, its contact with the metal can form a near-ideal ohmic contact.

Ohmic contact is commonly made by making a heavily doped region on an n-type or p-type semiconductor and then contacting with a metal to form a metal-$n^{+}n$ or metal-$p^{+}n$ structure. There are many ways to form a contact between a metal and a semiconductor, e.g., evaporation, sputtering, and electroplating.

When a metal and a semiconductor form a rectifying contact, it has a similar rectifying effect to the PN junction. A diode is formed by drawing leads at the p-side

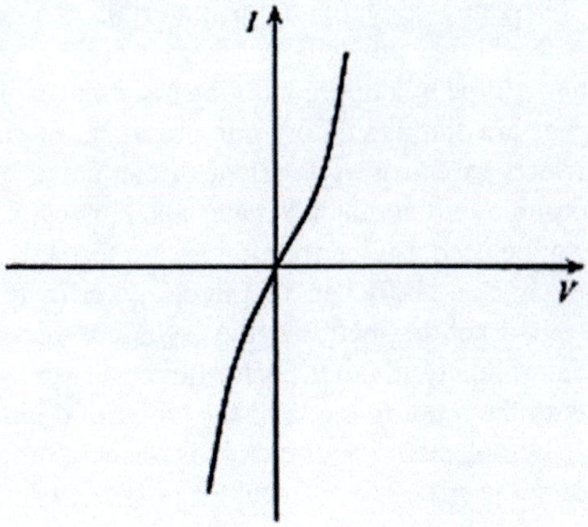

Fig. 2.16 I-V characteristics of metal–semiconductor ohmic contacts

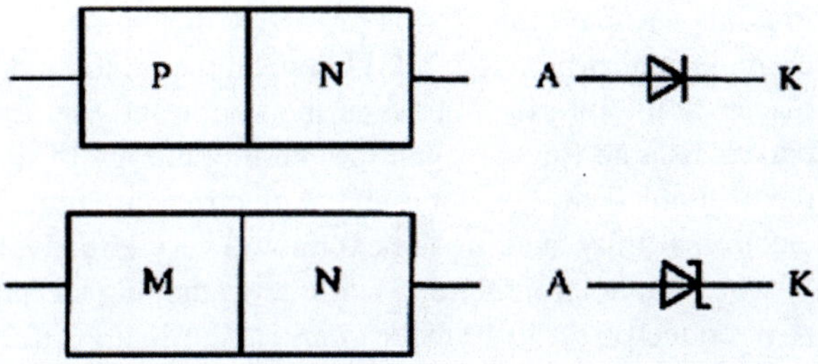

Fig. 2.17 Schematic diagram of a diode formed by a semiconductor junction and a metal–semiconductor junction

and the n-side of the PN junction, respectively. Likewise, diodes can also be made by using the rectifying effect of the metal–semiconductor contact and are termed as Schottky barrier diodes. The schematic diagram of the two diodes is shown in Fig. 2.17.

4 Junctions in organic semiconductor materials

With the progress of organic electronics, a variety of electronic devices with different functions have appeared frequently, e.g., organic photovoltaic cells, organic light-emitting diodes, and field-effect transistors. The mentioned devices composed of multilayer materials have heterogeneous junctions with different functions between the layers of their different materials. They can perform different functions, e.g., charge separation of excitons, energy transfer between excitons and other objects, charge transfer of different materials in the ground state, and the blocking effect of different carriers.

Heterojunctions in organic solar cells are often associated with the working principle of the battery, that is, how the heterojunction causes the separation of photogenic excitons (electron–hole pairs) that convert solar energy into chemical energy. How to effectively decompose excitons into electric energy in the battery involves the introduction of a heterojunction, otherwise, the ideal photoelectric conversion efficiency of cells cannot be obtained.

Organic semiconductor materials can also be classified into p-type and n-type, whereas they are different from inorganic semiconductor materials that form p-type and n-type by doping with intrinsic materials. Overall, organic semiconductor materials are conjugated organic polymers or small molecular compounds. Compounds which are electron-deficient or can act as electron acceptors are considered n-type organic semiconductor materials. However, compounds that are electron-rich or that can be used as electron donors are termed as p-type semiconductor materials. Most organic and polymeric semiconductor materials are p-type.

The heterojunction in the organic semiconductor materials can fall under a bilayer heterojunction, a bulk heterojunction, and a molecular D-A junction in accordance with the binding mode of the electron donor (D) and the electron acceptor (A).

A molecular D-A junction is formed by covalently attaching a unit with an electron donor property to an acceptor polymer or a small molecule.

The double-layer heterojunction forms a planar D-A interface by layering the organic materials of the donor and the acceptor. The driving force of charge separation in a bilayer heterojunction is primarily the energy difference of the lowest untaken up molecular orbital (LUMO) between the donor and the acceptor, i.e., the electron barrier at the interface between the donor and the acceptor. At the interface, if the barrier is large (greater than the binding energy of the exciton), the dissociation of the excitons is more favorable: the electrons are transferred to materials with higher electron affinity, making the exciton be very effectively dissociated.

The bulk heterojunction is similar to the bilayer heterojunction exploiting the D-A interface effect to transfer charge. The main differences between them are [11] as follows. ① the charge separation in the bulk heterostructure is generated in the whole active layer, while the charge separation in the double-layer heterostructure only occurs in the space charge region (several nanometers) at the interface. Thus, the exciton dissociation efficiency is higher in the bulk heterojunction, and the exciton recombination probability is lower, and as a result, the energy loss due to the small diffusion length of the organic exciton is reduced or avoided. ② For the presence of the interface in the whole active layer, the carrier transferring to the electrode in the bulk heterojunction device is primarily through the percolation between particles, and the carrier transfer medium in the double-layer heterojunction is the donor and the acceptor with continuous spatial distribution, so the carrier transfer efficiency of the double-layer heterojunction is relatively high. Due to being limited by the carrier transport characteristics, bulk heterostructure is sensitive to the morphology and particle size of materials, and the filling factor is correspondingly small. The morphology of the material and the mixing degree of donor/acceptor in a bulk organic heterostructure have a great influence on charge separation, photocurrent generation, and total energy conversion efficiency. If the particle size is too large, the separation efficiency of the charge will be down-regulated. If the particle size is overly small, the transfer of the charge will be blocked, and the probability of recombination of the free charge is relatively high. Accordingly, the particle size of materials should be optimized in the bulk heterostructure to satisfy higher charge separation efficiency and transport efficiency.

2.3 Upper Efficiency Limit of Solar Cells

In Chapter 10, the efficiency limit of solar cells is to be elucidated. Here, we introduce them briefly so that we can have a basic concept of the efficiency limit.

The photoelectric conversion process of solar cells consists of the exchange of energy among subsystems in the system composed of three parts (i.e., the sun, the surrounding environment, and the solar cells). The solar cell environment here is often considered the earth environment. In such a system, the exchange of energy between subsystems is mutual. There is the radiation of the sun, the absorption of the solar cell, and the environment, as well as the emission of the solar cell and the ambient light of the earth. However, the temperature of the solar cell and earth environment is lower, while the wavelength of emission photon is longer. Lastly, the macro-system consisting of three parts is in equilibrium. Statistical theory points out that the critical and sufficient condition of a system's macroscopic equilibrium is the detailed balance condition, which is the basis for discussing the macroscopic system [2].

2.3.1 *Detailed Balance Principle*

The detailed balance principle is the critical and commonly used means to evaluate the theoretical efficiency of solar cells. The concept of detailed balance was proposed in 1954 by Roosbroeck and Shockley in the Journal of Applied Physics. In 1961, William Shockley and Hans J. Queisser published an article entitled "Detailed balance limit of efficiency of p–n junction solar cells" in Applied Physics, in which they proposed the concept of detailed balance efficiency limit. By given several assumptions, a method is proposed to calculate the efficiency limit, which predicts that the efficiency limit of single-junction solar cells reaches 31%.

For a physical system in thermodynamic equilibrium, the principle of detailed balance can be derived from microscopic reversibility by classical mechanics or quantum mechanics. For systems far from equilibrium or non-physical systems, the detailed balance principle is also true under certain circumstances. In other words, the principle of detailed balance is a critical principle of broad significance. When this principle is employed in some systems, compared with traditional methods, some important laws in statistical physics can be obtained in a significantly easier way and we can gain insights into several physical processes [12].

2.3.2 *Maximum Efficiency of Solar Cells*

In 2001, Green put forward the development process of the solar cell falling into three stages. The first-generation solar cells are bulk silicon solar cells (e.g., single-crystal

Si and polycrystalline Si) and the second-generation solar cells are thin film solar cells (e.g., amorphous Si and GaAs,CdTe, CIGS, etc.), which are single cells and have been basically commercialized. Besides continuing to maintain the thin film property, the third generation of solar cells also use rich, non-toxic raw materials, and exhibits the highest photoelectric conversion efficiency.

If all factors are optimized (e.g., electrical, optical, and material), what is the ultimate efficiency of solar cells? This is one of the critical issues that people are concerned about. In accordance with the detailed equilibrium limit principle, Shockley and Queisser calculated the efficiency of ideal single-junction solar cells as a function of material forbidden bandwidth Eg. When Eg$\approx$ 1.3ev, the ultimate efficiency under 1sun irradiation reaches only 31%, and the ultimate efficiency under all-concentrated light is 40% [13]. The significance of the detailed balance principle lies in that it is the lowest theoretical limit identified, lower than the Carlo efficiency, and it is the highest efficiency that can be achieved objectively. The reason why this theory has such low efficiency is that there are several hypotheses:

(1) Only photons exhibiting energy greater than the forbidden bandwidth can be absorbed, which cannot be absorbed below the forbidden bandwidth.
(2) A photon can generate at most one electron–hole pair.
(3) The absorbed photon energy is adopted to stimulate and store the potential energy of the electron–hole pair.
(4) Radiation recombination is the only case.
(5) Semiconductor materials fully comply with the behavior of blackbody.

The idea is explained that the sun radiates photons of different wavelengths to the solar cell in a way abiding by the "Planck blackbody radiation distribution". The portion of the mentioned photons from the 0 wavelengths to the lower limit of the forbidden bandwidth of the semiconductor is absorbed by the solar cell. By integrating the photon energy distribution of solar radiation, the overall number of photons absorbed by the solar cell can be calculated. Each photon generates an electron–hole pair, so the total number of carriers generated can be obtained. Second, solar cells acting as a device with a certain temperature inevitably radiate energy, so the energy loss attributed to the radiation of the solar cell becomes an inevitable part. Accordingly, when the battery shortens:

The current density = (the number of total carriers a generated-the number of carriers radiated by the solar cell itself) $\times$ (electron power). The output voltage refers to the "potential difference" of electron–hole pair, so the output energy = current $\times$ voltage can be obtained. Subsequently, the theoretical limit of efficiency is output energy $\div$ input energy.

Overall, a solar cell gets a limited amount of energy from the sun; besides being able to convert it into electricity for use, it should be able to radiate some of it out. The last available efficiency is the highest limit. The efficiency limit of detailed balance is not high for its rigorous assumptions. For instance, "one photon can only excite at most one electron–hole pair" (actually, a photon by collision ionization can produce multiple electron–hole pairs). Moreover, photons with energy less than the bandgap width are not absorbed (as also revealed from the practical situation and experiments, they are not fully absorbed, many are absorbed by a type of "exciton", and energy transfer also exists) [14].

When discussing the maximum efficiency of solar cells, the detailed balance principle should be complied with, and many other assumptions should be proposed. For instance, based on the thermodynamic theory, the conversion efficiency of the solar cell can reach a very high level. In Chapter 10, the maximal efficiency of solar cells is to be presented more generally.

References

1. T. Wang, Research on silicon solar cells, The Master Degree Dissertation of National University of Defence Technology, 2006
2. S. Xiong, M. Zhu, et al., *Solar Cell Foundation and Application*, (Science Press, Beijing, 2009), p.10, (in Chinese)
3. Baidu Encyclopedia: black-body radiation. https://baike.baidu.com/item/%E9%BB%91%E4%BD%93%E8%BE%90%E5%B0%84/516583?fr=aladdin
4. Wikipedia: Black-body radiation. https://en.wanweibaike.com/wiki-Blackbody%20radiation
5. S. Yu, J. Zhong et al., *Solar Energy Photovoltaic Device Technology*, (University of Electronic Science and Technology Press, 2011), (in Chinese)
6. NREL Website: Reference Solar Spectral Irradiance
7. D.A. Neamen, *Semiconductor Physics and Devices Basic Principles*, ed. by Y. Zhao et al., 3rd edn. (Electronic Industry Press, USA, 2005)
8. A. Luque et al., *Handbook of Photovoltaic Science and Engineering* (Wiley, 2003)
9. J. Si, Conversion efficiency calculation of the third generation solar cell, The Master Degree Dissertation of Hefei University of Technology, 2007
10. D. Yang, *Solar Cell Materials* (Wunan Book Publishing Company, 2008)
11. S. Wu et al., Some probles on the organic semi-conductor Hetero-junction. Imaging Sci. Photochem. **28**(2), 147–149 (2010)
12. L. Fu et al., Analysisi of ARP based on the principle of detailed balance. J. Nat. Sci. Heilongjiang Univ. **16**(2), 86–88 (1999)
13. J. Zhao et al., Advances in novel high-efficiency solar cell research, Physics **40**(4), 233–240 (2011)
14. The ultimate efficiency of solar cells. https://www.china-nengyuan.com/tech/17443.html

Chapter 3
Crystalline Silicon Solar Cells

3.1 Silicon Manufacturing Process

3.1.1 Silicon Materials

The distinctive nature exhibited by silicon makes it critical in the modern electronic information industry. The development of silicon is considered a milestone in materials and electronic information worldwide in the twentieth century, and it is silicon that underpins the booming of information in the twenty-first century.

The silicon material originates from high-quality quartz sand (i.e., silica sand), and it is primarily composed of SiO_2. China enjoys abundant and wide silica sand reserves, which facilitate industrial production.

Silicon materials can be decomposed into semiconductor grade silicon and metal silicon in accordance with their purity; based on their crystal forms, they can be split into single crystalline silicon, polycrystalline silicon, and amorphous silicon. Single crystalline silicon can be split into Czochralski single crystalline silicon and Float-Zone single crystalline silicon following the process approach. Polycrystalline silicon consists of high-purity polycrystalline silicon, thin-film polycrystalline silicon, ribbon polycrystalline silicon, as well as cast polycrystalline silicon. Polycrystalline and single crystalline silicon pertain to crystalline silicon. In such type of silicon, polycrystalline silicon exhibiting high purity is refined from metal silicon, and single crystalline silicon can be obtained by purifying and smelting high-purity polycrystalline silicon.

In the 1980s, polycrystalline silicon materials began to be applied in solar cells, so they have acted as a partial substitute for single crystalline silicon materials. Though polycrystalline silicon material took up a large proportion in the market for its low price, it exhibits low product quality, which could not match its market share at that time. As the physical mechanism and optical properties of polycrystalline silicon materials are increasingly expounded, the quality of polycrystalline silicon materials has been significantly elevated. In combination with its natural price merit,

C. Zhang et al., *Semiconductor Photovoltaic Cells*,
https://doi.org/10.1007/978-981-15-9480-9_3

polycrystalline silicon lastly takes up a larger market share than single crystalline silicon.

At present, most of the studies on silicon solar cells highlight commercial applications. More prominent material properties are critical to enhancing the final performance of the solar cells, so the characteristics of silicon materials should be studied [1].

3.1.2 *Solar Grade Silicon Materials*

Crystalline silicon solar cells have dominated the photovoltaic industry in the first place. Today, they still dominate the photovoltaic market through their studies have declined with the emergence of some novel innovations. However, they tend to dominate the market for a foreseeable period.

One of the reasons why crystalline silicon takes up a critical proportion in the photovoltaic market is that silicon technology has been significantly developed as stimulated by the microelectronics industry. On the one hand, with the booming of the microelectronics industry, silicon raw materials and related equipment have increasingly reasonable prices; on the other hand, the development of microelectronic processing technology directly and technically underpins the photovoltaic industry, allowing photovoltaic processing production to be transformed and developed rapidly.

The most common and economical way to prepare solar grade silicon is to purify the metal silicon directly until the metal exhibits purity satisfying the application requirement for solar cells. During the purification process, the critical thing is to down-regulate the concentration of impurities with higher content in the metal silicon to below $5 \times 10^{16}/\mathrm{cm}^3$. In metal silicon, the impurity content usually reaches over 5%, in which the concentration of iron is 1700×10^{-6}–3000×10^{-6} on the whole, the concentration of aluminum is 1200×10^{-6}–4000×10^{-6}, and the concentration of boron and phosphorus reaches 20×10^{-6}–60×10^{-6} (1×10^{-6} denotes the number of impurity atoms per million molecules). Among the impurities covered in the mentioned metal silicon materials, except for the higher segregation coefficient of boron and phosphorus, the segregation coefficient of other metals is low, so metal impurities can generally be removed by directional solidification. However, boron and phosphorus cannot be effectively removed to reach the required standard, so subsequent processes are required [2].

3.1.3 *Manufacture of Single Crystalline Silicon*

Single crystalline silicon exhibits a cubic crystal system and a diamond structure, so it is considered a semiconductor material displaying prominent properties. Since the initiation of polycrystalline silicon application in the 1940s, the growth technology of silicon materials has been increasingly perfect, and it is extensively employed in infrared spectral frequency optical components, infrared and ray detectors, integrated circuits, solar cells, etc. Moreover, silicon does not exhibit toxicity, and its raw material quartz (SiO_2) takes up nearly 60% of the crustal composition, so its raw material supply can be fully ensured. For the merits and applications, silicon material has been considered the critical, most productive, fastest-growing, and most versatile semiconductor material.

Thus far, the solar photovoltaic industry basically complies with silicon materials, and solar photovoltaic devices worldwide are primarily constituted by single crystalline silicon. To be specific, single crystalline silicon solar cells were initially studied and adopted, and it remains a critical material for solar cells. Single crystalline silicon refers to an ideal material for solar cells for its excellent integrity, high purity, abundant resources, advanced technology, stable working efficiency, high photoelectric conversion efficiency, and long service life. Accordingly, it has been highlighted and favored by researchers at home and abroad. As single-crystal silicon solar cells have been increasingly demanded, the competition in the single-crystal silicon market is becoming progressively furious. To dominate the market, breakthroughs should be made in the following two aspects: one is to continuously reduce costs. To this end, the crystal diameter, the amount of feed, and the pulling speed should be elevated. The other is to enhance the photoelectric conversion efficiency. To this end, a breakthrough should be made in the crystal growth process to downregulate the content of oxygen and carbon in silicon. The mentioned two aspects are the proposed novel requirements for the production and research of single crystalline silicon. To gain insights into the growth conditions, growth defects, and their relationship to device performance of single crystalline silicon, the crystal quality of silicon should be improved.

Single crystalline silicon can be split into the Czochralski method and the Floating Zone method in terms of growth pattern. To be specific, the single crystalline silicon produced with the Czochralski method is primarily applied in solar cells and the integrated circuit industry; it exhibits high mechanical strength and low cost, taking up a significant share of the single crystalline silicon market. The growth equipment and the Czochralski process of single crystalline silicon are illustrated in Fig. 3.1.

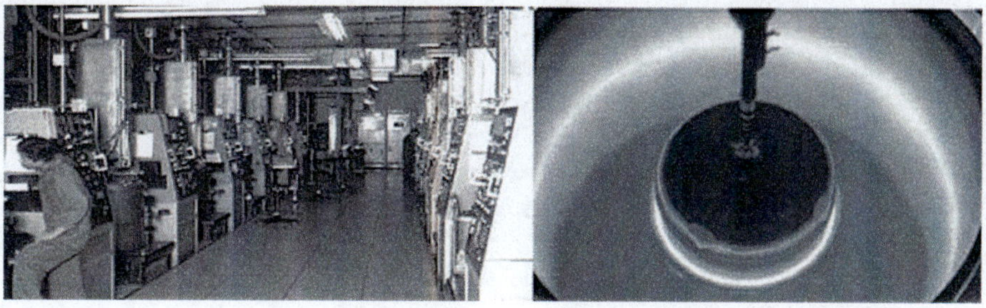

Fig. 3.1 Production equipment and the Czochralski process of single crystalline silicon

Fig. 3.2 Schematic diagram of the Czochralski method

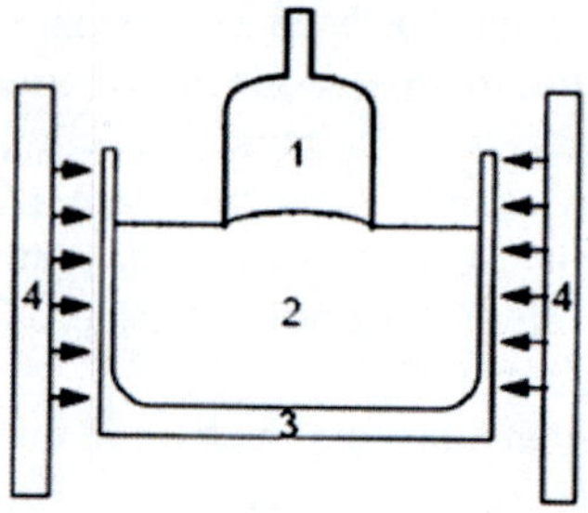

1 single crystalline silicon 2 molten silicon 3 crucible 4 heater

Frequently applied methods to prepare single crystalline silicon to consist of the Czochralski method, the Floating Zone method, the magnetron method, and the trimorphous silicon method.

(1) CZ method (Czochralski method)

The CZ method complies with the principle: above the quartz crucible that covers silicon raw material, there exists a clamping rod that can be pulled and rotated; subsequently, a seed crystal is clamped at one end of the rod. When the single crystal is prepared to be pulled, the seed crystal will be inserted into the molten silicon *liquid*; besides, the single crystal is produced by being pulled and rotated at a certain speed while pulling to the required length, and a schematic diagram is presented in Fig. 3.2.

On the whole, the process steps are relatively simple, primarily split into six steps: melting, seeding, necking down, shouldering, isodiametric growth, and winding up, as presented in Fig. 3.3.

(1) Melting. The bulk polycrystalline silicon is pulverized to an appropriate size; subsequently, the surface is cleaned and then placed in a quartz crucible. The single-crystal furnace is pumped to a certain pressure and then filled with protective gas; next, it is heated to make the temperature exceed the melting point of silicon (1412 °C). Indeed, high temperatures are conducive to reducing the melting cycle, they can damage the equipment as well. Thus, we should find a suitable temperature for melting.
(2) Seeding. After being treated, the selected seeds are sandwiched on the seed crystal clips. Subsequently, the seed crystal is moved down slowly and disrupted for a while when approaching the molten silicon surface, so the seed crystal can reach a higher temperature; otherwise, the direct entry into the liquid silicon causes a large thermal shock and damages the seed crystal. When a certain temperature is reached, the seed crystal head is placed into liquid silicon to be melted into one body; subsequently, the temperature is regulated, the seed crystal is rotated and then pulled at a certain speed; next, seeding is performed to form a single crystalline ingot.

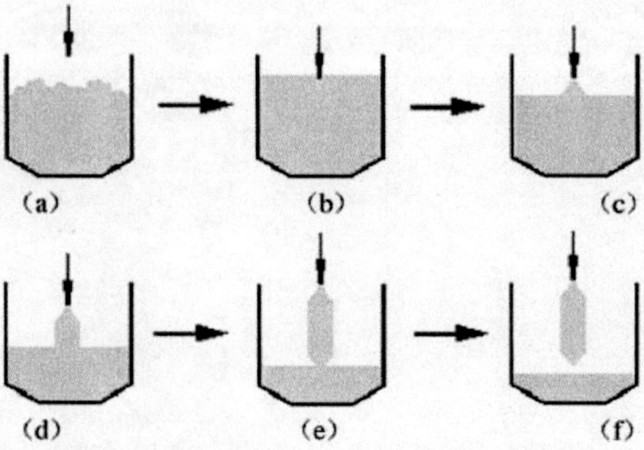

(a) Loading and melting of polycrystalline silicon. (b) Seeding. (c) Necking down and shouldering. (d) Isodiametric growth. (e) Winding up. (f) Completing.

Fig. 3.3 Process diagram of the CZ method

(3) Necking down. When the seed crystal comes into contact with the liquid silicon, thermal stress is generated, probably causing defects in the seed crystal. Thus, to avoid defects from being diffused significantly, it is necked. The main method is to lift the seed crystal, so the growth of single crystalline silicon can exhibit a certain speed; thus, the diameter of the silicon ingot can be reduced to the desired value.

(4) Shouldering. To reduce defects, the diameter of the crystal ingot is again elevated to the required size after necking, which is termed as shouldering. Then, the pulling speed decreases to up-regulate the diameter of the crystal ingot. The critical parameter here refers to the increased rate of diameter. If the increase is overly fast, it will affect the interface morphology while causing stress and defects.

(5) Isodiametric growth. When the desired diameter is reached after shouldering, the pulling and rotating speeds are accurately regulated to maintain the diameter of the crystal ingot at the desired size. With the crystallization of Czochralski, the liquid silicon residual in the crucible is increasingly less, so the rate of heating rises, thereby affecting the shape at the interface and causing stress and defects. Meantime, for the increasing size of silicon ingot, the heat dissipation is increasingly slower, so the pulling speed should be down-regulated during the subsequent crystallization to fully dissipate heat.

(6) Winding up. At the end of crystallization, the diameter of the crystal ingot will decrease and last to a point until leaving the interface, thereby completing the Czochralski process. The winding up is to primarily regulate the defect density of the material [3].

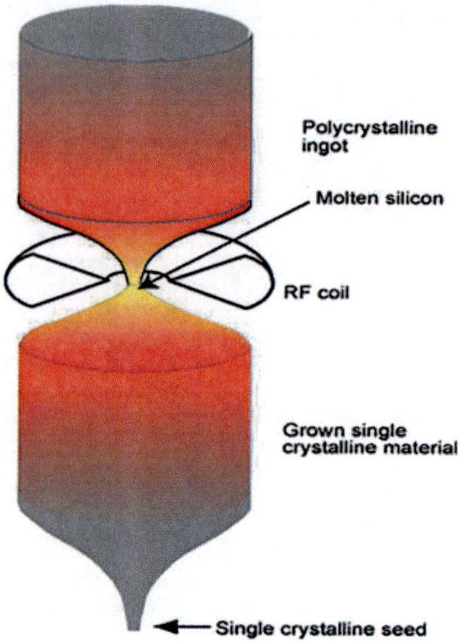

Fig. 3.4 Schematic diagram of the FZ method

(2) FZ method (Floating Zone method)

According to this method, the treated crack-free polycrystalline ingot and seed crystal are placed vertically between the vertical axes of the Float-Zone furnace, in which the seed crystal is underneath the polycrystalline ingot, and the bottom of the polycrystalline ingot is heated by the RF coil. The polycrystalline ingot begins to melt while it moves down, so more liquid silicon is attached to the seed crystal. By repeating the movement of the silicon ingot in such a way, the required single crystalline ingot can be produced, as presented in Fig. 3.4.

Compared with the CZ method, the FZ method exhibits a faster crystal growth rate, a shorter heating-cooling cycle, as well as a lower cost of consumables. Since the materials do not come into contact with the container (e.g., crucible) in the melting and crystallization processes, the contents of carbon, oxygen, metal impurities, and others are relatively low, the carrier lifetime is relatively long, and the resistivity at each position is comparatively uniform. However, the diameter of the single silicon ingot generated with the FZ method is slightly smaller than that produced with the CZ method, and the operation is more sophisticated, and the quality requirements of the raw materials are also slightly more rigorous. Single crystalline silicon obtained with the FZ method is generally applied in high-power devices and only takes up a small market share in the production of solar cells.

(3) Magnetron method

According to the Czochralski method, the oxygen content and its distribution act as parameters that are critical and difficult to control, primarily as the thermal convection in the molten mass aggravates the action of molten silicon and quartz crucible, i.e., O, B, Al, and other impurities in the crucible can easily enter the molten mass and crystal. Thermal convection also causes the temperature to fluctuate in the molten mass, so impurity stripes and vortex defects are formed in the crystal.

The molten semiconductor acts as an effective conductor. When the magnetic field is applied to the molten mass, the molten mass undergoes a Lorentz force opposite to its direction of movement, so the convection in the molten mass can be hindered, which is equated with elevating the viscosity in the molten mass. The effect of heat convection decreases, so the amount of impurities entering the liquid silicon is down-regulated. On the whole, horizontal magnetic field, vertical magnetic field, and other technologies are adopted in production.

In comparison with the Czochralski method, the magnetron technique exhibits the merit of hindering the temperature fluctuation in the molten mass. Overall, the temperature fluctuation in the molten mass near the solid-liquid interface in the Czochralski method reaches over 1 °C; with the magnetic field of 0.2 T applied, the temperature fluctuation reaches below 1 °C. Thus, the uniformities of the impurity distribution and the radial resistance distribution in the crystal can be significantly enhanced. Meantime, for the decreasing effect exerted by the molten silicon and the crucible as impacted by the magnetic field, the impurities in the crucible are hindered access into the molten silicon and crystal, thereby reducing the defect density in the single crystalline silicon, hindering the entry of impurities, and enhancing the purity of the crystal. Integrating the intensity of the magnetic field with the process parameters (e.g., crystal rotation and crucible rotation) can effectively regulate oxygen concentration variation in the crystal. For the magnetic viscosity, the thickness of the diffusion layer is elevated, and the uniformity of longitudinal distribution of impurities can be enhanced, which is conducive to improving productivity. High-quality crystals can be still obtained with magnetron technology (e.g., horizontal magnetic field), when the growth rate is twice that of the normal Czochralski method.

(4) Trimorphous silicon method

The trimorphous silicon method exhibits high mechanical strength, capable of cutting thinner silicon wafers instead of pure single crystalline silicon. The method to grow trimorphous silicon is identical to the CZ method. The only feature is that, when three seed crystals are applied simultaneously, the silicon ingot grows faster, and a silicon ingot exhibiting a diameter of 100–150 mm and a length of 700 mm can be grown. For the unique dislocation mode of trimorphous silicon, the dislocation density is significantly reduced. The trimorphous silicon products are presented in Fig. 3.5.

The solar cells composed of the trimorphous silicon material with the back-surface field technology achieve an average photoelectric conversion efficiency of 15.5% under standard test conditions, slightly higher than that achieved by the standard single crystalline silicon material. If the surface texturing process is applied, higher photoelectric conversion efficiency will be obtained [2, 4].

Fig. 3.5 Sample diagram of trimorphous silicon products

3.1.4 Manufacture of Polycrystalline Silicon

1. **Preparation of polycrystalline silicon**

Polysilicon indicates the development trend of silicon solar cells. In contrast to single crystalline silicon, the biggest merit of polycrystalline silicon lies in that raw materials are easier to obtain with lower manufacturing costs. Another merit is that rectangular or square polycrystalline silicon wafers are more efficient than circular or pseudo-square single crystalline silicon wafers. The photoelectric conversion efficiency of polycrystalline silicon solar cells is primarily affected by impurity atoms and extended defects (e.g., grain boundaries and dislocations). The critical issue in enhancing the photoelectric conversion efficiency of solar cells is to regulate the temperature variations in the ingot casting process and to regulate the flow in the cell production, which regulates the number of defects. Moreover, for polycrystalline silicon solar cells, the hydrogen passivation process is critical. The photoelectric conversion efficiency of industrial polycrystalline silicon solar cells is enhanced to 14–15% by the hydrogen passivation process via the silicon nitride deposition layer. Accordingly, the silicon solar cell industry has continuously transferred the raw material market to polycrystalline silicon [2].

Industrial production primarily complies with existing chlorination purification methods, while many difficulties remain in producing very cheap polysilicon materials. Thus, the development of two directions should be achieved: while maintaining and developing semiconductor polycrystalline silicon, novel technologies for producing cheaper solar grade polycrystalline silicon materials should be researched and developed. Over the past few years, novel technologies and novel processes to synthesize solar grade polycrystalline silicon has achieved great success and laid a solid foundation for subsequent development.

(1) Optimized Siemens method

Siemens of Germany initially developed the hydrogen reduction method of trichlorosilane ($SiHCl_3$) in 1954, also termed as the Siemens method. Such a method has been adopted by numerous large factories worldwide. This method is briefed below.

First, the silicon sand is crushed into metallurgical powder exhibiting a particle diameter of less than 0.5 mm; then, it is liquefied into metallurgical grade silicon in a reactor at 300–400 °C. The chemical reaction is expressed as

$$SiO_2 + 2C \rightarrow Si + 2CO_2 \uparrow$$

To satisfy the needs of high purity, subsequent purification should be made. Being placed into the reactor and then impacted by the catalysis of copper, silicon reacts with hydrogen chloride to synthesize $SiHCl_3$ and H_2. The reaction equation is expressed as follows:

$$Si + HCl \rightarrow SiHCl_3 + H_2 \uparrow$$

The reaction temperature is at 300 °C, and the heat will be released during the reaction [5]. Besides producing trichlorosilane, other by-products (e.g., H_2, $SiCl_4$, SiH_2Cl_2 gas, $FeCl_3$, BCl_3, and PCl_3) and other impurity chlorides should be purified by crude distillation and rectification. To obtain higher purity silicon, SiHC13 in the reactor is reduced with considerable hydrogen and precipitated into granular polysilicon on a silicon ingot under electric pressurization at 1000 °C. The chemical equation is defined as

$$SiHCl_3 + H_2 \rightarrow Si + 3\,HCl \uparrow$$

After a few days or more, the silicon core with only 8 mm diameter in the reduction furnace will grow to nearly 150 mm. The generated silicon ingots can act as raw materials for synthesizing single crystalline silicon with the Floating Zone method, which can also be crushed as raw materials for Czochralski single crystalline silicon [5].

$SiCl_4$ refers to the major pollutant generated by the Siemens process. It is a colorless, transparent, and smoky liquid, causing unpleasant asphyxiation and strongly stimulating the eyes and upper respiratory tract. A high concentration of $SiCl_4$ can cause several symptoms (e.g., corneal opacity, respiratory inflammation, and even pulmonary edema). Besides, other products also adversely affect the human body and the environment. Accordingly, the optimized Siemens method increases the absorption and reuse of tail gas, which protects the environment and human body while increasing the utilization rate and output of raw materials. The production process is illustrated in Fig. 3.6.

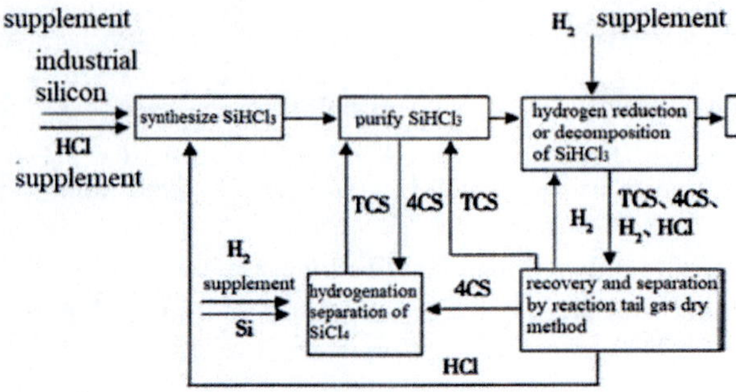

Fig. 3.6 Flowchart of the optimized Siemens method.

(2) Thermal decomposition method of silane

Silane refers to a material prepared by the hydrogenation of silicon tetrachloride, the decomposition of silicon alloy, direct hydrogenation of silicon, hydride reduction, etc. One of the methods is to exploit the reaction of magnesium silicide and ammonium chloride below zero degree Celsius, which was developed by Komatsu Company of Japan; the reaction equation is written as follows:

$$Mg_2Si + 4NH_4Cl \rightarrow 2MgCl_2 + 4NH_3 + SiH_4$$

Another approach is to react silicon tetrachloride, silicon metal, and hydrogen to synthesize silicon trichlorosilane; then trichlorosilane is decomposed, and lastly silane is generated. The reaction equation is expressed as follows:

$$3SiCl_4 + Si + 2H_2 \rightarrow 4SiHCl_3$$

$$2SiHCl_3 \rightarrow SiH_2Cl_2 + SiCl_4$$

$$3SiH_2Cl_2 \rightarrow SiH_4 + 2SiHCl_3$$

Subsequently, the generated silane gas is purified, and the rod-like polycrystalline silicon with high purity is synthesized in the pyrolysis furnace [2].

Silane can be exploited to purify silicon by decompression rectification, adsorption, and preheating decomposition. The reaction equation is written as follows:

$$SiH_4 \rightarrow Si + 2H_2 \uparrow$$

After the reaction, the silicon exhibiting high purity is generated [5].

Next, MEMC Pasadena Company in the United States produced granular polycrystalline silicon exhibiting high-purity silane gas as a raw material; a chlorine-free process was also developed with silicon tetrafluoride as a raw material in the

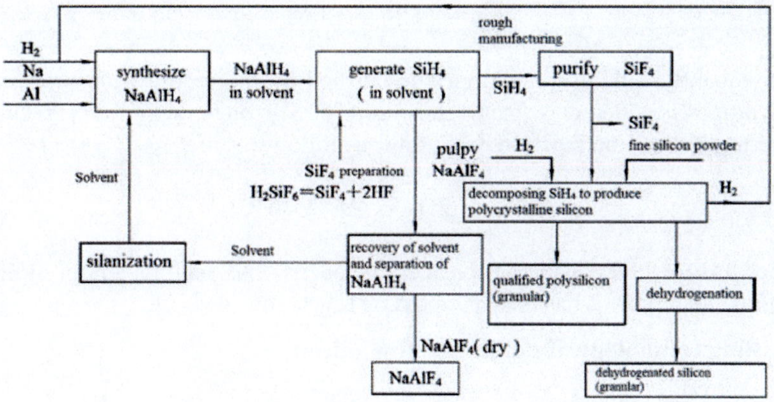

Fig. 3.7 Flowchart of chlorine-free n process.

silane production. This process can protect the product from the pollution of silicon tetrachloride [6]. The reaction flow is presented in Fig. 3.7.

Though the mentioned methods can effectively remove impurities from silicon raw materials, they consume considerable additional raw materials (e.g., Mg). Furthermore, SiH_4 is flammable and explosive and prone to danger when applied; thus, it is not very convenient to use them, and they are not adopted by considerable industrial production [5].

(3) Fluidized bed method

The Fluidized bed method pertains to a type of polycrystalline silicon preparation technology developed early by Union Carbide Company in the United States. In this method, $SiCl_4$, H_2, HCl, and industrial silicon act as raw materials to synthesize $SiHCl_3$ in a fluidized bed at a high temperature and under high pressure, thereby hydrogenating $SiHCl_3$ to synthesize SiH_2Cl_2; it in turn leads to the generation of silane gas. The generated silane gas is added into a fluidized bed reactor with small particle silicon powder to be continuously thermally decomposed, as an attempt to form granular polycrystalline silicon products. Since silicon involved in the reaction in the fluidized bed reactor covers a large surface area, this method is characterized by high production efficiency, low energy consumption, and low cost. The defects of such a method include poor safety, great risk, and relatively low purity of the product [6].

(4) Hydrogen reduction method of silicon tetrachloride

The hydrogen reduction method of silicon tetrachloride has been the most common technology in the early stage. According to such a method, silicon metal reacts with chlorine gas to synthesize silicon tetrachloride. The reaction equation is expressed as follows:

$$Si + 2Cl_2 \rightarrow SiCl_4$$

Subsequently, silicon tetrachloride is purified by the rectification process, and then hydrogen is added to react at 1100–1200 °C. Silicon tetrachloride is reduced to crystalline silicon. The reaction equation is as follows:

$$SiCl_4 + 2H_2 \rightarrow Si + 4HCl$$

Since this method consumes considerable energy and exhibits a poor utilization rate of raw materials, it has been adopted by few companies [2].

2. **Casting technology of polycrystalline silicon**

Two major casting technologies of polycrystalline silicon can be adopted, i.e., the Bridgman method and the casting method. Both methods are capable of synthesizing high-quality silicon ingots. Industrial companies adopting the Bridgman method to cast polycrystalline silicon consist of Kyocera of Japan and German Solar Company.

The technological difference between the two technologies is reflected as the process of silicon raw material melting and crystal growth is in the identical dry boiler, while the casting process is performed by another dry boiler during crystal growth. The primary difference is presented in Figs. 3.8 and 3.9.

In the Bridgman process, a layer of nitride (silicon nitride) should be applied inside the quartz crucible to melt silicon raw materials and the subsequent solidification process of polycrystalline silicon ingots. Since the volume expansion of silicon material inevitably extrudes or even damages the quartz crucible in the crystallization process, the nitride smeared is primarily to prevent the quartz crucible from sticking together with the silicon ingot, which causes the damage of the quartz crucible. For the casting method, the quartz crucible for melting the silicon raw material in the first step is not coated with silicon nitride; in the second step of crystal growth of the silicon ingot, however, the quartz crucible should be coated with a layer of nitride. Both steps have an identical effect.

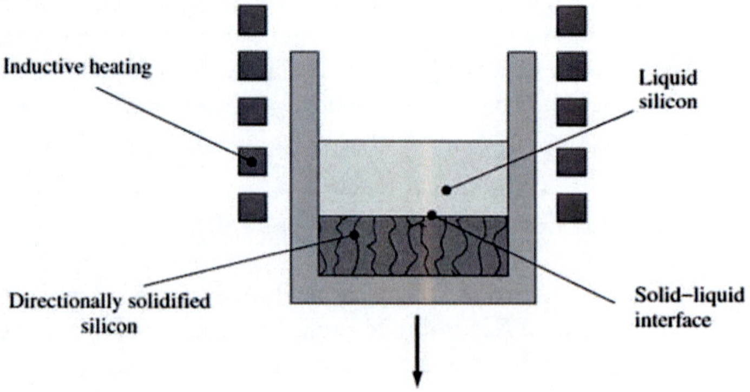

Fig. 3.8 Schematic diagram of casting polycrystalline silicon with the Bridgman method

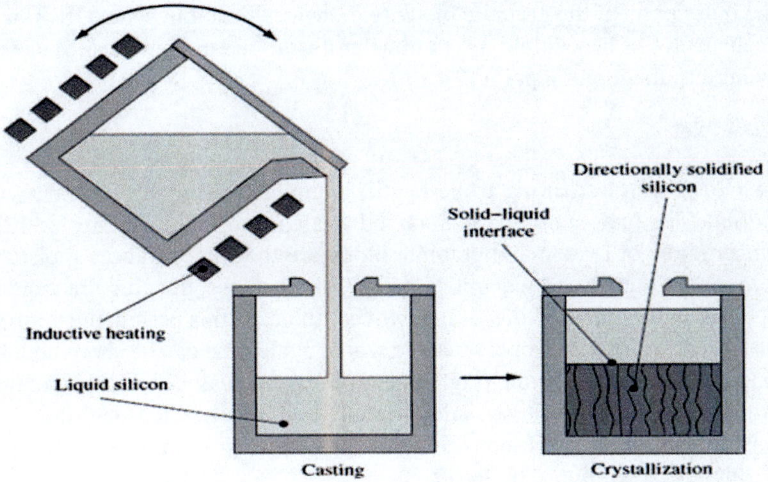

Fig. 3.9 Schematic diagram of casting polysilicon by the casting method

Overall, both production techniques begin to cool down via the bottom (below the melting point of silicon at 1410 °C), and then the liquid silicon at the bottom starts to be crystallized. The Bridgman method moves the quartz crucible supplemented by silicon solution slowly down from the high-temperature region for cooling. In the casting crystallization process, the temperature of the heater is regulated to cool down, and then the crucible itself is fixed.

Before the bottom begins to be crystallized, i.e., in the solid-liquid mixed state, the crystal surface is perpendicular to the crystal growth direction. After the columnar silicon ingot is obtained accordingly is cut out of the silicon wafer, the defect (e.g., grain boundary and dislocation) of the adjacent silicon wafer is highly similar [6].

The crystallization rate of the Bridgman method is usually about 1 cm/h. To enhance productivity, the crystallization rate should be elevated, whereas the cooling rate of silicon should be considered. Otherwise, an overly fast cooling rate and large temperature gradient may cause the internal fragmentation and even cracking of silicon ingots. In terms of the casting method, for a more flexible and advanced heating system, the crystallization rate can significantly increase, as well as the productivity [7].

3. **Main processes of casting polycrystalline silicon**

(1) Heating

Heating is primarily to heat graphite blocks and silicon material to as high a temperature as possible in the shortest possible time. At a temperature below 100 °C, the temperature regulated is not very stable, so it should be heated in the power mode. All stages of the heating period are completed in a vacuum, which can bake the moisture absorbed by the graphite block and the heat-shielding layer and then evaporate it from the surface of the silicon material. The power regulate mode is adopted to heat the inner parts of the graphite block (e.g., heater, crucible plate, DS-Block, and the inner surface of heat shielding) to transfer heat to the molten mass to elevate the temperature to the melting point [2].

(2) Melting

The first stage of the melting cycle is still completed in a vacuum, i.e., with the water dried. The temperature is maintained at a constant temperature for 1.5 h, so the temperatures of silicon and graphite block are the same, and the moisture, oil, and grease are drained. Subsequently, in a few short stages, the pressure increases to the specified value, and melting and growth continue. In this period, the temperature varies slightly at a certain slope; thus, the whole cycle time can be shortened. Lastly, the temperature reaches the melting temperature and remains at this temperature for a specified time; as a result, the silicon melts completely. Slow melting is better, whereas if the whole cycle time is limited, slow melting will not be allowed. Thus, the heating speed requires a trade-off.

(3) Growing crystal

First, the temperature is lowered slightly, and then the temperature is kept unchanged. A crystal nucleus is formed, and the crystal growth is initiated. The rate of growth is nearly 1 cm/h. After all melt materials are solidificated, the crystal growth ends.

(4) Anneal

When the last silicon block is solidified, an obvious temperature gradient is identified from the bottom to the top of the solidified silicon ingot. This temperature gradient generates stress or forms a very small stress crack in the silicon ingot. In this state, if the silicon ingot is cooled to ambient temperature, the mentioned cracks will be observed when it is cut on a band saw or a fretsaw. The stress can be eliminated by elevating the temperature of the whole silicon ingot. Besides, the stress can be easily transferred to the crucible. To eliminate the crystal growth stress, it is optimal to maximize the uniformity of the temperature gradient of the silicon ingot at a critical temperature for 1.5–3 h. Next, the furnace is switched to power regulate mode to reduce the power in accordance with a certain law. Thus, the temperature difference between the top and bottom of the silicon ingot will be more uniform in the cooling process.

(5) Cooling

The cooling rate of silicon ingots and internal components is the fastest in helium, followed by argon, and the slowest in a vacuum. With the use of helium, it is necessary to avoid stress cracks in the inner parts of the silicon ingot for the rapid cooling rate, and the outside of the ingot is notably cooler than the middle; thus, another harmful temperature gradient is likely to be produced. In the first step of cooling, set the power to 0 in argon, continue to cool the silicon ingots above 1000 °C gently, and prevent the top of the ingots from cooling faster than the bottom. Subsequently, argon is discharged in a vacuum; then, helium is added, and the pressure is gradually elevated to the specified value. Keep cooling until the temperature drops to 450 °C. The silicon ingots are taken out and cooled to ambient temperature.

3.1.5 Present Situation of Solar Grade Silicon in China

Compared with foreign countries, the study on solar grade silicon material in China has a relatively late start, and the technologies are relatively backward. For single crystalline silicon, most domestic manufacturers produce small-size wafers, and large-size single crystalline silicon production remains unmatured. For polycrystalline silicon, it has extensive domestic production. However, the process of its purification is not mature, resulting in high energy consumption, high pollution, and high cost, which hinders the development of industry and society. Over the past few years, the development of the silicon material industry in China has been in a semi-deformed state. Most of the raw materials are dependent on imports, and most of the sales of the products are determined by exports. Moreover, domestic demand is insufficient to support industrial development, resulting in serious overcapacity.

Thus, for the global energy shortage today, the photovoltaic industry should arouse more attention. At present, solar cells based on silicon materials still take up the dominant position in the photovoltaic industry. Accordingly, it is urgent to improve the independent intellectual property rights, reduce the defect density of silicon materials, and facilitate large-size wafer production. Meantime, the development of novel silicon materials on that basis is another way to subsequently develop the photovoltaic industry [8].

3.2 Principle and Basic Structure of Crystalline Silicon Solar Cells

3.2.1 Principle of Crystalline Silicon Solar Cells

In the second chapter, the basic principle of the solar cell is explained. Crystalline silicon solar cells are briefed here.

As mentioned, above, crystalline silicon solar cells are PN junction diodes under illumination. When a beam of light with the energy greater than the forbidden bandwidth of the semiconductor material is perpendicularly incident on the PN junction surface (Fig. 3.10), the incident photons will be absorbed in the range of $1/\alpha$ of a certain depth from the surface (α denotes the light absorption coefficient). If $1/\alpha$ exceeds the thickness of the PN junction, and incident photons excite electron-hole pairs in the junction region and the space near the junction, the photogenerated electrons and holes generated in the space charge region are separated via the junction

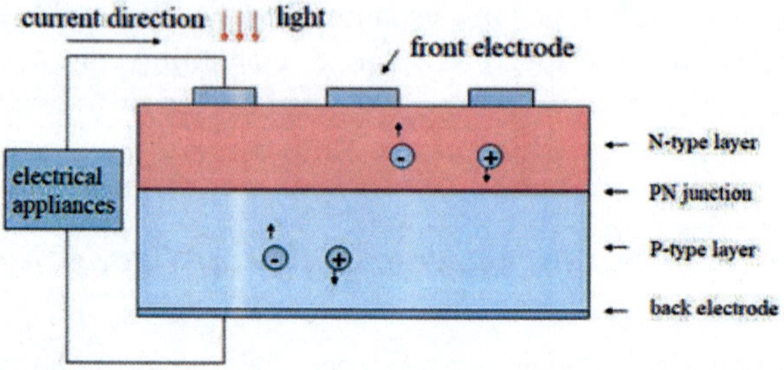

Fig. 3.10 Schematic diagram of crystalline silicon solar cells.

electric field. Photogenerated carriers generated in the vicinity of the junction and in the range of the charge diffusion length from the space charge region also exhibit diffusion into the space charge region, which is subsequently separated by the electric field. As impacted by the electric field, the electrons in the p region drift to the N region, and the holes in the N region drift to the p region, thereby forming the photogenerated current from the N region to the p region. The photogenerated carrier drift accumulates to form an electric field −qV opposite to the direction of the junction electric field under thermal equilibrium; meantime, a forward junction current opposite to the direction of photogenerated current is produced. The opposite electric field remedies the junction electric field and lowers the potential barrier. When the photogenerated current is equated with the forward junction current, a stable potential difference is established across the PN junction, which is the photogenerated voltage as mentioned above. The carrier concentration in the N region and that in the p region increases under illumination, so Fermi energy level is separated; besides, the difference between the energy levels on both sides is the open-circuit voltage V_{oc}. If the external circuit is short-circuited, the forward current of the PN junction is zero, and the current of the external circuit pertains to the short-circuit current, which is ideally the photocurrent [9].

3.2.2 Basic Structure of Crystalline Silicon Solar Cell

For ideal solar cells, four main assumptions are proposed:

(1) ideal light capture without reflection loss of light;
(2) the smallest composite model: defects and surface recombination are overlooked, and only Auger recombination is considered;
(3) ideal electrode: no light barrier and no series resistance loss;
(4) there exists no transport loss, and the body recombination is minimal.

Under the mentioned assumptions, the minimum Auger recombination and good free carrier collection can be obtained with the intrinsic substrate material. Meantime, the light absorption and recombination should be balanced to determine the substrate thickness. Over-thickness leads to the rise in recombination probability;

over-thinness leads to the poor light absorption effect. After weighing the two factors, a substrate of nearly 80 μm thickness should be employed theoretically. Under ideal conditions, a silicon cell can exhibit a photoelectric conversion efficiency of 29% (ambient temperature) based on the simulated sunlight of AM1.5 [10].

The ideal conditions discussed above do not indicate how to arrange the electrode position. To meet the mentioned conditions, the electrode can be placed on the front surface close to the place where the carriers are generated, as presented in Fig. 3.11a. The other way is to place the electrodes on the back surface to satisfy the requirements of a hypothesis (3) (Fig. 3.11b) [11]. This type of back contact electrode exhibits a significantly high photoelectric conversion efficiency. In most cells, different electrodes are placed on a variety of surfaces, as presented in Fig. 3.11c. Most of the collector electrodes of the minority carriers are placed on the front surface. However, for the low concentration of the minority carriers in the substrate, electrode collection is difficult to achieve, and most carriers can reach the back surface with rare loss; thus, some design forms put the minority collector electrode on both surfaces, as presented in Fig. 3.11d, which can elevate the absorption efficiency of photogenerated carriers [12].

Based on the mentioned analysis, a bifacial cell can be adopted, i.e., the front and rear surfaces of the cell can absorb the light and produce photogenerated carriers. Thus, the absorption of light per unit area of the cell can significantly increase, and the output power can be promoted. If both surfaces are transparent, any of the mentioned electrode design structures is applicable [4, 13]. The mentioned ideal cell structure models will be subsequently elaborated when the actual structure of a specific silicon solar cell is introduced later in the present chapter.

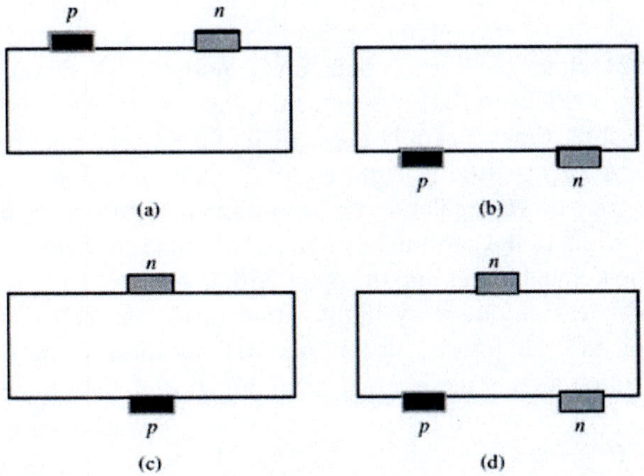

Fig. 3.11 Electrode structure

3.3 Electrical Characteristics and Limiting Factors of Crystalline Silicon Solar Cells

3.3.1 Electrical Characteristics of Crystalline Silicon Solar Cell

1. **Bulk characteristics of crystalline silicon solar cells**

The forbidden band of crystalline silicon falls into an indirect bandgap of $E_g =$ 1.12 eV and a direct bandgap of $E_g = 3$ eV [14]. Such bandgap structure determines the diversity of silicon at the wavelength of light absorption [15]. One photon can be absorbed under the light with a short ultraviolet wavelength to produce two electron-hole pairs, though the number can be overlooked [16]. Moreover, the generation and recombination of free carriers exist simultaneously [17]. The intrinsic carrier concentration is a vital parameter related to the bandgap structure. The concentration of non-equilibrium carriers can be regulated by external voltage [18].

In materials producing high concentration carriers by doping or external excitation, the bandgap structure varies. In solar cells, high-concentration doped regions tend to be transformed into so-called low-photovoltaic regions, i.e., dead layers [19] (elucidated later). The recombination in silicon materials is usually dominated by defect recombination. Materials exhibiting long minority carrier lifetime or large diffusion length are known as better materials. Auger recombination is another basic recombination form at high carrier concentration [20]. In accordance with the exciton effect (a series of effects exerted by the fact that the generated excitons are not separated into electron-hole pairs for the Coulomb force between electrons and holes), the Auger recombination coefficient increases at the corresponding concentration [21]. Inter-band recombination is another basic recombination form, whereas it can be overlooked [22]. When the doping concentration is low, the mobility of electrons is three times that of holes, and all of them are limited by phonon scattering [23, 24]. Impurity scattering is more prominent at high doping concentrations [25].

By down-regulating the recombination rate of the carriers, the reverse saturation current of the cell can be reduced, which is conducive to elevating the current output power and increasing the open-circuit voltage. The rise in the number of unbalanced carriers produced by illumination inside the cell or the extraction of the built-in electric field will cause a potential variation in the PN junction depletion region, which affects the internal recombination. Inside the entire cell, the net generation rate of the minority carriers and the recombination attributed to the bias effect of a few carriers are balanced. As suggested from the ideal open-circuit voltage formula 3.1, the open-circuit voltage more clearly indicates the internal recombination than the short-circuit current [9]:

$$V_{oc} = \frac{kT}{q} \ln\left(\frac{I_{sc}}{I_0} + 1\right) \tag{3.1}$$

where V_{oc} denotes the open-circuit voltage; I_{sc} is the short-circuit current; and I_0 is the reverse saturation current. Thus, reducing various internal recombination can significantly optimize crystalline silicon solar cells.

2. **Surface characteristics of crystalline silicon solar cells**

The surface of the substrate material falls into an upper surface and a back surface. An electrode connection is achieved on the surface, and the major function of the electrode is to act as the contact point between the internal carrier and the external circuit. The carriers in the substrate reach the surface as impacted by the concentration gradient field and the electric field; they are recombined at the other end of the substrate via the external circuit. Accordingly, the physical and chemical morphologies near the surface of the substrate noticeably impact the parameters of the device [4].

In the substrate, electrons reach the n-type surface by the action of the built-in electric field, and the holes reach the p-type surface via the built-in electric field. Minority carriers are recombined with interface defects and majority carriers before reaching the surface or on the surface, thereby increasing the saturation current and affecting the performance parameters of the device. Thus, it is generally required that the concentration of majority carriers cannot be overly high; otherwise, the recombination rate of minority carriers will increase, and the probability of survival will be significantly reduced. The number of defects on the surface can be reduced by surface passivation treatment, so the surface recombination rate of minority carriers is down-regulated [9].

However, it is noteworthy that a prominent device displays a small series resistance. Small series resistance can be obtained by elevating the surface doping concentration. However, for diffusion doping, the doping concentration near the upper surface increases as it is closer to the surface. The surface morphology of such high doping (close to the highest dissolution concentration) significantly increases the probability of recombination of minority carriers here, and there essentially exists no photovoltaic action; thus, this region is also referred to as the dead layer. The thickness of the dead layer is nearly several tens of nanometers. Short-wavelength light exhibits a relatively high absorption coefficient in silicon materials. In the presence of the short-wavelength illumination on the emitter of the dead layer for the excessive concentration, the photogenerated carriers are smoothly recombined, so the short-wave response is significantly poor, demonstrating that it is difficult for the substrate to achieve the required photogenerated carrier concentration for the high-speed recombination of carriers produced by short-wavelength light. Accordingly, excessive concentration of doping should be avoided to a great extent to prevent the appearance of an overly thick dead layer. Meantime, the doping will be treated by heat treatment, so some impurities will produce gaps, dislocations, and others at the original lattice, which will become the recombination center; besides, minority carriers are more likely to be recombined. Thus, the doping concentration should not be overly high, whereas the requirement of reducing contact resistance should be considered, so an optimal doping concentration should be developed [2].

High recombination probabilities are identified in the contact area between the cell surface and the electrode, and the recombination can be reduced in two ways. The first is to reduce the electrode contact area maximally. The second is to dope heavily in the contact region of the electrode, thereby effectively decreasing minority carriers near the electrode; as a result, the probability of surface recombination is reduced [9].

3.3.2 Factors Limiting the Photoelectric Conversion Efficiency of Crystalline Silicon Solar Cells

1. **Surface characteristics of crystalline silicon solar cells**

The first condition to synthesize high-efficiency solar cells is to choose high-quality silicon wafers as substrate. For instance, some laboratory-made high-efficiency solar cells use FZ silicon. However, the cost of this type of silicon wafer is high, and it does not apply to large-scale industrial production. A development direction of high-efficiency solar cells refers to down-regulating the cost of silicon wafers. Despite the low cost of Czochralski single crystalline silicon material, the cells made of boron-doped p-type silicon wafers are less efficient, and their performance is degraded when stored in light or even under dark conditions. The critical reason for the reduction of properties is the high oxygen content in this type of material, which reacts with the doped boron in the presence of light. Accordingly, the effective way to solve the performance degradation is to avoid considerable oxygen atoms and boron atoms existing in the silicon wafer at the same time.

Two effective solutions can be adopted: one is to synthesize crystalline silicon by the magnetic-field-applied Czochralski method, which has a low oxygen content in this silicon material. The other is to use other dopant sources instead of boron for doping in the silicon material (e.g., gallium). The performance of solar cells made by the mentioned two methods is significantly better than that of boron-diffused Czochralski crystalline silicon solar cells.

N-type material is another prominent choice, with some specific merits (e.g., long carrier lifetime, low boron oxidation reaction after sintering, good conductivity, and low saturation current). Some foreign manufacturers have employed n-type silicon to fabricate high-efficiency solar cells. As fueled by the advancement of solar cells and the increasing shortage of p-type single crystalline silicon materials, n-type materials are considered to be more extensively used [4].

2. **Diffusion depth of surface junction**

The quality of the PN junction displays tight relationships to the characteristics of solar cells. If the PN junction is overly shallow, short-wavelength light absorbed by the material near the surface cannot be significantly converted to carriers (for the existence of dead layers). Besides, in terms of screen-printed crystalline silicon solar cells, the last step of co-firing causes the metal gate lines on the front surface to form good ohmic contact coupled with the silicon body. For the overly shallow junction depth, the metal will diffuse into the n-type area of the cell, thereby reducing the quality of the PN junction. Thus, to avoid this effect, the PN junction should exhibit a certain depth. However, if the PN junction is overly deep and overly far from the surface, the metal electrode on the surface will not effectively collect the electron-hole pairs separated at the PN junction; thus, the junction depth of the PN junction cannot be overly deep. In brief, to more significantly improve the performance of the crystalline silicon solar cells, the junction depth of the PN junction should not be overly deep or overly shallow, instead it should have an optimal junction depth value.

3. **Silicon wafer thickness selection**

In terms of the electrical characteristics of the solar cell, the thickness selection of the substrate is taken given the numerous factors (e.g., substrate quality and device structure). When the diffusion length of the carrier exceeds the thickness of the substrate, surface recombination should be majorly considered. When some minority carriers can come into contact with the surface, reducing the substrate thickness accelerates the surface recombination speed, while down-regulating the generation probability of photogenerated carriers seriously reduces the device performance. Passivation treatment of the back surface down-regulates the recombination rate of the back surface. With the thickness of the substrate rising from 280 μm to nearly 400 μm, the performance of the device is significantly enhanced for the weakening of non-ideal factors (e.g., high surface recombination speed and small optical absorption rate).

Given the transport loss of minority carriers, when minority carriers are harvested from the illuminated surface, the substrate becomes thinner, and its transport loss is reduced. Under the back-surface electrode structure, a thin substrate is conducive to the collection of electrons and holes. Meantime, since photonic absorption cannot make the substrate overly thin, the thickness is set at 150–200 μm.

The diffusion depth of industrially produced solar cells is relatively shallow, and the performance of this cell exhibits poor sensitivity to the thickness of the substrate since most of the surface-recombined carriers are produced near the surface. The thickness of 200–300 μm refers to the range of choice for most manufacturers. However, given that the thin substrate can save raw materials to reduce costs, thin substrates have always been the direction followed by manufacturers [4].

3.4 Structure and Preparation of Crystalline Silicon Solar Cells in Industrial Production

3.4.1 Structure of Industrial Crystalline Silicon Solar Cells

There are a wide variety of crystalline silicon solar cell structures, especially those developed for high-efficiency solar cells. However, existing industrialized silicon solar cells exhibit simple structures. The single crystalline silicon with the Czochralski method or the polycrystalline silicon with the casting method has been adopted on a large scale. Generally, these silicon materials are boron diffusion doped, with a resistivity of 0.5–0.6 Ω cm. First, a textured surface is prepared by corrosion, and then a uniform emission zone is formed by the diffusion process. Lastly, the anti-reflective layer is deposited, and the upper positive and negative electrodes are developed to obtain the solar cells with the complete structure, as illustrated in Fig. 3.12 [2].

3.4.2 Basic Production Process for Industrialization

The fabrication of crystalline silicon solar cells consists of three main processes, i.e., preparing a junction by diffusion, vapor deposition of an anti-reflection film, and electrode preparation). The primary difference between solar cells and other semiconductor devices is that they require a large area of PN junction to realize the conversion from solar energy to electric energy. The electrodes are adopted to output electric energy, and the anti-reflection layer is employed for promoting the capture of light by the substrate; thus, the output power of solar cells is elevated. Overall, the characteristics of the PN junction are the major factors affecting the photoelectric conversion efficiency of cells. Besides affecting the electrical performance of the cell, the electrode is also associated with the reliability and life span of the cell [5]. The basic preparation of industrial silicon solar cells is presented in Fig. 3.13.

Silicon wafer detection should be performed before the process begins, since the quality of silicon wafer seriously affects the performance of the device. Most of

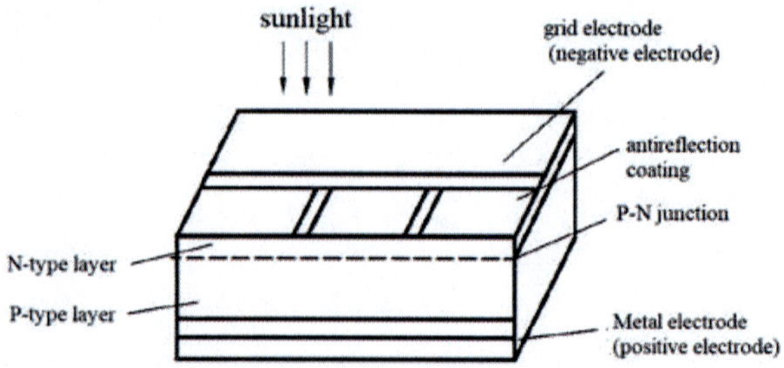

Fig. 3.12 Basic structure of crystalline silicon solar cells.

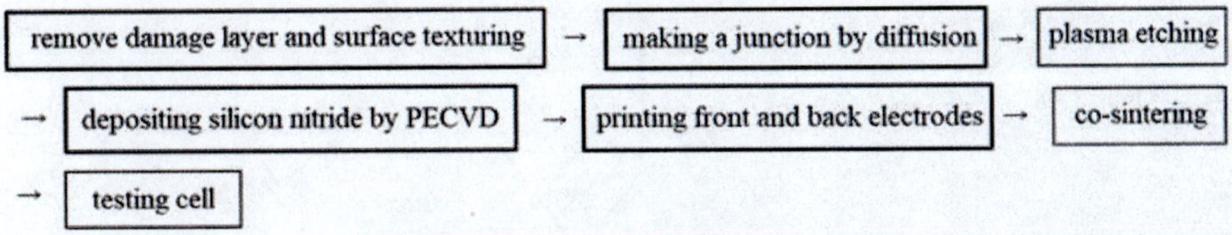

Fig. 3.13 Flowchart for the preparation of silicon solar cell.

the chips employed in industrial production include CZ–Si, shaped like a square, or square polycrystalline silicon wafer, as presented in Fig. 3.14. The major parameters of the test cover the flatness of the silicon wafer surface, minority carrier lifetime, resistivity, microcracks, etc.

1. **Removing the damaged layer**

Since the surface of the silicon wafer is stained with grease, metals, metal ions, a range of inorganic compounds, and mechanical damage left by cutting silicon wafer, it should be roughly thrown, as an attempt to remove the mechanical damage and oil pollution on the silicon wafer surface and to smooth and clean its surface. Surface treatment involves placing the wafer in an acidic or alkaline solution for corrosion. The surface treatment effect of the acidic solution outperforms that of the alkaline solution. Given the low cost of the alkaline corrosion and less pollution to the environment, however, the alkaline corrosion method is commonly adopted for industrial production. The alkaline corrosion method is to heat the 10% NaOH solution to 83 °C (as regulated between 80 and 85 °C), and then the prepared p-type silicon wafer is added into such solution for initial throwing (1–2 min, specific reaction time as determined by the thickness of the silicon wafer; the thicker the corrosion thickness, the longer the reaction time will be) to remove the mechanical damage layer of the silicon wafer. The chemical reaction equation is expressed as [26].

$$Si + 2NaOH + H_2O \rightarrow Na_2SiO_3 + H_2 \uparrow \ (80\,^{\circ}C)$$

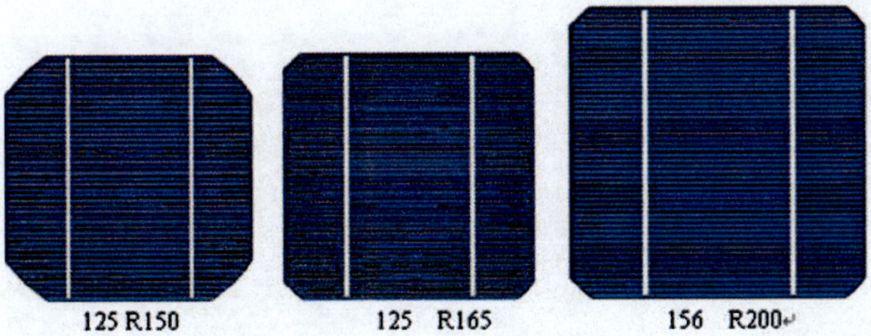

Fig. 3.14 Solar cells

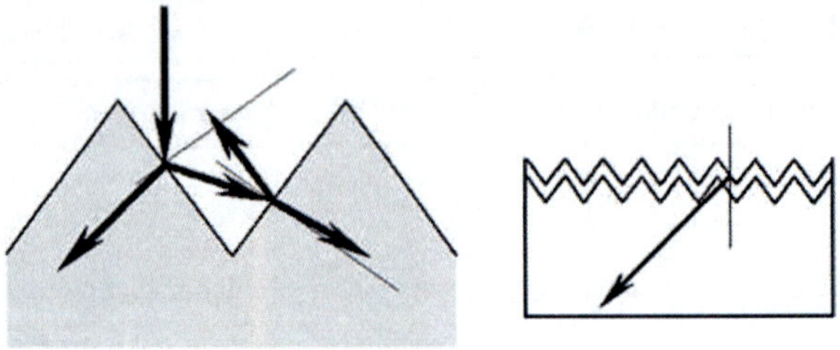

Fig. 3.15 Schematic diagram of the effect of surface texturing

2. **Surface texturing**

Surface texturing is to reduce the reflectivity of the cell surface, which facilitates the absorption of light in silicon materials, as presented in Fig. 3.15.

The texturing is formed by etching the silicon surface following a selective etching process. Such corrosion method causes the silicon lattice structure to corrode faster in one direction than in the other; as a result, some of the planes are exposed in the lattice. Figure 3.16 shows that the little pyramid-like vertebral bodies are formed by the intersections of the mentioned crystal planes. On the whole, the surface of the silicon wafer parallels the (100) plane, and the pyramid of the surface vertebra obtained by the texturing is formed by the intersection of the 111 planes.

The angle of the pyramid is dependent on the orientation of the crystal plane that gives the incident light at least two chances to access the cell. If 33% of the light is reflected at a respective incident point as is normally incident on the bare silicon surface, the overall reflection is 0.33×0.33, i.e., nearly 11%. If an anti-reflection film is adopted, the reflection of sunlight can be completely regulated at 3%. Another desirable feature is that the angle of light entering the silicon should ensure that the light is absorbed closer to the cells' surface, which allows the cells to more probably collect photogenerated carriers, especially for the long-wave part with weak absorption [27].

The texturing process is expressed as follows. In a constant temperature container, a solution of 1–2% NaOH in ethanol is heated to 70–80 °C (experimental results show that the texturing surface is better at a NaOH concentration of 1.5% and a

Fig. 3.16 Electron microscopy image of silicon surface after texturing [22]

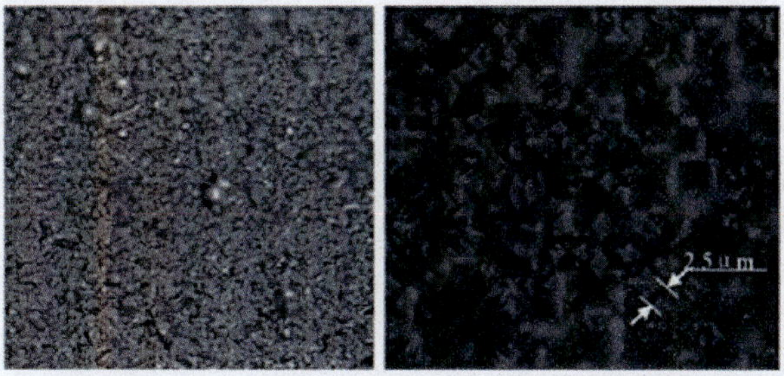

Fig. 3.17 Comparison of surface morphology before and after texturing [23]

(a) Electron microscopy photos before texturing.

(b) Electron microscopy photos after texturing.

temperature of 77 °C) to prepare a solution for texturing. The silicon wafer should be put (slowly and lightly) into the solution for texturing immediately to the texture after being roughly thrown, which hinders the surface of the silicon wafer from being oxidized rapidly. The duration of texturing is approximately 45 min, and the equation is identical to that of the rough throwing. Figure 3.17 shows the electron microscopy photos before and after texturing. As the NaOH solution has different corrosion rates against single crystalline silicon in various orientations, the pyramid texturing is formed as presented in Fig. 3.17b.

The solution of NaOH in ethanol used here is the most common chemical etchant to synthesize crystalline silicon solar cells. The formula below can be referenced to synthesize the texturing solution:

$$\begin{aligned}1\%\ \mathrm{NaOH} + 10\%\ \mathrm{CH_3CH_2OH} &= 13000\,\mathrm{mL}\ \mathrm{H_2O} + 140\,\mathrm{g}\ \mathrm{NaOH}\\ &\quad + 1400\,\mathrm{mL}\ \mathrm{CH_3CH_2OH}\end{aligned}$$

Besides adding ethyl alcohol absolute at the beginning of the texturing process, it should also be added at each later interval of 10 min, about 100 mL each time. NaOH reacts with Si to form H_2, as attached to the surface of the silicon wafer, thereby making the surface reaction uneven. When ethanol is added, on the one hand, since it is volatile, H_2 bubbles on the silicon surface can be removed timely to make the contact between the NaOH solution and the silicon surface more uniform, so a pyramid of similar size and a better texturing surface can be formed. On the other hand, for the large proportion of cold ethanol, the solution can be fully mixed and stirred to make the reaction on the silicon wafer surface more uniform.

The preparation of texturing exhibits sensitivity to the temperature and the concentration of the NaOH solution. Overly high or overly low temperature and concentration cannot lead to the formation of significant texturing. Accordingly, the concentration of the solution should be prepared, and the initial temperature is set prior to the operation.

After the texturing, the residual NaOH solution on the silicon wafer surface should be neutralized by the following two steps:

Step 1: soak in 0.5% HCl solution for 5 min + deionized water for cleaning;
Step 2: soak in 10% HF solution for 5 min + deionized water for cleaning.

The intuitive criterion to judge the quality of texturing is to observe whether the color of the texturing is uniform, as well as there are flower basket prints, white borders, or raindrop prints. Through theoretical calculations, the light-receiving area of texturing is obtained to be three times higher than that of unprocessed texturing, and the reflectivity of texturing is significantly reduced [23].

3. **Preparing a junction by diffusion**

Most solar cell manufacturers in China exploit tubular diffusion furnaces. The silicon wafer is placed vertically, and it is pushed into a furnace tube by pushing the boat, and oxygen and $POCl_3$ carried by nitrogen are introduced under a nitrogen atmosphere. The P_2O_5 generated by the reaction is deposited on the silicon wafer surface, and the phosphorus atoms diffuse into the silicon wafer to form a PN junction as impacted by the concentration gradient field and thermal motion [28]. The reaction equation is written as follows:

$$5POCl_3 \rightarrow 3PCl_5 + P_2O_5\ (600\,°C)$$

$$4PCl_5 + 5O_2 \rightarrow 2P_2O_5 + 10Cl_2\uparrow$$

$$2P_2O_5 + 5Si \rightarrow 5SiO_2 + 4P\downarrow$$

For the diffusion process of silicon solar cells, the major influencing factors consist of diffusion temperature, diffusion time, nitrogen flow rate carrying $POCl_3$, etc. The mentioned parameters impact the content of phosphorus diffused into the silicon wafer, the depth of diffusion, the surface impurity concentration, as well as the impurity distribution in the silicon wafer. The surface impurity concentration of the silicon wafer can be characterized by the square resistance, and the depth of diffusion can be obtained by experimentally determining the depth of the junction [29].

If the desired diffusion depth is deeper and the concentration is higher, it can be overall achieved by increasing the nitrogen carried and the temperature. The specific parameters are determined in accordance with the needs.

The Cl_2 generated during the reaction acts to clean the silicon surface and the furnace tube. The formed PN junction is relatively uniform, of which the square resistance is regulated in the range of 40–55 $\Omega/\square$, and the diffusion time is approximately 1 h. The yield per tube can reach 300 pieces in the case of single-sided diffusion. Phosphorus is capable of absorbing impurities during diffusion, which can extend the minority lifetime of the material. On the whole, the lifetime of the silicon wafer after diffusion is above 10 μs. Extending the diffusion time and decreasing the maximum diffusion temperature can increase the lifetime of the minority. Double-sided diffusion exhibits a longer minority lifetime and higher photoelectric conversion efficiency than single-sided diffusion. The existing closed-tube diffuser consumes less energy

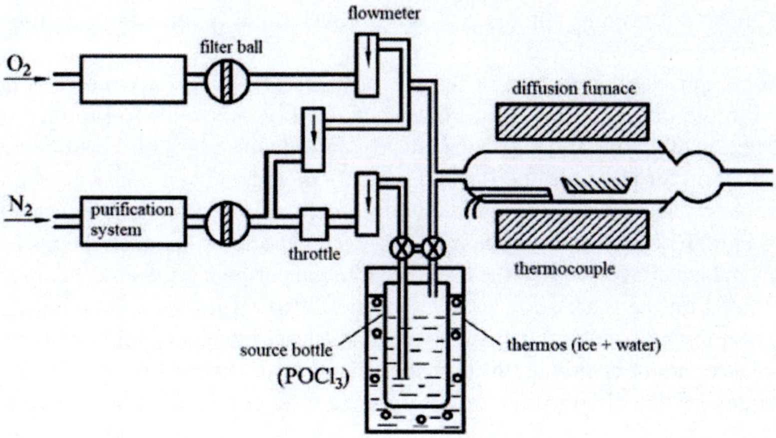

Fig. 3.18 Schematic diagram of diffusion device.

and produces less metaphosphoric acid, as compared with the open-tube diffuser. Numerous international companies have conducted research on closed-tube diffusion equipment and technology and made some achievements. The diffusion equipment is illustrated in Fig. 3.18.

Chain (multi-step continuous automation) diffusion is considered a critical industrialized production technology, exhibiting a high degree of automation, which hinders the process of human participation and avoids the process deviation attributed to the technical difference of operators. Besides, it can be well combined with texturing and edge etching techniques. The treated phosphoric acid is uniformly attached to the surface of the silicon wafer by means of a coating source or ultrasonic spray. Subsequently, a PN junction is generated by the chain diffusion furnace based on different temperature zones. The shortest diffusion time is only 5 s, no complexed loading and unloading device is required, and it is equipped with a non-contact sheet resistor for online detection, which underpins the automated production [28].

4. **Etching edge junction**

After the diffusion is achieved, the silicon wafer is taken out. After cooling, the square resistance is experimentally determined with the four-probe method. The common value of R□ in industrial production is 30–50 Ω/□. Overly high indicates that the diffusion is insufficient; otherwise, the concentration of diffused phosphorus is overly high. Afterwards, 10% HF is used to remove the phosphor-silicate glass (PSG) on the silicon wafer surface, and then considerable deionized water is adopted for rinsing.

As the PN junction is also formed around the diffused cells in the diffusion furnace, the diffusion area at the edge of the silicon wafer connects the electrodes on the upper and lower surfaces, thereby reducing the series resistance of the device and the performance of the device. To remove this part of the area, cryogenic dry etching technology is applied in production [26].

The most extensively applied dry etching technique is ion beam etching. The wafers are laminated and placed in a barrel-shaped etching chamber. Such an etching method can protect the surface of the silicon wafer; besides, only the edge area is exposed to the plasma. CF4 and SF6 are activated by RF fields to produce highly active species, and ions and electrons quickly corrode the edges of the silicon wafer. Such a method for mass production significantly increases the output of devices [4].

Moreover, laser cutting is another edge etching method. A fine laser beam exhibiting high energy density can melt and even vaporize silicon, so the purpose of cutting can be achieved. When the laser is cut to the periphery, the laser beam should be irradiated on the back surface. Overall, the diffusion layer of the silicon wafer exhibits a thickness of about 0.4 μm. Overall, the laser intensity should be regulated, and the laser cannot be broken through the silicon wafer; otherwise, it will affect the PN junction of the silicon solar cells. A pulsed laser source common for cutting a single crystalline silicon wafer displays a wavelength of 1.06 μm and a frequency of 2.0 kHz. When the solar cell is being sliced, its backside is cut upward, and the laser intensity is regulated by regulating the intensity of the laser current. The reference value of laser operating current is nearly 23 A, and the area of the solar cells will be narrowed slightly after the edge junction is removed [26]. The etching diagram is presented in Fig. 3.19.

5. **Deposition of anti-reflection film**

Early screen-printed solar cells adopt TiO_2 and SiO_2 as anti-reflection films, where TiO_2 cannot passivate the silicon wafers. Though SiO_2 can passivate the surface of silicon wafers well, it cannot passivate the polycrystalline silicon body, and its refractive index is overly low to perform a significant optical effect; besides, its oxidation is performed at high temperature, so the polysilicon material is subject to certain damage. For this reason, a better anti-reflection layer technology is urgently required. In 1984, Kyocera Corporation of Japan first applied the technology of

Fig. 3.19 Schematic diagram of laser edge etching

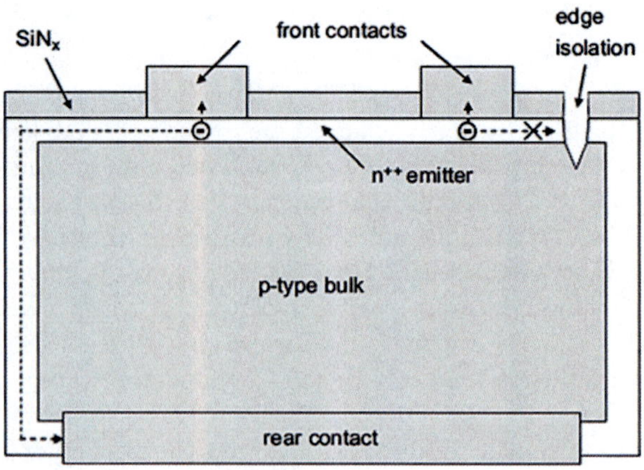

plasma-enhanced chemical vapor deposition (PECVD) to prepare silicon nitride anti-reflective film in commercial solar cell production, and they obtained significant anti-reflection and passivation effects.

Nowadays, PECVD technology is extensively applied in the commercial production of solar cells. SiH_4 reacts with NH_3 at $(0.1–1) \times 10^2$ Pa and 200–450 °C to deposit a layer of silicon nitride exhibiting a thickness of about 75 nm and a refractive index of 2.05 on the silicon wafer surface, and the reflectivity can be reduced to below 3%. The silicon nitride can exert a high passivation effect. PECVD is available in both tubular and flat versions. Flat PECVD exhibits higher productivity. Tubular PECVD can more significantly passivate polycrystalline silicon solar cells for the denser deposited silicon nitride film. As fueled by the advancement of technology, in VHF PECVD, microwave PECVD, remote PECVD, and other technologies, how to hinder the radiation damage to the cell surface, increase the hydrogen content in the film, enhance the surface and body passivation effect, and acquire the appropriate refractive index are critical to optimizing the process. Besides the preparation of silicon nitride with the use of PECVD technology, atmospheric pressure chemical vapor deposition (APCVD), low-pressure chemical vapor deposition (LPCVD), and magnetron sputtering can also lead to the formation of silicon nitride anti-reflective layers.

The refractive index n of the SiN_x:H film decreases with the decrease in NH_3/SiH_4 flow ratio, since the film formed is considered as Si-rich (small refraction coefficient) or N-rich (large refraction coefficient), and the N/Si ratio and refraction index of the film show a linear relationship. For the silicon nitride anti-reflection layer of the crystalline silicon solar cell, the passivation effect of the SiN_x:H thin film should be considered besides the optical factors (n value and thickness). The Si–N bonding density can affect the V_{OC} of cells. The optimal Si–N bonding density is about 1.3×10^{23} cm^{-3}. For silicon wafers exhibiting poor material quality, the optimization of Si–N bonding density is of higher significance to the results.

The end bonds of SiN_x:H film are formed as Si–H and N–H bonds. During heat treatment, the bonding forms and microstructure are varied. The hydrogen atoms on Si–H and N–H are likely to be incorporated into hydrogen molecules to escape from the film; meantime, novel Si–N bonds are generated. However, if the end bonds of the SiN_x:H film are only N–H bonds or Si–H bonds, Si–N bonds cannot be formed when heat treatment is being performed for dehydrogenation. The refraction coefficient of the SiN_x:H film is linearly related to the N/Si ratio. The SiN_x:H film with a low N/Si ratio shows a serious loss of hydrogen content after being sintered and releases considerable hydrogen. Since the sintering temperature is significantly higher than the terminal hydrogen cleavage of the Si–H/N–H bonds and dehydrogenation temperature (dehydrogenation starts at 500 °C and dehydrogenation peaks at 680 °C), the properties of the SiN_x:H film should cover the passivation effect of hydrogen released from sintering on the substrate body. Moreover, researchers should consider whether the variation in the properties of the film itself attributed to the structural variation complies with the optical design requirements and the surface passivation effect of the substrate [28]. The reaction equation is expressed as follows:

$$SiH_4 + NH_3 \rightarrow Si_3N_4 + H_2 \uparrow (200-500\ °C, 50-300\ Pa)$$

$$3SiH_4 + 4NH_3 \rightarrow Si_3N_4 + 12H_2 \uparrow$$

6. **Electrode preparation**

Solar cells (e.g., basic chemical cells) require electrodes to conduct electricity to power the load. The electrode should exhibit high solderability, low resistivity characteristics, etc. The electrode linked to the p-type area of the PN junction is termed as the positive electrode, and the electrode linked to the n-type area is termed as the negative electrode. Electrodes typically connected with highly doped portions are also commonly referred to as emitters (especially front surface electrodes), and electrodes that are lightly doped or undoped are referred to as bases (particularly back electrodes). However, the front surface is the part irradiated by light, and the electrode on the front surface blocks the incident light. Thus, to narrow the barrier area, the electrode lines are narrowed, and then the current collection is performed by several wider main electrodes. The back-surface electrode predominately covers the back surface, which is conducive to reducing the connection resistance [26].

Most positive existing surface electrodes are made by screen printing technology. Screen printing is to print a slurry containing metal powder particles into a defined electrode pattern, which is subsequently calcined in a high-temperature furnace to remove resins and other organic matters. In the subsequent rapid heating zone, the metal powder particles are melted and lead to the formation of effective contact with the silicon surface. The process is illustrated in Fig. 3.20.

At present, the fine grid line of screen printing in the industry is 110–150 μm, the main grid is 1.5–2 mm, and the loss of photoelectric conversion efficiency for shading is nearly 8%. How to optimize the current plane printing technology to more significantly reduce the electrode width is a hotspot in existing studies.

In terms of printing equipment, the automation degree of screen printing can satisfy the requirements of industrialization. Overall, the imported automatic printing line uses the CCD digital camera to detect the screen benchmark. The intelligent

Fig. 3.20 Schematic diagram of the screen printing technology.

Fig. 3.21 Electrode printing equipment [23]

technology automatically calibrates the chip position, and the output per hour is about 1000 pieces, as presented in Fig. 3.21. How to make the printing pressure more uniform and more significantly reduce the fragmentation rate under the trend of lamination should be considered by equipment manufacturers. In terms of the positive surface electrode silver slurry, how to selectively dissolve Si_3N_4 in the electrode sintering process to prevent electrode materials from entering overly deep into the silicon body, and how to facilitate slurry composition to make it apply to large-block resistance solar cells are the major problems that slurry preparers should tackle [28].

The conductive slurry used in screen printing primarily covers metal particles, resins, organic solvents, glass/ceramic materials, and other additives. For the problem of sedimentation and aggregation of the powder, it should be thoroughly stirred before use. Moreover, the slurry should be stored in the indoor shade to avoid deterioration from high-temperature exposure. The storage time is only a few months commonly, and the ambient humidity and temperature should be monitored during use. At present, the positive surface electrode of crystalline silicon solar cells exploits silver slurry, and the back electrode uses aluminum slurry or silver–aluminum mixed slurry. The compositions of different slurries are different, the printing conditions and subsequent sintering conditions are different as well, and the results significantly impact the cells' performance. To be specific, the front electrode of the cell pierces the anti-reflection layer to form an effective electrode contact with the emitter and the formation of the back-surface field of the cells is of the most critical significance. The characteristics of the slurry itself significantly impact the product. For instance, for the thermal expansion coefficient of the material and the difference in the composition of the slurry, the warping problem of the solar cell is caused, and the thinner silicon wafer exhibits even higher thinness, which is easy to cause debris, and therefore, to affect the yield. The adhesion of slurry after sintering should also be considered. Moreover, the sintered aluminum slurry is foamed, which should be solved in the material and the process [2].

To enhance the photoelectric conversion efficiency of the cell, the back surface also should reduce the reflectivity and passivation. In the industry, back-surface passivation is performed by screen printing technology to cover Al on the silicon wafer for alloying to form an aluminum backfield (as elucidated in 3.5.3). The thickness, uniformity, reflectivity of the aluminum backfield, and the curvature of the sintered cell are largely determined by the thickness of the aluminum slurry printed on the backside of the silicon wafer. Overall, the dry weight of the aluminum slurry (the weight of aluminum slurry after drying) is regulated at 6–10 mg/cm^2. The greater the dry weight of the aluminum slurry, the greater the curvature of the silicon wafer after sintering. Aluminum, silicon can form a eutectic structure at 577 °C. In accordance with the Al–Si binary phase diagram, a liquid phase of the Al–Si phase is produced in the heating process. With the decline of the temperature, recrystallization occurs in the silicon. In accordance with the solubility curve, a certain amount of aluminum is melted in the silicon to form a P^+ back-surface field layer. The thickness of the P^+-si region obtained after the optimized sintering process is nearly 6–7 μm [28].

7. **Sintering to form contact**

After screen printing, the cell substrate cannot be used directly. Since some impurities remain on the silicon wafer surface, it should be sintered to remove impurities, thus leaving pure electrode material. When the electrode material and the silicon wafer are at high temperatures, the silicon atoms diffuse into the electrode material in different proportions at a range of temperatures to form the desired ohmic contact. If the temperature drops at this time and the system begins to cool, the silicon atoms originally dissolved in the electrode metal material will be recrystallized in a solid state. In other words, an epitaxial layer is grown on the contact interface between the metal and the crystal. If the epitaxial layer contains sufficient impurities of the identical conductivity type as the original crystal material, an ohmic contact formed by the alloying process will be obtained. A PN junction formed by an alloying process is obtained if the crystal layer contains sufficient impurities opposite to the conductivity type of the original crystal material.

In terms of the device (as presented in Fig. 3.22), the mass production process requirements include a long preheating zone (100–250 °C) to sufficiently volatile organic solvents that exist in the last line of printing slurry, long baking area (500–600 °C) to sufficiently burn off the organic solvent and resin, the temperature in the instant heating area (700–850 °C) being set in accordance with the conditions of the anti-reflective layer, the characteristics of the slurry, and the speed of the belt, so the sintering temperature should consider the characteristics of the residual slurry [2].

8. **Photoelectric characteristics testing and sorting of solar cells**

The optical I-V characteristic of the solar cell was tested at ambient temperature 25 °C under standard solar simulator conditions. The major parameters describing the electrical performance of solar cells include open-circuit voltage V_{OC}, short-circuit current I_{SC}, fill factor FF, photoelectric conversion efficiency η, maximum output

Fig. 3.22 Sintering furnace [23]

power P_m, shunt resistance R_{SH}, series resistance R_S, and temperature coefficient, as elucidated in Chap. 2.

The photoelectric test characteristics of standard solar cells should comply with international norms. The test light source of the crystalline silicon solar cells is taken as the AM1.5 light source based on the spectrum near the surface, with the light intensity of 1000 W/m^2. At present, solar simulators exploit xenon arc lamps as light sources, primarily since their spectra are the closest to sunlight. Some areas above 800 nm are noticeably different from the standard spectrum, so a better sunlight simulator will be integrated with a filter to down-regulate the light intensity in this area.

The major factors affecting the test accuracy of the I-V characteristic curve include

(1) Stability of light source;
(2) Compensation ability of software and hardware of the photoelectric sensor;
(3) Contact resistance of the voltage probe;
(4) Temperature control accuracy of the test system.

After the test, the products will be sorted. Usually, the defective solar cells will be eliminated. Other cells will be classified by the unified classification standard, and the products with the closest maximum current at a fixed voltage are usually selected. The mentioned cells are categorized in accordance with performance levels so that the cells of the same level can reduce the mismatch loss when assembled. Automated test systems have significantly increased production.

3.5 Process Details and Improvement

3.5.1 Screen Printing

1. **Screen printing process**

(1) Grid

The grid is a stainless-steel wire or aluminum frame that is printed tightly on the device surface. The grid is coated with photoresist, and in the photoresist-coated portion, the aluminum is removed as in a photographic technique as presented in Fig. 3.23.

To make a prominent thin layer that meets the requirements of the front surface electrode, the printed lines should be thin and closely spaced. Moreover, the width of the cross line is many times larger than the width of the maximum printed particle. A typical grid line width is approximately 200 feet per foot. The line diameter is 10 μm, and the grid width is 30 μm [30].

(2) Pasting

Pasting is the use of an adhesive to adhere the active material onto the surface of the device. The components of the adhesive can optimize the printing effect. Pasting is primarily composed of the following characteristics:

(1) The organic solvent exhibits fluidity, which is conducive to printing.
(2) The organic binder is capable of bonding the active powders together before thermal evaporation.

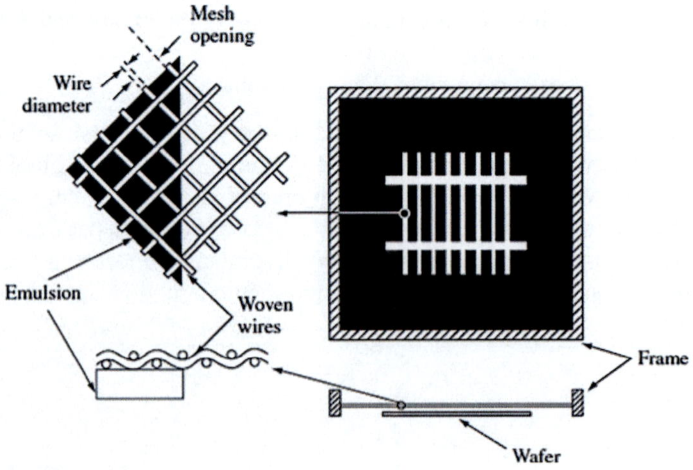

Fig. 3.23 Schematic diagram of grid printing

(3) The conductive material is composed of silver crystal grains having a size of 0.1 μm. For p-type electrode contact, Al material is applied as well, taking up 60–80% of the total mass of the slurry.
(4) The glass frit takes up 5–10% of the weight of the slurry. It is primarily composed of an oxide (e.g., oxide of lead, antimony, and silicon) powder.

The material has a lower melting point, high activity at the processing temperature, and enhanced mobility of the silver molecules, so it can diffuse into the silicon wafer to form an effective electrode contact. The composition of the slurry is important for the metallization effect and the slurry characteristics are very sensitive to temperature variations [4].

(3) Printing

A schematic diagram of the process of printing a latex map on a grid is presented in Fig. 3.24. The mesh and wafer surface are not in contact, and the intermediate separation distance is termed as the intermediate distance.

After aligning the slurry, the pressure is imposed on the blade made of rubber material. Subsequently, the mesh comes into contact with the wafer, and the blade is moved from one side to the other side. For this reason, the slurry is pressed against the surface of the wafer. When the blade leaves, the mesh bounces off the surface for elasticity and the paste sticks to the surface. The amount of slurry applied is determined by the thickness of the mesh material, the emulsion, the mesh area, as well as the area of the grid lines. The viscosity of the slurry is a critical factor: the slurry should exhibit certain liquidity, so it can fill the inside of the grid without void under pressure and will not spread on the surface after being printed. The critical

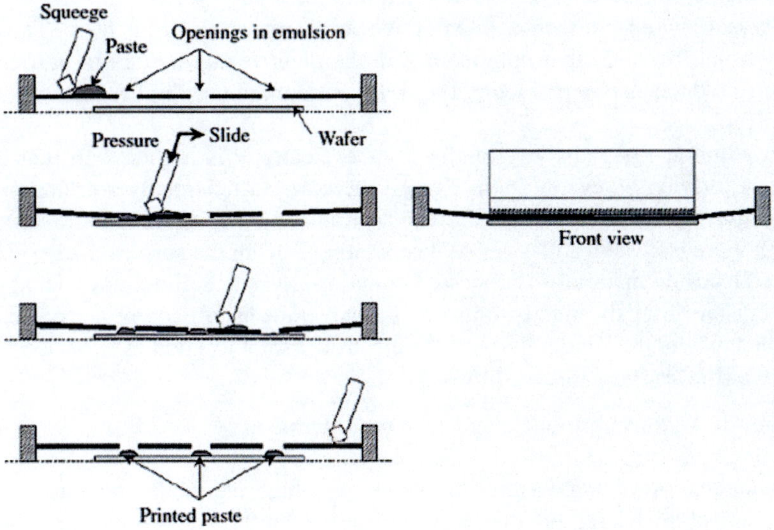

Fig. 3.24 Schematic diagram of the printing process

parameters of this process consist of the blade pressure, the blade travel speed, and the distance between the grid and the wafer [4].

(4) Drying

The solvent is evaporated at 100–200 °C, so the printed pattern is not damaged.

(5) Firing

The firing process in an infrared drum furnace basically falls under three steps. In the first step, the organic mixture combines with the particles and burns with the temperature rise. In the second step, it is maintained at higher temperature ranging from 600 to 800 °C for a few minutes. Higher temperatures are required if the printed layer should be heated sufficiently. Afterwards, the crystal orientation of the material and the composition of the slurry should also be considered. Third, the temperature is lowered.

In the process of firing, the phenomenon is sophisticated. The oxide is melted into a molten glass to sinter the silver particles to form a continuous conductor layer for the reduction of the surface resistance. When the melting point of silver is not reached, as well as the eutectic temperature of silicon, the two are sintered to form a solid microcrystalline structure in intimate contact. Meantime, the active molten frit corrodes a portion of the silicon and a portion of the silver particles to form a bond in the intimate contact with the surface of the cell. The silicon surface exhibits a corrosion depth of about 100 nm. When TiO_2 or SiN is present on the surface, the glass frit can corrode them. As a matter of fact, an effective connection is achieved by the homogenization of materials.

The electrode connection pattern at the electrode appears as two parts after cooling. On the inside, the silver crystal is inserted into the silicon surface to form an interfacial crystal, which constitutes an effective point contact electrode form. The silver particles, the frit, and other components in the outer region pertain to porous void structures, which also explain why the silver paste exhibits a higher resistivity than pure silver.

Furthermore, the resistivity of the printed electrode is higher than that of the electrode forming by evaporation. Even if there are sufficient silver grains to form an effective connection with the surface, not all the silver grains are linked to the surface; there is always a glass crystal separating it from the surface. Moreover, the surface electrode material to be pasted contains silver and aluminum. Though the Al–Si crystals inlay the metal atoms into the substrate, an effective electrode contact is formed, where the conductivity of the electrode is not as good as expected for the locality of the crystal particles.

2. **The future development trend of screen printing**

In the application of screen-printing techniques, high electrode contact resistance and etching of the frit require that the front surface emitting region should be heavily doped, and the emissive layer should not be overly thin. Such limitation can only be addressed by altering the formulation of the slurry and the process. A narrow and thin chip electrode connection is required, while the line width is specified to be significantly larger than the printed material particles. Increasing the line width refers to increasing the slurry used and expanding the light occlusion area of the front surface, which is limited by line requirements. Moreover, in this case, applying this method will deform the mesh, which causes the printed pattern to be blurry.

Metal grinders can also be exploited to make grids. To be specific, metal grinders can generate wide and high-quality lines, simplifying part of the process, without the consideration regarding degradation, maintenance, and cleaning.

During the grid printing process, the wafer is subjected to considerable pressure. Thin and irregular wafers (e.g., those obtained by sheet growth) are susceptible to chipping. Accordingly, metallization acts as an alternative process in production.

3.5.2 *Thin Wafer Process*

The sheet growth technique underpins the production of thin substrate materials. The sheet growth technique refers to a sheet growth technique regulating the pulling speed of the crystal to alter the thickness of the wafer with different seeding methods (Fig. 3.25). Wafers exhibiting a thickness of less than 200 μm have been available. When a thin substrate is adopted to fabricate a solar cell, some problems will arise accordingly.

During operation, the probability of fragmentation of thin wafers is significantly increased, especially when large-area wafers are operated. The corresponding operating tools should be designed, and the corresponding operating steps are of significance as well. For instance, in a chemical bath, convection creates a large force on the wafer surface. This type of problem will be discussed from the mechanical

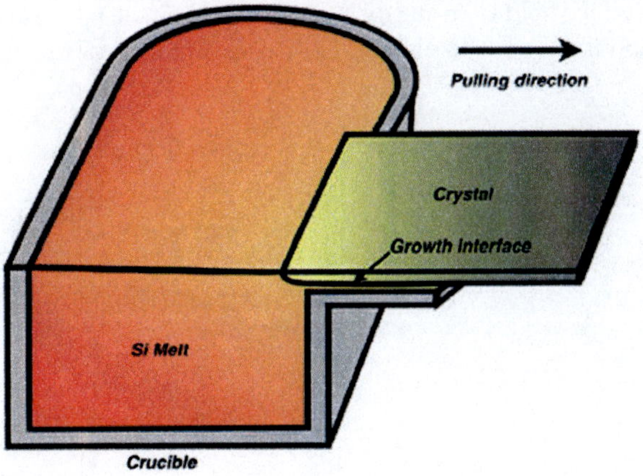

Fig. 3.25 Schematic diagram of sheet-like silicon growth technology

properties of silicon wafers, and it will also facilitate the emergence of novel crystal production [31].

The operation in the process is modified in accordance with the target of reducing heat demand. Moreover, in operation, the thin silicon wafer is extremely flexible, so some special operation processes should be developed for the thin silicon cells [32].

Thin silicon solar cells rely primarily on technologies (e.g., surface passivation and light confinement). If the technology reaches an acceptable level, the photoelectric conversion efficiency of the cells will be significantly enhanced. Thus, some novel cell structures and technologies should be developed [4].

3.5.3 *Surface Passivation Process*

Passivation on the surface critically impacts solar cell performance for the existence of multiple recombination mechanisms on the surface.

1. **Positive surface passivation**

Via a thermal oxidation process, a passivation layer can be formed on the front surface of the silicon solar cells to function as passivation and as an anti-reflective layer. However, if the oxide layer is overly thick, it will affect the function of the anti-reflective layer; thus, the oxide layer obtained by thermal oxidation should be thin. Meantime, electrodes are deposited over a portion of the thin oxide layer to form a metal-insulator-semiconductor (MIS) structure, which enables carrier collection by tunneling and makes minority carriers less likely to be recombined on the surface. The structure is presented in Fig. 3.26.

2. **Back-surface passivation**

Back-surface passivation is primarily based on back-surface field technology. Back-surface field is primarily heavily doped on the back surface or forms an alloy junction to form a built-in electric field. The direction of the built-in electric field is opposite

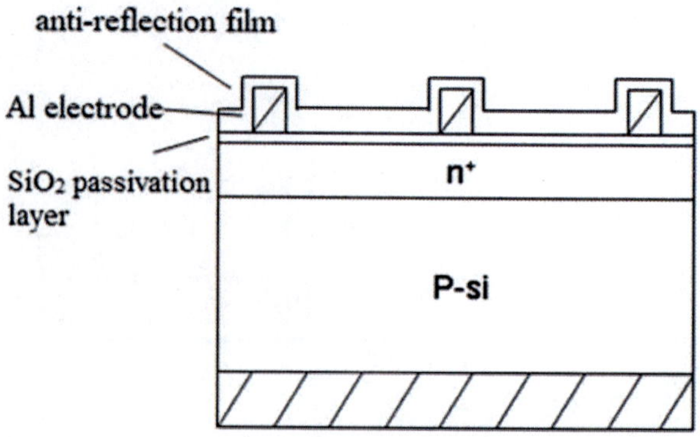

Fig. 3.26 Schematic diagram of the MINP structure.

to that of the minority carrier diffusion, so the photogenerated minority carriers generated in the body are far away from the back surface, thereby reducing the surface recombination probability of minority carriers. Accordingly, the reverse saturation current is reduced, which in turn increases the open-circuit voltage and short-circuit current density of the cells. Such technology is extensively employed in silicon solar cells and significantly enhances the photoelectric conversion efficiency of solar cells.

(1) Aluminum back-surface field

The aluminum back surface has a certain effect of absorbing impurities besides acting as an electrode, and a p-doped region exhibiting a high diffusion concentration on the back surface (for instance, a diffusion region of aluminum in silicon) forms easily in the process of screen-printing aluminum on the surface of the superalloy. For this reason, numerous silicon solar cell manufacturers have employed aluminum back field technology in their production lines. Aluminum backfield technology integrates the process. After the current surface is applied by screen printing technology for aluminum slurry printing, a sintering process is performed, so the aluminum and silicon can form an alloy in the sintering process. If a lower temperature process is applied in this process, damage to the substrate or the like can be hindered, and device performance can be enhanced. Meantime, the soldering of the back-surface silver electrode is another problem encountered in the development of thin solar cells [4].

(2) Boron back-surface field

The boron back-surface field is also primarily used to remove impurities and reduce the surface recombination rate. Boron can be surface-diffused as a diffusion source (e.g., phosphorus), so it can be combined with other steps in the basic process. Though it may be suggested to be beneficial to combine the processes in the heat treatment operation steps, the effect after the combination does not appear to be higher than that obtained by the separation step [4].

3.5.4 Selective Emitter Technology

The n-type semiconductor doping concentration in the PN junction is related to numerous critical characteristics of solar cells. To reduce the contact resistance, the conventional process is to up-regulate the doping concentration maximally. However, when the concentration increases, the holes generated by the irradiation of the short-wavelength light near the surface are easily recombined by the electrons in the highly doped n-type region, so the surface recombination probability increases. The carriers are not efficiently harvested by the electrodes, so the short-wave absorption effect near the surface is weakened; as a result, the energy conversion efficiency is lowered. Accordingly, adopting a conventional process to obtain a solar cell exhibiting a

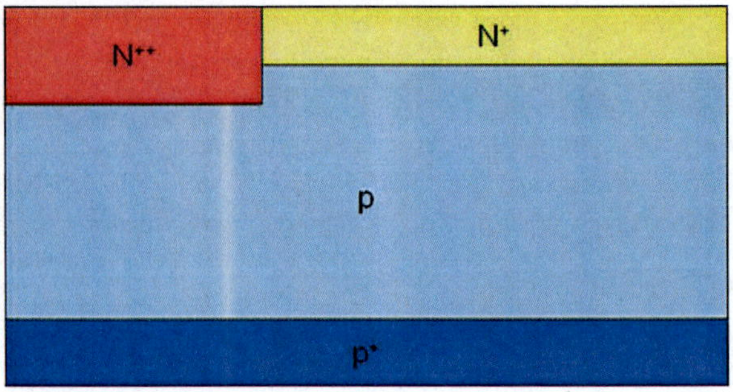

Fig. 3.27 Schematic diagram of selective emitter technology

very low contact resistance and a high photoelectric conversion efficiency is hard to achieve. To achieve the most satisfactory device performance, the optimization work must be carried out.

The so-called optimization refers to elevating the FF by reducing the contact resistance, and to increase the open-circuit voltage and short-circuit current by reducing the surface recombination. In this scenario, the selective emitter technique is obtained in two aspects, as presented in Fig. 3.27.

As shown in the above figure, the selective emitter technique is primarily heavily doped at the contact electrode, so a significant ohmic contact can be obtained, and the contact resistance is reduced, thereby increasing the FF; light doping at the non-electrode contact can lead to a small recombination rate of surface carrier, while the open-circuit voltage and short-circuit current of the solar cells can be enhanced without affecting the contact resistance at the electrode. With this selective doping, the higher photoelectric conversion efficiency can be achieved [2].

Selective emitter technology can be split into many methods in accordance with different difficulty and complexity.

Here are some of the methods used in the industry:

(1) Two diffusion methods

After the silicon wafer is textured, a dielectric layer (oxide or nitride of silicon) is thermally grown or deposited as a diffusion barrier. Photolithography or laser etching or localized etching on the dielectric layer is identical to metallization. High-density deep junction diffuses in a tubular or chain diffusion furnace, except for the pattern area; other parts are covered by the dielectric layer, thereby blocking the diffusion of impurities, and thus, impurities diffuse in the desired pattern area only. After the dielectric layer is removed, the second uniform diffusion of the shallow junction with low surface concentration is performed. The diffusion doping concentration is noticeably lower than the first diffusion, so different diffusion concentrations in various regions are achieved.

The merits of the two diffusions are the increased production efficiency, the easily regulated process, and the prominent surface passivation effect. However, the process is relatively complicated, and there are many steps that require heat treatment, which is harmful to the device.

(2) Partial diffusion method

The partial diffusion method is similar to the two diffusion methods, primarily with a non-complete masking method, with a thin dielectric layer to allow a portion of the impurity material to pass through the medium into the surface of the substrate. Thus, the doping concentration of the non-fully covered portion is relatively low, and the unmasked portion has a higher doping concentration. Thus, the diffusion of the heavily doped and lightly doped regions is completed at one time. The process is simple, the heat treatment is less, and the damage to the device is small. However, it is difficult to grasp the uniformity of diffusion and the accuracy of the concentration.

(3) Diffusion etch back method

The diffusion etch back method is first heavily doping and diffusing the surface, and then patterning the heavily doped portion with a printing slurry. Subsequently, the surface portion is etched with an etching solution, and the portion covered by the slurry is protected, and the uncovered portion is etched to a certain depth, and the diffusion junction becomes shallow. Thus, the doping concentration of the etched portion is relatively low. This method has fewer heat treatment steps, less damage, and better short-wave response. However, the etching and covering have higher requirements on the etching liquid and the slurry, and the diffusion concentration is not well understood, and the requirement of surface passivation is higher.

(4) Laser drive-in method

A shallow junction is spread on the silicon wafer surface, and then a diffusion source is spin-coated on the surface. Subsequently, the laser is locally driven in by the laser in accordance with the desired re-diffusion region. The steps being relatively simple, for the surface texturing, the local laser irradiation amount is different, resulting in uneven diffusion. Meantime, for the different line widths, some lines need multiple exposures to complete the diffusion. Moreover, the pulse interval and intensity of the laser are difficult to regulate [27].

3.5.5 *Rapid Heating Technology*

In a closed furnace or a conveyor-type furnace in a conventional process, the silicon wafer is heated, and the equipment itself (chamber and chassis) is in the heating process as well. Since most parts and materials are at an identical temperature and the whole temperature is high, and the period from heating to cooling is long, some impurities will diffuse into the substrate, causing pollution. Moreover, high temperature for a long time causes high energy consumption.

Moreover, in the past few years, the microelectronics industry has evolved to produce devices that can be heated quickly: a device that heats only silicon wafers to high temperatures without heating other objects. Selective heating is achieved by intense semiconductor ultraviolet radiation. Its main benefit is to reduce the heating-cooling cycle from the time required for conventional equipment to a few minutes. Accordingly, the production of solar cells will be significantly improved. Moreover, the extra parts are not at high temperatures, reducing potential sources of contamination and reducing energy consumption [4].

Laboratory-scale rapid heating equipment has been applied. The rapid heating technology is well applied in the screen printing sintering process and the metallization process of aluminum, as well as in the rapid surface passivation technology of nitrogen and oxygen. It can be said that any process that requires heat treatment in the manufacturing of solar cells can be achieved by rapid heating technology [33, 34].

This technique may cause defects in the substrate for a rapid heating-cooling process, which causes the substrate to age to a certain extent relative to conventional processes, shortening the lifetime [35].

A major impediment to the industrialization of rapid heating processes is the inability to produce suitable industrial equipment since microelectronic heating equipment requires batch reactors with large capacity or requires a continuous transfer line. The uniformity of heating temperature is another problem to be considered if a conveyor-style industrial rapid heating device is used [36].

3.6 Some Special Methods for Manufacturing Polycrystalline Silicon Solar Cells

Polycrystalline silicon solar cells may not apply to standardized processes for certain special properties. Some alternatives to the standard process have been proposed, while they have not been adopted for their relatively high cost. People are still looking for a solution, two of which are not the same as the single crystalline silicon process.

First, for the presence of metal impurities (dissolution or precipitation) and crystal defects (e.g., grain boundaries and misalignment), the material quality of polycrystalline silicon is relatively poor, which affects the lifetime of carriers in the material, thereby reducing the photoelectric conversion efficiency of the cells. To solve this problem, two methods of impurity removal and hydrogen passivation are primarily used.

Second, the surface texturing of polycrystalline silicon is also difficult to develop since different crystal planes are exposed from the surface, making the standard process uncommon. Thus, to improve the capture and absorption of light, the corresponding method should be considered [4].

3.6.1 Removal of Impurities from Polycrystalline Silicon Solar Cells

As is generally said, the impurity removal technique can also be applied to a single crystalline silicon process, whereas for polycrystalline silicon materials, the impurity removal technique is particularly important. The removal techniques of P and Al in the process steps of polycrystalline silicon solar cells are generally integrated into one step. The impurity removal conditions (e.g., temperature and process duration) of polycrystalline silicon are different from those of single-crystal silicon, primarily for the interaction of crystal defects, metal impurities, and other impurities (primarily O and C) in the material.

It is experimentally revealed that the efficiency of impurity removal depends largely on the quality of the material itself [37, 38]. Polycrystalline silicon ingots obtained with different techniques have different numbers and distributions of defects, and even different distributions of defects can be identified in the identical silicon ingot [39].

Moreover, on a silicon wafer, the surface and the inside also exhibit unevenness in the distribution of impurities and defects. Accordingly, the impurity removal has different effects in different regions, which probably eventually affect the electrical characteristics of the cells [40, 41].

3.6.2 Hydrogen Passivation Process

Silicon nitride is extensively used as a masking film in the field of microelectronics [42]. For solar cells, silicon nitride can also act as an effective anti-reflective layer. Overall, silicon nitride is achieved by several deposition methods, where the most common technique is chemical vapor deposition (CVD), and exploiting silane gas and ammonia gas. Plasma-enhanced chemical vapor deposition is a method taken by most people compared with other chemical vapor deposition methods (the atmospheric pressure CVD method or the low-pressure CVD method) since it is a low-temperature operation (less than 500 °C). This means that complexity is reduced, so the aging speed of the device is reduced.

The critical characteristic of a significant PECVD material is that it can produce a hydrogenation reaction, which is very conducive to silicon [43, 44]. Hydrogen atoms react with most of the impurities and defects in the silicon, neutralizing the negative charge center. Such effect is termed as bulk passivation. An amorphous silicon nitride film usually contains hydrogen atoms (it is generally considered to be pure silicon nitride, while its composition is formed as SiN_x:H). The subsequent operation is heating to activate the contained hydrogen atoms, which is often done simultaneously during the metal sintering process [45].

Moreover, surface passivation techniques for depositing silicon nitride based on PECVD methods have been verified [46]. The effect of the phosphor-doped emitter

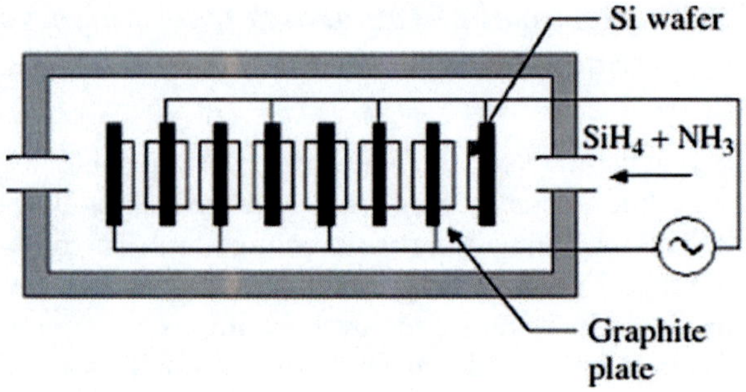

Fig. 3.28 Direct plasma reaction

surface passivated with the mentioned method is equivalent to the effect of high-quality oxygen surface passivation. The recombination speed was 4 cm/s obtained on a polished silicon wafer exhibiting a resistivity of 1.5 Ω cm [47].

The effects of the three processes of anti-reflective layer, bulk passivation, and surface passivation are related, and the process of any one of them cannot be arbitrarily altered. Thus, the processing parameters (e.g., temperature, the excitation power and frequency of plasma, and the flow rate of gas) should be optimized to address the problem to obtain a balanced process method. Moreover, different PECVD methods give different results [48, 49].

Most of the PECVD processes in the industry are direct PECVD, as presented in Fig. 3.28.

The gas to be employed in the process is excited by an electromagnetic field to form a plasma. Subsequently, the wafer is directly placed in a plasma atmosphere for processing. However, placing the wafer directly in a plasma atmosphere for an overly long time can affect the effect of surface passivation. Moreover, exposure to ultraviolet light can cause wafer aging.

Direct PECVD is split into high frequency (13.56 MHz) and low frequency (10–500 kHz). The high frequency method is better for the surface passivation effect, and the stability of the ultraviolet irradiation is higher as well. However, such a method is more difficult to regulate, and it cannot effectively obtain a relatively uniform passivation layer.

Another method refers to separating PECVD, with the aim to place the wafer outside of the plasma atmosphere. Applying such a method can hinder the surface damage of the silicon wafer, thereby achieving better surface passivation, whereas the body passivation effect may be worse. After years of development, such technology has been applied to industrial production. A schematic diagram of a separate passivation system is presented in Fig. 3.29.

The separate PECVD method achieves a continuous passivation process, and the wafer is continuously fed into the apparatus, which is evidently superior to the intermittent direct PECVD method [4].

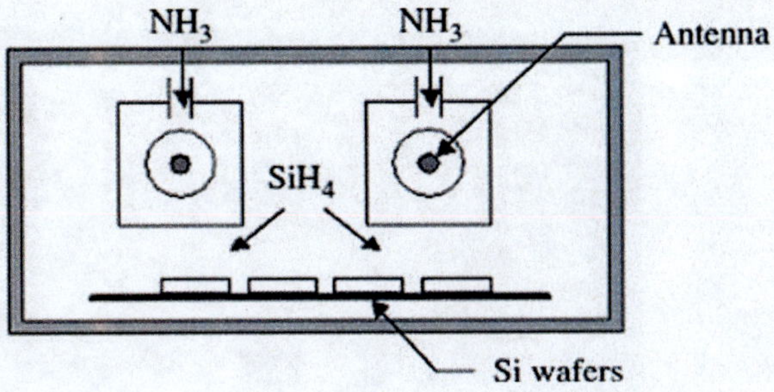

Fig. 3.29 Schematic diagram of a separate passivation system

3.6.3 Light Capture Process

The standard alkaline texturing process for single crystalline silicon can also be used for polycrystalline silicon wafers, where the satisfaction of the etching results is not as great as that of single crystalline silicon.

For the different etch rates of different crystal orientations, the surface texturing pattern is not very regular, resulting in relatively high surface reflectance. Moreover, there may be vacancies between the crystal grains so that electrode breaks occur when screen printing achieves electrode contact.

The existence of the above problems is another reason for considering novel alternative processes. The novel process considers the problem of surface reflectance of light, as well as surface damage and the compatibility of the metallization process. Some of the mentioned processes have already been applied and are being put into industrial production, and some processes require subsequent verification. Geometric optics will no longer be applicable when the geometry size produced by texturing is comparable to or less than the wavelength of light. Then, the structure of the texturing surface will become a diffraction grating, a scattering medium, or be limited to a small geometrical dimension as a graded refractive layer [4].

1. **Chemical etching**

Some chemical etching techniques have been verified. Some of them can lead to the production of inverted pyramid-shaped structures, where lithographic masks are required, which is not consistent with some industrial considerations, so it is difficult to industrialize [50]. Approximately 20% of the surface of the polycrystalline silicon solar cell is textured by acid etching with robust oxidation [51]. One of the simple methods refers to exploiting an etchant containing nitric acid, hydrofluoric acid, and some additives to achieve isotropic etching. The structures etched with the mentioned method are uniformly distributed pits of 1–10 μm. This results in uniform reflectivity on the surface and where it is missing between crystal grains, as presented in Fig. 3.30.

The solar cell obtained by isotropic etching of the surface has a rise in short-circuit current of about 1 mA/cm^2 compared with that obtained by anisotropic etching [52]. However, some technical problems have arisen (e.g., the dry junction loss of the solution and the exothermic effect). The mentioned problems may make processes

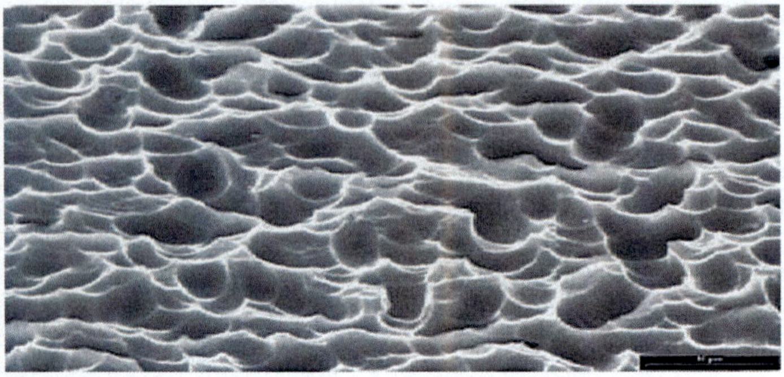

Fig. 3.30 Micrograph of an acid etched surface [4]

in industrial production incompatible. In the face of the mentioned problems, automatic wetting table, automatic temperature regulation system for etching liquid, and automatic chemical replenishment system have been developed.

To reduce the absorption loss attributed to reflection, porous silicon materials have been developed [53]. The analysis shows that if both absorbed and reflected lights are considered, porous silicon can achieve an optimum light absorption loss of about 5–6%. Though porous silicon exhibits its merits in development, process incompatibility problems occur in the screen printing process, so the manufacturing process of porous silicon solar cells requires further improvement.

2. **Mechanical texturing**

Mechanical texturing obtained a V-groove of about 50 μm deep by a mechanical etching method with a conventional dicing saw and a bevel blade and is then etched with an alkaline solution to reduce surface damage. With this technology, the photoelectric conversion efficiency of the packaged solar cell can be elevated by 5% [54]. The direction in which the finger electrodes are printed parallels the direction of the V-groove. By using the mechanical texturing method, there can be no texturing on the high platform, making it easier to print the electrodes and facilitates some process operations that require calibration. Besides the grid printing process, other electrode contact techniques (e.g., roll printing [55] and buried layer electrodes [56]) have been applied as well. Some automated production processes are being currently verified.

Another method is surface texturing by laser engraving [57]. Two sets of orthogonal parallel grooves are laser-etched, and then silicon is removed by chemical etching, so a pyramid-shaped etch pit having a height of about 7 μm can be obtained. Combined with a single-layered anti-reflective layer process, the laser texturing method is capable of decreasing light reflection to 4%. The improvement is primarily achieved by anisotropic etching and anti-reflective coatings. Some processes have been modified to achieve smoother and smaller grooves to accommodate the screen printing process.

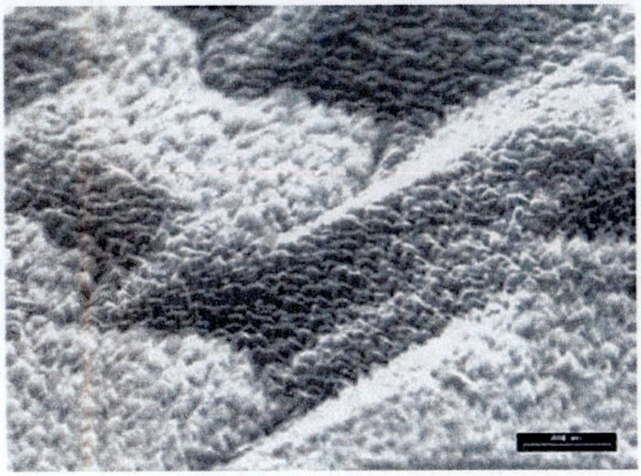

Fig. 3.31 Effect diagram of reactive ion etching [4]

3. **Reactive ion etching (RIE)**

This reactive ion etching method refers to etching the silicon wafer in a chlorine plasma gas. Such a method pertains to a dry etching method capable of achieving a large density of corrosion pits exhibiting a feature size of less than 1 μm on the surface (Fig. 3.31) [58]. The short-circuit current of the solar cell obtained with the mentioned method is 1.4 mA/cm^2 larger than that of the solar cell obtained by the anisotropic non-mask method [59]. Reactive ion etching methods can also be integrated with mask etching techniques to develop more regular corrosion effects. However, if they are applied to industrial production, the production efficiency will be reduced [60], and the application of toxic and corrosive chlorine gas is considered as another problem to be addressed in industrial production [61].

4. **Reflective coating and packaging**

The reflective properties characterized with different texturing methods are different since they are fabricated in conjunction with conventional anti-reflective coating processes on the whole (e.g., precipitation of TiO_2 or TiO_2/SnO_2 by the atmospheric pressure CVD method, and deposition of silicon nitride by the PECVD method) and cell packaging processes. Accordingly, the gap between the merits and defects of some etching methods is narrowed (Table 3.1).

Table 3.1 Comparison of reflectivity after several surface treatments [69]

Reflectivity (%)	Acid etching	Alkaline etching	Non-mask reactive ion etching
Die	34.4	27.6	11.0
Depositing silicon nitride	9.0	8.0	3.9
Depositing silicon nitride and packaging	12.9	9.2	7.6

3.7 Technology and Structure of High-Efficiency Crystalline Silicon Solar Cells

3.7.1 Strip-Shaped Silicon Technology

Since the strip-shaped silicon technology eliminates the slicing process, it is very conducive to production costs. The edge film growth process has become the most advanced technology to produce strip-shaped silicon, in which the banded and dendritic roll process has been adopted for industrial production.

For the high density of defects (e.g., dislocations, grain boundaries, and impurities) of the strip-shaped silicon substrate, the strip silicon solar cell processing should comply with some specific processes. In particular, a relatively deep back-surface field is printed with an aluminum slurry to facilitate the absorption of impurities; silicon nitride is deposited by PECVD for bulk passivation and as an anti-reflective coating.

For the strip-shaped silicon solar cell, it cannot be metallized with the screen printing process for the uneven surface of the material, whereas it is realized by pad printing through directly extruding the silver slurry and the ink. Strip-shaped silicon solar cells exhibit an average photoelectric conversion efficiency of 14% in industrial production; in some scenarios, the photoelectric conversion efficiency reaches 14.7% [62]. The growth of large-area strip-shaped silicon will be another way to lower production costs in the future, and the growth of large-area strip-shaped silicon will facilitate the reduction of thermoelastic stress, as well as the generation of more uniform and thinner wafers. Moreover, thin curved wafer solar cells require novel manufacturing techniques. The photoelectric conversion efficiency of banded silicon has reached 14.7% [63], and photovoltaic modules of 50 and 100 W have already been commercialized [64].

3.7.2 High-Efficiency and Low-Resistance Silicon Solar Cell

A high-efficiency low-resistance silicon solar cell (RESC) is a solar cell developed with melted silicon exhibiting a resistivity of 0.2 and 0.3 Ω cm in the p-type region. The major feature is to make a passivation layer at the emitter of the cell; as a result, the surface recombination rate of the photogenerated carriers on the surface can be down-regulated, and the position of the PN junction is deep. For the optimization process described above, the cell exhibits a higher open-circuit voltage, a higher short-circuit current, as well as a higher fill factor. Under the illumination of a standard solar simulator, a photoelectric conversion efficiency of 21.6% is obtained [5].

3.7.3 *Passivated Emitter and Rear Cell*

Since the aluminum electrode covers the back surface of the entire solar cell, it is critical to absorb impurities and it acts as a P+ layer to prevent minority carriers from migrating to the back surface and reduce surface recombination. However, under the trend of thinning of the silicon wafer, the cell will bend, resulting in a rise in the surface recombination rate and a decrease in the long-wave absorption efficiency. Passivated emitter and rear cell (PERC) can more effectively solve such problem. The PERC silicon solar cell is considered the initial high-efficiency solar cell researched at the Photovoltaic Device Laboratory of the University of Novel South Wales, Australia. The cell structure is illustrated in Fig. 3.32.

PERC exploits an oxide layer to passivate the surface of the cell and replaces the all-aluminum structure with a point-contacting electrode. The passivation of the oxide layer decreases the interface state density of the surface, metal impurities, and surface faults, which significantly extends the lifetime of the minority carriers. The oxide layer is located between the metal layer and the silicon wafer, thereby reducing the amplitude of the silicon wafer bending, and it also acts as an anti-reflective layer to up-regulate the photon absorption rate.

The point-contacting electrode is to connect some of the separated small holes through the passivation layer and the substrate, and the photoelectric conversion efficiency of the cell can reach 23.2%. Since the electrode comes into direct contact with the substrate, it is not passivated. To decrease the recombination rate of the surface, the design distance of the holes exceeds the thickness of the substrate. However, if the distance is overly large, the surface resistance will be enhanced, and the open-circuit voltage and other parameters will be lowered. Thus, a substrate material of low resistivity should be exploited, and an appropriate distance between holes should be set.

The electrode of PERC penetrates the dielectric layer to achieve good ohmic contact, and the process step is critical. At present, there are primarily the following methods for electrode fabrication, i.e., photolithography, mechanical methods, inkjet printing, and laser sintering. The first three methods are to open a hole in the surface,

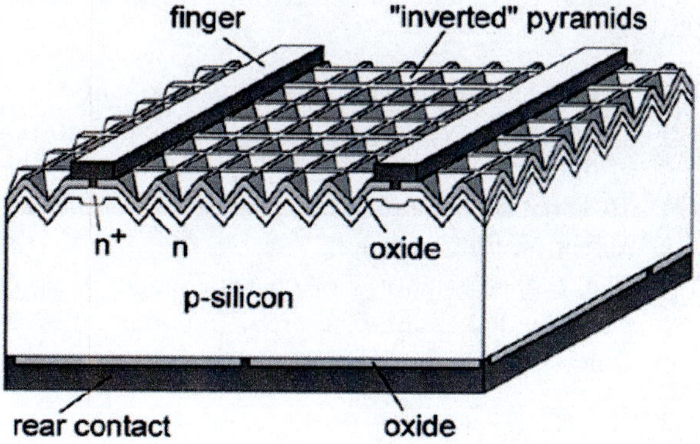

Fig. 3.32 Schematic diagram of PERC structure

Fig. 3.33 Image of laser sintering

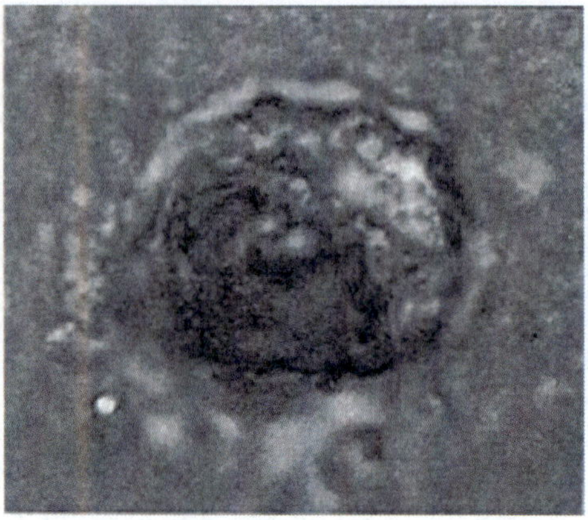

then deposit metal into the small hole, and form an effective electrode contact by sintering. The laser sintering method refers to depositing a metal material on the surface and then penetrating the metal into the substrate to form an electrode contact as impacted by a laser. Figure 3.33 is an image of laser sintering [65].

3.7.4 *Passivated Emitter and Rear Locally Diffused Cell*

Passivated emitter and rear locally diffused cells (PERL) are primarily designed for the defect of the large contact resistance of the PERC. The main improvement is the doping of the boron material at the point electrode contact, where the doping of the boron material impacts the lifetime of the minority carrier. Next, researchers used boron bromide for diffusion to make the effect much better. Meantime, the surface resistance can be reduced by reducing the hole pitch to improve the photoelectric conversion efficiency. Next, after a series of improvements, the photoelectric conversion efficiency of the PERL reaches 24.7%, which is marked as a record of the photoelectric conversion efficiency.

In brief, the outstanding performance of PERL is primarily for the following design:

(1) With a substrate material with better quality;
(2) The surface texturing technology is adopted to enhance the absorption efficiency of light;
(3) To enhance the secondary absorption efficiency of light, two layers of anti-reflection film are used;
(4) The point-contacting electrode design is applied, and selective doping is performed meantime, thereby reducing the surface recombination and the contact resistance [5].

3.7.5 Notching and Burying Grid Technology

Notching and burying grid technology refers to a technology of high-efficiency solar cells developed by the University of Novel South Wales. Such technique involves forming a deep trench on the silicon surface and then depositing the metal in the trench by electroless plating (Fig. 3.34).

A higher longitudinal electrode length and a smaller metal shading rate can be obtained with a structure developed by notching and burying the grid. For the notching and burying grid technology, several techniques have been extensively proposed; to be specific, the laser etching method is generally considered the most suitable process for large-scale production. The etched trench is approximately 40 μm deep and 20 μm wide, as presented in Fig. 3.35. Other processes have also been employed to the

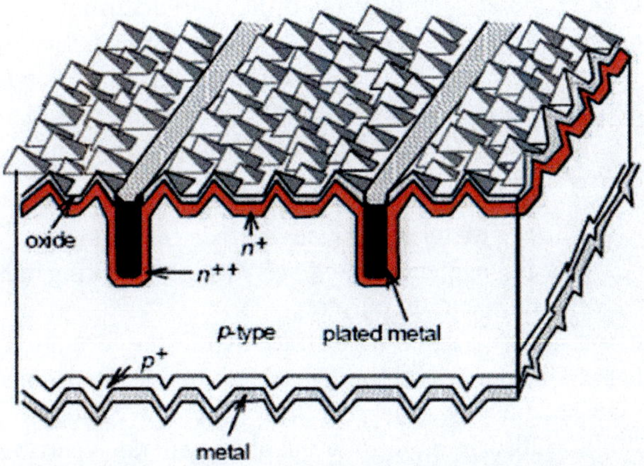

Fig. 3.34 Schematic diagram of a structure fabricated by notching and burying grid technlogy

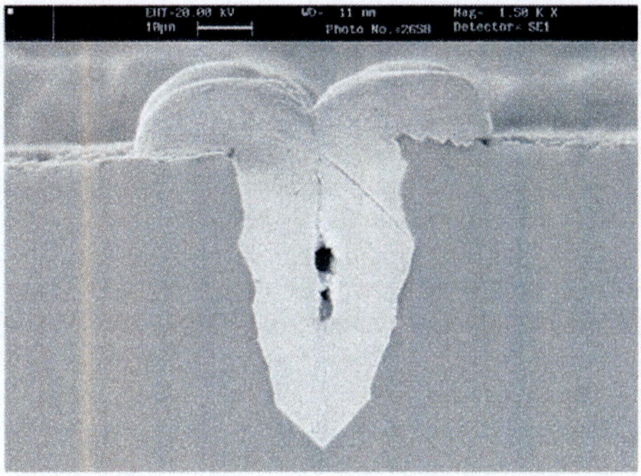

Fig. 3.35 Actual microscopic view of notching and burying grid technlogy [4]

structure of notching and burying grid technlogy (e.g., selective emitter technology, low doping technology, finger electrode technology, as well as back-surface field technology).

The primary process flow of the notching and burying grid technology is expressed as follows: ① Cleaning, surface texturing; ② Cleaning; ③ Diffusion of light phosphorus; ④ Thermal oxidation passivation; ⑤ Notching; ⑥ Corrosion in the notch; ⑦ Cleaning; ⑧ Diffusion of concentrated phosphorus in the notch; ⑨ Evaporation of aluminum on the back side; ⑩ Burning the back field; ⑪ Electroless plating buried gate; ⑫ Fabrication of the back electrode; ⑬ Evaporation of the reflective layer; and ⑭ De-sintering.

With this technique, it is more difficult to etch a thin notch. At present, there are primarily two etching methods: dry etching and wet etching. The most frequently applied method for dry etching is laser notching. The etch line width of this method is easier to control, and the notch shape is etched accurately. However, the etching speed is slow, the etching depth is difficult to determine accurately, and substrate defects are easy to form. When wet etching is applied, the etching depth is easy to control, where an etching protection layer should be formed on the surface, and for deep notch etching, the speed is slow, and it is difficult to meet industrial requirements.

The reasons why solar cells with buried grid exhibits higher photoelectric conversion efficiency are presented as follows:

(1) Surface texturing reduces surface reflectance;
(2) The finger electrode reduces the light blocking rate of the electrode and increases the short-circuit current;
(3) Light doping on the surface avoids the formation of "dead layers";
(4) Selective heavy doping reduces contact resistance;
(5) The burying gate process reduces the contact resistance of the electrode, and the absorption ratio of the photogenerated carriers is significantly increased for the increased contact area between the electrode and the substrate [2].

3.7.6 *Obliquely Evaporated Contact Solar Cell*

Obliquely evaporated contact solar cell (OECO cell) is a novel cell structure developed by the ISFH Institute in Germany. The cell structure design is creative, low-cost, simple to operate, and applies to mass production. The cell is primarily designed based on the MIS structure, the silicon wafer is inclined at a certain angle, and then a layer of Al is deposited as an electrode on the side of the groove. A very simple process is required to obtain better electrode contact, and electrode evaporation can be performed in large quantities. The structure of the cell is presented in Fig. 3.36.

The OECO cell surface consists of several well-arranged square trenches with a shallow emitter n+ doping on the trench surface and a thin layer of tunnel oxide on it. An aluminum electrode material is deposited on the side of the trench, and silicon nitride is subsequently evaporated on the surface as a passivation layer and an anti-reflective layer.

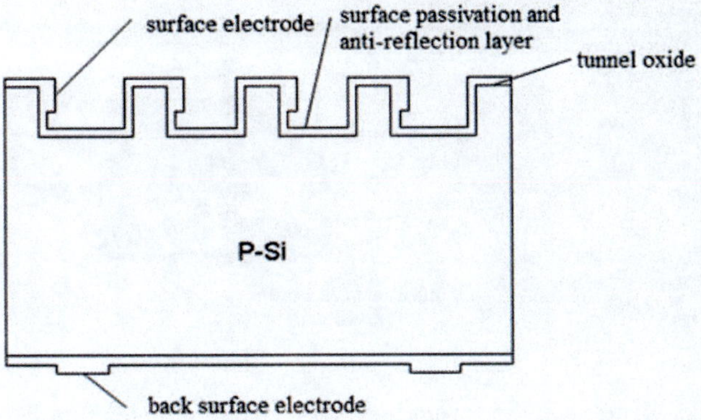

Fig. 3.36 Diagram of cell structure and electrode fabrication.

The reasons why OECO cells exhibit higher performance include the following:

(1) Since the electrode is vapor-deposited on the side surface of the groove, light is less blocked by the surface electrode, which is conducive to enhancing the short-circuit current;
(2) Based on the design of the MIS structure, no direct contact is identified between the electrode and the substrate; as a result, the surface recombination is reduced, and the open-circuit voltage and the fill factor can be improved;
(3) Since the method of oblique vapor deposition is employed, an effective electrode contact is formed;
(4) Since the electrode does not directly contact the substrate, the emitter region can be shallower;
(5) The surface treatment process at a lower temperature does not cause excessive substrate damage.

For the application of mechanical grooving, the process is relatively simple, and the electrode can be vapor-deposited on a large scale and applies to a thin silicon process, which can save cost. The critical thing is that the electrode material Al replacing Ag and does not require a mask. It can significantly reduce expenses, is very much applicable to large-scale industrial production, and the photoelectric conversion efficiency can reach 21.2% [5].

3.7.7 Metal Wrap Through

Metal wrap through (MWT) is a novel solar cell structure developed by Solland Solar, the largest solar cell manufacturer in the Netherlands, as presented in Fig. 3.37. This technique preserves the electrode on the front surface, where the lines are very thin. The current collected by the front surface is connected to the main gate on the back surface through a metallized hole. Since the electrode lines on the front surface become thinner, the blocking effect on light is reduced. Moreover, the back-surface recombination of the electrodes can reduce the design difficulty of the components

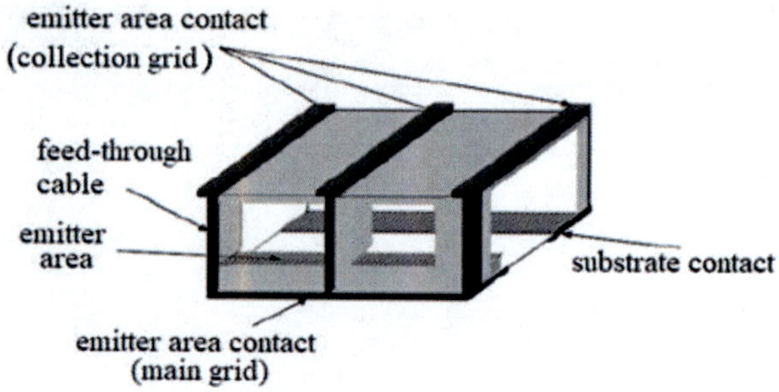

Fig. 3.37 Schematic diagram of MWT structure.

and save costs. Meantime, the through-type electrode reduces the distance that the carrier reaches the electrode, improves the carrier collection efficiency, and increases the short-circuit current of the cell [66].

The main process flow of the metal wrap through technology is expressed as follows: ① Cleaning; ② Laser drilling; ③ Surface texturing; ④ Diffusion of emitter; ⑤ Removal of phospho-silicate glass; ⑥ Preparation of anti-reflective layer; ⑦ Screen printing of front electrode lines and filling holes; ⑧ Printing of back aluminum electrodes; ⑨ Sintering; and ⑩ Front and back isolation.

The punching technique is an important part of the preparation of the MWT cell since the surface of the hole is in good ohmic contact with the electrode. Overall, the punching technology can be split into mechanical punching, wet etching punching, and laser drilling. But the first two methods take a long time. Accordingly, the laser etching method is generally used in the industry. Though the etching is better in morphology and speed, the temperature at the aperture is higher for laser etching, which probably damages the substrate material. Thus, after the perforation, it undergoes lye treatment to reduce the damage, where the existence of the damage cannot be avoided. Accordingly, solving the problem of punching is a direction of process research in the future [66].

Moreover, the filling and sintering of the electrode slurry of the holes to form the electrode contact also require the quality of the slurry. Positive and negative electrodes on the back surface may also cause punch-through, resulting in current loss, so measures to isolate the electrodes are required.

3.7.8 *Interdigitated Back-Sided Contact Solar Cell*

Interdigitated back-sided contact (IBC) solar cell is a novel high-efficiency solar cell structure developed by SunPower, as presented in Fig. 3.38.

The IBC cell structure removes all the front metal gate electrodes, and the positive and negative electrodes are arranged crosswise at the back surface. The front surface is completely unobstructed. The benefits of this cell structure are primarily as follows:

Fig. 3.38 Schematic diagram of IBC structure

(1) There is no electrode shielding on the front surface, so the absorption rate of light increases, and the short-circuit current increases;
(2) The design of the finger-type electrode on the full back surface makes the design of the external component simple, and the splicing between the cell substrates is easier, which improves the production efficiency and reduces the process difficulty, so the cost can be significantly reduced.

However, for the relatively novel structural design, there are some process problems.

First, since the front surface electrode is moved to the back surface, the light absorption rate of the front surface increases. Short-wavelength light is absorbed at a position very close to the surface, generating excitons, and then the excitons diffuse to the PN junction to separate into free carriers. However, for the smaller distance from the front surface, the probability of being recombined by the front surface is significantly increased. Thus, if the photogenerated carriers are to be collected by the back electrode, they should spread a long distance to reach the back surface. Hence, it is necessary to increase the lifetime of photogenerated carriers and seek a better process for manufacturing high-quality silicon. Usually, n-type silicon has high carrier mobility, so it is a significant choice for the substrates of IBC solar cells. Meantime, reducing the thickness of the silicon wafer also contributes to the absorption of carriers by the electrodes, and the production cost can be reduced. Surface passivation can also reduce the surface recombination rate, thereby increasing the absorption efficiency of carriers by the back electrode [2]. On the whole, the surface passivation method can fall into surface oxidation to form a SiO_2 layer, deposition of a silicon nitride layer, or formation of a surface field. The method of forming the surface field can form a passivation layer to reduce defects, while facilitating the carrier migration, i.e., expanding the lifetime of the minority and up-regulating the collected number of the minority. If an n-type silicon substrate is employed, passivation can be achieved by surface p-doping [66].

Second, since the positive and negative electrodes are overall located on the back surface, the position and arrangement among the electrodes directly affect the performance of the cell. First, a wider electrode is required to obtain a smaller contact resistance, while the surface recombination rate increases. Accordingly, heavy doping at the electrodes and a point-contact-type electrode process are employed. Meantime, the electrodes should be passivated for isolation to prevent current loss. To increase the absorption effect of the electrodes, the distance between the electrodes is shortened, where a smaller distance between the electrodes makes the etching more difficult to achieve. Given the requirements of cost, operation difficulty and efficiency in industrial production, electrodes are generally prepared by laser etching and sintering [66].

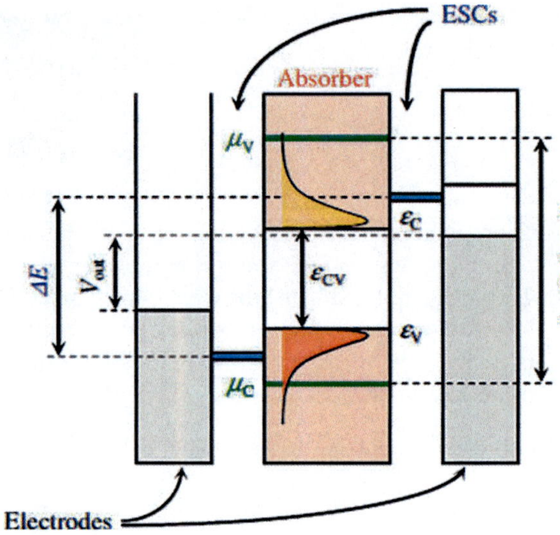

Fig. 3.39 Schematic diagram of HCSC cell

3.7.9 *Hot-Carrier Solar Cell*

After the photon excitation generates photogenerated carriers, the energy of the carriers may be lowered by Auger recombination, phonon scattering, photon radiation, and others, so it is lost. Thus, hot-carrier solar cells (HCSC) are to avoid loss of photogenerated carriers, so they maintain high energy [2].

The HCSC cell aims to collect the carrier energy before it is lost to the electrode. Accordingly, one approach is to enhance the carrier mobility and shorten the time it takes to be collected, so it is absorbed by the electrode timely. Another way is to make a superlattice that quantifies carriers, capable of extending the lifetime of photogenerated carriers. Moreover, strong light irradiation can be performed to increase the survival time. The schematic diagram of the HCSC cell is presented in Fig. 3.39 [67].

3.7.10 *High-Efficiency Solar Cell with Back-Surface Reflection*

High-efficiency solar cell with back-surface reflection is a device primarily with a material acting as a highly effective reflective layer that evaporates between the electrode on the back surface and the passivation layer, which is generally Al. The solar cell can absorb light in a range of wavelengths, where the light that is not in this range is generally lost through the cell. Thus, a reflective layer is designed on the back surface to reflect the unabsorbed light back into the cell for secondary absorption; as a result, the light absorption rate of the cell can be elevated, and the short-circuit current of the cell can increase. In the trend of cells becoming thinner and thinner, the utilization of light will decrease as the substrate becomes thinner; thus, the reflective layer on the back surface is critical. Meantime, the reflective layer on the back surface can also reflect the light near the infrared area with a longer wavelength out of the cell to prevent it from absorbing the part of the light and generate heat, since the variation of temperature may cause degradation of the cell performance.

To enhance the photoelectric conversion efficiency, p-type heavy doping can also be performed near the electrode on the back surface to form the back-surface field, thereby reducing surface recombination and facilitating short-circuit current. Moreover, the potential difference of the formed back-surface field is imposed on the open-circuit voltage to increase it [2].

3.7.11 Heterojunction with Intrinsic Thin-Layer Solar Cell

The heterojunction with intrinsic thin-layer (HIT) solar cell refers to a high-efficiency solar cell developed by Sanyo Corporation of Japan. This solar cell structure is applicable to making large-area solar cells. A photoelectric conversion efficiency of 23% is obtained on a silicon wafer of 100.4 cm^2 size, and a photoelectric conversion efficiency of 22.8% is also obtained on a 98 μm thick Cz–Si. The cell structure is presented in Fig. 3.40.

As presented in Fig. 3.40, intrinsic amorphous silicon is deposited on the front surface of the surface-textured n-type crystalline silicon, and p-type amorphous silicon is formed thereon, thereby forming a PN junction structure. Moreover, a layer of intrinsic amorphous silicon is deposited on the back surface, while a layer of n-type amorphous silicon is formed to establish the back-surface field (BSF). Subsequently, a transparent electrode and a metal gate are formed on the amorphous silicon on both sides. The mentioned processes are all completed below 200 °C, which can effectively mitigate cell damage and lower power consumption [4].

A significant feature of the solar cell of the HIT structure is that amorphous silicon exhibiting high quality is deposited on the single crystalline silicon as the passivation layer. Such significantly thin intrinsic amorphous silicon remedies the dangling bonds on the heterojunction surface, reducing the defect density, which allows an open-circuit voltage of more than 720 mV to be obtained. Meantime,

Fig. 3.40 Schematic diagram of a HIT structure

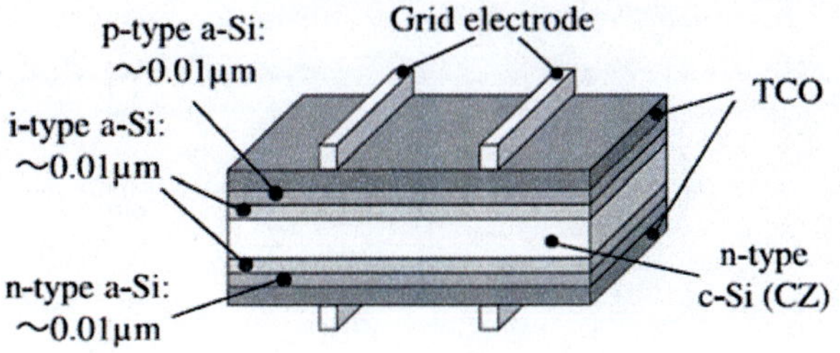

the cell also has a significant variation of the photoelectric conversion efficiency with temperature (0.23%/°C). Moreover, since the structure of the HIT solar cell has symmetry and the surface electrode is transparent, it is very much applicable to be used as a double-sided cell. The mentioned good cell characteristics make it more economical than ordinary crystalline silicon solar cells at the identical output of power generation.

In 2009, the photoelectric conversion efficiency of HIT solar cells in the laboratory reached 23.0%. To gain greater competitive merit, several technologies have shifted from the laboratory stage to the mass production stage.

To reduce the production cost of HIT cells, the use of thinner silicon wafers is a noticeable choice, while there are also some unsolved problems. First, the curling of thin silicon cells is a problem. Secondly, the insufficiency of the light absorption and the reduction of the photocurrent attributed to the thinning of the silicon wafer is another urgent problem to be solved. The third problem is that as the carriers are closer to the surface, the increased surface recombination rate affects the open-circuit voltage. Accordingly, the application of thin silicon wafers requires a corresponding process.

The 58 μm thick HIT cell originating from the experiment did not have any curl since the low temperature technique is applied in the fabrication process and the cell exhibits a symmetrical structure.

As presented in Fig. 3.41, the decrease in photocurrent attributed to the decrease in the thickness of the cell resulting in insufficient absorption of light is true. However, since the solar cell is thinned, the body resistance of the solar cell is reduced, thereby increasing the open-circuit voltage, which overly remedies the decrease in the short-circuit current. Thus, thin silicon HIT solar cells are a viable solution to lower production costs.

Figure 3.42 gives the variation of open-circuit voltage for crystalline silicon HIT cells under different surface recombination rates. As suggested from the figure, when the surface recombination rate is larger than 100 cm/s, the open-circuit voltage

Fig. 3.41 Curve of open-circuit voltage and short-circuit current versus thickness of silicon wafer [9]

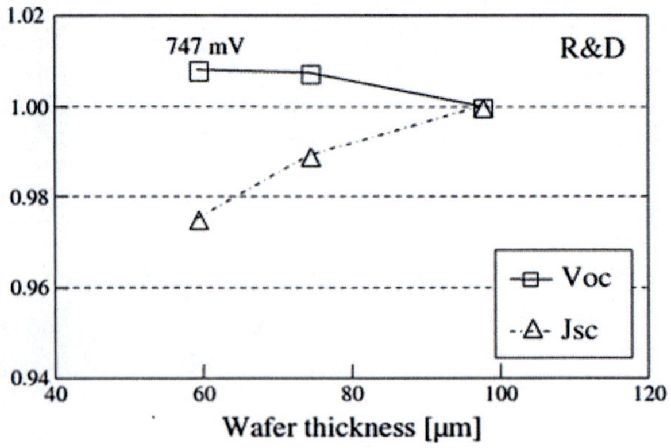

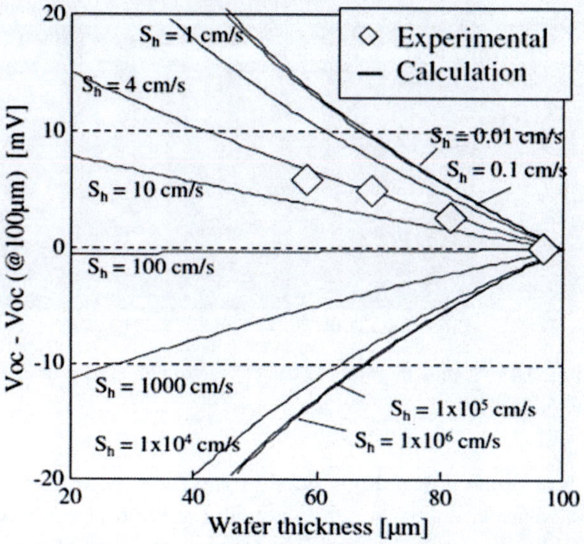

Fig. 3.42 Curve of open-circuit voltage versus thickness of silicon wafer at different surface combination rates [68]

decreases as the silicon wafer becomes thinner. However, at the surface recombination rate is smaller than 100 cm/s, the result is reversed. For a well surface passivated HIT cell, the surface recombination rate can be low as only 4 cm/s. Accordingly, as the thickness of the silicon wafer is reduced, a higher open-circuit voltage can be obtained [68].

3.8 Outlook of Crystalline Silicon Solar Cell

Lowering the cost of solar photovoltaic power generation has been constantly pursued by research and development in the crystalline silicon solar cell industry. After years of development, great progress has been achieved in this aspect: over the past few years, with the emergence of advanced production processes and emerging cell structures, the photoelectric conversion efficiency of commercial single crystalline silicon solar cells have reached 16–19%, and that of the polycrystalline silicon solar cells have reached 15–17%. Figure 3.43 presents the trend of photoelectric conversion efficiency of crystalline silicon cells over the past few years. It is revealed that the ideal photoelectric conversion efficiency of silicon solar cells still cannot be achieved. Moreover, many points require further improvement (e.g., surface passivation, texturing, and electrode preparation of the cell). Meantime, the quality of crystalline silicon material also seriously affects the photoelectric conversion efficiency, so elevating the quality of industrial silicon is another direction of future development.

Subsequent developments also require lower costs reduced from material costs and processing costs. First, the thickness of the silicon wafer is reduced, i.e., the

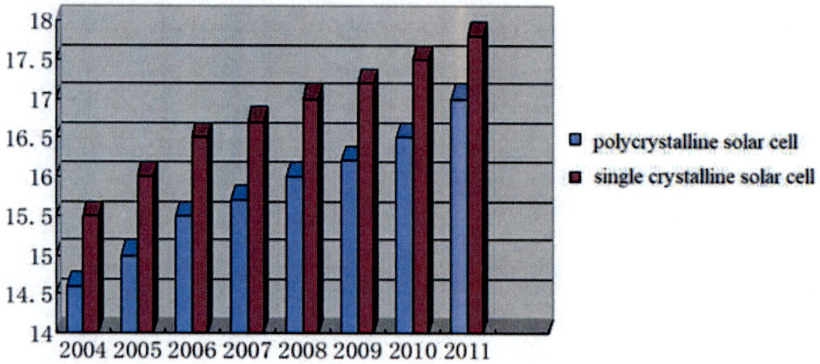

Fig. 3.43 Trends in photoelectric conversion efficiency of crystalline silicon solar cells over the past few years.

development of a thin silicon process. From cells with some structures, thin silicon saves raw materials, while enhancing photoelectric conversion efficiency (e.g., the HIT cell). The existing thickness of crystalline silicon wafers for solar cells is nearly 180 μm, which will decrease in the future. However, for the existence of shallow junctions, the thin silicon process increases the surface recombination rates, so good surface passivation and other related processes are essential. Moreover, technologies (e.g., rapid heating devices and laser etching) are highly useful to enhance production efficiency and lower costs. With the emergence of novel processes, the applications of equipment and the emergence of novel solar cell structures, silicon solar cells are considered to have a better prospect based on advanced silicon technology.

References

1. S. Xiong, M. Zhu et al., *Solar Cell Foundation and Application* (Science Press, Beijing, 2009) (in Chinese)
2. M. Green, *Silicon Solar cells. Advanced Principles and Practice*, Chap. 7 (Centre for Photovoltaic Devices and Systems, University of New South Wales, Sydney, 1995)
3. P. Verlinden et al., *Proceedings of 14th European Conference on Photovoltaic Solar Energy Conversion* (1997), pp. 96–100
4. H. Ohtsuka, M. Sakamoto, K. Tsutsui, Y. Yazawa, Prog. Photovolt. **8**, 385–390 (2000)
5. A. Luque, J. Ruiz, A. Cuevas, M. Agost, *Proceeding of 1st European Conference on Photovoltaic Solar Energy Conversion* (1977), pp. 269–277
6. J.L. Gray, *Handbook of Photoltaic Science and Engineering* (Steven Hegedus Wiley, 2002)
7. R. Turton, Band structure of Si: overview, in *Properties of Crystalline Silicon*, ed. by R. Hull (INSPEC, Stevenage, UK, 1999)
8. M. Green, M. Keevers, Prog Photovolt. **3**, 189–192 (1995)
9. S. Kolodinski, J. Werner, T. Wittchen, H. Queisser, Appl. Phys. Lett. **63**, 2405–2407 (1993)
10. D. Clugston, P. Basore, Prog. Photovolt. **5**, 229–236 (1997)
11. A. Sproul, M. Green, J. Appl. Phys. **70**, 846–854 (1991)
12. P. Altermatt et al., *Proceedings of 16th European Conference on Photovoltaic Solar Energy Conversion* (2000), pp. 102–105

13. J. Dziewior, W. Schmid, Appl. Phys. Lett. **31**, 346–351 (1977)
14. P. Altermatt et al., *Proceedings of 16th European Conference on Photovoltaic Solar Energy Conversion* (2000), pp. 243–246
15. M. Green et al., Nature **412**, 805–808 (2001)
16. W. Thurber, R. Mattis, Y. Liu, J. Filliben, J. Electrochem. Soc. **127**, 1807–1812 (1980)
17. W. Thurber, R. Mattis, Y. Liu, J. Filliben, J. Electrochem. Soc. **127**, 2291–2294 (1980)
18. D. Kane, R. Swanson, *Proceedings of 20th IEEE Photovoltaic Specialist Conference* (1988), pp. 512–517
19. B. Dai, H. Zheng, *Solar Cells Technical Manuals* (Posts & Telecom Press, Beijing, 2012)
20. Y. Ma, industrial crystalline silicon solar cell technology. Chin. J. Nat. **32**(3), 161–165 (2010)
21. J. Yu, *Solar Energy Photovoltaic Device Technology* (University of Electronic Science and Technology Press, Chengdu, 2011). (in Chinese)
22. J. He, Discussion on preparation technology of crystal silicon solar cells. J. Chuxiong Normal Univ. **27**(6), 26–30 (2012)
23. R. Zhao, Z. Liang, J. Li, J. Jin, H. Shen, New development of manufacture technology of crystalline silicon solar cells. Mater. Rev. **23**(5), 25–29 (2009)
24. S. Wu, J. Guo, X. Cai et al., High efficiency and low cost wafer crystalline silicon solar cell technology. China National invention patent, CN200710063643.0, 2008-08-13
25. J. Nijs et al., *Proceedings of 1st World CPEC* (1994), pp. 1242–1249
26. B. Finck von Finckenstein et al., *Proceedings of 28th IEEE Photovoltaic Specialist Conference* (2000), pp. 198–200
27. K. M¨unzer, K. Holdermann, R. Schlosser, S. Sterk, IEEE Trans. Electron Dev. **46**, 2055–2061 (1999)
28. S. Sivoththaman et al., *Proceedings of 14th European Conference on Photovoltaic Solar Energy Conversion*, 400–403(1997)
29. P. Doshi et al., *Proceedings of 25th IEEE Photovoltaic Specialist Conference* (1996), pp. 421–424
30. P. Doshi et al., Sol. Energy Mater. Sol. Cells **41**(42), 31–39 (1996)
31. D. Biro et al., Sol. Energy Mater. Sol. Cells **74**, 35–41 (2002)
32. I. P´erichaud, F. Floret, S. Martinuzzi, *Proceedings of 23rd IEEE Photovoltaic Specialist Conference* (1993), pp. 243–247
33. S. Narasimha, A. Rohatgi, IEEE Trans. Electron Dev. **45**, 1776–1782 (1998)
34. D. Macdonald, A. Cuevas, F. Ferraza, Solid-State Electron. **43**, 575–581 (1999)
35. J. Gee, B. Sopori, *Proceedings of 26th IEEE Photovoltaic Specialist Conference* (1997), pp. 155–158
36. C. del Ca˜nizo, I. Tob´ıas, R. Lago, A. Luque, J. Electrochem. Soc. **149**, 522–525 (2002)
37. S. Gandhi, *VLSI Fabrication Principles*, Chap. 8 (Wiley, New York, 1994)
38. J. Johnson, J. Hanoka, J. Gregory, *Proceedings of 18th Photovoltaic Specialist Conference* (1985), pp. 1112–1115
39. B. Sopori, X. Deng, S. Narayanan, S. Roncin, *Proceedings of 11th European Conference Photovoltaic Solar Energy Conversion* (1992), pp. 246–249
40. J. Szulfcik et al., *Proceedings of 12th EC Photovoltaic Specialist Conference* (1994), pp. 1018–1021
41. C. Leguijt et al., Sol. Energy Mater. Sol. Cells **40**, 297–345 (1996)
42. A. Aberle, R. Hezel, Prog. Photovolt. **5**, 29–50 (1997)
43. W. Soppe et al., On combining surface and bulk passivation of SiNx:H layers for mc-Si solar cells, in *Proceedings of 29th IEEE Photovoltaic Specialist Conference* (New Orleans, 2002) (in press)
44. D. Ruby, W. Wilbanks, C. Fieddermann, *IEEE 1st WPEC 1335–1338* (1994)
45. K. Shirasawa et al., *Proceedings of 21st IEEE Photovoltaic Specialist Conference* (1990), pp. 668–673
46. J. Zhao, A. Wang, P. Campbell, M. Green, IEEE Trans. Electron Dev. **46**, 1978–1983 (1999)
47. S. De Wolf et al., *Proceedings of 16th European Conference Photovoltaic Solar Energy Conversion* (2000), pp. 1521–1523

48. R. Bilyalov, L. Stalmans, L. Schirone, L´evy-clement C. IEEE Trans. Electron Dev. **46**, 2035–2040 (1999)
49. M. Spiegel et al., Sol. Energy Mater. Solar Cells **74**, 175–182 (2002)
50. F. Huster et al., *Proceedings of 28th IEEE Photovoltaic Specialist Conference* (2000), pp. 1004–1007
51. W. Joos et al., *Proceedings of 16th Europe Conference Photovoltaic Solar Energy Conversion* (2000), pp. 1169–1172
52. L. Pirozzi et al., *Proceedings of 12th Europe Conference Photovoltaic Solar Energy Conversion* (1994), pp. 1025–1028
53. D. Ruby et al., *Proceedings of the 2nd World CPEC* (1998), pp. 39–42
54. Y. Inomata, K. Fukui, K. Shirasawa, Sol. Energy Mater. Sol. Cells **48**, 237–242 (1997)
55. S. Winderbaum, O. Reinhold, F. Yun, Sol. Energy Mater. Sol. Cells **46**, 239–248 (1997)
56. R. L¨udemann, B. Damiani, A. Rohatgi, G. Willeke, *Proceedings 17th Europe Conference Photovoltaic Solar Energy Conversion* (2001), pp. 1327–1330
57. D. Macdonald et al., Texturing industrial multicrystalline silicon solar cells. *ISES Solar World Congress* (2001)
58. W. Schmidt, B. Woesten, J. Kalejs, Prog. Photovolt. **10**, 129–140 (2002)
59. V. Yelundur, A. Rohatgi, J. Jeong, J. Hanoka, IEEE Trans. Electron Dev. **49**, 1405–1410 (2002)
60. R. Janoch et al. *Proceedings of IEEE 28th Photovoltaic Specialist Conference* (2000), pp. 1403–1406
61. M. Fu, Performance analysis of several high efficiency crystalline silicon solar cells. Mater. Reports **26**(3), 1–5 (2012)
62. Yasuhiko Takeda, Highly efficient solar cells using hot carriers generated by two-step excitation. Sol. Energy Mater. Sol. Cells **95**, 2638–2644 (2011)
63. Satoshi Tohoda, Future directions for higher-efficiency HIT solar cells using a Thin Silicon Wafer. J. Non-Cryst. Solids **358**, 2219–2222 (2012)
64. S. Tie, Development status on solar energy silicon material. J. Qinghai Univ. (Nat. Sci.) **27**(1), 33–38 (2009)
65. W. Koch et al., Solid State Phenom. **401**, 57–58 (1997)
66. C. H¨aßler et al. *Proceegdins of 2nd World Conferences on Photovoltaic Solar Energy Conversion, 1886* (1998)
67. L. Liu, Growth principle and technique of single crystal silicon. J. Changchun Univer. Sci. Technol. **32**(4), 569–573 (2009)
68. F. Huang, Research and development status of silicon materials for solar cell. Spec. Cast. Nonferrous. Alloys **28**(12), 925–930 (2008)
69. Y. Meng, Research on crystalline silicon solar cells with low cost and high efficiency. Semicond. Optoelectron. **32**(2), 151–157 (2011)

Chapter 4
High-Efficiency III-V Single-Junction and Multi-junction Solar Cells

4.1 III-V Semiconductor Materials

4.1.1 Nomenclature of III-V Semiconductor Materials

The III-V semiconductor is an essential branch of compound semiconductors. The understanding of III-V semiconductors must be based on the understanding of compound semiconductors. Compound semiconductors are defined as substances with semiconductor properties composed of two or more elements. It is worth noting that the concept of "compound" in compound semiconductors is slightly different from that of "compound" in chemistry, which is defined as a pure substance formed by two or more elements following a fixed composition ratio. Compounds have fixed physical and chemical properties such as fixed melting point, boiling point, and density at room temperature. According to the law of determining composition or proportion, their composition is the same and the compound semiconductor is closer to the alloy, no matter what method is used to prepare compounds. The required characteristics can be obtained by adjusting the proportion of constituent elements. Compound semiconductors can be classified into III-V, IV-IV, and II-VI groups according to the composition elements. It can be found from Table 4.1, that compounds consisting of III and IV elements such as GaN are called III-V compound semiconductors. Similarly, compounds consisting of II and VI elements such as zinc selenide (ZnSe), cadmium sulfide (CdS), cadmium telluride (CdTe), or zinc oxide (ZnO) are called II-VI compound semiconductors.

Besides, the compound semiconductor can also be divided into binary, ternary, quaternary, and even quinary compound semiconductors according to the types of elements constituting the compound semiconductor. The common binary compound semiconductors include silicon carbide (SiC), gallium arsenide (GaAs), indium phosphide (InP), gallium nitride (GaN), and zinc oxide (ZnO). There are various nomenclature methods for compound semiconductors while there is no unified nomenclature method at present. A simpler nomenclature is regarded as A-based B according to different elements; among them, A is a group V element, B is a group III element.

C. Zhang et al., *Semiconductor Photovoltaic Cells*,
https://doi.org/10.1007/978-981-15-9480-9_4

Table 4.1 Partial periodic table related to semiconductors [1]

II	IIIA	IVA	VA	VIA
	B 5	C 6	N 7	O 8
Mg 12	Al 13	Si 14	P 15	S 16
Zn 30	Ga 31	Ge 32	As 33	Se 34
Cd 48	In 49	Sn 50	Sb 51	Te 52
Hg 80	Tl 81	Pb 82	Bi 83	Po 84

For example, AlGaAs is generally written as $Al_xGa_{1-x}As$ and read as aluminum gallium arsenide; GaAsP is generally written as $GaAs_{1-x}P_x$ and read as phosphorus gallium arsenide. $Al_xGa_{1-x}As$ can be obtained by adding the aluminum source to the growth process of GaAs material during the growth of gallium arsenide material, in which both aluminum and gallium are group III elements; "x" represents the molar ratio of aluminum in the group III element of the entire III-V compound semiconductor. $Al_xGa_{1-x}As$ with different bandgaps can be obtained by changing the ratio of gallium source to aluminum source in the growth process. Therefore, $Al_xGa_{1-x}As$ can be regarded as the alloy of GaAs and AlAs. Since $Al_xGa_{1-x}As$ is composed of three elements, it is called a ternary compound semiconductor. Common ternary compound semiconductors include InGaN, GaAsP, and GaInP. Similarly, the compound semiconductors consisting of four elements are called quaternary compound semiconductors. Common ones are InGaAsN, GaAlInP, and InGaAsP. Therefore, InGaAsNSb consisting of five elements is called a quinary compound semiconductor.

4.1.2 *Properties of III-V Semiconductor Materials*

Single crystal, polycrystalline, and amorphous silicon can be applied in silicon-based solar cells. III-V compound semiconductor in III-V compound semiconductor solar cells is a single crystal. The common III-V compound semiconductor GaAs is generally obtained by the Bridgman method and Czochralski method, and then cut into substrates with appropriate thickness. At present, the size of the GaAs substrate used in the semiconductor industry can reach 150 mm. Substrates of this size are mainly used in high-speed electronic components. The maximum size of general photoelectric components is usually 75 mm. Different from larger mainstream silicon substrates, there are two main reasons for using smaller GaAs substrates: first, the larger size substrates are easy to break down during transportation because the GaAs substrates are brittle; second, photoelectric components require a more stringent surface of the substrates, and it is difficult for large size substrates to obtain uniform epitaxial layers, resulting in a complication in the current GaAs material epitaxy.

III-V semiconductors have advantages over common silicon materials in solar cell applications. The main reason is that most III-V semiconductors are direct bandgap semiconductors, suggesting that the change in momentum (P) is not involved in the energy process of transferring photons to electrons. For silicon with an indirect bandgap semiconductor, the transition of electrons from the conduction band to the valence band is accompanied by momentum change and energy conservation; thus, the absorbed photon energy is the sum of potential energy and phonon energy. The phonon energy caused by the lattice vibration will eventually be converted into heat dissipation. Therefore, the highest theoretical conversion efficiency of solar cells based on GaAs and other direct bandgap semiconductors is higher than that of silicon-based solar cells.

Moreover, the lattice constant of materials is also a significant parameter for semiconductor devices. The lattice constants of common III-V materials are illustrated in Fig. 4.1.

The lattice constant denotes the length of the unit cell formed by the constituent atoms in the material. Taking $Al_xGa_{1-x}As$ as an example, the material is GaAs when the molar ratio of aluminum is x = 0; its lattice structure is sphalerite structure, the lattice constant is 0.56533 nm, and the bandgap is 1.42 eV. The material is AlAs when x = 1; its lattice constant at room temperature is 0.56605 nm, and the bandgap is 2.17 eV. It is noteworthy that GaAs would change from direct bandgap to indirect bandgap when the molar ratio of Al exceeds 0.45. The bandgap of GaAs increases with the molar ratio of Al while the lattice mismatch is only 0.1%. This indicates that materials with different bandgaps can be obtained by changing the molar ratio of Al. Therefore, gallium arsenide solar cells can absorb light of different wavelengths in

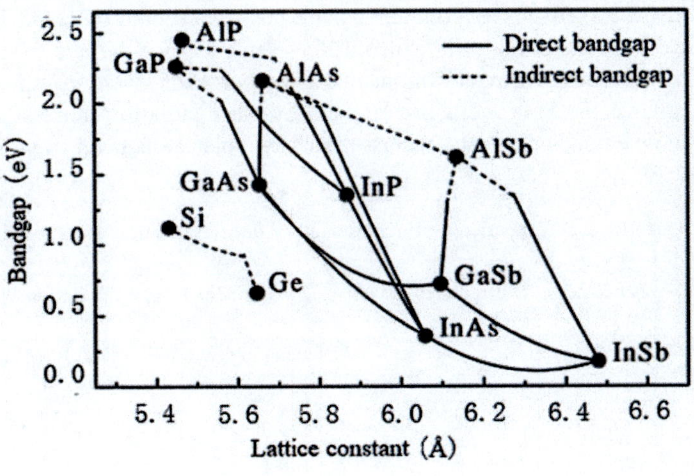

Fig. 4.1 Lattice constants of common III-V materials [2]

the solar spectrum by using a multi-layer structure, leading to higher photoelectric conversion efficiency.

4.2 Application of III-V Semiconductor Solar Cells

4.2.1 Space Application

Semiconductor solar cells used in space have been developed for three generations: the single-junction silicon-based solar cells represented by silicon materials, the single-junction heterojunction solar cells represented by GaAs/Ge, and the multi-junction tandem solar cells represented by GaInP/GaAs/Ge materials. The performance of typical space solar cells under different irradiation and temperature conditions are illustrated in Table 4.2. It can be observed from Table 4.2, that the efficiency of solar cells applied in space is increased from 14 to 21.5% by replacing materials and adopting tandem structure. Simultaneously, the power of solar cells decreases as irradiation or temperature increases when other conditions remain unchanged.

The application of III-V semiconductor solar cells in space can be traced back to the 1950s. Spectrolab, an investment subsidiary of Boeing, USA, invested in the development and manufacture of GaAs solar cells in 1956. Spectrolab plays a significant role in the development of solar cells used in space. The development process of solar cells used in space is exhibited in Fig. 4.2.

Explorer 6 launched in 1959, was equipped with Spectrolab solar cells as the main power source. In the 1960s, researchers explored that GaAs PN junction solar cells fabricated by zinc diffusion technology had excellent high-temperature operating characteristics, including good stability at a high-temperature and good radiation resistance. Besides, III-V semiconductor materials represented by GaAs have been widely used in space solar cells Since 1965. Venera-2 and Venera-3 spacecrafts launched by the former Soviet Union, in November 1965, were equipped with 2 m^2 GaAs solar cells. Solar cells are affected by solar radiation more significantly because the detector is closer to Venus; therefore, solar cells need to work at high

Table 4.2 Performance comparison of representative three-generation solar cells for space applications [3]

Solar cells materials	Efficiency (%)	Unirradiated power (W)		Power under 1 MeV irradiation (W)			
				3.4×10^{14} e/cm^2		1×10^{15} e/cm^2	
		28 °C	50 °C	28 °C	50 °C	28 °C	50 °C
Si	14.8	170.9	149.5	129.0	112.2	113.0	98.8
GaAs/Ge	18.5	218.1	208.2	188.1	179.6	166.8	159.3
GaInP/GaAs/Ge	21.5	253.5	242.8	223.0	211.9	192.7	183.0

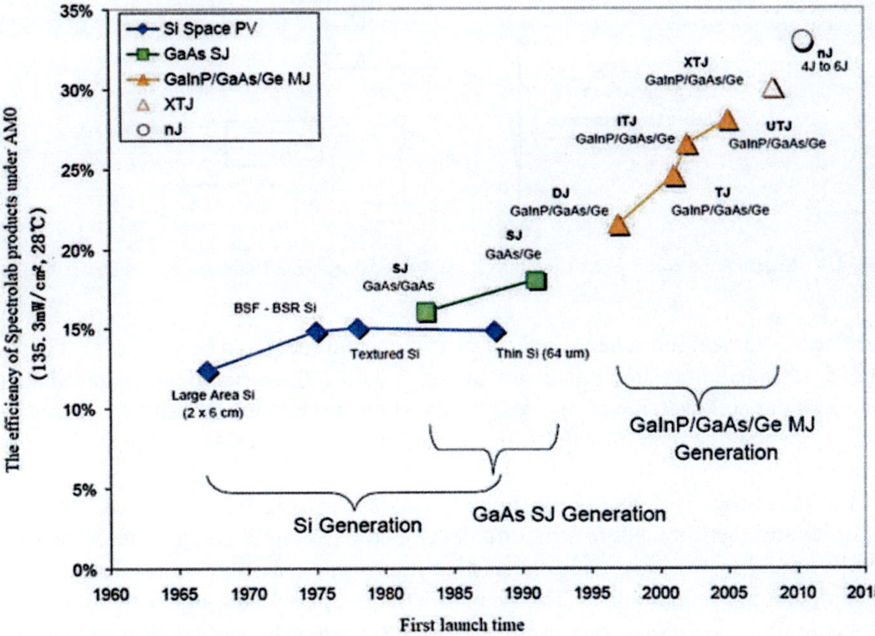

Fig. 4.2 Development of Spectrolab solar cells [4]

temperatures; GaAs solar cells just meet this demand. Subsequently, the former Soviet Union launched the lunar rovers Lunokhod-1 and Lunokhod-2 equipped with gallium arsenide solar cells of 4 m^2 in 1970 and 1972, respectively. Simultaneously, the fabrication technology of silicon-based solar cells has also been improved; Surface Texturing and Anti-Reflection Coating Technologies have been proposed and developed to increase the photoelectric conversion efficiency to 14%. However, the surface temperature of the moon can reach 130 °C. At this temperature, the intrinsic excitation degree of silicon-based materials has approached the ionization degree of impurities, leading to the failure of PN junctions and the failure of the silicon-based solar cell. Then, AlGaAs/GaAs tandem junction solar cells had been developed in the 1980s; their photoelectric conversion efficiency could reach 18–19%. Besides, the area of single heterojunction solar panels assembled on MIR Space Station launched by the former Soviet Union reached 70 m^2. The space station was originally designed and used for three years while it crashed at the end of the 2001 mission cycle. The space station had been in operation for 15 years; the AlGaAs/GaAs solar cells were equipped with an efficiency attenuation of less than 30% under the conditions of the great space temperature difference and high irradiation. As illustrated in Fig. 4.3, the open-circuit voltage V_{OC} and short-circuit current I_{SC} are functions of the irradiation density. With the increase of irradiation, the open-circuit voltage V_{OC} remains unchanged and the short-circuit current I_{SC} increases linearly. The solar cell would

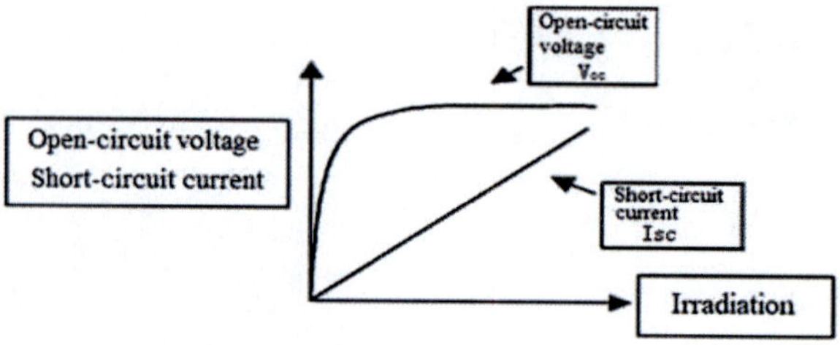

Fig. 4.3 Relations between open-circuit voltage V_{OC}, short-circuit current I_{SC}, and irradiation density [5]

fail due to short-circuit when the short-circuit current reaches a certain level. Therefore, a wide-bandgap semiconductor material with a direct bandgap is introduced into the solar cell used in space to reduce the short-circuit current caused by irradiation. It can be illustrated in Fig. 4.1, that GaInP/GaAs has a larger bandgap compared to silicon.

The photoelectric conversion efficiency of solar cells is further improved by the multi-heterojunction tandem structure developed based on a single heterojunction. As illustrated in Fig. 4.4, GaInP/GaAs/Ge multi-junction solar cells fabricated by Spectrolab have a photoelectric conversion efficiency of 41.6% under the 340 solar concentrating conditions; this is independently verified by the US Department of National Renewable Energy Laboratory (NREL) under the test conditions of 340 solar, 25 °C, and AM1.5D. The typical GaInP/GaAs/Ge multi-heterojunction tandem structure is exhibited in Fig. 4.4. The structure consists of $Ga_xIn_{1-x}P$ top cell with a bandgap of 1.8–1.9 eV, heavily doped PN tunnel junction, GaAs intermediate cell, PN tunnel junction, and Ge bottom cell. To date, the fabrication technology of semiconductor solar cells based on GaInP/GaAs/Ge multi-heterojunction tandem structure has been mature. Recent reports indicated that the photoelectric conversion efficiency of semiconductor solar cells under 340 solar concentrating conditions is over 41.6%, which is close to the theoretical upper limit of 45% [16].

4.2.2 *Earth Energy Applications*

At present, III-V semiconductor materials have completely replaced silicon materials in the application of space solar cells. However, silicon-based solar cells still play a dominant role in earth energy applications because silicon is cheaper and can obtain larger size substrates than GaAs. However, the power generation cost of III-V solar cells with high power concentrating system and GaInP/GaAs/Ge multi-heterojunction tandem structure can be greatly reduced [7]. Then, the development of concentrating III-V semiconductor solar cells become one of the effective means to alleviate the energy crisis in the twenty-first century due to its high efficiency, expected low-cost, and the feasibility of the concentrating system.

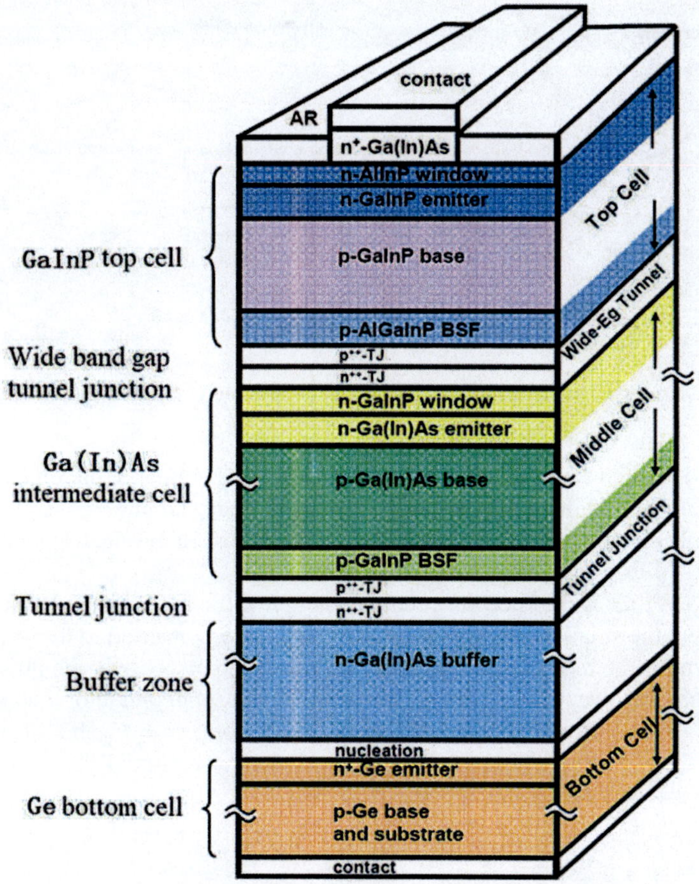

Fig. 4.4 GaInP/GaAs/Ge multi-heterojunction tandem structures [4]

4.3 The Foundation of III-V Single-Junction and Multi-junction Solar Cells

4.3.1 Direct Bandgap and Indirect Bandgap

The basic factors limiting the photoelectric conversion efficiency of single-junction III-V semiconductor solar cells need to be revealed to better understand the structure of III-V semiconductor solar cells and further analyze their performance. The bandgap of III-V semiconductors is recorded as E_g. The electrons in the semiconductor conduction band will absorb the energy of the photon to the valence band if the energy of the incident photon $hv > E_g$. Besides, the energy of the photons will be converted into potential energy if the material is a direct bandgap semiconductor (such as GaAs and InP). Moreover, the photon energy will be converted into potential

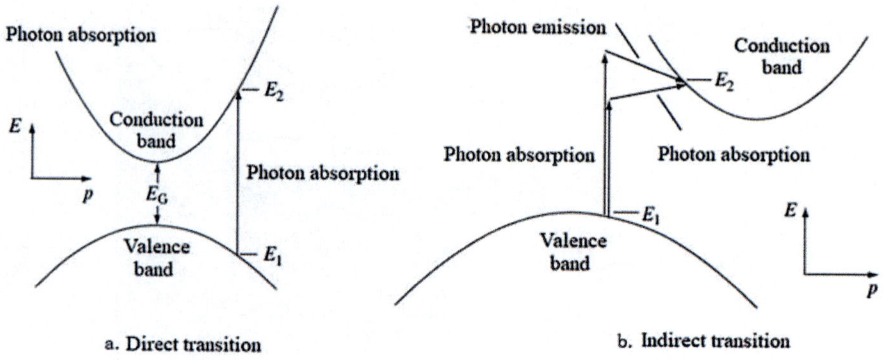

Fig. 4.5 Schematic diagram of electronically excited transitions [6]

energy and phonon energy if the material is an indirect bandgap semiconductor (such as Si or Ge). The phonon energy does not contribute to the photogenerated current and is finally converted into heat dissipation. The process of the electronically excited transition is illustrated in Fig. 4.5.

The greater the difference between the energy of the incident photon hv and the bandgap E_g, the smaller the ratio of the photon energy converted to the photogenerated current to the incident photon energy. From one perspective, the photons with incident energy lower than E_g are not absorbed by the semiconductor material; these photons do not contribute to the photocurrent. Therefore, $hv = E_g$ should be ensured to obtain the maximum photon absorption efficiency; it should be noted that this efficiency is always less than 100%.

4.3.2 Principle Limits for Single-Junction and Multi-junction Solar Cell Efficiency

The principal limitations of single-junction and multi-junction solar cell efficiency will be briefly introduced in this section to better understand the III-V solar cells. Before the introduction, the energy distribution of the solar spectrum is reviewed. The solar spectrum in different environments is illustrated in Fig. 4.6. Since the solar spectrum is wider, the photon energy in sunlight is 0–4 eV. Essentially, the efficiency of a single solar cell for sunlight is much lower than that of a single solar cell for monochromatic light because a single semiconductor material cannot absorb sunlight of all wavelengths. The way to solve this problem, in theory, is very simple, that is, the spectrum is divided into several parts, and semiconductor materials with different bandgaps are used to absorb different spectral ranges of photon instead of a single cell absorbing all the photons in the spectrum. For example, the spectrum can be divided into three parts: $h\upsilon_1 \sim h\upsilon_2$, $h\upsilon_2 \sim h\upsilon_3$, and $h\upsilon_3 \sim \infty$; $h\upsilon_1 < h\upsilon_2 < h\upsilon_3$. The photons in the corresponding spectral range are absorbed separately when the

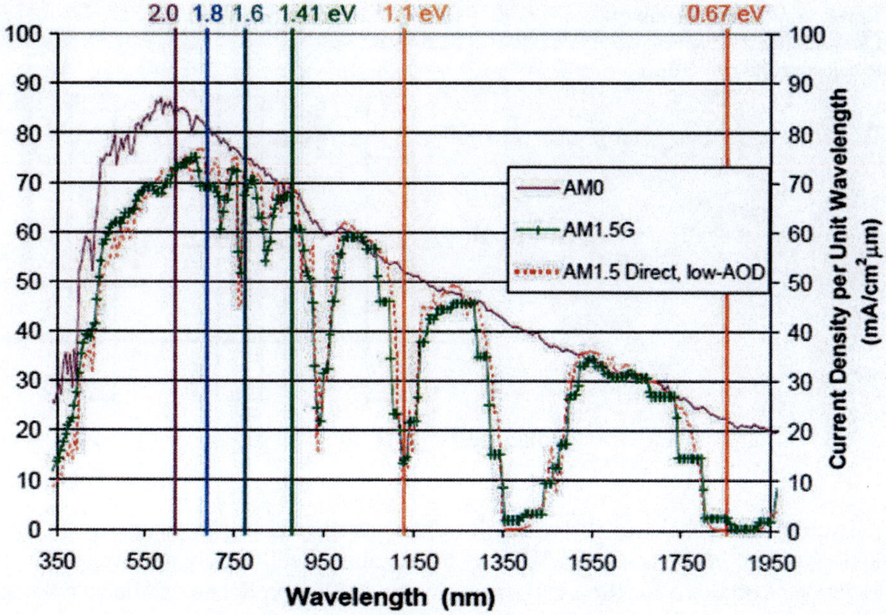

Fig. 4.6 Solar spectrum in different environments

bandgap is $E_{g1} = h\upsilon_1$, $E_{g2} = h\upsilon_2$, and $E_{g3} = h\upsilon_3$. The more the spectral regions, the higher the photon absorption efficiency; this is the theoretical foundation for improving the efficiency of tandem solar cells.

Henry calculated the absorption efficiencies of 1, 2, 3, and 36 forbidden band semiconductor materials under standard atmospheric mass (AM 1.5); they were 37%, 50%, 56%, and 72%, respectively [8]. The absorption efficiency increases significantly when the number of forbidden bands of semiconductor materials increases from 1 to 2; however, this trend gradually disappears as the number of forbidden bands continues to increase [9]. It is very difficult to manufacture a tandem solar cell with 4–5 junctions. More importantly, multi-junction solar cells cannot be achieved without careful consideration of the lattice constants of the materials. Besides, the number of forbidden bands that can be used in solar cells is also limited by the existence of a lattice mismatch.

The working mechanism of the tandem solar cell is illustrated in Fig. 4.7; the forbidden bandwidth is reduced from left to right. Since a low-pass reflective layer is placed between the cells and the reflection threshold of the reflective layer is the bandgap of the cell material, incident photons are prevented from being injected into the unabsorbable cell. With this configuration, each cell has its own load circuit; therefore, they can be biased at different voltages. It is worth noting that in principle, the absence of the reflective layer results in a lower absorption efficiency if the number of tandem cells is limited. Besides, the placement of the reflective layer will have no effect on the absorption efficiency if the number of tandem cells is infinite.

Fig. 4.7 Schematic diagram of the working mechanism of the tandem solar cell [10]

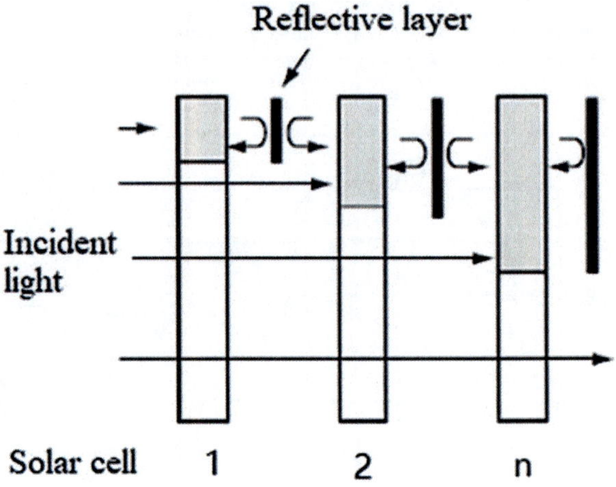

The efficiency limit of the cell will be discussed in detail in Chap. 10. In this section, a brief analysis of the efficiency of the tandem cell is only performed. For a tandem cell of a two-junction cell structure, the energy produced can be calculated by the following equation.

$$W = qV_l[N(T_s, 0, E_{gl}, E_{gh}, Hs) - N(T_a, qV_l, E_{gl}, E_{gh}, H_r)] + qV_h[N(T_s, 0, E_{gh}, \infty, Hs) - N(T_a, qV_h, E_{gh}, \infty, H_r)]$$

$$\mathrm{H}_{\mathrm{s(r)}} = \pi \mathrm{A}\sin^2\theta_{\mathrm{s(r)}} \quad (4.1)$$

where subscripts l and h denote low and high; V is the quasi-Fermi level difference between n-type and p-type materials; E_g represents the bandgap of the semiconductor material; N is the photon flow entering or leaving the cell; T_s is the sun temperature (6000 K); T_a is the environment temperature of the solar cell (300 K); H_s and H_r are related to the solid angle θ of the solar cell with solar radiation (indicated by subscript s) and solar cells radiating to the surrounding space (indicated by subscript r). The function for maximum efficiency can be optimized by changing the variables V_l, V_h, E_{gl}, and E_{gh}. Figure 4.8 illustrates that the optimum efficiency of a tandem cell consisting of two cells is a function of the bandgap of the two materials making up the cell. It can be observed in Fig. 4.8, that the area where the maximum efficiency occurs corresponds to a very wide area, suggesting that materials can be combined to form a cell in a wide range.

This fact has also been verified by a series of experiments. For example, the InGaP/GaAs/Ge multi-junction cell prepared by Spectrolab achieved an efficiency of 34% under the test conditions of AM1.5G and 210 times sunlight [12, 13]. When the number of layers of the cell is infinite, the maximum efficiency of the cell can be calculated by the following equation if each layer of the cell has independent

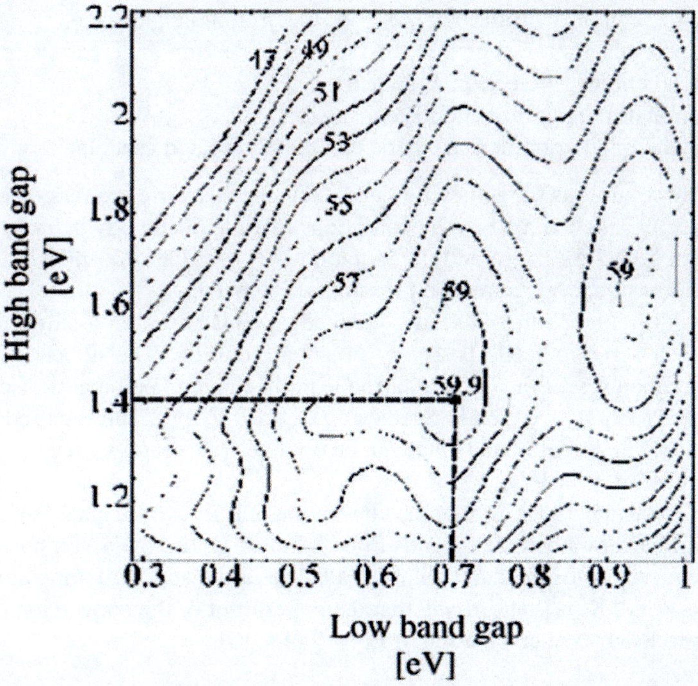

Fig. 4.8 The ideal efficiency of a double-junction cell is a function of the material bandgap E_l and E_h [11]

monochromatic illumination and bias voltage.

$$\eta = \frac{1}{\sigma_S T_s^4} \int_0^\infty I(E, V) V_{\max} dE \tag{4.2}$$

where σ_s denotes the Stefan constant; $I(E, V)$ represents the current of the cell under the monochromatic illumination; E is the energy; V denotes the difference between the quasi-Fermi level of the p-type material and the n-type material; the remaining parameters are constant. The upper limit of the maximum efficiency of an ideal solar cell under certain conditions can be calculated by Eq. 4.2 [13]. Tandem solar cells emit luminescent radiation at room temperature. Besides, the entropy change of these cells during the photoelectric conversion process is also positive due to the positive entropy change of each monochromatic photocell during the photoelectric conversion process. The conditions of Landsberg photoelectric conversion efficiency are listed as follows:

(1) Only can the photons with energy greater than the bandgap be absorbed, while the photons with energy lower than the bandgap cannot be absorbed.

(2) A photon can only produce at most one electron–hole pair.
(3) The absorbed photon energy is used to excite electron–hole pairs and store the potential energy of electron–hole pairs.
(4) Only radiation recombination is considered.
(5) The semiconductor material can be completely treated as a blackbody.

Since the conditions for achieving Landsberg photoelectric conversion efficiency cannot be fully satisfied, that is, the conditions that the blackbody behavior entropy of condition 5 becomes zero and the radiation is free radiation cannot be satisfied, the upper limit of energy conversion cannot be achieved.

It is not difficult to integrate multiple bandgap solar cells on a single substrate. Connecting a series of cells is a very practical solution. The upper limit of the conversion efficiency of the GaInP/GaAs/Ge multi-junction cell mentioned above is determined by Eq. 4.1, which is related to E_{gl}, E_{gh}, V_l, V_h, and thus reducing the maximum conversion efficiency that can be obtained. The total voltage obtained by the tandem cell is $V = V_l + V_h$.

Then, what is the upper limit of the conversion efficiency of the tandem cell when the cell number of tandem cells tends to be infinite? Under the assumption that the surface temperature of the sun is $T_s = 6000$ K and the surface temperature of the earth is $T_a = 300$ K, it is surprising that the upper limit of the conversion efficiency can reach 86.8% by the calculation of Eq. 4.2 [14, 15].

4.3.3 *Spectral Separation*

Tandem solar cells require incident photons to be directly absorbed by materials corresponding to the bandgap. This requirement is well-satisfied by the concept of spectral separation. Photons of different energies are separated and irradiated into different regions using scattering optics such as prisms; besides, the cell material in this region can absorb photons. This process can be represented in Fig. 4.9a.

Although the concept of prism separation in most cases is very simple, it is difficult to apply in practice due to the optical complexity of a mechanical knot conforming to this concept. Another possible method is to stack the individual cells; therefore, the sunlight is first absorbed by the uppermost wide-bandgap solar cell and then gradually absorbed by the low-bandgap solar cell, as illustrated in Fig. 4.9b. Therefore, in Fig. 4.9b, the photon with a photon ($h\upsilon_3 > E_{g3}$) is first absorbed by the upper wide-band solar cell, and the bandgap of the upper cell material has a bandgap of E_{g3}. Then, the photon with the photon energy ($E_{g2} < h\upsilon < E_{g3}$) is absorbed by the solar cell in the middle, and the bandgap of the cell material is E_{g2}. Finally, the photon with photon energy $h\upsilon < E_{g2}$ is absorbed by the lower layer of the low-bandgap solar cell, and the bandgap of the cell material is E_{g1}. In short, different junctions such as optical devices can separate the solar spectrum and absorb the photons of the corresponding energy. Besides, the bandgap of the tandem solar cell material decreases from top to bottom. This structure can better avoid the use of optics

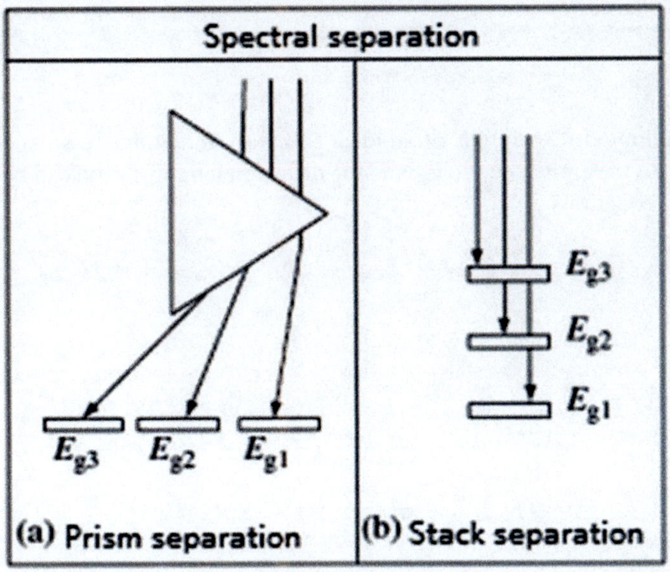

Fig. 4.9 Comparison of different spectral separation methods

devices; the individual cells are physically separated while they can be mechanically stacked onto the same substrate. Apparently, all junctions in this stacked structure need to be optically transparent concerning the underlying junction in addition to the bottom cell, resulting in increasing the difficulty of substrate selection and back contact metallization to some extent.

In this section, the effects of different parameters on a typical double-junction tandem cell structure will be analyzed. Meanwhile, this analysis can be extended to GaInP/GaAs/Ge triple-junction solar cells widely used in space. Our analysis is based on four preconditions [16, 17]:

(1) The interconnected tunnel junction is completely transparent and has a resistance of 0;
(2) No reflection loss;
(3) The series resistance is 0;
(4) The incident photons are completely absorbed by the material, and the J-V curve of the cell can be described by an ideal diode model.

It is worth noting from the analysis that the power conversion efficiency of high-quality III-V material semiconductor cells is 90% of the predicted power conversion efficiency.

Each junction constitutes a separate sub-cell; the short-circuit current density J_{SC} of the sub-cell is mainly determined by the quantum efficiency $QE(\lambda)$ of the cell and the incident spectrum $\Phi_{inc}(\lambda)$.

$$J_{sc} = e\int_0^\infty QE(\lambda)\Phi_{inc}d\lambda \tag{4.3}$$

The quantum efficiency *QE* of an ideal cell with finite thickness depends on the base width x_b, the emitter width x_e, and the diffusion length *W*, and can be expressed by standard equations.

$$QE = QE_{\text{emitter}} + QE_{\text{depl}} + \exp(-\alpha(x_e + W))QE_{\text{base}} \tag{4.4}$$

where

$$QE_{\text{emitter}} = f_\alpha(L_e)\left(\frac{l_e + \alpha L_e - \exp(-\alpha x_e)\times [l_e\cosh(x_e/L_e)+\sinh(x_e/L_e)]}{l_e\cosh(x_e/L_e)+\sinh(x_e/L_e)} - \alpha L_e\exp(-\alpha x_e)\right) \tag{4.5}$$

$$QE_{\text{depl}} = \exp(-\alpha x_e)[1 - \exp(-\alpha W)] \tag{4.6}$$

$$QE_{\text{base}} = f_\alpha(L_b)\left(\begin{array}{c} l_b\cosh(x_b/L_b) + \sinh(x_b/L_b) \\ -\alpha L_e - \frac{-(\alpha L_b - l_b)\exp(-\alpha x_b)}{l_b\sinh(x_b/L_b)+\cosh(x_b/L_b)}\end{array}\right) \tag{4.7}$$

$$l_b = S_b L_b/D_b, l_e - S_e L_e/D_e, D_b = kT\mu_b/e, D_e = kT\mu_e/e \tag{4.8}$$

$$f_\alpha(L) = \frac{\alpha L}{(\alpha L)^2 - 1} \tag{4.9}$$

The wavelength dependence of photons is classified into the wavelength-dependent absorption efficiency $\alpha(\lambda)$ rather than being considered separately. Besides, $\mu_{b(e)}$, $l_{b(e)}$, and $S_{b(e)}$ denote the basic region (emitter region) mobility, diffusion length, and the minority carrier recombination velocity of the surface, respectively. T represents the absolute temperature. Assuming that each photon is absorbed by the material and forms a photogenerated current. It can be found by first-principle approximation that this assumption is correct for most of the III-V junctions, in which the case quantum efficiency is related to the thickness of the device; $x = x_e + W + x_b$, and it can be obtained

$$QE(\lambda) = 1 - \exp[-\alpha(\lambda)x] \tag{4.10}$$

where part of the incident light $\exp[-\alpha(\lambda)x]$ passes directly through the cell without being absorbed. Apparently, Eq. 4.10 is valid, where $S = 0$, $L \gg x$, and $L \gg 1/\alpha$; Eq. (4.10), is derived from Eqs. (4.4)–(4.7).

For incident photons with energy below the bandgap, $\alpha(\lambda) = 0$, thus $\exp[-\alpha(\lambda)x] = 1$. The incident spectrum $\Phi_{inc}(\lambda)$ is the solar spectrum Φ_s. Besides, the spectrum reaching the bottom cell is filtered by the top cell; the solar spectrum can be expressed as $\Phi_s \exp[-\alpha_t(\lambda)x_t]$, where x_t denotes the thickness of the top cell and α_t represents the absorption coefficient of the top cell. Assuming that the bottom cell is thick enough to absorb all incident photons with energy greater than its bandgap, the short-circuit current density J_{SCt} of the top cell and the short-circuit current density J_{SCb} of the bottom cell can be derived.

$$J_{\mathrm{SCt}} = e\int_0^{\lambda_t} (1 - \exp[-\alpha_t(\lambda)x_t])\Phi_s(\lambda)d\lambda \tag{4.11}$$

$$J_{\mathrm{SCb}} = e\int_0^{\lambda_b} \exp[-\alpha_t(\lambda)x_t])\Phi_s(\lambda)d\lambda \tag{4.12}$$

where $\lambda_b = hc/E_{gb}$ and $\lambda_t = hc/E_{gt}$ are the wavelength corresponding to the bandgap. The minimum integration limit of the short-circuit current density J_{SCb} of the bottom cell is 0 instead of λ_t. No photons in the shortwave range will pass to the bottom cell unless the top cell is infinitely thick. Since the photon reaching the bottom cell is filtered by the top cell, the short-circuit current density J_{SCb} of the bottom cell depends on not only the bandgap E_{gb} of the bottom cell material, but also the bandgap E_{gt} of the top cell. However, the short-circuit current of the top cell density J_{SCt} depends only on the bandgap E_{gt} of the top cell. When the top cell material is sufficiently thick and fully absorbed, the light of $h\upsilon \geq E_{\mathrm{g}}$, Eqs. (4.11) and (4.12), can be simplified to the following form:

$$J_{\mathrm{SCt}} = e\int_0^{\lambda_t} \Phi_S(\lambda)d\lambda \tag{4.13}$$

$$J_{\mathrm{SCb}} = e\int_{\lambda_t}^{\lambda_b} \Phi_S(\lambda)d\lambda \tag{4.14}$$

For a double-ended multi-junction solar cell with m junctions, the J-V curve of a single junction can be expressed as $V_i(J)$. Besides, the J-V curve of the whole multi-junction solar cells is a simple superposition of the J-V curve of each single-junction solar cell when the short-circuit current density of all sub-cells is the same; it can be expressed as

$$V(J) = \sum_{i=1}^{m} V_i(J) \tag{4.15}$$

This means that for the case of constant current, the voltage is the superposition of the voltage of each sub-cell. Each single-junction sub-cell has its own independent maximum power point $\{V_{mpi}, J_{mpi}\}$, where the product of voltage and current density is the largest. However, the current of each junction in series multi-junction solar cells is limited to a fixed value. The sub-cells will work at the maximum power point only when the maximum circuit density of each junction is equal, that is, $J_{mp1} = J_{mp2} = J_{mp3} = \cdots J_{mpm}$. If this condition holds, the maximum power output of multi-junction solar cells is the sum of the product of V_{mpi} and J_{mpi}. Moreover, some sub-cells will deviate from the maximum power point when the maximum circuit density of each sub-cell is not equal.

For high-quality III-V cells, the maximum power point is crucial, determining whether the solar cell will leak under normal bias or breakdown immediately under reverse bias. In Fig. 4.10, the J-V curve of a series multi-junction cell, in this case, is illustrated. The curves in the figure are the J-V curve of GaInP top cell, the J-V curve of a GaAs bottom cell, and the J-V curve of a double-junction cell formed by a series of two sub-cells. In this sample, the short-circuit current J_{SCb} of the bottom cell is higher than the short-circuit current J_{SCt} of the top cell; the J-V curve of the top cell is slightly shifted to better illustrate the change of the short-circuit current J_{SC} in the multi-cell. For a given current value, the voltage of the multi-junction cell satisfies $V_{series} = V_{top} + V_{bottom}$; this can be also seen in Fig. 4.10. The short-circuit

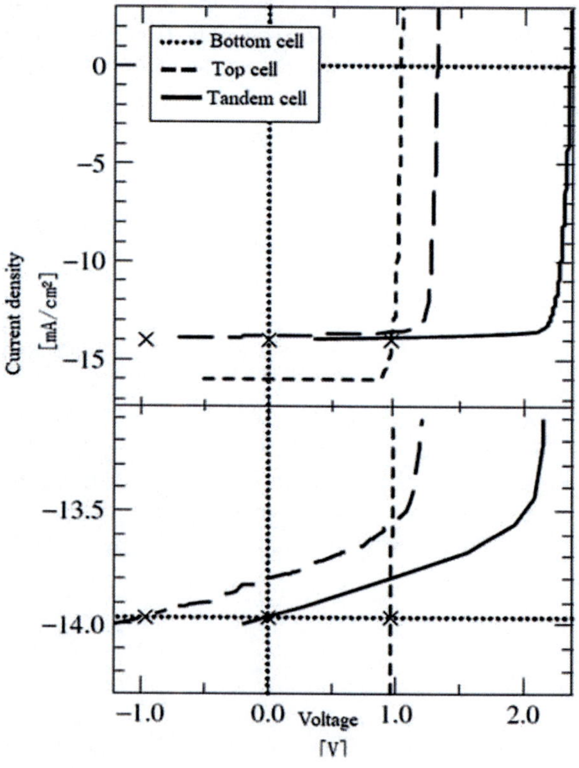

Fig. 4.10 Schematic diagram of the J-V curve of the double-junction series sub-cell

current density at the interface in the multi-junction series cell is $J_{SC} = -14$ mA/cm^2. The enlarged portion in the lower part of Fig. 4.10, is the J-V curve when the current density is close to the upper limit of the current density of the top cell. The black cross represents the voltage of the top cell, the bottom cell, and the double-junction cell when the current density is -14 mA/cm^2. All sub-cells are forward biased when the current density is $J = -13.5$ mA/cm^2; meanwhile, the corresponding bias voltage is only slightly lower than its open-circuit voltage V_{OC}. The bias voltage of the bottom cell is still near the open-circuit voltage V_{OC} and remains forward biased when the current density value exceeds -14 mA/cm^2. The bias voltage of the top cell is still about 1 V when the current density value exceeds -14 mA/cm^2, while the bias state changes from the forward bias to reverse bias. At this current density, the double-junction cell is zero biased, suggesting that there is no significant leakage current or reverse breakdown for each sub-cell. The short-circuit current density J_{SC} in a multi-junction cell is limited to a value lower than the short-circuit current density of each sub-cell; it is worth noting that this current-limiting effect makes the performance of the sub-cell in a multi-cell cell lower than that of a single-junction cell.

An expression for the J-V curve is necessary for constructing a quantitative analysis model for multi-junction solar cells. As mentioned earlier, the J-V curve expression of the classic photodiode [19] can be used

$$J = J_0[\exp(eV/kT) - 1] - J_{sc} \tag{4.16}$$

where e denotes the charge of the electron. Assuming that the ideal factor of the diode is 1. Then, the open-circuit voltage V_{OC} can be expressed as

$$V_{oc} \approx (kT/e)\ln(J_{sc}/J_0) \tag{4.17}$$

Since $J_{SC}/J_O \gg 1$ in practice, the dark current density J_O can be expressed as

$$J_0 = J_{0,\text{Base}} + J_{0,\text{Emitter}} \tag{4.18}$$

where

$$J_{0,\text{Base}} = \mathrm{e}\left(\frac{D_b}{L_b}\right)\left(\frac{n_i^2}{N_b}\right)\left(\frac{(S_b L_b/D_b) + \tanh(x_b/L_b)}{(S_b L_b/D_b)\tanh(x_b/L_b)+1}\right) \tag{4.19}$$

A similar expression can also be used to describe the dark current $J_{0,\text{Base}}$. The intrinsic carrier concentration n_i can be expressed as

$$n_i^2 = 4N_c N_v (2\pi kT/h^2)^3 (m_e^* m_h^*)^{3/2} \exp(-E_g/kT) \tag{4.20}$$

where m_e^* and m_h^* denote the effective masses of electrons and holes, respectively; Nc and Nv represent the effective state density of the bottom of the conduction band and the effective state density of the top of the valence band, respectively.

Each junction in the multi-junction cells can be described by Eqs. (4.16)–(4.20). The dark current $J_{0,i}$ of the i-th junction and short-circuit current $J_{SC,i}$ corresponding to the current density–voltage relationship is $V_i(J)$. The current density–voltage relationship $V(J)$ of the entire cell is obtained by adding the current density–voltage relationship $V_i(J)$ of the individual junctions. The maximum power point $\{V_{mpi}, J_{mpi}\}$ can be found by iterating over the points on the J-V curve. The performance parameters of different solar cells can be derived from the J-V curve, $V_{OC} = V(0)$, $FF = J_{mp}V_{mp}/(V_{OC}\,J_{SC})$.

A single variable needs to be controlled for discussion to determine the effect of different parameters on the performance of multi-junction solar cells. Literature [16], proposed a model of a double-junction N/P solar cell; the properties apart from the material bandgap of the bottom cell are the same as those of a gallium arsenide (GaAs) solar cell model. The light absorption efficiency also varies with the bandgap to ensure that the light absorption rate is zero when the photon energy is lower than the bandgap. Similarly, in addition to the variable bandgap, the properties of the top cell material are consistent with those of gallium indium phosphide (GaInP). For gallium arsenide (GaAs) cells, the carrier diffusion length is $L_h = 17$ μm, $L_e = 0.8$ μm at 300 K, and $L_h = 3.7$ μm, $L_e = 0.6$ μm for gallium indium phosphide (GaInP) cells. With the purpose of facilitating the understanding of the optimal performance of multi-junction solar cells, it is assumed that there is no recombination on all surfaces. For all the sub-cells, the thickness of the emitter region is $x_e = 0.1$ μm, the ionization impurity concentration is $N_e = 2 \times 10^{18}/cm^3$, and the base ionization impurity concentration is $N_b = 10^{17}/cm^3$. These values are very close to the actual values when the multi-junction solar cell is operating. Specifically, the efficiency profile of an infinitely thick double-junction solar cell in a solar AM1.5 standard spectrum is illustrated in Fig. 4.11a. Nell and Barnett [20] and Wanlass et al. [17] calculated the efficiency of infinitely thick double-junction solar cells in different spectra. Under the optimized bandgap combination of $E_{gt} = 1.75$ eV, $E_{gb} = 1.13$ eV, the predicted efficiency of the double-junction solar cell is 38%, which is higher than that of the best single-junction cell (29%).

The bandgap of $E_{gt} = 1.95$ eV and $E_{gb} = 1.42$ eV is calculated to be higher than the efficiency of the best single-junction cell even with a non-optimal bandgap combination. The bandgap of $E_{gt} = 1.95$ eV and $E_{gb} = 1.42$ eV is chosen because the bandgap of the bottom cell material is the bandgap of gallium arsenide (GaAs), while the bandgap of the top cell is only slightly higher than the bandgap of gallium indium arsenide (GaInP) prepared by the normal method (1.85 eV). However, the power conversion efficiency rapidly decreases from 35 to 30% when the bandgap E_{gb} of the bottom cell maintains the bandgap of gallium arsenide (GaAs) and the bandgap E_{gt} of the top cell drops from 1.95 to 1.85 eV. This drop is due to the dependence of the photocurrent of the top and bottom cells on the bandgap E_g of the top cell material. As mentioned above, the short-circuit current J_{SCb} of the bottom

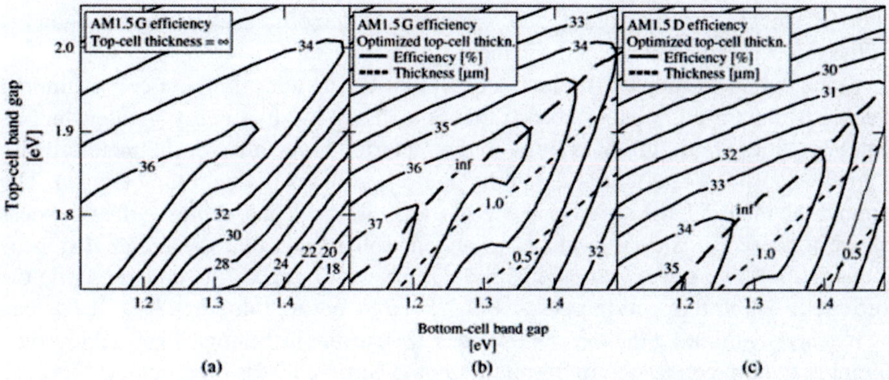

Fig. 4.11 Relationship between efficiency and bandgap of a double-ended double-junction tandem cell [17]

battery can be reduced and the short-circuit current J_{SCt} of the top battery can be increased by keeping the bandgap E_{gb} of the bottom cell unchanged and lowering the bandgap E_{gt} of the top cell. In Fig. 4.12a, the short-circuit current densities J_{SCb} and J_{SCt} being a function of the top cell bandgap E_{gt} when the bottom cell bandgap is constant ($E_{gb} = 1.42$ eV) is illustrated.

The short-circuit current J_{SC} of the multi-junction cell is lower than the short-circuit current density J_{SCt} of the top cell and the short-circuit current density J_{SCb} of the bottom cell. It can be observed in Fig. 4.12a, that the current matching effect is the best when the bandgap of the top cell is $E_{gt} = 1.95$ eV. Besides, the current matching degree drops rapidly when the bandgap of the top cell is $E_{gt} < 1.95$ eV. The decrease in the short-circuit current J_{SC} is caused by the limitation of the photocurrent of the low-bandgap bottom cell, as illustrated in Fig. 4.11a. The dependence of the short-circuit current density J_{SC} and the power conversion efficiency of the multi-junction cell on the bandgap E_{gt} of the top cell make it difficult for GaInP/GaAs dual-junction cell to achieve high power conversion efficiency. Fortunately, the dependence of the

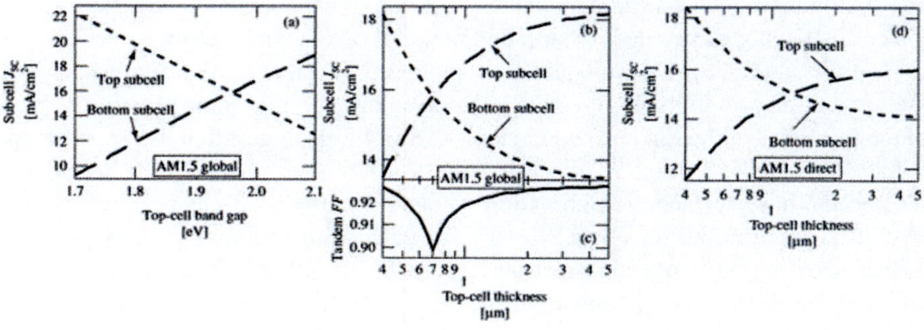

Fig. 4.12 The changes in tandem cell performance with the cell thickness

short-circuit current density J_{SC} and power conversion efficiency on the bandgap E_{gt} of the top cell can be alleviated by thinning the top cell.

Since the absorption coefficient α of a solar cell is not infinite, a cell of limited thickness cannot completely absorb the incident light with energy greater than its bandgap. Some light will pass through the material, especially for the material with a small absorption coefficient α and the photons with energy near the bandgap. The thinner the material, the stronger the light transmission. Thus, thinning the thickness of the top cell for a double-junction tandem cell means that the incident light is redistributed between the two cells; the current of the top cell decreases while the current of the bottom cell increases. If $J_{SCb} < J_{SCt}$ before thinning, $J_{SCb} = J_{SCt}$ can be reached by thinning the top cell. Besides, the current matching can be achieved by thinning the top cell to obtain optimum photoelectric conversion efficiency because the short-circuit current J_{SC} of the cell in the multi-junction tandem cell depends on the smaller one of J_{SCb} and J_{SCt}. The relationship between the thickness of the top cell and the current matching when the bandgap of the top cell is $E_{gt} = 1.85$ eV and the bandgap of the bottom cell is $E_{gb} = 1.42$ eV is illustrated in Fig. 4.12b. In this double-junction tandem cell, $J_{SC} \approx \min(J_{SCt}, J_{SCb})$; the current matching is acquired between the sub-cells when the top cell thickness is 0.7 μm. At this thickness, $J_{SC} = J_{SCt} = 15.8$ mA/cm^2, which is 85% of the short-circuit current of the infinite thickness top cell. The reason why such a thin cell can absorb a relatively high proportion of incident light is that the cell material is a direct bandgap semiconductor. Therefore, the short-circuit current J_{SC} of the double-junction cell with the infinite thickness top cell needs to depend on the short-circuit current $J_{SCb} = 13.4$ mA/cm^2 of the bottom cell in order to compare with the double-junction cell of the thinned top cell. The short-circuit current J_{SC} is increased by thinning the top cell, leading to the improvement of the power conversion efficiency of the double-junction cell; meanwhile, the power conversion efficiency of the double-junction cell is increased from 30% to 15.8/13.4 * 30% ≈ 35%. Apparently, this is only an approximate estimate. The effect of top cell thinning on the fill factor FF and the open-circuit voltage V_{OC} will be discussed below. However, these are the secondary effects. Only considering the influence of short-circuit current JSC on power conversion efficiency can be a good approximation. As illustrated in Fig. 4.12, the dependence of the power conversion efficiency on the bandgap E_{gt} of the top cell is reduced by the thinned top cell, allowing us to select the top cell material over a wide range.

The fill factor FF of a multi-junction solar cell depends on the photocurrent of the top and bottom cells. Figure 4.12c indicates that the fill factor FF is a function of the thickness of the top cell. Equations (4.18)–(4.20) suggest that the open-circuit voltage V_{OC} is also a function of cell thickness. Therefore, how the finite base width $\mathbf{X_b}$ and the base surface recombination velocity S_b affect the open-circuit voltage V_{OC} of a gallium indium (GaInP) cell is revealed. Since doping has an essential influence on the surface recombination velocity S_b, the doping should be considered when designing the power conversion efficiency of the GaInP/GaAs cell.

Light is distributed between the individual sub-cells; therefore, photocurrent is generated in each sub-cell and determined by the spectrum of the incident light. Thus, the optimal top cell bandgap and top cell thickness depend on the spectrum. The relationships between the power efficiency and the bandgap of the top and bottom cells in the AM1.5D and AM1.5G spectra are illustrated in Fig. 4.11c, b, respectively. For a double-junction solar cell with a given bottom cell bandgap, the optimal bandgap E_{gt} of the top cell in the AM1.5D spectrum is lower than that in the AM1.5G spectrum because the AM1.5D spectrum contains less blue light, which reduces J_{SCt}/J_{SCb}. Lower E_{gt} allows more light to be absorbed by the top cell. Similarly, for a double-junction solar cell with a fixed top cell bandgap E_{gt} and a bottom cell bandgap E_{gb}, the optimal top cell thickness in the AM1.5D spectrum is greater than that in the AM1.5G spectrum. Figure 4.12d illustrates that the top cell short-circuit current J_{SCt} and the bottom cell short-circuit current density J_{SCb} are functions of the cell thickness when the top cell bandgap and the bottom cell bandgap are 1.85 eV and 1.42 eV, respectively. The other conditions are kept unchanged in Fig. 4.12a, and AM1.5G is replaced by the AM1.5D spectrum; therefore, the top cell thickness required for current matching is about 1.2 μm higher than that in the AM1.5G spectrum (0.7 μm). In contrast, the top cell thickness in the AM0 spectrum containing more blue light components was also calculated, and the required thickness was about 0.5 μm.

The above analysis provides a method for determining the thickness of the top cell based on the spectrum, and no single spectrum can accurately describe the light incident on the surface of the solar cell. The spectrum fluctuates with time and place. The specific design of multi-junction solar cells suitable for different spectra is more complicated, referring to the literature [18].

In the previous discussion, it was all assumed that there was no reflection of incident light on the surface of the cell. However, without an anti-reflective layer coverage, III-V solar cells may reflect up to 30% of light in a specific spectrum. This reflection can be reduced to 1% by covering the anti-reflective layer, while it is only effective for a part of the spectrum. This limitation is crucial for the current matching of multi-junction solar cells. Therefore, the detection of the anti-reflection layer problem is significant for the III-V multi-junction solar cell.

Therefore, a numerical model of surface reflection needs to be established in order to analyze the effect of anti-reflection layer parameters on the performance of multi-junction solar cells. In this section, the relatively simple Lockhart-King model [19] can be used. This model calculates the normal reflection of the three-layer coating. Assuming that each layer exists, the j-th layer reflection index is n_j, and the layer thickness is d_j. The value of j is from 1 to 4; $j = 1$ represents the substrate, and $j = 4$ represents the top layer of the cell. For example, the layer 4/3/2/1 for a two-layer magnesium difluoride/zinc sulfide (MgF_2/ZnS)-covered gallium indium phosphide (GaInP) cell with an aluminum indium phosphide (AlInP) window is MgF_2/ZnS/AlInP/GaInP. The reflection R is a function of the wavelength λ and can be expressed as

$$R = |(X - 1)/(X + 1)|^2 \tag{4.21}$$

where

$$X = \begin{bmatrix} n_2(n_3n_4 - n_2n_4t_2t_3 - n_2n_3t_2t_4 - n_3^2t_3t_4) + in_1(n_3n_4t_2 + n_2n_4t_3 + n_2n_3t_4 \\ n_3^2t_2t_3t_4]/[n_1n_4(n_2n_3 - n_3^2t_2t_3 - n_3n_4t_2t_4 - n_2n_4t_3t_4 + in_2n_4(n_2n_3t_2 \\ +n_3^2t_3 + n_3n_4t_4 - n_2n_4t_2t_3t_4 \end{bmatrix} \quad (4.22)$$

$$t_j = \tan(2\pi n_j d_j/\lambda) \quad (4.23)$$

Although this method incorrectly believes that there is no absorption of light from top to bottom and completely ignores the reflection and refraction of light in the tandem cells, the result is easier to understand and is in good agreement with the more rigorous calculations [20]. The effect of the antireflection layer can be discussed by Eq. (4.21).

The reflection model of a GaInP cell covered with an anti-reflective coating of MgF_2/ZnS is illustrated in Fig. 4.13a; the optical constants reported in the literature [22, 23], is used by changing the thickness of the coating [21].

The dependence of reflectivity on coating thickness can be discussed from two aspects: the ratio of the thickness of the two coatings and the total thickness of each coating. A low, flat trough-shaped reflection curve can be obtained by choosing the right coating ratio. Besides, the total thickness of the coating determines the position

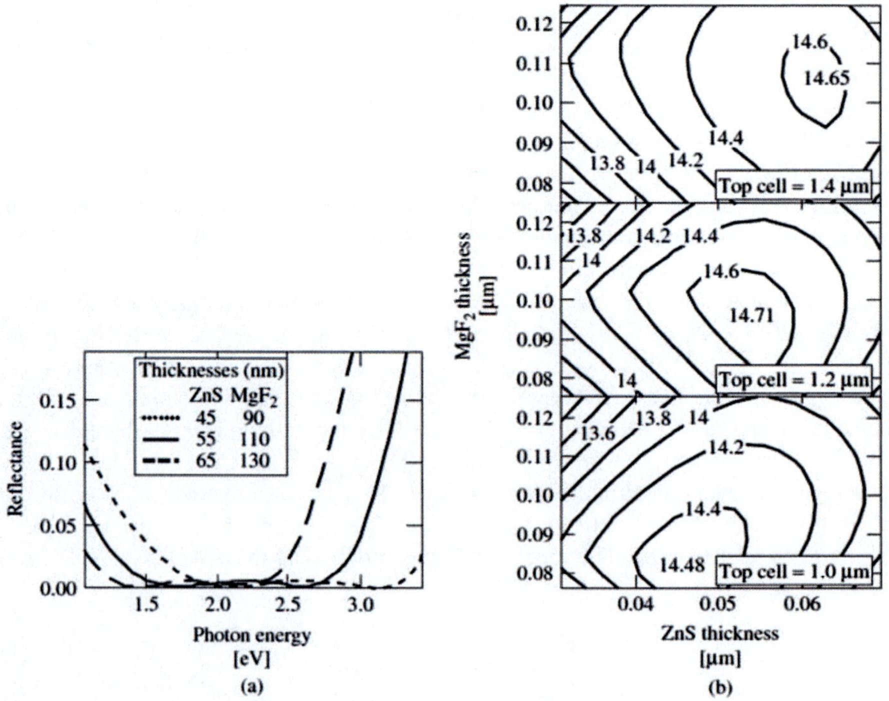

Fig. 4.13 The relationship between cell performance and anti-reflective layer

of the lowest value when the coating ratio is constant; meanwhile, the bottom of the groove moves to the position of low photon energy as the thickness of the coating increases. Since the width of the trough is smaller than the solar spectral range, the photocurrent of the sub-cell is smaller than the ideal photocurrent without reflection, no matter where the trough bottom is. Moving the bottom of the trough to high-energy photons causes more light to be absorbed by the top cell and less light to be absorbed by the bottom cell, and vice versa. Thus, the current matching of the solar cell is affected by the anti-reflective layer. Figure 4.13b illustrates that the photocurrent of GaInP/GaAs solar cells in the AM1.5D spectrum is a function of the thickness of the MgF_2/ZnS anti-reflective layer. With the increase in the thickness of the top cell, the thickness of the optimal anti-reflective layer also increases to allow more light to reach the bottom cell.

Solar cells in terrestrial applications are generally used in conjunction with concentrating systems; in this way, the cost of solar cells is also increased to some extent. III-V semiconductor multi-junction solar cells are well suited for concentrating systems because their high power conversion efficiency can remain good not only under a single solar illumination, but also under the concentration of 1000 suns. In specific applications, it is necessary to adjust the multi-junction solar cell structure operating under monochromatic light to make it suitable for use in concentrating systems, referring to the literature [24], for a detailed discussion. The relationship between different concentrating conditions and the power conversion efficiency of solar cells is displayed in Fig. 4.14.

It can be observed from Fig. 4.14, that the power conversion efficiency increases as the degree of concentrating increases. However, this relationship is nonlinear, and the degree of concentration increases the cell cost to a certain extent. Therefore, it is necessary to balance the concentrating conditions and the cell cost in the further application of solar cells.

Because the temperature at which the solar cell operates is not constant, the dependence of the main parameters of the solar cell on temperature requires to

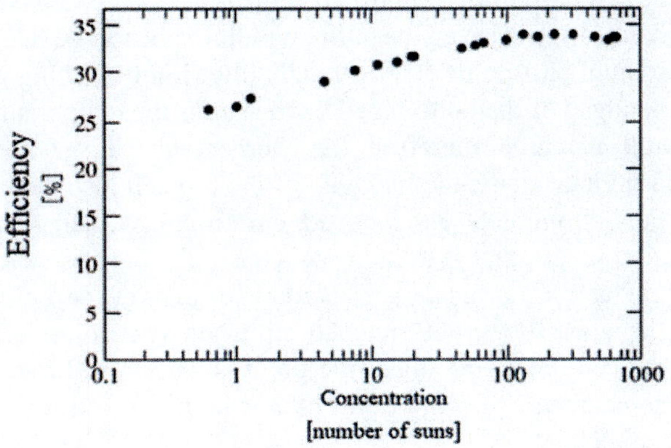

Fig. 4.14 Relationship between different concentrating conditions and power conversion efficiency of solar cells

Table 4.3 Temperature correlation coefficient of multi-junction cell and its sub-cells under 300 K [30]

Solar cell structure	Open-circuit voltage V_{OC} (mV)	dV_{OC}/dT (mv/K)	$1/V_{OC}dV_{OC}/dT$ (%/K)
Ge	200	−1.8	−0.90
GaAs	1050	−2.0	−0.19
GaInP	1350	−2.2	−0.16
GaInP/GaAs	2400	−4.2	−0.17
GaInP/GaAs/Ge	2600	−6.0	−0.23
GaInP/GaAs/Ge (500 suns)	3080	−6.0	−0.19
PERL Si	711	−1.7	−0.24

be discussed to further evaluate the performance of the solar cell. Since the open-circuit voltage V_{OC} of the series multi-junction cell is the sum of the open-circuit voltages V_{OCs} of the sub-cells, the cell temperature correlation coefficient dV_{OC}/dT is also the sum of the temperature correlation coefficients of the sub-cells. Taking a gallium phosphide indium/gallium arsenide (GaInP/GaAs) solar cell as an example, the temperature correlation coefficient of a gallium indium arsenide (GaInP) cell and a gallium arsenide (GaAs) cell is $dV_{OC}/dT \approx -2$ mV/°C. Therefore, the temperature correlation coefficient of the multi-junction cell is $dV_{OC}/dT \approx -4$ mV/°C [29]. The temperature correlation coefficients for different types of cells are listed in Table 4.3. Assuming that the ideal factor of the junction is n = 1. Unless otherwise stated, the temperature correlation coefficients of the devices in the table are in solar illumination, compared with the temperature correlation coefficient of the locally diffused silicon cell on the backside with the passivated emitter [30].

For the short-circuit current J_{SC}, the short-circuit current of the cell depends on the one with the lowest short-circuit current in the sub-cells. Taking the gallium phosphide indium/gallium arsenide (GaInP/GaAs) solar cell as an example, the short-circuit current of gallium arsenide (GaAs) sub-cells depends not only on the GaAs bandgap but also on the gallium indium phosphide (GaInP) bandgap because gallium indium phosphide (GaInP) cells filter light reaching gallium arsenide (GaAs). The bandgap of the bottom cell decreases as the temperature of the multi-junction series cell increases; therefore, the short-circuit current J_{SC} has an increasing tendency. Simultaneously, the bandgap of the top cell also decreases; thus, the light reaching the bottom battery is reduced, contributing to alleviating the tendency of the short-circuit current J_{SC} to increase. Generally, the short-circuit current density J_{SCs} does not have a significant temperature correlation coefficient. There is an inflection point for a nearly current-matched multi-junction solar cell. The short-circuit current of the cell below the inflection point is determined by one of the sub-cells, while the short-circuit current of the cell above the inflection point is determined by the other sub-cell.

Particularly, dE_{ff}/dT is a relatively smooth function of temperature T because the power conversion efficiency is proportional to $V_{OC} \times J_{SC} \times FF$, and dJ_{SC}/dT and dFF/dT change when the temperature tends to the current matching temperature.

4.3.4 *Device Structure*

There are many ways to take the power cord out of the cell, including multi-junction tandem solar cells. Each structure provides varying degrees of electrical isolation. In a four-ended lead-out structure, each sub-cell has its own two separate ports and is electrically isolated from other sub-cells. One benefit of this cell is that the structure has no limitation on the polarity (p/n or n/p) or current and voltage polarity of the junction in the sub-cell. However, the complicated cell structure and preparation process of the four-ended cell makes it difficult to realize electrical isolation and port extraction between the sub-cells in a monolithic structure. Generally, the four-ended cell is mechanically stacked, which is not a good choice compared to the monolithic structure.

Conversely, each sub-cell in a three-ended cell configuration is not electrically isolated, and the upper cell is electrically coupled to the underlying cell. The preparation of a three-ended cell is easier, even though the difficulty is increased relative to a two-ended cell. Therefore, the cell material must have semiconductor characteristics in order to introduce an intermediate port in the subsequent process and form an ohmic contact with the intermediate port. Consequently, different sub-cells have different photocurrents due to the introduction of the intermediate port. Besides, the introduction of the intermediate port allows the sub-cells to have different polarities; for example, the upper cell is the p/n junction and the lower cell is the n/p junction. A more detailed three-ended and four-ended cell module interconnection can refer to the literature [25].

The serial connection at both ends has the most stringent requirements for the cell interconnection. This structure requires that different sub-cells have the same polarity and the photogenerated current must match because the minimum photogenerated current of the sub-cell in this structure determines the current of the cell. Although the limitation of current matching has a serious impact on the selection of cell materials, the presence of a high-quality single-layer tunnel junction as the middle port allows only the uppermost and lowermost layers of the cell to be metalized like a single cell. This advantage allows the two-terminal tandem cell to be realized in a simple process like a two-ended single cell. A typical double-ended triple-junction gallium phosphide indium/gallium arsenide/germanium semiconductor solar cell is illustrated in Fig. 4.4.

4.4 Development and Problems of GaInP/GaAs/Ge Multi-junction Cell

4.4.1 GaInP Solar Cell

The main advantage of gallium phosphide indium/gallium arsenide/germanium (GaInP/GaAs/Ge) is the high lattice matching of the materials. The main preparation method of the monolithic cell is a heterojunction epitaxy, which is generally achieved by metal–organic chemical vapor deposition (MOCVD). Similar lattice constants make heteroepitaxial simple, such as epitaxially AlGaAs on GaAs; epitaxial lattice-mismatched materials make it more difficult. Lattice mismatch can be alleviated by nucleation and dislocations; the dislocation density is related to the degree of lattice mismatch and the thickness of the material. These dislocations become non-radiative recombination centers and affect the lifetime and diffusion length of minority carriers, resulting in further affecting the photoelectric conversion efficiency of the cell.

The bandgap of the semiconductor alloy $Ga_xIn_{1\text{-}x}P$ is linearly related to the molar ratio x of gallium; the bandgap can be expressed by Eq. 4.24.

$$a_{Ga_xIn_{1-x}P} = xa_{GaP} + (1-x)a_{InP} \tag{4.24}$$

where $a_{GaP} = 0.54512$ nm and $a_{InP} = 0.58686$ nm are the lattice constants of GaP and InP, respectively. The parameters of more III-V binary compounds are illustrated in Table 4.4.

Table 4.4 The lattice constant, force constant and Poisson ratio of binary III-V compound semiconductor [26]

Materials	Lattice constant/(nm)	$C_{11}/(10^{10}$ N/m$^2)$	$C_{12}/(10^{10}$ N/m$^2)$	Poisson ratio/ν
AlP	0.546354			
GaP	0.54512	14.12	6.25	0.307
		14.05	6.203	0.306
InP	0.58686	10.22	5.76	0.360
		10.11	5.61	0.357
$Ga_xIn_{1\text{-}x}P$[x = 0.516]				0.333
GaAs	0.565318	11.81	5.32	0.311
		11.91	5.951	0.333
InAs	0.60583	8.239	4.526	0.352
Ge	0.5657906	12.89	4.83	0.273
		12.40	4.13	0.250

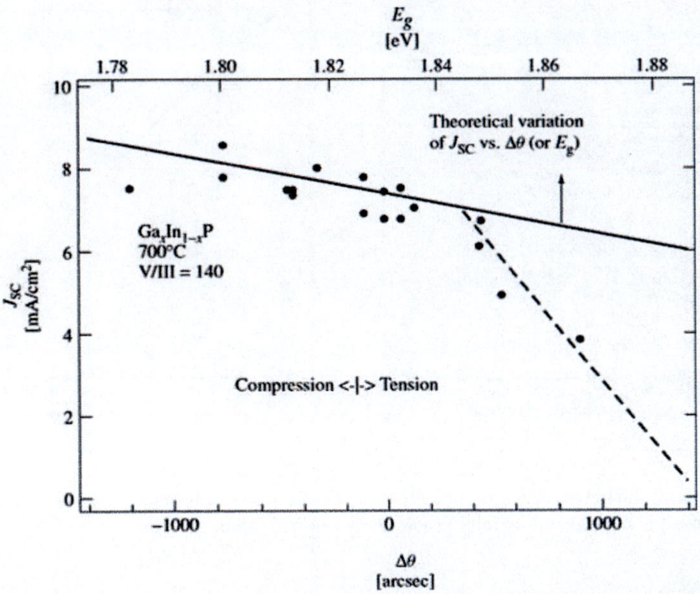

Fig. 4.15 Short-circuit current of gallium phosphide indium solar cells with different forbidden bandwidths

The lattice constant of the gallium arsenide substrate $a_{GaAs} = 0.565318$ nm; the material under 25 °C can achieve lattice matching when the epitaxial gallium indium arsenide ($Ga_xIn_{1\text{-}x}P$) satisfies the condition of $X = 0.516 = X_{LM}$. The quality of the epitaxial gallium indium phosphide ($Ga_xIn_{1\text{-}x}P$) material is still good when the epitaxial material fluctuates slightly near the X_{LM}.

The short-circuit current of gallium phosphide indium solar cells with different bandgaps is displayed in Fig. 4.15. This current is a function of $\Delta\theta$; $\Delta\theta$ can be measured by the twin-crystal X-ray rocking curve diffraction and is a parameter to measures the quality of an epitaxial layer [27].

If the thickness of the gallium indium phosphide ($Ga_xIn_{1\text{-}x}P$) material is much lower than the critical thickness associated with x, then

$$\Delta\theta = \tan\theta_B \left(\frac{x a_{GaP} + (1-x)a_{InP} - a_{GaAs}}{a_{GaAs}}\right)\left(\frac{1 + [\nu_{GaP}x + \nu_{InP}(1-x)]}{1 - [\nu_{GaP}x + \nu_{InP}(1-x)]}\right) \tag{4.25}$$

where θ_B is the Bragg angle and $\nu_{GaP}x + \nu_{InP}(1-x)$ is the Poisson ratio of gallium indium phosphide ($Ga_xIn_{1\text{-}x}P$) obtained by the Poisson ratios of GaP and InP, which can be found in Table 4.4 (Poisson ratio is defined as the negative value of the ratio of transverse stress to longitudinal stress in the case of uniaxial stress). If the material is free of stress, the value of the last factor of Eq. (4.25), becomes 1. The relationship between the value of x and the angular change m of gallium indium phosphide on gallium arsenide is illustrated in Fig. 4.16a.

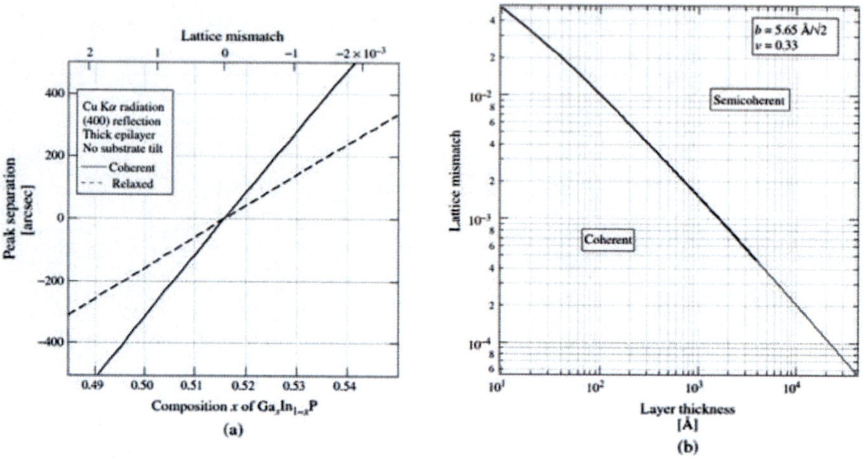

Fig. 4.16 Relationship between composition X, lattice mismatch, and the material thickness of gallium indium phosphide epitaxially grown on gallium arsenide

The critical thickness is determined by the polymerization energy produced by the strain and the self-consistent energy caused by the mismatch. The lowest energy state of the system is an epitaxial layer with the same lattice constant as the substrate when the thickness of the epitaxial layer is lower than the critical thickness. Besides, the lowest energy state is the epitaxial layer stress and the dislocations alleviating the stress of the epitaxial layer when the thickness of the epitaxial layer is greater than the critical thickness. Mattews and Blakeslee first researched this field [28]. The relationship between lattice mismatch and epitaxial layer thickness is illustrated in Fig. 4.16b. It can be observed from Fig. 4.15, that the critical thickness is infinite and there is no dislocation when the angle change $\Delta\theta$ is 0, suggesting that the short-circuit current density J_{SC} is only related to the mobility of minority carriers in the epitaxial material. The solid line with a negative reciprocal represents the relationship between the theoretical short-circuit current density J_{SC} and the angular change $\Delta\theta$. For $\Delta\theta < 0$, the In content of the epitaxial layer is higher ($x < x_{LM}$), and the bandgap is lower than the lattice-matched gallium indium phosphide ($Ga_xIn_{1-x}P$); therefore, the short-circuit current density J_{SC} increases as the angle change $\Delta\theta$ decreases. For $\Delta\theta < 0$, the In content of the epitaxial layer is higher ($x > x_{LM}$). In the portion of the curve where the In content is high, the short-circuit current density J_{SC} first decreases as the angle change $\Delta\theta$ increases and then decreases rapidly as the angle change $\Delta\theta$ increases. Critical angle changes are a function of a range of kinetic factors including film thickness, growth temperature, and growth rate. Epitaxial layers with higher In content tend to be subjected to compressive strain. Compared with the material with stretched strain, the material with compression strain is less likely to produce slip dislocation.

Table 4.5 Important parameters of Ge, GaAs, and GaInP at 298 K

Material	Ge	GaAs	$Ga_xIn_{1-x}P$	$Al_xIn_{1-x}P$
Atom/cm^3	4.42×10^{22}	4.44×10^{22}		
Lattice constant/Å	5.657906	5.65318	When x = 0.516, equal to the value of GaAs	When x = 0.532, equal to the value of GaAs
Bandgap/eV	Indirect bandgap 0.662	1.424	Mismatch 1.91	Indirect bandgap 2.34
	Direct bandgap 0.803			Direct bandgap 2.53
Conductor band state density Nc/cm^{-3}	1.04×10^{19}	4.7×10^{17}		
Valence band state density Nv/cm^{-3}	6.0×10^{18}	7.0×10^{18}		
Intrinsic carrier concentration/cm^{-3}	2.33×10^{13}	2.1×10^{6}		
Coefficient of thermal expansion/K^{-1}	7.0×10^{-6}	6.0×10^{-6}	5.3×10^{-6}	

$Ga_xIn_{1-x}P$ and GaAs are lattice-matched at room temperature, while the two materials are not lattice-matched at the growth temperature; this is caused by the difference in thermal expansion coefficients between $Ga_xIn_{1-x}P$ and GaAs (See Table 4.5). From a kinetic point of view, the lattice constant matching of the two materials at the growth temperature is more essential. The epitaxial layer lattice-matched to the substrate at the growth temperature of 625 °C will have the lattice mismatch of $\Delta\theta = -200°$ at room temperature [29], or the material lattice-matched to the substrate at room temperature will have the lattice mismatch of $\Delta\theta = 200°$ at the growth temperature of 625 °C. Since defects and dislocations are more easily introduced at high temperatures, the quality of the material lattice-matched to the substrate at the growth temperature is better. The error at the growth temperature is ±50 °C while it will become $-250° < \Delta\theta < -150°$ at room temperature.

As mentioned earlier, materials subjected to compressive strain are less prone to dislocations and slip dislocations than materials subjected to tensile strain. Therefore, the allowable error range is slightly larger when $\Delta\theta < 0$.

Due to the existence of the dynamic scattering effect, the measured $\Delta\theta$ of the thin epitaxial layer (thickness less than 0.1 μm) is smaller than that of the thinner material [30].

The value of $\Delta\theta$ in the epitaxial layer is not uniform and it is related to the crystal orientation of the substrate and the incident X-ray beam. The effective x is generally taken as the average of the two measurements, the first measurement using the traditional measurement method and the second rotation of the sample 180° [31]. For the substrates with a crystal orientation close to <100>, this effect is not significant and produces an error of about 10% or 6°. Besides, the error caused by the substrate with crystal orientation of <511> is as high as 50%.

Most publications before 1986, considered that gallium indium phosphide ($Ga_xIn_{1-x}P$) matched the lattice constant of gallium arsenide (GaAs) and had the same bandgap (1.9 eV). However, Gomyo et al. [32], reported in 1986, that the bandgap of gallium indium phosphide ($Ga_xIn_{1-x}P$) obtained by Metal–Organic Chemical Vapor Deposition (MOCVD) was less than 1.9 eV, and the bandgap was related to growth conditions. Subsequent studies [33], indicated that the shift in the bandgap is related to the orientation of Ga and In in the sublattice. The crystal is a $CuPt_B$ structure (referred to as a CVD grown lattice structure); the (111) crystal plane is replaced by $Ga_{0.5+\eta/2}In_{0.5-\eta/2}P$ and $Ga_{0.5-\eta/2}In_{0.5+\eta/2}P$, where η is the long-range order parameter. The perfectly arranged GaInP ($\eta = 1$) replaces the (111) plane with GaP and InP. The theoretical calculation of the indium gallium phosphide indium ($Ga_xIn_{1-x}P$) orientation was first obtained by Kondow and his collaborators by using the tight-binding method [34]. Afterward, Kurimoto and Hamada performed similar calculations using the Linear Augmented Plane Wave (LAPW) method in the first principle [35]. The orientation of the indium gallium phosphide ($Ga_xIn_{1-x}P$) and the change in the energy gap ΔE_g is first published by Capaz and Koiller [36], as expressed in Eq. (4.26).

$$\Delta E_g = -130\eta^2 + 30\eta^4 \tag{4.26}$$

The further calculation is [37]

$$\Delta E_g = -484.5\eta^2 + 435.4\eta^4 - 174.4\eta^6 \tag{4.27}$$

The effects of different growth conditions on the orientation and bandgap of gallium indium phosphide ($Ga_xIn_{1-x}P$) have been extensively investigated. The bandgap of indium gallium phosphide ($Ga_xIn_{1-x}P$) depends on its growth temperature T_g, the growth rate R_g, phosphorus partial pressure (PH_3), lattice mismatch, and the doping level. The influence of these effects on the energy gap is illustrated in Fig. 4.17. In the higher growth temperature range (700–725 °C), the bandgap increases as the lattice mismatch increases. In the intermediate growth temperature range (600–675 °C), the bandgap decreases first, reaches the minimum value, and then increases as the lattice mismatch increases.

Although the influence of these factors on the bandgap of gallium indium phosphide ($Ga_xIn_{1-x}P$) is very complicated, only a few factors have been specifically characterized. For example, a bandgap of 1.8–1.9 eV for a substrate having a lattice orientation close to 100 is calculated by using common growth temperature T_g, growth rate R_g, and phosphorus partial pressure (PH_3). The calculated energy gap is 1.9 eV when the growth temperature T_g, the growth rate R_g, and the phosphorus partial pressure (PH_3) are all limited. However, the properties of the material are also related to other factors, such as minority diffusion length, elemental composition, and surface topography. The most direct way to achieve a higher bandgap width is to use a substrate deviating from the <100> crystal orientation toward the (111)

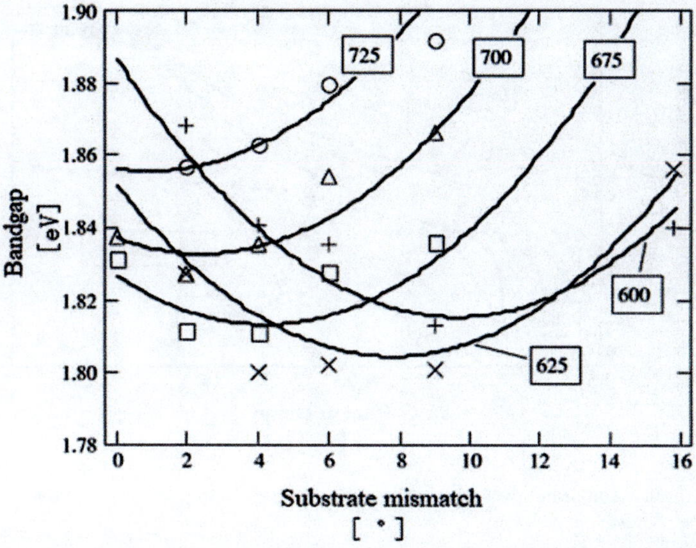

Fig. 4.17 Effect of material growth temperature T_g on the bandgap and substrate mismatch of gallium indium phosphide ($Ga_xIn_{1-x}P$)

crystal plane. The direction of (111) crystal plane towards the III group element in the sphalerite system is defined as the <111> A crystal orientation; the direction of (111) crystal plane towards the V group element is defined as the <111> B crystal orientation. A substrate deviating from the <111> crystal orientation and towards the <111> B crystal orientation can contribute to the increase in the order of the element arrangement. Besides, there is no difference between the <111> A and the <111> B crystal orientation when the gallium indium phosphide ($Ga_xIn_{1-x}P$) is grown on the surface of Ge; meanwhile, it is very difficult for III-V semiconductor (GaAs or GaInP) to control the crystal direction of A or B. Thus, using a substrate with a lattice orientation deviation angle greater than 15° is the simplest method of growing gallium indium phosphide ($Ga_xIn_{1-x}P$) with a high bandgap on the Ge surface. The growth process is accompanied by a high growth rate R_g, a moderate growth temperature T_g, and a partial pressure PH_3.

Apparently, there are other factors affecting the orientation of gallium indium phosphide ($Ga_xIn_{1-x}P$) materials, such as optical anisotropy [38–41], transport anisotropy [42, 43], and surface topography [44, 45].

It is necessary to establish a precise light transmission model of gallium indium phosphide ($Ga_xIn_{1-x}P$) to further characterize the properties of gallium indium phosphide ($Ga_xIn_{1-x}P$) materials and evaluate the cell performance. The optical constants of indium gallium phosphide ($Ga_xIn_{1-x}P$) can be obtained by optical ellipsometry experiments models [46, 47] and light transmission experiments [48]. These results are summarized in Fig. 4.18. In most cases, there is no uniform model to describe the transmittance of indium gallium phosphide ($Ga_xIn_{1-x}P$) epitaxial on different crystal

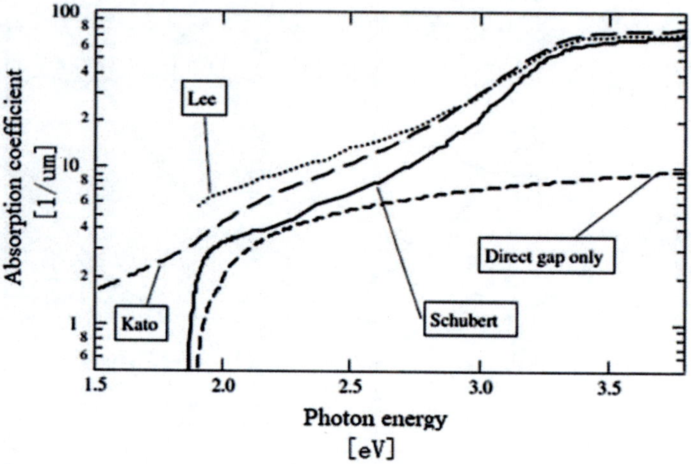

Fig. 4.18 Comparison of light absorption rates of indium gallium phosphide obtained by different methods

orientation substrates. For example, Kato's model [47], successfully described light transmission in the shortwave range; unfortunately, it was not suitable for the wavelength range where the energy is close to the bandgap of gallium phosphide. The models established by Schubert and Kurtz are only suitable for the wavelengths in which the energy is close to the bandgap of gallium phosphide. Besides, Kurtz et al.'s model provided an expression for the photo-absorption rate of indium gallium phosphide [48], as expressed in Eq. (4.28).

$$\alpha = 5.5\sqrt{E - E_g} + 1.5\sqrt{E - (E_g + \Delta_{so})}\,[\mu m^{-1}] \tag{4.28}$$

where E denotes the photon energy and E_s represents the bandgap. Δ_{so} is the spin–orbit energy and generally set to 0.1 eV; it is independent of the degree of alignment ηs. Apparently, the absorption rate varies with the value of the energy gap E_g. This model is relatively accurate about E_g and $E_g + \Delta_{so}$, which is very beneficial for deriving the minority carrier diffusion length through photo-response experiments. However, this model is not applicable to higher energy photons.

The material must be doped in order to form a PN junction. According to the type of doping, Gallium indium phosphide ($Ga_xIn_{1-x}P$) has n-type and p-type. The main n-type dopants are selenium (Se) and silicon (Si).

The selenium (Se) element is a common III-V semiconductor donor element, and the usual doping method is to decompose H_2Se. The doping behaviors of H_2Se have been investigated by many research groups [49–54]. Under most growth conditions, the concentration of electrons increases with increasing H_2Se flow or partial pressure and reaches a maximum at $2 \times 10^{19}/cm^3$. This maximum concentration, like many semiconductors, also depends on the growth temperature Tg. The doping of selenium

(Se) also depends on the partial pressure PH_3 (for GaInP) or partial pressure AsH_3 (for GaAs). This relationship can be better described by the following equation:

$$n^{-1} = (1 + \alpha P_V)(\beta P_{Se})^{-1} + k^{-1} \tag{4.29}$$

where n denotes the electron concentration; P_V and P_{Se} denote the partial pressures of the group V element and Se, respectively. The correlation coefficients a and b depend on the temperature T_g and the residence time (carrier flow rate), respectively; therefore, the values of these correlation coefficients vary with different reaction systems. This doping behavior can be described by the modified Langmuir model, in which selenium (Se) and the group V elements absorbed by the constant k have competitive absorption, depending on the concentration gradient of the two elements. This model is much better than the common $n \propto P_V P_{Se}$ model.

When the electron concentration exceeds $2 \times 10^{18}/cm^3$, the bandgap of indium gallium phosphide ($Ga_xIn_{1\text{-}x}P$) increases, the crystal orientation decreases, and the surface becomes smooth [55]. Besides, the surface begins to rough again when the H_2Se flow is large enough. Meanwhile, the electron concentration also begins to decrease [49], and the precipitated selenium atoms are observed by transmission electron microscopy [56]. The molar ratio of Al in the compound at the connection of the selenium atoms such as the DX center in ($Al_xGa_{1\text{-}x}$) $In_{0.5}P$ (i.e., the deep level associated with the donor impurity in the compound semiconductor) is higher than 0.4[57].

Silicon (Si) elements are widely used as p-type doping of III-V semiconductors, and the common dopant is Si_2H_6. Hotta and his collaborators first discovered the doping effect of Si_2H_6 [58], revealing that the pyrolysis rate of Si_2H_6 also decreased as the growth temperature decreased under the condition of the growth temperature $T_g < 640$ °C; meanwhile, the electron concentration n also decreased as expected. Under the condition of $T_g > 640$ °C, the electron concentration reached saturation status at $n = 5 \times 10^{18}/cm^3$; this is in line with expectations. The reason for the saturation of the electron concentration is the appearance of a non-ionizing complex such as $Si_{III}^+ - Si_V^-$ or $Si_{III}^+ - V_{III}^-$. This result [57], is close to the results of the quantitative analysis of silicon-doped gallium arsenide. Besides, Scheffer and his co-workers [59], used Si_2H_6 as a dopant; no significant electron concentration saturation was observed when the electron concentration was as high as $8 \times 10^{18}/cm^3$. Moreover, Minagawa and his collaborators [59], found that the saturation status occurred when the electron concentration reached $1 \times 10^{19}/cm^3$ [60]; this saturation is independent of substrate orientation and growth temperature. It is reported that for a uniform gallium indium phosphide ($Ga_xIn_{1\text{-}x}P$) epitaxial layer not only improves the concentration of electrons, but also the mobility of electrons, compared with the δ doping of silicon (Si) (i.e., doping only occurs in a single layer or several layers instead of the entire epitaxial layer) [61, 62]. It can be concluded from this study that the δ doping of silicon (Si) introduces shallow-level acceptor defects. Silicon (Si) appears to be a silicon atom existing as a deep-level impurity in $(Al_xGa_{1-x})_{0.5}In_{0.5}P$ with a molar ratio of $x > 0.3$ rather than a deep-level impurity in gallium indium phosphide ($Ga_xIn_{1-x}P$) [63]. Similar to selenium (Se), gallium indium phosphide ($Ga_xIn_{1-x}P$) begins to exhibit lattice mismatch when the doping concentration of silicon (Si) exceeds a critical value; meanwhile, its bandgap also increases. However, there are some differences in the effect of the doping concentration of silicon on lattice mismatch. Gomyo and his co-workers investigated that the concentration of doped silicon (Si) is lower than that of doped selenium (Se) when lattice mismatch appears in gallium indium arsenide ($Ga_xIn_{1-x}P$) [50]. Meanwhile, Minagawa and his collaborators explored that gallium indium phosphide ($Ga_xIn_{1-x}P$) began to appear lattice mismatch [51], when the concentration of doped silicon (Si) was as high as $1 \times 10^{19}/cm^3$.

Zinc (Zn) is often used for p-type doping of gallium indium phosphide ($Ga_xIn_{1-x}P$); the common zinc sources are dimethyl zinc (DMZ) and diethylzinc (DEZ). The doping characteristics of zinc haven investigated by many researchers [49, 55–57]. The relationship between the doping ratio and the injection flow is generally quasi-linear while the linear level is improved at a low growth temperature T_g and a high growth rate R_g. An analytical model of this phenomenon was proposed by Kurtz et al. [64].

Some problems in p-type doping of gallium indium phosphide ($Ga_xIn_{1-x}P$) can be caused by the high zinc (Zn) concentration. Specifically, the order of the gallium indium phosphide ($Ga_xIn_{1-x}P$) crystal is destroyed and the bandgap increases when the carrier concentration is $1 \times 10^{18}/cm^3$ [65]. Simultaneously, a high zinc (Zn) concentration, or more precisely, a large flow of dimethylzinc (DMZ) leads to a lattice mismatch of gallium indium phosphide ($Ga_xIn_{1-x}P$) and aluminum gallium indium phosphide (AlGaInP) [26]. This problem may be caused by the parasitic gas-phase reactions occurring between diethylzinc (DEZ), trimethylindium (TMIn), and phosphine (PH_3). The growth rate of the material and the molar ratio of gallium to indium in the material can be significantly affected by this parasitic reaction. The mismatch in the ratio of gallium to indium makes it difficult to achieve a lattice match between the two materials, further affecting the morphology of the material. Moreover, the doping of gallium (Ga) is also inhibited by the high diethylzinc (DEZ) flow to some extent.

The diffusion of zinc during the growth of the epitaxial layer may lead to the degradation of solar cell performance [67]. The substrate, the back surface, and the tunnel junction layer can store the zinc dopant and diffuse to the base region in the n-p-structured cells. This diffusion is largely affected by the point defects introduced during the growth of the n-type material. Zinc diffusion can be increased by reducing the doping level of n-type or p-type materials, such as increasing the diffusion barrier or using selenium (Sn) instead of silicon (Si) for n-type doping [67]. Particularly, the effect of the coating and cooling atmosphere on the hole concentration in zinc (Zn) doped $(Ga_{1-x}Al_x)_{0.5}In_{0.5}P(x = 0.7)$ was investigated by Minagawa and his co-workers [68]. The concentration of holes in the material can be reduced by cooling the temperature in hydrogen (H_2) containing arsine (AsH_3) and phosphine (PH_3). The hydrogen radicals generated by the hydride decomposition of the V group element can diffuse to the epitaxial layer and activate the zinc acceptor. Besides, the internal diffusion of hydrogen radicals can be hindered by the cover layer and promoted by the material below the cover layer, leading to a problem in p-n-structured cells [66].

Similarly, magnesium (Mg) can also be used for p-type doping of gallium indium phosphide ($Ga_xIn_{1-x}P$). The commonly used magnesium source is cyclopentadienyl magnesium, and there have been certain results [64, 68–71]. Maintaining a high hole concentration in the AlGaInP and AlInP materials contributes to an improvement of the cell performance. However, the doping rate of magnesium decreases as the temperature decreases; therefore, a high material growth temperature is more favorable for the doping of magnesium, being an advantage for AlGaInP. However, the growth rate of the material increases rapidly as the temperature increases, resulting in the higher difficulty of tunnel junction preparation and an unsuitable GaAs/Ge interface. For GaInP, the use of magnesium doping is no better than using zinc doping, and the use of magnesium doping also produces a memory effect [69]. Moreover, good zinc-doped aluminum indium phosphide (AlInP) can be obtained by the careful selection of the zinc source and the improved system cleanliness [72].

The emission surface window is used to reduce the surface states on the emitting surface, which are minority traps. The surface state is generally characterized by the surface recombination rate S. The value of the un-activated gallium indium phosphide (GaInP) emissive layer S is as high as 10^7 cm/s while the value of the high-quality AlInP/GaInP interfaces can be as low as 10^3 cm/s. The photo-response of the gallium indium phosphide (GaInP) solar cell is reduced by the high surface recombination rate; this reduction in this response is mainly concentrated in the blue light range. A valid window layer material of n-p-structured cell should have the following characteristics:

(1) The lattice constant is similar to gallium indium phosphide;
(2) The energy gap E_g is greater than that of the emission layer material;
(3) Compared with the material of the emissive layer, a larger energy band shift os needed to form a barrier to block the minority carriers;
(4) The relatively higher hole concentration (the order of hole concentration should exceed $10^{18}/cm^3$);
(5) It should have the property of easy preparation of low interfacial recombination rate.

Aluminum indium phosphide (AlInP) has most of the above characteristics. $Al_xIn_{1-x}P$ has the same lattice constant as gallium arsenide (GaAs) when $x = 0.532$. The indirect bandgap is 2.34 eV, which is 0.4–0.5 eV higher than that of gallium indium phosphide (GaInP). The bandgap arrangement of the $Al_xIn_{1-x}P$/GaInP alloy tends to i-type; $\Delta E_c \sim 0.75E_g$, $\Delta E_v \sim 0.25E_g$ [73]. This means that the holes in the emitter of the n-p-structured cell are well-limited by the structure. At photon energy of 3.5 eV, the internal quantum efficiency of a gallium indium phosphide (GaInP) cell with a good $Al_xIn_{1-x}P$ window layer exceeds 40%. However, aluminum (Al) in $Al_xIn_{1-x}P$ is easily bonded to oxygen (O) atoms, while oxygen (O) atoms become deep donor impurities in $Al_xIn_{1-x}P$. Thus, the quality of the $Al_xIn_{1-x}P$ material will decrease if the reaction chamber or raw material is contaminated with water vapor or other oxygenates. The low-quality $Al_xIn_{1-x}P$ material reduces the photo-response of

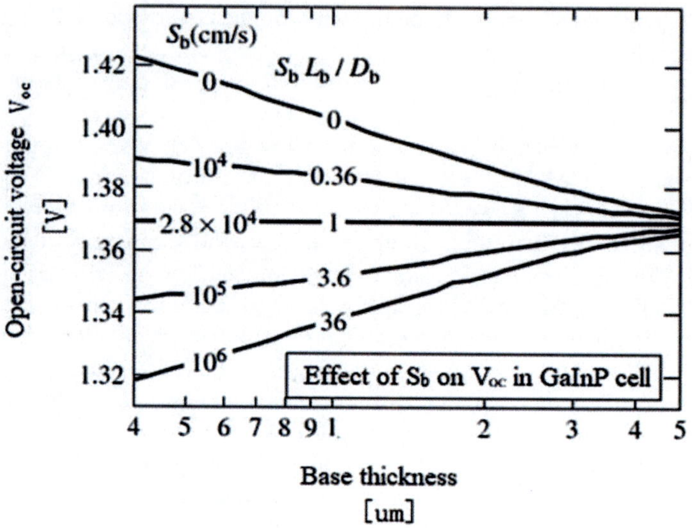

Fig. 4.19 The effect of surface recombination rate on open-circuit voltage V_{OC}

the gallium indium phosphide (GaInP) cell in the blue light range and also reduces the fill factor FF by affecting the contact resistance [74].

The back surface barrier layer is used to reduce the surface state density of the top cell and interconnect tunnel junction interfaces and the out-diffusion of dopants in interconnected tunnel junctions in some cases [75]. The high surface recombination rate at this interface not only affects the photo-response (especially the photo-response in the red range), but also the open-circuit voltage V_{OC}. The effect of surface recombination rate on the open-circuit voltage V_{OC} is illustrated in Fig. 4.19.

It is worth noting that there is an obvious effect of the surface recombination rate on the open-circuit voltage V_{OC}, which is also affected by the minority diffusion length and the base thickness. Similar to the surface window layer, the back surface barrier layer for the n-p-structured cell should have the following characteristics:

(1) The lattice constant is similar to gallium indium phosphide;
(2) The energy gap E_g is greater than that of the emission layer material;
(3) The relatively high hole concentration (the order of hole concentration should exceed $10^{18}/cm^3$);
(4) The good minority transport performance;
(5) The good transparency for photons to reach the underlying gallium arsenide cell.

The earliest calculation of the back surface was performed by Friedman et al. [76]. This calculation indicates that the gallium indium arsenide (GaInP) with irregular or high bandgap width is the best back surface of the low-bandgap gallium indium arsenide (GaInP) cell compared to an aluminum gallium indium phosphide (AlGaInP). This is due to the oxygen contamination in AlGaInP. As a deep donor, oxygen is an essential complication in the application of p-type aluminum gallium indium phosphide (AlGaInP). Other researchers have revealed that the strained, gallium-rich gallium indium arsenide ($Ga_xIn_{1-x}P$) as a back surface material performs better than disordered lattice-matched gallium indium phosphide ($Ga_xIn_{1-x}P$) and aluminum gallium arsenide (AlGaAs) [77]. However, recent reports suggest that aluminum gallium indium phosphide (AlGaInP) [13] or aluminum indium phosphide (AlInP) [78], is used in the commercial laminated cells with the best performance. There are also many reports on the growth methods and quality evaluation of zinc (Zn) doped aluminum gallium indium phosphide (AlGaInP) [71, 73, 79].

Since the bandgap E_g of gallium indium phosphide ($Ga_xIn_{1-x}P$) changes dramatically with growth conditions, the material quality of gallium indium phosphide ($Ga_xIn_{1-x}P$) characterized only by the power conversion efficiency of gallium indium phosphide ($Ga_xIn_{1-x}P$) solar cells is meaningless. With the increase in the bandgap E_g, the open-circuit voltage V_{OC} should also increase, while the short-circuit current density J_{SC} and the power conversion efficiency PCE should decrease. Simultaneously, the thickness of the gallium indium phosphide ($Ga_xIn_{1-x}P$) material in many optimized tandem structures is very thin and almost optically transparent, leading to low power conversion efficiency. Therefore, the thickness of the material and the bandgap are two parameters that must be considered when the power conversion efficiencies of two single-junction solar cells are compared. Generally, it is useful to describe the behavior of a cell with the open-circuit voltage V_{OC} and the energy gap E_g.

4.4.2 GaAs Solar Cells

Although the lattice constant of epitaxial gallium arsenide (GaAs) matches the lattice constant of the germanium (Ge) substrate, the qualities of the gallium arsenide epitaxial layers on the germanium substrates are not the same. The standard for measuring the quality of heteroepitaxial gallium arsenide (GaAs) materials is the power conversion efficiency of gallium arsenide (GaAs) cells and gallium indium arsenide (GaInP) cells. Generally, the surface of a good epitaxial layer is close to the mirror surface, while pits, hillocks, or slips are relatively few. For a mirror epitaxial gallium arsenide (GaAs) surface, faint "cross" patterns caused by a dislocation array of gallium arsenide and germanium interfaces can be observed. Sometimes, the absence of the cross pattern means that the screw dislocations can alleviate the lattice mismatch, while the transport properties of the minority carriers in gallium arsenide (GaAs) cells and gallium indium arsenide (GaInP) cells would be affected by the higher screw dislocation density. Therefore, it is necessary to avoid high screw dislocation density.

The epitaxial gallium indium phosphide ($Ga_xIn_{1-x}P$) material morphology is more sensitive to the quality of the original gallium arsenide (GaAs) material. Tiny surface defects of gallium arsenide (GaAs) caused by the different adsorption of gallium and indium on differently oriented surfaces are modified by gallium phosphide indium ($Ga_xIn_{1-x}P$) materials.

Gallium arsenide that is almost lattice-matched to the ruthenium substrate can be obtained by adding about 1% of indium (In). Simultaneously, the addition of indium (In) also eliminates the "cross" pattern on the surface of a well heteroepitaxial GaAs material but increases the difficulty of heteroepitaxial growth. It is reported that the performance of $Ga_{0.99}In_{0.01}As$ solar cells is better than that of gallium arsenide (GaAs) solar cells on germanium (Ge) substrate when all batteries have the best results [80].

For the detailed optical model parameters of gallium arsenide (GaAs), the various optical parameters of gallium arsenide (GaAs) researched by Aspnes and his collaborators can be referred [81]; the dielectric function model for gallium arsenide (GaAs) and aluminum gallium arsenide ($Al_xGa_{1-x}As$) was proposed [82].

Indium gallium arsenide ($Ga_xIn_{1-x}P$) and aluminum indium phosphide ($Al_xIn_{1-x}P$) are well suited as window and back surface electric field materials for gallium arsenide (GaAs) solar cells [83, 84]. Both gallium indium phosphide ($Ga_xIn_{1-x}P$) and aluminum indium phosphide ($Al_xIn_{1-x}P$) have the same i-type band arrangement as gallium arsenide (GaAs) and have suitable conduction band and valence band offset relative to gallium arsenide (GaAs). Ideally, aluminum indium phosphide ($Al_xIn_{1-x}P$) is more suitable as a window layer compared to indium gallium phosphide ($Ga_xIn_{1-x}P$) because the indium phosphide ($Al_xIn_{1-x}P$) has a larger bandgap. However, aluminum indium phosphide ($Al_xIn_{1-x}P$) can hardly form a good interface with gallium arsenide (GaAs) like gallium indium phosphide ($Ga_xIn_{1-x}P$) due to the sensitivity to oxygen pollution of aluminum indium phosphide ($Al_xIn_{1-x}P$) (forming a good AlGaAs/GaAs interface is a major problem in current single-junction gallium arsenide (GaAs) solar cells) [84]. The undoped $Ga_xIn_{1-x}P$/GaAs interface has the lowest recombination rate ($S < 1.5$ cm/s) among all known heterojunction structures including Si/SiO2 heterojunction [83]. Furthermore, it will be difficult to achieve p-type doping of aluminum indium phosphide ($Al_xIn_{1-x}P$) when the doping hole concentration $p > 1 \times 10^{18}/cm^3$. Therefore, in a GaInP/GaAs tandem cell structure, gallium indium phosphide ($Ga_xIn_{1-x}P$) is often used as a window layer and a back surface electric field layer in a gallium arsenide (GaAs) cell.

4.4.3 Ge Solar Cells

Germanium (Ge) has excellent optical and electrical properties [26], as well as a diamond structure and a lattice constant similar to that of gallium arsenide (GaAs). It is often used as the substrate for growing gallium arsenide (GaAs) due to its better mechanical properties compared to gallium arsenide (GaAs). The bandgap of germanium (Ge) is 0.67 eV, which can form current matching with the gallium arsenide (GaAs) top cell [16]. Besides, it is also the best choice for the bottom cell in a four-layer tandem cell [86]. However, some properties of germanium (Ge) make it have some disadvantages:

(1) The open-circuit voltage V_{OC} is limited by the indirect bandgap structure of germanium (Ge) and is generally around 300 mV; besides, V_{OC} is also sensitive to temperature [87].
(2) Relatively, the price of germanium (Ge) is high. In addition to being used in aerospace, it is difficult to become an ideal solar cell material.
(3) The germanium (Ge) element is an n-type impurity for gallium arsenide (GaAs) and gallium indium arsenide (GaInP). Therefore, germanium (Ge) in gallium indium phosphide (GaInP) has amphoteric behavior when the compensation ratio is $N_a/N_d = 0.4$ [88]; this may be related to the deep acceptor state of the germanium (Ge) element [89].
(4) Gallium (Ga), arsenic (As), phosphorus (P), and indium (In) are shallow-level impurities relative to germanium (Ge). Therefore, the control of junction formation is complicated by the combination of heteroepitaxial epitaxy.

The diffusion of group III and V elements in the germanium (Ge) sub-cell is very common. In fact, the diffusion of Group III and Group V elements to germanium (Ge) substrates is inevitable due to the high-temperature required for III-V epitaxial layers and heteroepitaxial epitaxy. Obtaining the germanium (Ge) sub-cell with good photovoltaic characteristics and a defect-free gallium arsenide (GaAs) epitaxial layer with requested conductivity type and conductance level at the same time by controlling the growth process has been always the major challenge to manufacture GaInP/GaAs/Ge tandem cell.

Since the thermal diffusivity is positively correlated with temperature, the doping elements are generally less active with a more stable heterojunction at lower growth temperatures.

Tobin et al. [90] proposed that at 700 °C, the diffusion coefficient of arsenic (As) in germanium (Ge) is higher than that of gallium (Ga) while the solid solubility of gallium (Ga) in germanium (Ge) is higher than that of arsenic (As).

For a triple-junction gallium indium arsenide/gallium arsenide/germanium (GaInP/GaAs/Ge) tandem cell, the only germanium (Ge) sub-cell parameter associated with the overall performance of the cell is the open-circuit voltage V_{OC} because the current density J_{SC} of the Ge sub-cell is much higher than that of gallium indium phosphide (GaInP) or gallium arsenide (GaAs) sub-cells.

The currently open-circuit voltage V_{OC} of the germanium (Ge) sub-cell is reported as 0.239 V; V_{OC} is very sensitive to process conditions, especially the III-V material germanium (Ge) interface and its formation [86].

Arsenic (AsH_3) has an etching effect on germanium (Ge); the etching rate increases as the temperature increases and the partial pressure of arsine (AsH_3) increases. The etched germanium (Ge) surface is very rough under the microscope [86]. Therefore, it should be avoided to expose the germanium (Ge) to arsine (AsH_3) for a long time.

The etch rate of phosphine (PH_3) is relatively low, and the exposure of germanium (Ge) to phosphine (PH_3) has little effect on surface roughness [90]. The diffusion coefficient of phosphorus (P) at 600 °C is two orders of magnitude lower than that of arsenic (As) [85]. Thus, phosphine (PH_3) is a better group V n-type dopant compared to arsine (AsH_3).

There are various reports of gallium arsenide (GaAs) growing on a (100) crystal grain germanium (Ge) substrate, and mirror morphology and low stacking fault density have been obtained. However, these reports have some contradictions. For example, Pelosi et al. [92], revealed that when the ratio of the V group source to the III group source is close to 1, a high-quality gallium arsenide(GaAs) epitaxial layer can be obtained under a middling growth rate ($R_g \approx 3.5$ μm/h) and a low growth temperature ($T_g = 600$ °C). From another perspective, Li et al. [93], suggested that a better gallium arsenide (GaAs) epitaxial layer can be acquired by the high V group source to III source ratio, low growth rate R_g, and high growth temperature T_g. Besides, Chen et al. [94], prepared the gallium arsenide (GaAs) epitaxial layer with a "good" surface only when the growth temperature R_g was between 600 °C and 630 °C.

The reason for this difference is still unclear, probably due to the different reaction chamber designs, the purity of the raw materials, or the purity of the germanium (Ge) substrate. Besides, some researchers [91], believed that this phenomenon is related to the pre-nucleation state or the surface state of germanium (Ge) before the nucleation of gallium arsenide (GaAs).

There are various reports about the structure of (100) crystal orientation germanium (Ge) [86–91], while the surface in most reports was obtained under a high vacuum (UHV) or molecular beam epitaxy (MBE) environment. However, it has been reported that the surface states of arsenic (Ge) treated by arsine (AsH_3) in the metal–organic chemical meteorological deposition (MOCVD) reaction chamber are not the same in most cases. Similar to the surface of (100) arsenic gallium (GaAs), arsenic (As) aggregates on the (Ge) surface terrace of the (100) crystal orientation [91, 95, 96].

4.4.4 Tunnel Junction

The tunnel junction between gallium indium phosphide (GaInP) sub-cell and gallium arsenide (GaAs) sub-cell serves as a low resistance interconnection between the p-type back surface electric field layer of gallium indium phosphide (GaInP) sub-cell and the n-type window layer of gallium arsenide (GaAs) sub-cell. The PN junction will have a polarity or forward turn-on voltage opposite to the top or bottom cell if there is no interconnected tunnel junction. Under illumination, this forward turn-on voltage will reduce the photovoltage generated by the top cell. The tunnel junction is composed of heavily doped or degenerate p^{++} and n^{++} materials. The space charge regions formed by heavily doped p^{++} and n^{++} materials are very narrow with a typical value of 10 nm. The thermal current characteristics of the common PN junction under the forward bias are not obvious. The narrow space charge region of the tunnel junction causes the "short-circuit" current characteristic of the tunnel junction. Thus, below the critical peak current density J_p, the IV characteristics of the tunnel junction are similar to the resistance. The critical peak current density J_p is presented in exponential form

$$J_p \propto \exp\left(-\frac{E_g^{3/2}}{\sqrt{N^*}}\right) \tag{4.28}$$

where E_g is the energy gap and $N^* = N_A N_D/(N_A + N_D)$ is the effective doping concentration [97]. The critical peak current density J_p must be higher than the photogenerated current of the tandem cell. If $J_p < J_{SC}$, tunnel current tends to be a thermal emission current, and the voltage drop across the tunnel junction increases to the voltage drop of the normal PN junction.

The ideal tunnel junction for high power conversion efficiency solar cells should have no defect. Life limits and in-band defects often result in the extra current. The lower critical peak current density J_p can be covered by the high extra current, while the conductivity of the tunnel junction is quite low. From another perspective, the space charge region can be widened by high-density point defects or extended defects, leading to a decrease in the tunneling current. Besides, the quality of the tunnel junction and the material extended on the tunnel junction can be also degraded by defects. Thus, the tunnel junction in tandem solar cells should be defect-free.

The earliest reported high-efficiency double-junction gallium indium phosphide/gallium arsenide (GaInP/GaAs) tandem cell was an optically transparent gallium arsenide (GaAs) tunnel junction. High-quality tunnel junctions are achieved by doping carbon (C) and selenium (Se) in gallium arsenide (GaAs). Therefore, the tunnel junction is relatively stable in a hot environment; it can meet the temperature requirements of the top cell growth and also work under the concentration of 1000 suns. The tunnel junction with a thickness of 30 nm only blocks about 3% of the light expected to reach the bottom cell. For optically transparent and unannealed tunnel junctions, the critical peak current density J_p can reach 14 A/cm^2, while the extra current is close to zero [75].

4.5 Development of High-Efficiency Multi-junction Solar Cells

4.5.1 *Problems in High-Efficiency Multi-junction Solar Cells*

The current problems in multi-junction solar cells need to be revealed to understand the development of solar cells and further optimize the structure of solar cells in order to develop solar cells with better performance. The common problems in an n-p-structured solar cell are listed in Table 4.6.

Table 4.6 Common problems in multi-junction solar cells [98]

Problems	Phenomena	Verification
Thin window layer	Poor blue light response	QE model
High front surface recombination	Poor blue light response (lower V_{OC})	QE model
The diffusion length $L_{emitter}$ of the emitter region is smaller than the thickness d of the emitter region	Poor blue light response (lower V_{OC})	QE model
Heavy doping of the emitter region	Poor blue light response (lower V_{OC}, low block resistance of the emitter region)	Measure the doping of the emitter region
Sight doping of the emitter region	Lower V_{OC}, high sheet resistance of the emitter region	Depletion of the emitter region
Sight doping of the base region	lower V_{OC}	Measure the doping of the base region
Heavy doping of the base region	Lower V_{OC}, reduce red light response	The dark I-V curve: $n = 1$
The diffusion length L_{base} of the base region is smaller than the thickness of the base region	Lower V_{OC}, reduce red light response	The dark I-V curve: $n = 1$
High back surface recombination	Lower V_{OC}, reduce red light response	The dark I-V curve: $n = 1$
Line dislocation	Low V_{OC}, FF and J_{SC}	Larger dark current, in most cases $n = 2$
Too thin metallization	Low FF—series resistance	Measure grid line resistance
Poorer tunnel junction	Low FF—series resistance or low V_{OC}	Measuring tunnel junction
Extra junction	Low FF—(Difficult to form ohmic contact) series resistance	I-V curve is dependent on the spectrum
Poor metallization contact	Low FF—(Difficult to form ohmic contact) series resistance	Measure the transmission line resistance on the front surface and the two square resistances on the back surface
Resistive window layer	Low FF—series resistance	Measure transmission line resistance of the emitter
Granulation	Low FF—Leakage	Luminescence under the forward bias is related to surface topography
Incomplete step isolation	Low FF—Leakage, sometimes there is a higher J_{SC}	Etching step
The front metal layer is in contact with the lower layer	Low FF—Leakage	Microscopic observation

(continued)

Table 4.6 (continued)

Problems	Phenomena	Verification
Severe color deviation	Low *FF* under the spotlight	I-V curve deviates from the normal curve

The epitaxial layer needs to be characterized to further analyze the performance of the device. For alloys such as gallium indium phosphide (GaInP), the lattice constant can be determined by X-ray diffraction in a rocking mode [98]. The carrier concentration, the bandgap, and the minority carrier diffusion length in the epitaxial layer can be measured by the improved Polaron analyzer [99]. Besides, the minority carrier lifetime and surface recombination velocity in different thickness materials can be obtained by the double heterojunction time-varying photoelectric measurement [100, 101]. In addition to C-V and photocurrent (QE) measurements, transmission line measurement is also a means of characterizing contacts and is generally used for device analysis [102]. Moreover, various electrical and physical characterization methods are widely used in III-V solar cells [103, 104]. The application of these methods has enabled us to understand the material and device characteristics more deeply and to find related problems, playing an essential role in the development of III-V solar cells.

4.5.2 Development of High-Efficiency Multi-junction Solar Cells

The gallium phosphide indium/gallium arsenide/germanium (GaInP/GaAs/Ge) multi-junction cell has become more mature in space applications. The power conversion efficiency based on this structure is still increasing and can achieve 30% in the case of AM0 and the concentration of 500 suns. The power conversion efficiency of ground-concentrated GaInP/GaAs/Ge multi-junction cells prepared by Spectrolab is expected to be further improved [105, 106]. The theoretical conversion efficiency of a concentrating GaInP/GaAs/Ge multi-junction cell is 45% in the case of AM1.5 and the concentration of 500 suns [16]. The theoretical power conversion efficiency of multi-junction solar cells composed of III-V materials can reach as high as 80% [16].

According to reports, with the increase of the bandgap of gallium phosphide indium (GaInP), the power conversion efficiency of GaInP/GaAs/Ge multi-junction cells will be promoted under the case of AM0. However, the bandgap of gallium indium phosphide (GaInP) can be increased by the addition of aluminum (Al) in gallium indium phosphide (GaInP); it cannot improve the power conversion efficiency of the top cell because the short-circuit current J_{SC} is reduced by 10% and the increase in the open-circuit voltage V_{OC} is very small. Aluminum (Al) is easily contaminated by oxygen element, which affects the characteristics of minority carriers and thus reduces the performance of the device [75].

In GaInP/GaAs/Ge multi-junction cells, the germanium (Ge) sub-cell collects about twice as much current as of the other two cells. Theoretically, the power conversion efficiency of the multi-junction cell can be improved by reducing the bandgap of gallium arsenide (GaAs), increasing the bandgap of germanium (Ge), or adding a fourth cell between gallium arsenide (GaAs) and germanium (Ge) cells.

The most common method is to dope indium (In) into a gallium arsenide (GaAs) layer or a gallium phosphide indium (GaInP) layer. Without the buffer layer, the matching degree of the material and the germanium (Ge) can be improved by adding indium (In), contributing to the improvement of power conversion efficiency [107]. Then, the method of doping a higher ratio of indium (In) (such as more than 12%) was also tested in order to achieve greater power conversion efficiency; the power conversion efficiency was increased by 2% [108]. Particularly, high-efficiency multi-junction solar cells require a good quality epitaxial layer. This epitaxial layer advantageously relieves strain at the interface and reduces the screw dislocations and other dislocations in the active layer of the solar cell. It is exciting that the experiment meeting these demands has been successful while the performance is still similar to that of cells with lattice matching [108–111]. Researchers from Varian tried to prepare a lattice-mismatched material on the back of the wafer and then prepare a material free of lattice mismatch on the "sunny side" [112, 113]. However, the impact of lattice mismatch on the mechanical properties and lifetime of solar cells remains unclear.

Simultaneously, a method of adding a fourth cell with a bandgap of 1 eV between gallium arsenide (GaAs) and germanium (Ge) cells has also been widely concerned. The power conversion efficiency of this four-junction solar cell can exceed 50% theoretically [16], and reach 40% in practice with the right material. Unfortunately, it is still very difficult to obtain materials with a bandgap of around 1 eV and lattice matching with gallium arsenide (GaAs) [114]. At present, although some materials have been developed, the most suitable material is indium gallium arsenide nitrogen (GaInAsN). Zinc germanium arsenide ($ZnGeAs_2$) is difficult to grow (especially at low pressures) and can also cause cross-contamination (for example, zinc can affect the subsequent growth). Thallium gallium phosphide (Ga0.5Tl0.5P) has been reported to have a lattice constant similar to gallium arsenide (GaAs) and a bandgap of 0.9 eV. However, the original experiment cannot be repeated [115]. Indium gallium boron arsenide (BGaInAs) is lattice-matched with gallium arsenide (GaAs), and its bandgap is also regulated below 1.35 eV while the bandgap is still higher than 1 eV so far [116]. Simultaneously, the material quality of Indium gallium boron arsenide (BGaInAs) is not good, leading to a decrease in photovoltage and photocurrent of solar cells [117].

When $x = 3y$, Gallium indium arsenide nitrogen ($Ga_{1-x}In_xAs_{1-y}N_y$) can be obtained, which is lattice-matched with gallium arsenide (GaAs) and has the bandgap of around 1 eV [118]. However, the minority carrier diffusion length of this material is small [119–121]. With about 3% nitrogen (N) being doped in gallium arsenide nitrogen (GaAsN), the bandgap was reduced from 1.4 to 1 eV. Scattering in this alloy is more severe than conventional alloys; this can explain, to some extent, why the mobility of majority carriers decreases. However, the more serious problem is the relatively short life of the minority, which has not been well-explained so far. Compared with GaInAsN obtained by molecular beam epitaxy (MBE), GaInAsN obtained by metal chemical vapor deposition (MOCVD) is contaminated seriously by carbon (C) and hydrogen (H) elements [122].

Mechanical stacking is another method to obtain a high-efficiency multi-junction solar cell. The biggest benefit of this method is that the problem of lattice mismatch cannot be considered. Currently, GaInP/GaAs, GaInAsP(1 eV)/GaInAs(0.75 eV), and gallium antimonide (GaSb) solar cells are the most suitable cells for mechanical stacking [123, 124]. The difficulty in completing this mechanical stacking is that the top cell must be transparent to the light absorbed by the bottom cell (using a transparent gallium arsenide substrate, using unconventional back contact, and adding an anti-reflective layer on both sides of the top cell), and the good heat dissipation and electrical isolation must be achieved at the same time. These problems become especially serious when using high power concentrating systems. The advantage of this method is that the photocurrent separation of the two junctions (when using four-ended measurement) makes the material selection more flexible and higher power conversion efficiency can be achieved when the spectrum changes [16].

Since the performance of a solar cell depends on its work environment, it is difficult to predict the performance of the solar cell outdoors. Moreover, the performance prediction of series multi-junction solar cells is more difficult under varying spectra. The predicted energy loss of the two-ended three-junction and four-junction cells is much higher than that of the six-ended or eight-ended structure while the energy loss at any temperature is lower than that of the silicon solar cell [54]. Mechanical stacking is the easiest for two-ended cells, just by bonding the two semiconductor materials together. The reflection losses and the use of transparent substrates are avoided in the bonding process because wafer bonding is a common process in the semiconductor industry, reducing the difficulty of heat dissipation and electrical isolation. Besides, the bonding process can also reduce substrate costs if the overhead of reusing the substrate continues to decrease.

There are many processes for improving the performance of multi-junction solar cells. However, most solar cells are based on the structure of Fig. 4.4. The weight of the cell can be reduced (this is very advantageous for space applications) by bonding III-V solar cells to a silicon (Si) substrate using the bonding process. If the substrate can be reused, the cost of the solar cell will be further reduced. Moreover, a method of bonding gallium arsenide (GaAs) and silicon (Si) and forming an ohmic contact on a silicon substrate has been reported [125]. Wafer bonding has not been applied to large-scale solar cell production, and wafer bonding of commercially available 8-inch silicon-on-insulator substrates has been achieved. Although the current price of this type of substrate is still relatively high, the price is expected to further decrease in the future with the improvement and development of the process.

The growth of group III-V materials on silicon (Si) substrates is also a method of bonding. The biggest problem with epitaxial gallium arsenide (GaAs) on silicon (Si) substrates is the severe lattice mismatch between the two materials. However, the growth of lattice-matched gallium arsenide (GaAs) on a silicon (Si) substrate is similar to epitaxial gallium arsenide (GaAs) on a germanium (Ge) substrate. Tu and his collaborators prepared aluminum-phosphorus gallium nitride (AlGaNP) lattice-matched with silicon (Si) on a silicon (Si) substrate; the material had a bandgap of 1.4–1.95 eV. The use of silicon (Si) as a plug-in cell material between high-efficiency tandem cells can reduce the weight of the cell while the photocurrent of the silicon cell in the red light range is not ideal. Although high power conversion efficiency is not achieved, the low cost and lightweight still make silicon (Si) be an attractive option.

Consequently, an indium phosphide/indium gallium arsenide (InP/GaInAs) double-junction solar cell can be prepared on an indium phosphide (InP) substrate to further improve the power conversion efficiency of the solar cell [126]. The three-junction or four-junction solar cell based on indium phosphide (InP) can achieve higher power conversion efficiency.

At present, gallium indium arsenide/gallium arsenide/germanium (GaInP/GaAs/Ge) multi-junction cells occupy a large share of space solar cells; a few improvements are only required to make them suitable for ground applications. First, the series resistance must be reduced in the case of concentrating light, as mentioned earlier. Second, the AM1.5 direct spectrum contains less blue light than AM0; hence, the gallium indium phosphide (GaInP) cell should be thinner to produce a photogenerated current similar to that of a gallium arsenide (GaAs) cell to achieve the current matching between the two cells.

Generally, the lifetime of GaInP/GaAs/Ge multi-junction cell is not a problem. However, packaging and the use of complex structures (such as tunnel junctions, in which high defect density is caused by using germanium substrates) in practical applications bring challenges for the preparation of high-reliability devices; this is also the focus of future research on GaInP/GaAs/Ge multi-junction cells.

References

1. B. Dai, H. Zheng, in *Solar Cells Technical Manuals*, Chapter 7 (Posts & Telecom Press, Beijing, 2012), p. 153
2. A.W. Bett et al., Appl. Phys. A **69**, 119–129 (1999)
3. T.V. Torchynska et al., Semiconductor physics. Quantum Electron & Optoelectron **5**(1), 63–70 (2002)
4. R.R. King et al., in *IEEE 4th World Conference on Photovoltaic Energy Conversion*, vol. 2 (2006)
5. A. Naser et al., in *ISECS International Colloquium on Computing Communication, Control, and Management*, vol. 1 (2009), pp. 373–378
6. A. Luque et al., in *Handbook of Photovoltaic Science and Engineering* (John Wiley & Sons Ltd, 2003)
7. R. Swanson, Prog. Photovolt. Res. Appl. **8**, 93–111 (2000)
8. C. Henry et al., J. Appl. Phys. **51**, 4494–4500 (1980)
9. J. Nelson, The physics of solar cells (Properties of semiconductor materials), Imperial College (2003)
10. A. Mart et al., Sol. Energy Mater. Sol. Cells **43**, 203–222 (1996)

11. B. Dai, H. Zheng, in *Solar Cells Technical Manuals*, Chapter 4 (Posts & Telecom Press, Beijing, 2012), p. 133
12. M. Green, Prog. Photovolt. **9**, 137–144 (2001)
13. N. Karam et al., Sol. Energy Mater. Sol. Cells **66**, 453–466 (2001)
14. A. Brown et al., Prog. Photovolt. Res. Appl. **10**, 299–307 (2002)
15. I. Tobıas et al., Prog. Photovolt. Res. Appl. **10**, 323–329 (2002)
16. S. Kurtz et al., J. Appl. Phys. **68**, 1890 (1990)
17. M. Wanlass et al., in *Proceedings of 22nd IEEE Photovoltaic Specialists Conference* (1991), pp. 38–45
18. P. Faine et al., Sol. Cells **31**, 259 (1991)
19. L. Lockhart et al., Opt. Soc. Am. **37**, 689 (1947)
20. G. Bader et al., Appl. Opt. **34**, 1684–1691 (1995)
21. D. Friedman et al., in *Proceedings of 12th NREL Photovoltaic Program Review*, vol. 306 (1993), p. 521
22. E. Palik, A. Addamiano, in *Handbook of Optical Constants of Solids*, vol. I, ed. by E. Palik (Academic Press, 1998), pp. 597–619
23. T. Cotter, M. Thomas, W. Tropf, in *Handbook of Optical Constants of Solids*, vol. II, ed. by E. Palik (Academic Press, 1998), pp. 899–918
24. V. Andreev, V. Grilikhes, V. Rumyantsev, in *Photovoltaic Conversion of Concentrated Sunlight* (Wiley, 1997)
25. J. Gee, Sol. Cells **24**, 147–155 (1988)
26. O. Madelung, in *Semiconductors: Group IV Elements and III-V Compounds* (Springer, 1991)
27. J. Olson et al., in *Proceedings of 19th IEEE Photovoltaic Specialists Conference* (1987), p. 285
28. J. Matthews et al., J. Cryst. Growth **27**, 118 (1974)
29. I. Kudman et al., Appl. Phys. **43**, 3760–3762 (1972)
30. C. Wie, J. Appl. Phys. **66**, 985 (1989)
31. B. Tanner et al., Mater. Lett. **7**, 239 (1988)
32. A. Gomyo et al., J. Cryst. Growth **77**, 367–373 (1986)
33. A. Gomyo et al., Appl. Phys. Lett. **50**, 673 (1987)
34. M. Kondow et al., J. Cryst. Growth **93**, 412 (1988)
35. T. Kurimoto et al., Phys. Rev. B **40**, 3889 (1989)
36. R. Capaz et al., Phys. Rev. B **47**, 4044–4047 (1993)
37. Y. Zhang et al., Phys. Rev. Lett. **63**, 201312 (2000)
38. A. Mascarenhas et al., Phys. Rev. B **41**, 9947 (1990)
39. A. Mascarenhas et al., Phys. Rev. Lett. **63**, 2108 (1989)
40. J. Luo et al., J. Vac. Sci. Technol. B **12**, 2552–2557 (1994)
41. J. Luo et al., Phys. Rev. B **51**, 7603–7612 (1995)
42. D. Friedman et al., Proc. Mater. Res. Soc. Symp. **280**, 493 (1993)
43. L. Chernyak et al., Appl. Phys. Lett. **70**, 2425–2427 (1997)
44. D. Friedman et al., Appl. Phys. Lett. **63**, 1774–1776 (1993)
45. D. Friedman et al., Appl. Phys. Lett. **65**, 878–880 (1994)
46. M. Schubert et al., J. Appl. Phys. **77**, 3416–3419 (1995)
47. H. Kato et al., Jpn. J. Appl. Phys. **33**, 186–192 (1994)
48. S. Kurtz et al., in *Proceedings of 1st World Conference on PV Energy Conversion* (1994), p. 2108
49. T. Iwamoto et al., J. Cryst. Growth **68**, 27 (1984)
50. M. Ikeda, K. Kaneko, J. Appl. Phys. **66**, 5285 (1989)
51. A. Gomyo et al., Jpn. J. Appl. Phys. **28**, L1330–L1333 (1989)
52. S. Kurtz et al., J. Electron. Mater. **19**, 825–828 (1990)
53. J. Goral et al., J. Electron. Mater. **19**, 95 (1990)
54. S. Kurtz et al., J. Electron. Mater. **23**, 431–435 (1994)
55. S. Kurtz et al., Sol. Cells **30**, 501 (1991)
56. J. Olson et al., in *Proceedings of 18th IEEE Photovoltaic Specialists Conference* (1985), pp. 552–555

57. M. Watanabe et al., J. Appl. Phys. **60**, 1032 (1986)
58. H. Hotta et al., J. Cryst. Growth **93**, 618–623 (1988)
59. F. Scheffer et al., J. Cryst. Growth **124**, 475–482 (1992)
60. S. Minagawa et al., J. Cryst. Growth **152**, 251–255 (1995)
61. C. Wang et al., Jpn. J. Appl. Phys. **34**, L1107–L1109 (1995)
62. L. Malacky et al., Appl. Phys. Lett. **69**, 1731–1733 (1996)
63. M. Suzuki et al., J. Cryst. Growth **115**, 498–503 (1991)
64. S. Kurtz et al., in *Proceedings of the InP and Related Materials Conference* (1992)
65. T. Suzuki et al., Jpn. J. Appl. Phys. **27**, L1549–L1552 (1988)
66. Y. Nishikawa et al., J. Cryst. Growth **100**, 63–67 (1990)
67. S. Kurtz et al., in *Proceedings of 25th IEEE Photovoltaic Specialists Conference* (1996), pp. 37–42
68. S. Minagawa et al., J. Cryst. Growth **118**, 425–429 (1992)
69. I. Hino et al., Inst. Phys. Conf. Ser. **79**, 151–156 (1985)
70. M. Kondo et al., J. Cryst. Growth **141**, 1–10 (1994)
71. G. Bauhuis et al., J. Cryst. Growth **191**, 313–318 (1998)
72. S. Stockman et al., J. Electron. Mater. **28**, 916–925 (1999)
73. K. Bertness et al., J. Cryst. Growth **196**, 13–22 (1999)
74. Y. Ishitani et al., J. Appl. Phys. **80**, 4592–4598 (1996)
75. K. Bertness et al., Appl. Phys. Lett. **65**, 989–991 (1994)
76. D. Friedman et al., in *Proceedings of 22nd IEEE Photovoltaic Specialists Conference* (1991), pp. 358–360
77. N. Rafat et al., in *Proceedings of First World Conference on Photovoltaic Energy Conversion* (1994), pp. 1906–1909
78. P. Chiang et al., in *Proceedings of 28th IEEE Photovoltaic Specialists Conference* (2000), p. 1002
79. K. Kadoiwa et al., J. Cryst. Growth **145**, 147–152 (1994)
80. T. Takamoto et al., in *Proceedings of 28th IEEE Photovoltaic Specialists Conference* (2000), p. 976
81. D. Aspnes, A. Studna, Phys. Rev. B **27**, 985–1009 (1983)
82. C. Kim et al., Phys. Rev. B **47**, 1876–1888 (1993)
83. J. Olson et al., Appl. Phys. Lett. **55**, 1208 (1989)
84. S. Kurtz et al., in *Proceedings of 20th First IEEE Photovoltaic Specialists Conference* (1990), pp. 138–140
85. H. Hovel, *Solar Cells*, vol. 11 (Academic Press, New York, 1975)
86. S. Kurtz et al., in *Proceedings 26th IEEE Photovoltaic Specialists Conference* (1997), pp. 875–878
87. D. Friedman et al., in *Proceedings of 28th IEEE Photovoltaic Specialists Conference* (2000), p. 965
88. J. Lee et al., Appl. Phys. Lett. **62**, 1620–1622 (1993)
89. I. Yoon et al., J. Phys. Chem. Solids **62**, 607–611 (2001)
90. S. Tobin et al., in *Proceedings of 20th IEEE Photovoltaic Specialist Conference* (1988), pp. 405–410
91. J. Olson et al., in *Proceedings of 2nd World Conference on Photovoltaic Energy Conversion* (1998)
92. C. Pelosi et al., J. Electron. Mater. **24**, 1723–1730 (1995)
93. Y. Li et al., J. Cryst. Growth **163**, 195–202 (1996)
94. J. Chen et al., J. Electron. Mater. **21**, 347–353 (1992)
95. W. McMahon et al., Phys. Rev. B Condens. Matter **60**, 2480–2487 (1999)
96. W. McMahon et al., Phys. Rev. B Condens. Matter **60**, 15999–16005 (1999)
97. S. Sze, in *Physics of Semiconductor Devices* (Wiley, 1981)
98. O. Peterson et al., Phys. Rev. **150**, 703 (1966)
99. S. Kurtz et al., in *Proceedings of 19th IEEE Photovoltaic Specialists Conference* (1987), pp. 823–826

100. P. Blood, Semicond. Sci. Technol. **1**, 7–27 (1986)
101. R. Ahrenkiel, Solid State Electron. **35**, 239–250 (1992)
102. R. Ahrenkiel, in *Minority Carriers in III-V Semiconductors: Physics and Applications*, vol. 39, ed. by R. Ahrenkiel, M. Lundstrom (Academic Press, 1993), pp. 39–150
103. H. Berger, J. Electrochem. Soc. **119**, 507–514 (1972)
104. S. Kurtz et al., in *Proceedings of First World Conference on Photovoltaic Energy Conversion* (1994), pp. 1733–1737
105. R. King et al., in *Proceedings of 28th IEEE Photovoltaic Specialists Conference* (2000), p. 998
106. R. King et al., in *Proceedings of 29th IEEE Photovoltaic Specialists Conference* (2002) pp. 776–779
107. H. Cotal et al., in *Proceedings 28th IEEE Photovoltaic Specialists Conference* (2000), p. 955
108. T. Takamoto et al., Sol. Energy Mater. Sol. Cells **66**, 511–516 (2001)
109. R. King et al., in *Proceedings of 28th IEEE Photovoltaic Specialists Conference* (2000), p. 982
110. A. Bett et al., in *Proceedings of 28th IEEE Photovoltaic Specialists Conference* (2000), p. 961
111. F. Dimroth et al., J. Electron. Mater. **29**, 42–46 (2000)
112. S. Sinharoy et al., in *Proceedings of 28th IEEE Photovoltaic Specialists Conference* (2000), p. 1285
113. L. Partain et al., in *21st IEEE Photovoltaic Specialists Conference* (1990), pp. 184–189
114. K. Volz, J. Solar Energy Eng. **129**, 269 (2007)
115. D. Friedman et al., in *Proceedings of NREL/SNL PV Program Review Meeting* (1998), pp. 401–405
116. M. Antonell et al., in *Proceedings of InP and Related Materials Conference* (1997), pp. 444–447
117. J. Geisz et al., Appl. Phys. Lett. **76**, 1443–1445 (2000)
118. J. Geisz et al., in *Proceedings of 28th IEEE Photovoltaic Specialists Conference* (2000), p. 990
119. M. Kondow et al., Jpn. J. Appl. Phys. **35**, 1273–1275 (1996)
120. J. Geisz et al., J. Cryst. Growth **195**, 401–408 (1998)
121. S. Kurtz et al., Appl. Phys. Lett. **74**, 729–731 (1999)
122. D. Friedman et al., J. Cryst. Growth **195**, 409–415 (1998)
123. T. Miyamoto et al., J. Cryst. Growth **209**, 339–344 (2000)
124. A. Moto et al., Sol. Energy Mater. Sol. Cells **66**, 585–592 (2001)
125. L. Fraas et al., in *Proceedings of 28th IEEE Photovoltaic Specialists Conference* (2000),p. 1150
126. Y. Hong et al., in *Photovoltaic for the 21st Century, Proceedings of 199th Meeting Electrochemical Society, 2001–10* (2001), pp. 415–422

Chapter 5
Thin-Film Solar Cells Based on Amorphous Silicon

5.1 Introduction of Solar Cells Based on Amorphous Silicon

Several square meters of solar cells often need to generate hundreds of watts of electrical power because of the diffusivity of sunlight. Therefore, it is necessary to develop large-area thin-film solar cells to reduce costs.

Thin-film solar cell based on amorphous silicon is an essential component of the thin-film solar cell family, including thin-film solar cells based on amorphous silicon (a-Si), microcrystalline silicon, and nanocrystalline silicon. Traditionally, hydrogenated microcrystalline silicon (uc-Si:H) or abbreviated microcrystalline silicon (uc-Si) is classified as thin-film materials based on amorphous silicon, which has its historical origins. In the early 1980s, it was found in the process of preparing amorphous silicon by decomposing silane using glow discharge that a film with much higher conductivity could be prepared by appropriately increasing the hydrogen dilution and the plasma power of silane. Since the electron diffraction spectrum exhibits some ring characteristics of crystals, it was called microcrystalline silicon. It is formed by self-embedding silicon grains with sizes ranging from several nanometers to tens of nanometers into a hydrogenated amorphous silicon substrate. The term nanocrystalline silicon (nc-Si) emerged in the mid-1980s and was originally referred to as polycrystalline silicon or microcrystalline silicon films deposited by chemical transport from hydrogen plasma. Afterward, this silicon film is called nanocrystalline silicon to conform to the definition of nanomaterials in the field of metals and ceramics (where the nanoscale size covers the entire range of 1–100 nm). With the rise of nanotechnology, the concept of "nanocrystalline silicon" has become more and more widely used. Therefore, there has been a tendency worldwide in recent years to regard microcrystalline silicon as nanocrystalline silicon. In fact, no strict distinction between nanocrystalline silicon and microcrystalline silicon is found in most cases. The two names refer to the same structure of thin films based on silicon. Hydrogenated nanocrystalline silicon or abbreviated nanocrystalline silicon is a silicon-based film with a grain size of only a few nanometers only on certain occasions. At that moment, the grain size of silicon can be compared with the de

C. Zhang et al., *Semiconductor Photovoltaic Cells*,
https://doi.org/10.1007/978-981-15-9480-9_5

Broglie wavelength of the electron, suggesting the significant quantum size confinement effects and quantum transport phenomena [1], which are beyond the scope of this book.

Thin-film solar cells based on amorphous silicon have been widely concerned because of their advantages as follows.

(1) The material cost is low and the amount of silicon material used is small. The substrate material of the amorphous silicon film mainly includes glass, stainless steel, and plastic with a low price. Besides, the thickness of silicon can be very thin, only about 0.5 um. This is highly related to the optical absorption coefficient of amorphous silicon materials. Meanwhile, single crystalline silicon cells need a relatively thick thickness of about 200 um to fully absorb sunlight. Moreover, amorphous silicon cells are not required to be sliced like single crystalline silicon, and the material waste is very little.
(2) The manufacturing process is simple, and it can be mass-produced continuously, in a large area, and automatically. Compared with amorphous silicon, the manufacturing of crystalline silicon solar modules requires trivial processes such as screening and welding of solar cells with a lot of manpower input. It is difficult to control the quality in the manufacturing process and realize the automated mass production.
(3) The manufacturing process consumes less electricity. Energy has a short repayment period. Conventional amorphous silicon thin-film solar cells are prepared by a gas decomposition method, with the substrate temperature of only 200–300 °C. Thus, the consumption of electricity is much less compared with single crystalline silicon repeatedly melted at 1,412 °C or more. The crystalline silicon solar cells have an energy repayment time of 2–3 years while the amorphous silicon solar cells have an energy repayment time of only 1–1.5 years.
(4) The temperature coefficient is positive, that is, the photoelectric conversion efficiency of amorphous silicon cells will increase as the temperature rises. Its principle will be mentioned later in the introduction of the S-W effect. For crystalline silicon, the increase of temperature will reduce the mobility of the carrier, resulting in reducing the energy conversion efficiency.
(5) Weak-light performance is strong. Amorphous silicon cells still have good photoelectric conversion efficiency under low light due to the low-energy level of valence electron of amorphous silicon.
A comparison of the monthly average power generation of power plants based on amorphous silicon and single crystalline silicon cells in Thailand from 1998 to 1999 is illustrated in Fig. 5.1. It can be observed that the power generation of amorphous silicon is higher in 11 months of 12 months due to the advantages mentioned in (4) and (5). The sun is not very strong on rainy days, and the weak-light performance of amorphous silicon is better than that of crystalline silicon, leading to a higher power generation. However, the higher temperature in a sunny or hot summer will reduce the photoelectric conversion efficiency of crystalline silicon cells and increase the photoelectric conversion efficiency of

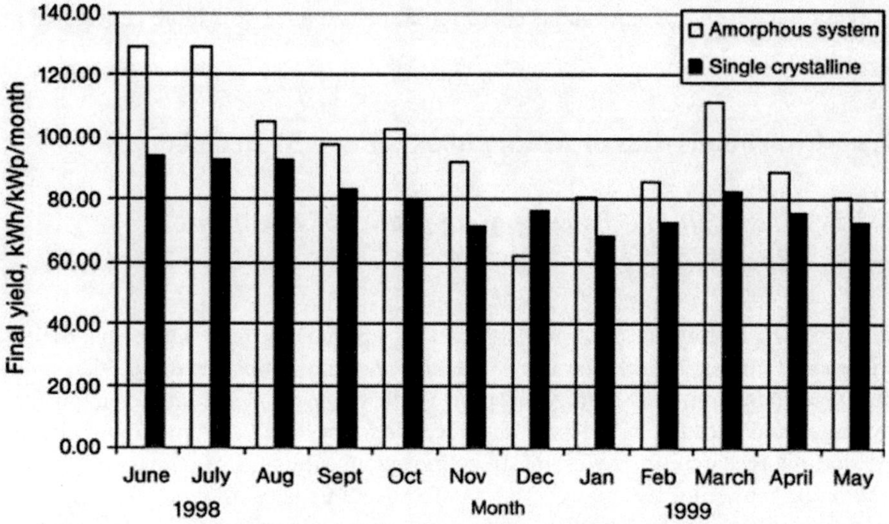

Fig. 5.1 Comparison of monthly average power generation of power plants based on amorphous silicon and single crystalline silicon cells in Thailand [2]

amorphous silicon cells. Generally, the power generation of amorphous silicon cells is still high.

(6) It is beautiful and generous, and easy to be integrated with the building. Amorphous silicon films can be deposited onto flexible substrates such as plastics, aluminum foil, and stainless-steel sheets. Solar cells on flexible substrates can be mounted on uneven surfaces of buildings; this is especially significant in the areas with less available space. Simultaneously, the cells deposited on the flexible substrate have lightweight and are suitable for the installation of buildings. There is no special load requirement for the design of buildings, making flexible silicon thin-film cells widely used in the integrated design of photovoltaic buildings. Besides, the color of modules is easily matched with the color of the building to beautify the indoor and outdoor environment when thin-film solar modules based on amorphous silicon is used as a roof and a wall surface; meanwhile, the fine and neat laser cutting line can make the building more beautiful, generous, and attractive.

(7) The hot spot effect is not obvious. Under certain conditions, the shaded solar modules in the series branches will be treated as loads, consuming the energy generated by other solar modules exposed to sunlight. Meanwhile, the obscured solar module will heat up. This phenomenon is known as the "hot spot effect", which could severely destroy the output power of the solar cell. Some or even all the energy produced by a working solar cell may be consumed by a shaded cell. It is inevitable that some modules are in the shadow when the area of the solar cell array is large. However, the hot spot effect is not obvious due to the small

current density of amorphous silicon solar cells; hence, it is more convenient to use and has better reliability [3].

5.2 Characteristics of Amorphous Silicon Materials

5.2.1 Research and Development Status of Amorphous Silicon Materials

The research history of amorphous silicon as a promising material for photoelectric energy conversion can be traced back to the late 1960s. In the British Standard Communication Laboratory, the glow discharge method was applied to prepare hydrogenated amorphous silicon films, revealing that it had certain doping effects. Before this research, the hydrogen-free amorphous silicon film prepared by evaporation or sputtering technology had a defect state density of about 10^{19} cm^{-3} or more with little application value. Nowadays, amorphous silicon film also refers to hydrogenated amorphous silicon film while amorphous silicon film containing no hydrogen has been rarely studied. In 1975, W. E. Spear et al. realized substitutional doping in a-Si:H materials and prepared a PN junction. They explored that hydrogen had a function of saturating dangling bonds inside amorphous silicon, and a-Si:H material had a low density of defect states (about 10^{16} cm^{-3}) and superior photosensitivity [4]. In 1976, D. E. Carlson of RCA (Radio Corporation of America) of the United States developed a p-i-n structured amorphous silicon solar cell with a photoelectric conversion efficiency of 2.4% [5]. In the context of the energy crisis and rising oil prices in the 1970s, various researches on thin-film solar cells based on amorphous silicon rapidly emerged worldwide. Meanwhile, amorphous silicon cells of various new materials and structures had sprung up and the efficiency had been increasing. The efficiency of amorphous silicon cells was increased to 8% in 1980 by Carlson, making it possible to industrialize. In 1997, the three-junction amorphous silicon cell made by Yang, Banerjee, and Guha [6] had an efficiency of 14.6%, a stable efficiency of 13.0% after illumination, and an effective area of 0.25 cm^2, as illustrated in Fig. 5.2.

Following the significant discovery that the characteristics of amorphous silicon can be improved by hydrogen doping, another important progress is the discovery that materials based on amorphous silicon can adjust their bandgaps by forming alloys, contributing to further improving the efficiency of amorphous silicon solar cells. For example, it has been revealed that the amorphous silicon-carbon (a-SiC:H) alloy thin film has a wide bandgap, and the open-circuit voltage and short-circuit current of the cells can be significantly increased by the p-type window layer used as a p-i-n amorphous silicon cell. The alloy film of amorphous silicon and germanium has a narrow bandgap and can be used to form a laminated cell with a-Si:H material to significantly expand the long-wave absorption spectrum of the cell. On this basis, the a-Si:H/a-SiGe:H laminated cell and the a-Si:H/a-SiGe:H/a-SiGe:H three-junction cell were developed to not only significantly improve the long-wave absorption of

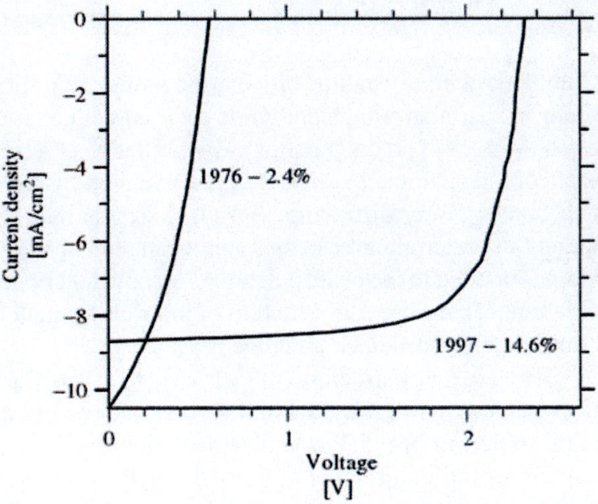

Fig. 5.2 I-V Curves of amorphous silicon cells with two early structures [6, 7]

the cell but also reduce the thickness of the intrinsic layer of each sub-battery, leading to the improvement of the illumination stability of the cell.

Besides, it is necessary to improve the growth rate of the film under the premise of ensuring the quality of amorphous silicon film in order to meet the requirements of reducing cost and achieving large-scale production. Over the past decades, various processes have been invented to achieve this goal; among them, ultra-high-frequency plasma chemical vapor deposition realized high-speed deposition of a high-quality film of amorphous silicon and microcrystalline silicon with a narrow bandgap under high pressure and high-power excitation and is expected to realize large-scale industrial production in the future. Moreover, the hot filament chemical vapor deposition, another promising preparation process, has also been continuously developed. These two growth processes will be discussed in detail later. In the process of preparing amorphous silicon films, the stability of amorphous silicon materials can be significantly improved by diluting silane with hydrogen. This hydrogen dilution technique has been widely applied to improve the microstructure and stability of amorphous silicon materials and solar cells. Besides, the formation of hydrogenated nanocrystalline silicon and hydrogenated microcrystalline silicon can be promoted by the dilution of a large amount of hydrogen. Moreover, significant progress has been made in the solid-phase crystallization of amorphous silicon to prepare film materials and cell devices of microcrystalline silicon or polycrystalline silicon.

At present, dozens of universities, national laboratories, and companies around the world are engaged in the research on thin-film solar cells based on silicon; their industrialization technology is maturing.

5.2.2 Atomic Structure

Compared with crystalline silicon, the atoms of amorphous silicon lose long-range order in spatial arrangement while their constituent atoms are not completely scattered. Structural characteristics very similar to crystals can be seen on a very small scale of a few atoms because they are bounded by chemical bonds, especially covalent bonds. X-ray diffraction (XRD) indicates that Si atoms in amorphous silicon largely maintain their structure and arrangement in crystalline silicon, and each Si atom is bonded to the nearest Si atoms by covalent bonds to form a regular tetrahedral structure. Therefore, the structure of amorphous materials is generally described as "long-range disorder, short-range order".

There are many structural models used to describe amorphous silicon. The continuous random network model (CRN model) is one of them, and its schematic diagram is illustrated in Fig. 5.3. It can be seen that there are still four atoms bonding with any atom; the bond angle and bond length have changed. Therefore, no periodic arrangement of atoms in amorphous silicon can be found in a large range.

It is difficult to make a model of a-Si if a ball-and-stick model for a-Si is established using little sticks as covalent bonds and little wooden balls with four holes as Si atoms. Therefore, small sticks must be bent to avoid the formation of a structure like c-Si. Then, it can be found that a fourth stick cannot be plugged into some Si atoms, and an amorphous structure with lots of imperfect "dangling bonds" emerges, as illustrated in Fig. 5.4.

Since a large number of defects such as dangling bonds and vacancies exist in amorphous silicon materials, there is a high density of defect states, which provides a place for the recombination of electrons and holes. Generally, amorphous silicon is not suitable for making electronic devices. The density of defects in the material will be reduced and then used to make solar cells only when these dangling

Fig. 5.3 Schematic diagram of the CRN model of amorphous silicon

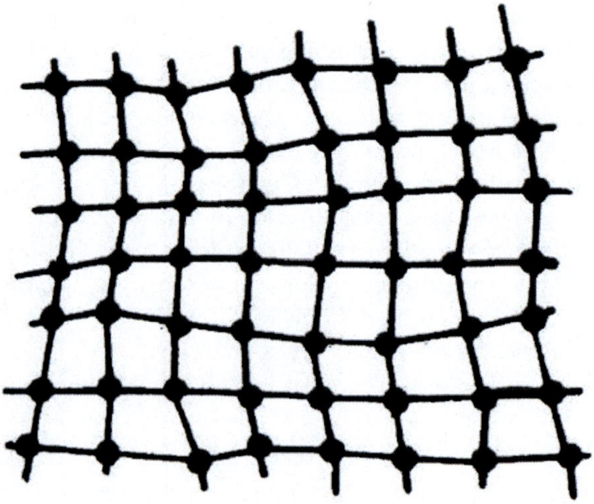

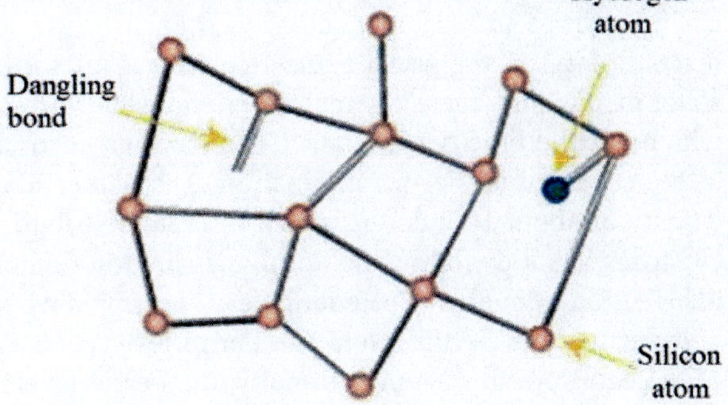

Fig. 5.4 Ball-and-stick model of a-Si:H material

bonds are filled with hydrogen atoms. Nowadays, amorphous silicon cells mean a-Si:H cells in most cases. However, there is little research on amorphous silicon cells without hydrogenation. The materials grown in the conventional preparation processes of amorphous silicon film such as plasma chemical vapor deposition and hot-wire chemical vapor deposition are all a-Si:H materials.

5.2.3 Electronic States of Amorphous Silicon Materials

Since the atomic arrangement of amorphous silicon maintains the sp^3 bond structure and short-range order, the electronic state in amorphous silicon maintains the basic characteristics of the band structure of crystalline silicon. Besides, amorphous silicon also has the valence band and conduction band; the forbidden band is between the other two bands. Thus, it is necessary to introduce the comparable and relatively simple knowledge of the band structure and electronic states of crystalline silicon materials before introducing the electronic states of amorphous silicon materials.

The band theory of solids is deduced by applying the principle of quantum mechanics to solid-state multibody systems, suggesting that the Schrödinger equation is solved under a specific lattice and corresponding potential distribution to obtain the distribution of electron states according to the distribution of energy in the system. In the crystal structure, the lattice structure has a periodicity in space; the corresponding potential also exhibits periodic distribution. Under the adiabatic approximation of lattice and single-electron approximation, the accurate electronic energy state distribution, namely, the electronic band structure, can be obtained. The fundamental characteristic of a crystal band is the existence of a conduction band and valence band; a forbidden band (or bandgap) separates the conduction band and the valence band. There is a single energy value at the bottom of the conduction band and the top of the valence band. The energy states above the bottom of the conduction band are extended states; the electrons in these states are free electrons with high mobility. The energy states below the top of the valence band are localized states of holes; the mobility of the free electrons is also very high. There is no electronic energy state in an ideal semiconductor bandgap. However, it is hard to avoid defects and impurities in semiconductors, and each has its own energy state. The lattice discontinuity on the surface and interface leads to the surface state and interface state; meanwhile, these anomalous energy states often fall into the forbidden band, which is called the defect state. The electrons occupied on it are localized and act as the center of carrier recombination. The defect states of the crystal have a small density of about 10^{15} cm^{-3}, with a discrete distribution. From another perspective, electrons in solids follow the Fermi distribution function according to the statistical distribution of energy. The energy level occupied by electrons with a probability of 1/2 is called the Fermi level. The Fermi level is also called the chemical potential of the equilibrium system. Normally, the Fermi level of a semiconductor is located in the forbidden band. The electrons are majority carriers and the material is n-type if the Fermi level is close to the bottom of the conduction band; the electrons are majority carriers and the material is p-type if the Fermi level is close to the top of the valence band. The position of the Fermi level can be adjusted by appropriate doping, indicating that the amount and type of semiconductor conductance can be regulated by doping.

Since the atomic arrangement of amorphous silicon maintains the sp^3 bond structure and short-range order, the electronic state in amorphous silicon maintains the basic characteristics of the band structure of crystalline silicon. However, it is a long-range disorder with small random changes in bond length and bond angle between atoms. Its actual structure is a network structure composed of silicon atoms. Besides, there are a large number of dangling bonds in the network. The electronic band structure of amorphous silicon is complex due to its disordered structure, which has some similarities and great differences with the crystal band. Their influences on the electronic states of amorphous silicon materials are mainly manifested in the following aspects:

(1) The loss of the periodicity and long-range order of atomic arrangement in amorphous silicon makes the electronic wave vector no longer a good quantum number to describe the electronic state. The energy state density distribution function N(E) of the electron can only be used to describe the band characteristics of amorphous silicon due to the uncertain dispersion relation between energy E and the wave vector K. Therefore, it is impossible to figure out whether amorphous silicon is an indirect-bandgap material or a direct-bandgap material.
(2) Another effect of the disordered structure is to obscure some sharp eigenstructure of the valence band and the conduction band; thus, the sharp band edge extends to the bandgap in a localized band-tail state. Moreover, in the middle of the bandgap, a continuously distributed defect state caused by structural defects such as dangling bonds is formed. Compared with the electronic band structure of crystalline silicon, the electronic energy states in the valence band of amorphous silicon can also be divided into the conduction band, valence band, and forbidden band while both conduction band and valence band have band-tail states extending to the forbidden band. Tail states are related to the random variation of bond length and bond angle. The bottom of the conduction band and

the top of the valence band are replaced by the corresponding fuzzy mobility edge, respectively; the extended state and localized state change continuously at the mobility edge. The high-density defect states in the bandgap are introduced by the high-density dangling bonds; its density is higher than 10^{17} cm^{-3}; the excess carriers recombine through the defect states. Therefore, the photoconductivity of the amorphous materials is very low, and the doping has little effect on the position of the Fermi level. This a-Si material has no useful electrical properties. However, most of the dangling bonds in the hydrogenated amorphous silicon materials are compensated by hydrogen to form a silicon-hydrogen bond, contributing to reducing the density of defect states to below 10^{16} cm^{-3}. Only in this way can such materials exhibit good device-level electronic characteristics.

It can be seen by comparing Figs. 5.5 and 5.6 that it is difficult to accurately determine the width of the forbidden band for amorphous silicon because the band-tail state extends to the forbidden band, and the position of the bottom of the conduction band and the top of the valence band is difficult to be determined due to the tail state

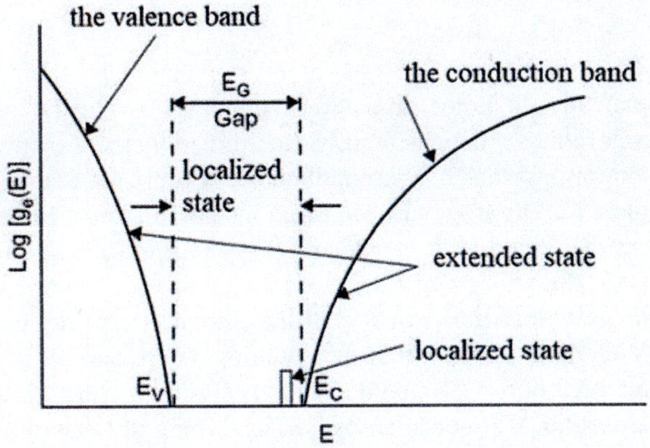

Fig. 5.5 Schematic diagram of electronic state density $g(E)$ in crystalline silicon [8].

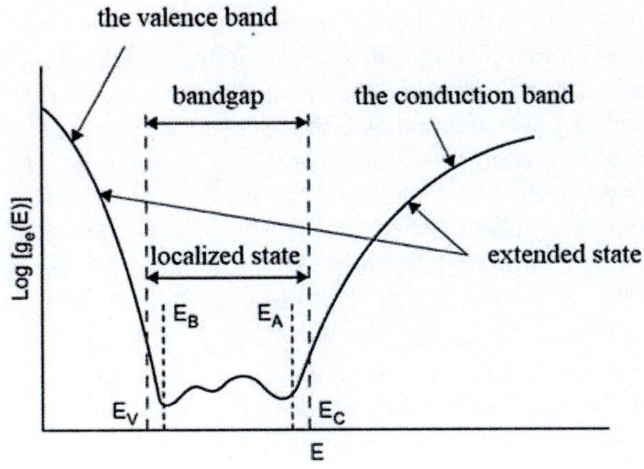

Fig. 5.6 Schematic diagram of electronic state density $g(E)$ in amorphous silicon [8].

being related to the random change of bond length and bond angle. However, the bandgap of amorphous silicon material is an essential parameter in the application of solar cells and is the main factor affecting the absorption of light. There is no doubt that it is significant to establish a set of normative standards to compare the bandgap of various materials. The most popular method is to analyze the measurement of optical absorption coefficient $\alpha(h\nu)$, which can be used to obtain an optical bandgap E_g [9].

Besides, the changes in band structure caused by hydrogen typing in a-Si:H can mainly reduce the density of the defect state and shift the valence band downwards. Although the bottom of the conduction band also moves up, the amplitude is much smaller, and the overall effect is the increased width of the bandgap. It can be found through experiments that there is a linear proportional relationship between the optical bandgap E_g of the a-Si:H film and its hydrogen content C_H, that is, $E_g = 1.48 + 0.019\ C_H$.

5.2.4 Doping and Electrical Properties of Amorphous Silicon

The mobility of electrons in amorphous silicon [about 1 $cm^2/(V{\cdot}s)$] is much larger than that of holes [about 0.01 $cm^2/(V{\cdot}s)$]. The DC dark conductivity of intrinsic amorphous silicon is mainly determined by the transport characteristics of electrons, exhibiting weak n-type conductance. Compared to amorphous silicon without hydrogenation, hydrogenated amorphous silicon has a lower density of defect states and can be doped with n-type and p-type impurities to control the conductivity. Therefore, the change in the conductivity at room temperature can reach 10 orders of magnitude. Similar to crystalline silicon, n-type doping is obtained by adding group V element phosphorus; meanwhile, p-type doping is obtained by adding group III element boron. Doped amorphous silicon material is an indispensable part of p-i-n amorphous silicon solar cells. The doping of crystalline silicon changes the position of the Fermi level of the material by intentionally introducing other atoms such as P and B; however, the effect of doping in amorphous silicon is quite different from that in the crystal. For example, the P atoms in crystalline silicon (c-Si) can replace the original position of Si atoms in the lattice after P is doped. Besides, the outermost four electrons of each Si atom in the crystal structure of crystalline silicon can form a covalent bond with the nearest four Si atoms. Moreover, the extra fifth “free” electrons occupy the energy level at a point below the bottom of the conduction band because there are 5 electrons in the outermost layer of P, leading to the increase in the Fermi level.

In a-si materials, most of the P atoms form covalent bonds with only the three nearest Si atoms. This structure is more stable from a chemical perspective; the 3-coordinate state has lower energy, which is more favorable chemically. Therefore, most of the P or B atoms are in the 3-coordinate states, and their energy levels are in the valence band of silicon, which does not function as doping. Only are a small part of P or B atoms in 4-coordination states, and their energy levels are in a certain range of the band tail of amorphous silicon, which acts as shallow donors or shallow acceptors. Why does this more favorable structure appear in amorphous silicon rather than crystalline silicon? The reason is that amorphous silicon does not have the same tight lattice structure as crystalline silicon. During the growing process of a piece of the amorphous silicon film, the material automatically adjusts the mesh of chemical bonds to form a near-ideal chemical arrangement when the impurity atoms are introduced. In crystalline silicon, the positions of several Si atoms must be redistributed and a large number of suspended bonds must be produced to make P atoms be bonded as in amorphous silicon. The energy required is higher than the direct replacement of Si atoms to form four covalent bonds.

Two essential inferences can be deduced from the doping mechanism of the P atom in amorphous silicon. First, the doping efficiency in amorphous silicon materials is very low. Most of the impurity atoms do not contribute to free electrons and thus cannot change the position of the Fermi energy level. Second, the electrons contributed by the impurity atoms are trapped by the dangling bonds in the amorphous silicon. Since the defect level is located below the conduction band, the doped P atoms in amorphous silicon are not as effective in raising the Fermi level as in c-Si. Moreover, the negatively charged dangling bonds introduced by doping are very effective hole traps, leading to greatly reducing the mobility of holes. The transmission of electrons and holes is a basic condition for photovoltaic devices, and the photons absorbed in the doped a-Si layer contribute little to the output power of the solar cell.

The substitutional doping of B in amorphous silicon is similar to that of P, and most of them are in the state of 3-coordination. The doping efficiency is also low, and defects are introduced. They can be used to establish the built-in potential and ohmic contact in amorphous silicon cells instead of as the light absorption layer because of the high density of defects and high recombination rate of the photogenerated carrier in p-type and n-type amorphous silicon. Therefore, amorphous silicon solar cells rely on the intrinsic layer to absorb sunlight and must adopt a p-i-n structure rather than a PN structure like crystalline silicon solar cells.

5.2.5 *Bandwidth Adjustment of Amorphous Silicon Alloy*

The structural and optical properties of amorphous silicon will vary with process conditions. For example, changing the temperature of the substrate in the process of growth or the concentration of silane in the plasma deposition will result in a change in the a-Si:H bandgap of at least 1.6–1.8 eV, which is caused by the microstructure of hydrogen in a-Si:H. The effect on the bandgap will be greater if other impurities such as Ge, C, O, and N are added. Besides, amorphous alloy materials with different bandgap widths can be obtained when a-si and other elements, such as Ge, C, O, and N are alloyed. The doping process is simple if appropriate proportions of GeH_4, CH_4, O_2, NO_2, or NH_3 gases are mixed in the plasma-deposited silane gas. therefore, the bandgap of a-Si:H can be adjusted over a wide range to meet the requirements

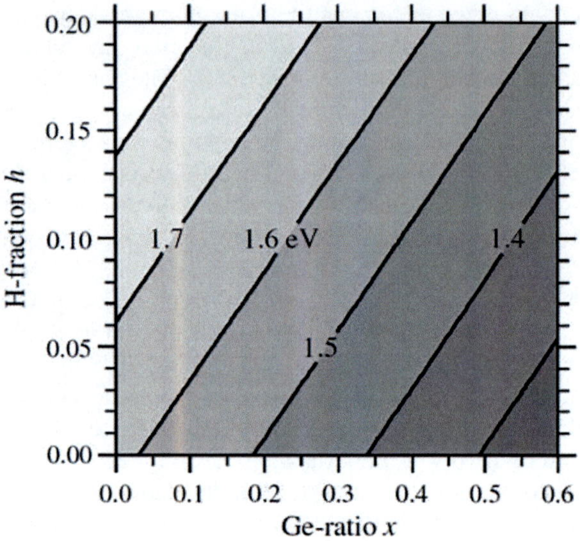

Fig. 5.7 Relationship between optical bandgap of a-$Si_{1-x}Ge_x$:H alloy and component Ge, component H [10, 11]

of devices. For example, a-SiGe alloy, whose bandgap can be continuously adjusted between 1.1 and 1.7 eV by adjusting the composition of Ge, can be used to fabricate absorption-layer cells with a narrow bandgap in multi-junction solar cells. The relationship between the optical bandgap of a-$Si_{1-x}Ge_x$:H alloy and component Ge, component H is illustrated in Fig. 5.7.

Another important a-Si:H alloy material is a-SiC:H with a bandgap of 1.7–2.2 eV, depending mainly on the doping amount of C [12]. The bandgap of a-SiC:H layer can be widened by the upward shift of the bottom of the conduction band. Therefore, the a-SiC:H material is suitable as a p-type window layer of a p-i-n a-Si:H solar cell to limit the reverse diffusion of photogenerated electrons, instead of as an n-layer of a cell because it will hinder the flow of photogenerated electrons to the n-layer and make them be collected. However, it is worth noting that the C atoms in the p-type a-SiC:H layer diffuse through the interface to the intrinsic a-Si:H layer, resulting in the reduced lifetime of intrinsic carriers and the degradation of cell performance. Therefore, an interface buffer layer should be added on the a-SiC:H/a-Si:H heterojunction interface to improve the performance of solar cells. Besides, most scientists believe that a-SiC material is not suitable for the i-type structure of the top layer in multi-junction solar cells. Particularly, a-SiC:H materials with appropriate band gaps have higher defect density compared to a-Si:H materials and must be prepared into thin films after the photodegradation effect occurs, impeding the absorption of the sunlight. At present, the wide bandgap material required in triple-junction solar cells is also a-Si:H, in which the composition of H is relatively high. The a-Si:H with a high composition of H can only be prepared by high-hydrogen dilution at relatively low temperatures [13].

5.2.6 *Optical Properties of Thin-Film Materials Based on Amorphous Silicon*

As mentioned earlier, the high absorption coefficient is an essential advantage of thin-film solar cells based on amorphous silicon. Only can a very thin layer of amorphous silicon absorb most of the sunlight, suggesting that the solar cells can be made very thin. The absorption coefficients of photons with different energies in a-Si:H and c-Si materials are illustrated in Fig. 5.8. Experimental data indicates that an a-Si:H film with a thickness of 500 nm can absorb most of the photons with energies greater than 1.9 eV, as exhibited in the shaded part of Fig. 5.8.

Moreover, a-Si:H can also form an alloy; the bandgap of the alloy can be easily adjusted by controlling the composition to further enhance the absorption of the incident light. It has been revealed through experiments that a-SiGe alloy with a bandgap of about 1.45 eV is very suitable for the i-type absorption layer in multi-junction p-i-n cells. Compared with a-Si, a-SiGe has a narrower bandgap and can absorb photons with lower energies, resulting in improving optical absorption efficiency. The relationship between the absorption coefficient of light with different wavelengths and component Ge in a-SiGe alloy is illustrated in Fig. 5.9. The number on the curve is

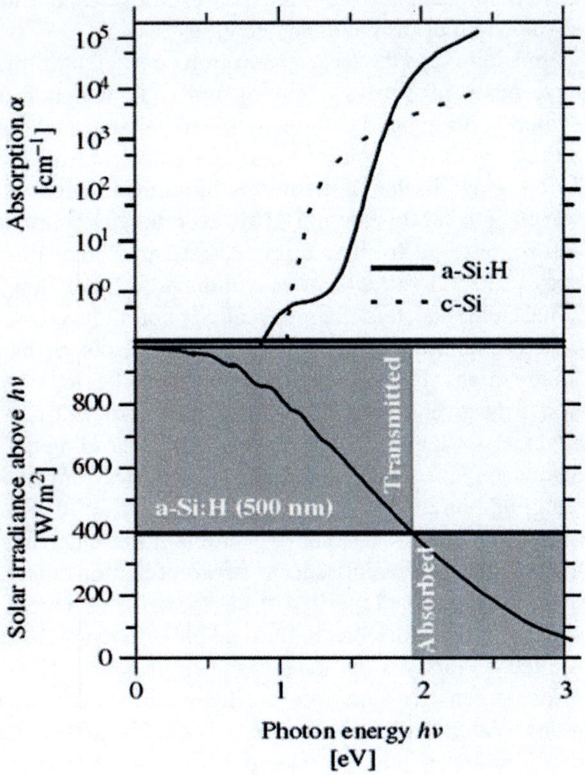

Fig. 5.8 Absorption coefficients of photons with different energies in a-Si:H and c-Si materials [14]

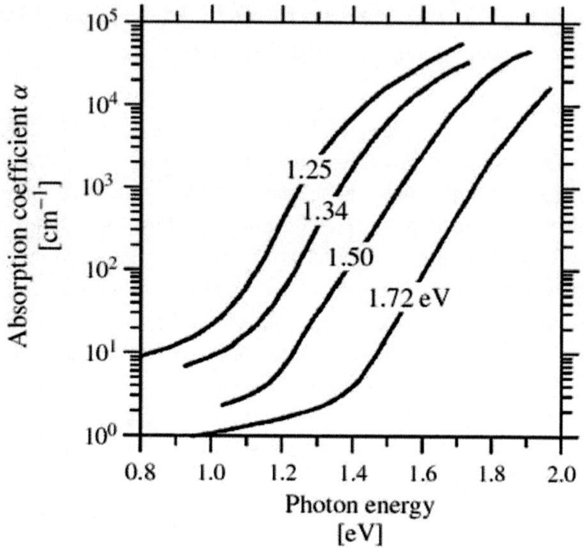

Fig. 5.9 Absorption coefficient of light with different wavelengths in the a-SiGe alloy

the optical bandgap, and the corresponding compositions of Ge are 0.58–1.25, 0.48–1.34, 0.30–1.50, and 0.0–1.72 eV [15]. Besides, the absorption coefficient increases as the width of the bandgap decreases.

In addition to the large absorption coefficient, films based on amorphous silicon have many properties changing under light conditions. In 1977, D. L. Staebler et al. first discovered that the dark conductance and photoconductivity are gradually reduced with time going by and tend to be saturated when the a-Si:H film prepared by the glow discharge method is illuminated (light intensity is 200 $mW{\cdot}cm^{-2}$, the wavelength is 600–900 nm). However, the dark conductance and photoconductivity can be restored to their original state after annealing for 1–3 h at a temperature above 150 °C. Since it can be eliminated by annealing, the degraded state is called a metastable state; this change is called photoinduced metastable change, which is later called a photoinduced change effect. It is also known as the Staebler–Wronski effect (hereinafter referred to as the S-W effect) because Staebler and Wronski were the first to discover this effect. Besides, the photoelectric conversion efficiency of cells is adversely influenced by the S-W effect. According to the techniques of electron spin resonance and sub-bandgap absorption spectrum, the light produces a metastable dangling bond defect state in a-Si:H material; its saturated defect concentration is about $10^{17}/cm^{-3}$. The energy position of these defect states is close to the middle of the bandgap and mainly acts as a recombination center, leading to the degradation of the photoelectric properties of the a-Si:H film material and the performance of the solar cell; therefore, the highest stable efficiency that the a-Si:H cell can achieve is limited. In fact, a significant factor limiting the large-scale application of thin-film solar cells based on amorphous silicon is how to reduce the degradation of cell performance caused by the S-W effect. In addition to the decreased photoconductivity and dark conductance of a-Si:H and the increased density of metastable dangling bond, the illumination also causes a series of changes in the physical properties of a-Si:H, such as the Fermi level shifting toward the center of the bandgap, the shortened lifetime of carriers, the reduced diffusion length, the increased density of band-tail states, the decreased intensity of the main peak of photoluminescence, the increased intensity of the luminescence peak of defects, and the obtained fatigue effect of photoluminescence.

The S-W effect has a lot to do with the season. It was revealed by the Advanced Photovoltaics Systems of Switzerland in an experiment examining the daily energy conversion efficiency and the ambient temperature of triple-junction solar cells based on amorphous silicon that the solar cells have the best average performance during the hot summer. The photoelectric conversion efficiency of solar cells changes with temperature as $+5\times10^{-3}$/K when the temperature is lower than 20 °C. It is worth noting that the solar cells did not exhibit permanent degradation during the later three years of testing in addition to the first 1000 h. Therefore, it was concluded from the experiment that the amorphous silicon solar cells reached a stable state after 1000 h of illumination [16], and the efficiency in this state was called stable efficiency.

This positive correlation between energy conversion efficiency and the temperature is not suitable for solar cells of other materials. For example, the correlation coefficient between energy conversion efficiency and temperature of a polycrystalline silicon cell is -4×10^{-3}/K. Interestingly, the coefficient of efficiency as a function of temperature is negative, about -1×10^{-3}/K if the temperature dependence of the a-Si cell is measured very quickly, that is, there is no time for the S-W effect [17]. Then, this phenomenon can be understood. There are two mechanisms affecting the efficiency of a-Si cells as a function of temperature. One is the negative temperature correlation coefficient of the cell itself; the other is that the rising temperature is equivalent to a slow annealing process to reduce the S-W effect, resulting in improving the photoelectric conversion efficiency of cells and making the efficiency and temperature positively correlated. The role of the latter is more apparent. The cell generally exhibits a positive correlation with temperature [18, 19].

Consequently, the light-induced degradation of amorphous silicon cells is caused by the S-W effect, resulting in a significant decrease in efficiency during the first few hundred hours of using the cell. Product testing by United Solar Systems [20, 21] indicated that the photoelectric conversion efficiency of a single-junction a-Si solar cell lost 30% after 1000 h of illumination; the condition of the triple-junction cell was much better while it also lost 15%. Moreover, the development of amorphous silicon solar cells is severely restricted by the light-induced degradation, leading to the most studied area recently.

As mentioned above, the electron spin resonance and the sub-bandgap absorption spectrum indicate that the S-W effect is mainly caused by the fact that the light produces a metastable dangling bond defect state in the amorphous silicon material with a saturated defect concentration of about 10^{17} cm^{-3}. The energy position of these defect states is close to the middle of the bandgap and mainly acts as a recombination center, leading to the degradation of the photoelectric properties of the a-Si:H film material and the performance of the solar cell; therefore, the highest stable efficiency that the a-Si:H cell can achieve is limited. Investigating the mechanism of the photo-induced metastable state and exploring the ways to overcome the light-induced degradation are significant for improving not only the basic theory of developing amorphous silicon materials but also the performance of solar cells.

All the laboratories engaged in the research and application development of a-Si in the world are investigating the problem of light-induced degradation, conducting various experimental observations, and proposing various theoretical models. However, there is still no convincing unified model to explain various major experimental facts. A more consistent view is that light-induced degradation is related to the motion of hydrogen in a-Si materials. The relatively important models are

(1) The non-radiative recombination of photogenerated carriers causes the breakage of weak Si–Si bonds, creating dangling bonds. Nearby hydrogen compensates for one of the dangling bonds by diffusion; meanwhile, a metastable dangling bond is added.
(2) The dangling bond in amorphous silicon requires more energy to obtain the second electron compared to the first electron. This energy difference is the correlation energy of the electron. D. Adler believes that the amorphous silicon network in some areas may be slack due to the non-uniformity and disorder. Besides, the accompanying lattice relaxation reduces the total energy and makes the effective correlation energy of the electron to be negative when the dangling bond captures the second electron. In these regions, the energy of the dangling state with two electrons is lower than that of the dangling state with one electron. Therefore, it is a positively charged dangling bond with no electron and a negatively charged dangling bond with two electrons instead of a stable neutral dangling bond with one electron. Moreover, these charged dangling bonds may capture electrons or holes and turn into metastable neutral dangling bonds when the light excites the carriers.
(3) The new model proposed by H. Branz in 1998 indicated that photogenerated electrons collide with each other to produce two movable hydrogen atoms. The diffusion of hydrogen atoms forms two immobile complexes with Si–H bonds. The metastable dangling bond appears at the position of excited hydrogen. This model can quantify the mechanism of photogenerated defects and explain some of the main experimental phenomena.

The relatively convincing models mentioned indicate that light-induced degradation is related to the movement of hydrogen in a-Si materials. The a-Si material is obtained by chemical decomposition of a silane-based gas plasma at a lower temperature and then depositing on a substrate. There are inevitably dangling bonds of silicon in its silicon network structure. The a-Si film deposited by a widely used PECVD method contains a hydrogen content of 10–15%; this makes the dangling bonds of silicon better compensated and far exceeds the density of the dangling bonds of silicon. Therefore, hydrogen occupies various positions with different activation energies in the a-Si material, one of which compensates for the position of the dangling bond, and the other is located at the position with lower activating energy. The ideal amorphous silicon material should have neither defects such as micro-voids nor bonding bodies such as SiH_2, $(SiH_2)_n$, and SiH_3. The density of amorphous silicon material should be as close as possible to the ideal density of crystalline silicon; besides, the dangling bonds of silicon are fully compensated by the proper amount of hydrogen, resulting in a low density of defect states and high stability of the structure.

5.3 Preparation Technology of Amorphous Silicon Film and Industrialization of the Amorphous Silicon Cell

5.3.1 Common Preparation Techniques of a-Si Film

Many methods of preparing thin-film materials based on silicon have been developed in the past few decades, mainly including chemical vapor deposition (CVD) and physical vapor deposition (PVD). The amorphous silicon film prepared by the physical vapor deposition such as the sputtering and the electron beam evaporation has no hydrogen and is deficient in the passivation effect of hydrogen on dangling bonds in amorphous silicon. The density of the grown film is very high and there is no application value for cells. Therefore, most of the current processes in the laboratory and industrial production are chemical vapor deposition. Generally, the chemical vapor deposition decomposes a silicon-containing gas (such as silane SiH_4 and disilane Si_2H_6) in a reaction chamber and then deposits the decomposed silicon atoms or radical groups containing silicon on the substrate. Phosphine (PH_3) is required in the preparation of the n-type doping material, and the p-type doping material requires the addition of diborane (B_2H_6), trimethyl borane $[B(CH_3)_3]$, or boron trifluoride (BF_3). Therefore, the silane is usually diluted with hydrogen or an inert gas such as helium (He) and argon (Ar) to improve the quality of the material.

Table 5.1 compares the common preparation techniques of a-Si films.

The first a-Si cell in the world was manufactured by Chittick et al. [23] and Spear, LeComber et al. [24] using a deposition process of glow discharge induced by radiofrequency voltages. This method is now called plasma-enhanced chemical vapor deposition (PECVD). Since then, researchers have started exploring new growth processes to improve the quality and growth rate of the material. Among all these growth processes, PECVD technology is still the most widely used to prepare a-Si materials in most of the laboratory research and factory production. Other emerging thin-film deposition technologies are obtaining more and more attention due to their high growth rate and the ability to grow high-quality microcrystalline silicon. The common processes for preparing a-Si and their advantages and disadvantages are summarized in Table 5.1. Next, plasma-enhanced chemical vapor deposition (PECVD) and hot filament chemical vapor deposition (HWCVD) will be introduced in the following sections.

Table 5.1 Comparison of various deposition processes of amorphous silicon films [22]

Process name	Maximum growth rate (Å/s)	Advantage	Disadvantage	Manufacturers
Direct current plasma-enhanced CVD method	3	The quality and consistency of the film is good	The growth rate of the film is slow	Most
Radio frequency plasma-enhanced CVD method	3	The quality and consistency of the film is good	The growth rate of the film is slow	BP solar
Ultra-high frequency plasma-enhanced CVD method	15	The growth rate of the film is fast	The consistency of the film is poor	None
Microwave plasma-enhanced CVD method	50	The growth rate of the film is very fast	The quality of the film is poor	Canon
Hot filament CVD method	50	The growth rate of the film is very fast	The consistency of the film is poor	None
Photoinduced CVD method	1	The quality of the film is good	The growth rate is slow	None
Sputtering method	3		The growth rate is slow, and the quality of the film is poor	None

Note The standard for the maximum growth rate is that the quality of the grown film will degrade rapidly once this rate is exceeded. All the data in the table are empirical values, representing the technology level at that time

5.3.2 *Reaction Kinetics During the Growth of Amorphous Silicon Thin Films*

The deposition of intrinsic amorphous silicon is generally performed using silane (SiH_4) or disilane (Si_2H_6). Due to the presence of various ions in the plasma, the gas-phase chemical reaction process is a complicated process with limited understanding. As an example of silane decomposition, the decomposition process is various. In Fig. 5.10, the particles and ions produced by the decomposition of silane in the plasma are illustrated, as well as the energy required to produce the various particles and ions.

Some of the possible primary chemical processes and the energy required are displayed as follows:

$$SiH_4 + e^-(8.75\ eV) \rightarrow SiH_3 + H + e^-$$

$$SiH_4 + e^-(9.47\ eV) \rightarrow SiH^* + H_2 + H + e^- \quad SiH_2 + H_2 + e^-$$

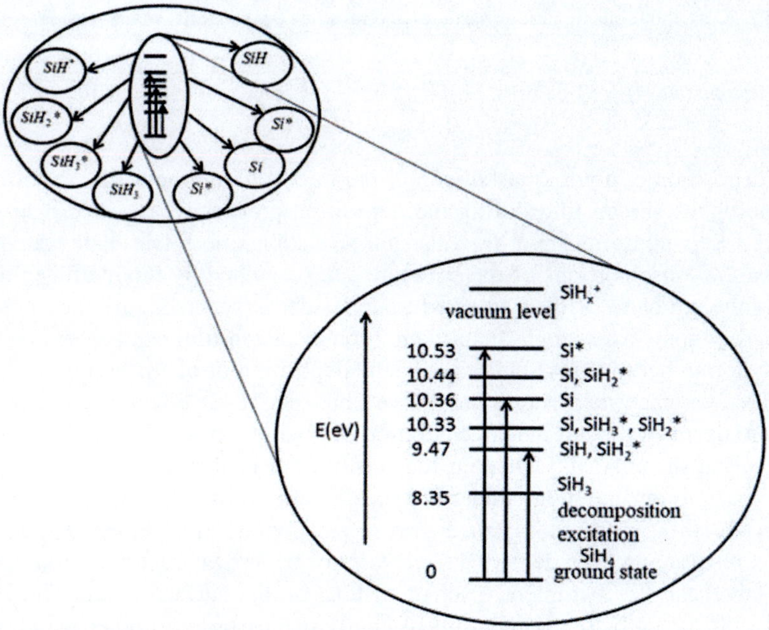

Fig. 5.10 Particles and ions produced by the decomposition of silane in the plasma and the energy required to produce various particles and ions [25].

$$SiH_4 + e^-(\sim 10\ eV) \rightarrow SiH_x^- + (4 - x)H$$
$$SiH_4 + e^-(10.33\ eV) \rightarrow SiH^* + H_2 + H + e^- \quad Si + 2H_2 + e^-$$
$$SiH_4 + e^-(10.53\ eV) \rightarrow Si^* + 2H_2 + e^-$$
$$SiH_4 + e^-(> 13.6\ eV) \rightarrow SiH_x^* + (4 - x)H + 2e^-$$

where SiH_x^* is a positively charged ion; SiH^* and Si^* are particles in an excited state, returning to the ground state by releasing photon energy:

$$SiH^* \rightarrow SiH + hv(414\ nm)$$
$$Si^* \rightarrow Si + hv(288\ nm)$$

The characteristics of the plasma can be investigated by measuring the luminescence spectrum of the plasma. The concentrations of various particles and ions in the plasma are different because the decomposition of silane into different particles and ions requires different energies, and their lifetimes are also different. The concentrations of various particles and ions in a conventional silane plasma are listed in Table 5.2.

It can be seen from the table that the main component in the plasma is SiH_3. Neutral SiH_3 particles are generally considered to be precursors for the growth of high-quality amorphous silicon. Besides, atomic hydrogen plays an essential role

Table 5.2 Concentrations of various particles and ions in a conventional silane plasma

Radical group and ion	SiH_x^+, H^+	SiH^*, Si^*	Si	SiH	SiH_2	SiH_3
Concentration (cm^{-3})	10^8–10^9	10^5	10^8–10^9	10^8–10^9	10^9	10^{12}

in the deposition of amorphous silicon. First, the dangling bonds on the surface of the amorphous silicon film during the deposition process require hydrogen to be saturated. Second, atomic hydrogen also has an etching effect. During the deposition process, the loose portions of the structure can be etched by the hydrogen atoms, making the structure of the deposited material dense and reducing the density of the micro-voids; thus, a high-quality amorphous silicon film can be obtained. The role of atomic hydrogen is significant in the deposition of microcrystalline silicon. Compared with neutral particles, they have a non-negligible effect on the deposition of materials, even though the concentration of charged ions is lower. The negative effect is that the positively charged ions diffuse out of the plasma zone, enter the dark zone, and obtain energy under the acceleration of the electric field. From one perspective, these ions with certain energy have a bombardment effect on the growth surface, resulting in high-density defect states of the grown material. From another perspective, the bombardment of charged ions on the surface of the film is also positive. Specifically, the bombardment can transfer energy to other particles and increase the local temperature of the growth surface, resulting in increasing the diffusion coefficient of the particles and ions on the surface; therefore, it is easy for particles to find low-energy regions, and the quality of the material is improved. This effect is especially significant during high-speed deposition. Therefore, controlling the bombardment of high-energy charged ions is an essential means to optimize the deposition of thin-film silicon materials at high speed. Since the potential of the plasma is positive, the negatively charged ions are trapped within the plasma. These negatively charged ions trapped in the plasma and neutral particles are combined to form large particles, leading to affecting the quality of the deposited material.

Generally, the rapid growth of high-quality amorphous silicon films should meet the following requirements:

(1) There must be a certain distance between the formation position of the SiH_3 atomic group and the growth surface of the film.
(2) The concentration and flow rate of atomic H should be maintained at a high level to passivate the growth interface film.
(3) The bombardment energy of the plasma should be reduced to improve the quality of the film.
(4) The atomic groups with a short lifetime should be prevented from reaching the growth surface of the film. The internal defects in the film and the deterioration of the quality of the film are often caused by these atomic groups.
(5) The formation of complex atomic groups and powders should be avoided.

SiH_2 is generally considered to harm the stability of the material. Since the production of SiH_2 requires high-energy electrons, the materials deposited under high-power conditions have poor stability. High silane or large-mass particles are prone to be produced during secondary or advanced chemical reactions. Besides, high silanes also hurt the quality and stability of the material. The high silane leads to the presence of SiH_2 and poly-hydrogen groups in the material, making the material tend to generate defect states in the light. The large-mass particles cause micro-voids and a high density of defects in the material, leads to the accumulation of dust in the reaction chamber, and increases the maintenance cost of the reaction system. Therefore, both SiH_2 and high silane produced in the reaction chamber harm the quality of the material, which should be considered in the process of optimizing material.

Chemical reactions on the surface are an important part of the process of depositing amorphous silicon. After the neutral particles and charged ions coming out of the plasma reach the growth surface, part of them combine with the chemical bonds on the surface to form a solid material while part of them return from the surface to the gas. During the growth process, most of the silicon atoms on the surface of the silicon material are saturated by hydrogen atoms while part of the silicon atoms on the surface form dangling bonds. Neutral particles (such as SiH_3) and charged ions reaching the surface diffuse on the growth surface. They can be bonded to dangling bonds on the surface and remove hydrogen atoms from the surface to bond to the silicon atoms on the surface. The main factor affecting the reaction on the surface is the temperature of the substrate. Therefore, the temperature of the substrate needs to be increased to raise the diffusion coefficient of the particles on the surface. However, the coverage of hydrogen on the growth surface is reduced by the substrates with excessively high temperatures; besides, hydrogen in the film also diffuses out. Therefore, there is a relatively low content of hydrogen and excessively high defect states in the amorphous silicon grown on a substrate with excessively high temperature. At low temperatures, the degree of disorder in the material is too high and the density of defect states is also increased due to the low diffusion coefficient of the particles on the surface, even though amorphous silicon contains a sufficiently high content of hydrogen. Therefore, optimizing the temperature of the substrate is significant for optimizing the quality of amorphous silicon materials. Generally, the temperature of the substrate for growing amorphous silicon is 150–350 °C according to other deposition parameters. The optimized temperature of the substrate can be relatively low in the processes of low-speed deposition. For example, at a deposition rate of less than 1 Å/s, the temperature of the substrate can be less than 200 °C. Under conditions of high-speed deposition, particles reaching the surface of the substrate require a faster diffusion rate; thus, the optimized temperature of the substrate is relatively high. For example, at a growth rate of greater than 10 Å/s, the temperature of the substrate needs to be greater than 300 °C. Besides, the coverage of the hydrogen on the growth surface is related to the dilution of the hydrogen in the reaction chamber. Under a condition of high dilution of hydrogen, the coverage of the hydrogen on the growth surface is relatively high; therefore, the temperature of the substrate can be relatively low.

The content of the hydrogen in amorphous silicon is directly related to the temperature of the substrate. Particularly, the content of hydrogen decreases as the temperature of the substrate increases under certain conditions. Since the content of hydrogen has a direct effect on the bandgap of amorphous silicon, the bandgap of the material can be adjusted by optimizing the temperature of the substrate. The influence of the temperature of the substrate on amorphous silicon–germanium alloy material is more apparent. Moreover, the deposition temperature of amorphous silicon–germanium material is generally higher than that of amorphous silicon because GeH_3 is heavier than SiH_3 and has a small diffusion coefficient on the growth surface. Furthermore, the density of defect states in amorphous silicon–germanium material is higher as the bond energy of Ge–H is lower than that of Si–H. Therefore, reducing the density of defect states in amorphous silicon-germanium alloy materials is an essential step to improve the photoelectric conversion efficiency of multijunction solar cells based on amorphous silicon.

5.3.3 *Plasma Enhanced Chemical Vapor Deposition*

A schematic of the chamber and related parts of the RF PECVD process are illustrated in Fig. 5.11. First, the chamber is evacuated by a vacuum pump. Then, a silicon-containing gas such as a mixed gas of silane and hydrogen is introduced into the vacuum chamber. There are two electrodes inside the chamber, through which a radio-frequency voltage is applied to the chamber to generate a plasma in the chamber. The generated plasma excites and decomposes the gas to produce atomic groups and ions. Besides, the desired substrate material is preliminarily deposited on the electrodes, and a layer of a-Si is deposited on the substrate as the atomic groups and ions diffuse toward the electrodes. Moreover, the quality of the grown a-Si film can be optimized by adjusting the temperature of the substrate because the adsorption of the film surface to atoms and the diffusion of atoms on the film surface are related to temperature.

A complete PECVD system usually consists of the following main parts: (1) Gas transmission part (gas cylinder, pressure regulator, large flow controller, gas valve corresponding to each gas); (2) deposition chamber, equipped with electrodes (there

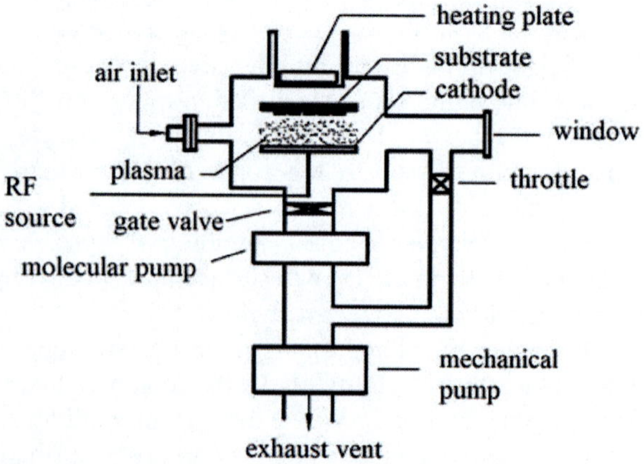

Fig. 5.11 Structure of equipment for RF plasma chemical vapor deposition [22].

Table 5.3 Process parameters of optimal conditions for growing a-Si:H film

Range of variation	Gas pressure (Torr)	Density of RF source (mW/cm^2)	Temperature of substrate (°C)	Spacing of electrodes (cm)	Flow rate of effective gas current ($sccm/cm^2$)	Dilution of H_2
Maximum	2	100	350	5	0.02	100
Median	0.5	20	250	3	0.01	10
Minimum	0.05	10	150	1	0.002	0

Note (1) The flow rate of effective gas current (such as the mixed gas of SiH_4, GeH_4, or Si_2H_6) is defined as the gas flow per unit substrate area (electrode + substrate + inner wall of chamber); (2) The dilution of hydrogen is defined as the proportion of H_2 in the effective gas current (e.g. H_2/SiH_4)

are pre-deposited substrate materials on the electrode.), substrate heating device, and RF power unit; (3) vacuum pump device, usually including a mechanical pump and a molecular pump; (4) the control system of air pressure, including capacitor pressure gauge, ionization gauge, thermocouple vacuum gauge, and reduction valve, used to control and monitor the air pressure in the chamber; (4) exhaust system for the removal of waste gases (typically using chemical detergents to neutralize the waste gases or using a combustion chamber to crack the waste gases). Besides, transfer devices are also required in multi-chamber systems to transfer substrate material in the vacuum system among chambers through different gate valves. These processes can be performed by controlling the panel of the instrument; meanwhile, the vacuum degree of the chamber can be read and controlled.

There are several procedures for preparing a film using the PECVD process: the diffusion of source gas, electron collision dissociation, gas-phase chemical reaction, and the diffusion of atomic groups and deposition. The optimum empirical values of various process parameters during growth to prepare a high-quality a-Si film are illustrated in Table 5.3. The data in the table are empirical data, which correspond to the experimental conditions at the time of publication.

The gas pressure during deposition is generally 0.05–2 Torr. A lower gas pressure facilitates the formation of a uniform film while a higher voltage is generally used to prepare a microcrystalline silicon film. Researchers tend to deposit a-Si films using a pressure of 0.5–1 Torr. The power density of the RF power source is typically set to 10–100 mW/cm^2 through a capacitive coupling reactor. RF power supplies below 10 mW/cm^2 are not sufficient to generate and sustain the plasma required for growth while RF power supplies above 100 mW/cm^2 form the polymeric hydrides of silicon, which in turn contaminate the a-Si film. Moreover, low-pressure deposition or reducing the concentration of hydrogen can alleviate this problem if a high-power RF power supply must be used due to the conditions (such as rapid growth).

Generally, the temperature of the substrate is set to 150–350 °C. The hydrogen component in the a-Si:H film is relatively high when the temperature of the substrate is low, resulting in slightly widening the bandgap of a-Si:H. However, the formation of the polymeric hydrides of silicon could be accelerated by lower temperatures

(<150 °C) of the substrate unless the high-hydrogen doping is employed. Besides, the hydrogen component in the a-Si:H film is relatively low and the bandwidth of a-Si:H is slightly reduced when the temperature of the substrate is high. This is because the diffusion coefficient of the adsorbed atoms on the surface of the grown film has an effect of thermotropic enhancement. It can be inferred that the lattice network of a-Si:H is more desirable at higher temperatures because only a small amount of hydrogen can fill the internal dangling bonds. Then, the device was fabricated by exploiting this effect of the temperature of the substrate on the width of the bandgap. A material with a wider bandgap is suitable as the top layer of a three-junction solar cell [26, 27] while a material with a narrow bandgap is suitable as the i-type absorber layer in an a-Si/a-SiGe tandem-junction cell to improve the absorption efficiency of light. However, the quality of the film will be seriously degraded if the temperature of the substrate is too high (>350 °C) because the high temperature affects the passivation of hydrogen to the dangling bonds in the amorphous silicon.

The pitch between the electrodes in the RF glow discharge reactor is generally 1–5 cm. The smaller the pitch, the better the uniformity of the grown film; the larger the pitch, the easier the maintenance of the plasma. The flow rate of the incoming gas current depends on the rate of deposition and the area of the reactor panel. A portion of the silicon atoms in the gas current is deposited directly on the substrate or the walls of the chamber. Besides, the remainder is pumped out as a waste gas by a vacuum pump. Generally, manufacturers are more willing to increase the utilization of silicon by controlling growing conditions (such as using lower flow rates of gas current and increasing the power of RF power supplies); however, this is finished at the expense of the quality of the film.

The standard frequency of the RF power supply for RF glow discharge reactors is $f = 13.56$ MHz, which is also the frequency of industrial production marked in the United States and internationally. However, the frequency range of the RF source used in the process of actual application is very wide, from DC ($f = 0$), low frequency (f is in the order of kHz), very high frequency (VHF, $f \approx 20$–150 MHz), to the microwave frequency (MW, $f = 2.45$ GHz). In RCA laboratories, the DC glow discharge deposition technology was used to deposit a-Si films in the early period. However, it is generally not used to prepare an a-Si:H cell now since the cell prepared by this method has a too poor performance. AC glow discharge deposition technologies, including RF, VHF, and MW PECVD, are becoming more widely used. Moreover, plasmas with a high ionization efficiency can be generated and sustained due to these technologies. Among them, VHF and MW technologies are especially concerned by people because their growth rates are particularly high and they can be used to prepare films of microcrystalline silicon and polycrystalline silicon in addition to a-Si. These two techniques for depositing a-Si are described as follows.

Universiťe de Neuchatel [28] first proposed that VHF plasma can be used to prepare a-Si films at high rates (greater than 10 Å/s). In addition to its high preparation rate, another advantage is that it does not produce a powder of the polymeric hydride of silicon, leaving a problem in the past. Specifically, the powder of the polymeric hydride under low-frequency conditions increases to raise the growth rate and the frequency of the RF source, resulting in contaminating the a-Si film. Researchers

Table 5.4 Comparison of performance of p-i-n cells, of which i-type layers are prepared by RF and VHF processes at different growth rates

Frequency of power supplies (MHz)	Growth rate (Å/s)	Initial efficiency (mW/cm^2)	Degree of degradation (%)
RF (13.56)	0.6	6.6	14
VHF (70)	10	6.5	10
RF (13.56)	16	5.3	36
VHF (70)	25	6.0	22

have used the VHF process to produce high-quality a-Si films and cells with the excellent performance [29]. The comparison of the four single-junction a-Si:H cells are exhibited in Table 5.4; among them, the i-type layers are prepared by RF and VHF technologies (other conditions are the same). It can be found that the performance of the cell prepared by the VHF process is superior to that prepared by the RF process, both in terms of photoelectric conversion efficiency and stability.

It can be observed from Table 5.4 that the VHF process is superior to the RF process in terms of the growth rate, and the initial efficiency of the cell is not significantly reduced when the growth rate is greatly increased.

Although the VHF process has various advantages as described above, it has not been widely used in practical large-scale production for the following two reasons. Firstly, a uniform layer of a-Si film needs to be deposited on a large area in practical production. However, the fluctuation effect such as diffraction and interference may affect the film growth on the electrode when the size of the electrode is large enough to compare with the wavelength of the RF electromagnetic wave, resulting in ultimately affecting the uniformity of the a-Si film. Secondly, VHF coupling is difficult. It is difficult to couple VHF power from a generator to a large-area electrode while many researchers are performing research in this area and have made great progress [30].

Since the microwave (MW) plasma deposition processes use higher frequency RF sources, a-Si films faster than the VHF processes can be prepared. If the plasma generated by the microwave plasma deposition process is directly in contact with the substrate material, the optical and electrical properties of the grown a-Si film are very poor compared with the a-Si film prepared by the VHF process; thus, it cannot be applied to the absorbing layer of the cell. Therefore, researchers invented the noncontact microwave plasma deposition process [31] by improving the process and successfully applied this process to grow high-quality a-Si films. During the noncontact microwave plasma deposition process, the substrate material is separated from the plasma; then, the microwave plasma is used to excite and decompose intermediate gases such as He, Ar, and H_2; next, these high-energy intermediate gases pass through the surface of the substrate. These high-energy intermediate gases decompose SiH_4 or Si_2H_6 to form an a-Si film on the substrate and the walls of the chamber. Besides, the concentration of SiH_3 radicals can be maintained through this noncontact process while the concentration of other atomic groups (such as SiH_2 and SiH) can be minimized. The only defect is that the growth rate of the a-Si film is reduced by this method reduces.

Both the United Solar Corporation of the United States and the Canon of Japan have investigated the microwave plasma deposition processes. Generally, the a-Si film prepared by the MW process is generally inferior to the a-Si film prepared by the RF process in terms of structural and photoelectric properties while its ultra-high growth rate is unmatched by other processes (such as RF and VHF process).

5.3.4 Hot-Wire Chemical Vapor Deposition

The hot-wire chemical vapor deposition (HWCVD) process has a simple basic principle. The silane gas can be thermally decomposed after the wire is heated to 1500–2000 °C. Initially, it was thought that no ions were generated in the process, and thus there was no influence on the quality of the amorphous silicon film due to the bombardment of the film surface by ions. However, Sung-Soo Lee et al. [32] proposed in 2008 that there are charged particles playing a significant role in the deposition process. However, the bombardment energy of ions in the hot-wire chemical vapor deposition process is much lower than other deposition processes, and the bombardment effect on the film is relatively small, even though there are ions generated. Therefore, the flow rate of the silane, the temperature of the hot wire, the distance between the hot wire and the substrate, and the number of filaments are all the key points to increase the growth rate.

A few years after the first HWCVD, Mahan et al. [33] improved the deposition process to greatly promote the quality of a-Si films grown; besides, they established a worldwide research boom in this process technology to rapidly prepare high-quality films of amorphous and microcrystalline silicon.

The device of the HWCVD process is illustrated in Fig. 5.12. In the HWCVD process, firstly, a mixture of silane gas or other gases (such as H_2 or He) is introduced into the chamber; then, the reaction gas is heated by the heating filament to 1800–2000 °C. At high temperatures, the source gas is decomposed into atomic groups and ions under the action of hot-wire catalysis. Afterward, these silicon-containing radicals diffuse inside the chamber and are deposited on the substrate material heated to 150–450 °C; a few centimeters from the heating filament exist inside the chamber. Besides, Mahan et al. revealed that the a-Si film prepared by the HWCVD process has a relatively low hydrogen composition and a relatively small photodegradation effect compared with the a-Si film prepared by the RF PECVD process. It has been reported that the p-i-n cells with an i-type layer of a-Si material prepared by the HWCVD process have an initial efficiency of ~10% [34, 35]. Therefore, the HWCVD process is very popular due to these advantages. Although it has not been widely used in industrial production, its ultra-high growth rate of a-Si:H and a-SiGe:H films [36] is still worthy of attention.

The growth rates and characteristic parameters of a-Si:H films prepared by the HW CVD process under different process conditions are listed in Table 5.5. As indicated

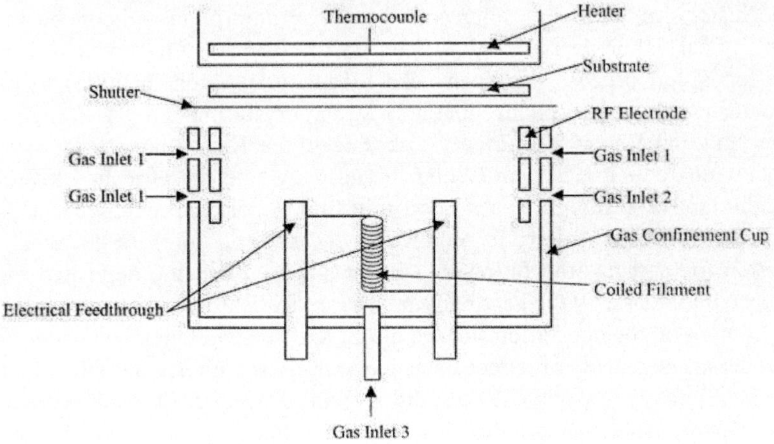

Fig. 5.12 Schematic diagram of HW CVD device

Table 5.5 Growth rate and characteristic parameters of a-Si:H film grown by HW CVD under different process conditions

Sample number	SiH_6 current (sccm)	H_2 current (sccm)	Temperature of substrate (°C)	Growth rate (Å/s)	Content of component H	Width of bandgap E_g (eV)
A1	105	0	100+	700	>10	~1.9
A2	105	0	200+	460	12	1.74
A3	105	0	300+	240	10	1.63
A4	105	0	400+	240	6	1.61
B2	105	100	200+	180	19	1.78
B3	105	100	300+	200	7	1.62
B4	105	100	400+	150	3	1.63
C	70	0	175+	200	8	1.63
D	140	0	100+	800	>8	~1.9

Note The temperature of the hot wire during the growth of all samples is 2,000 °C

in the table, (1) the influence of the temperature of the substrate on the growth rate is apparent; it can be seen by comparing A1, A2, A3, and A4 or comparing B2, B3, and B4 that the higher the temperature of the substrate, the faster the growth rate; (2) the growth rate is also affected by the flow rate of the source gas; it can be revealed by comparing A1 and D that the higher the flow rate of SiH_4, the faster the growth rate; besides, the comparison of A2 and B2 suggests that the higher the flow rate of H_2, the slower the growth rate; (3) the component H in the grown amorphous silicon film can be controlled by adjusting the ratio of the flow rates of SiH_4 and H_2, thereby changing the width of the forbidden band.

The history of preparing solar cells based on amorphous silicon by the hot-wire chemical vapor deposition has been up to 20 years. The hot-wire chemical vapor deposition process has been used by many research institutes to prepare solar cells based on amorphous silicon and microcrystalline silicon. Some of them are the National Renewable Energy Laboratory, the Kaiserslautern University in Germany, the Juelich Research Center in Germany, and the Utrecht University in the Netherlands. Although it is a good thin-film deposition technique, it has not been widely used, especially in production because the hot-wire chemical vapor deposition method has the following disadvantages. First, the deposited material may be contaminated by the heating wire. Theoretically, there is no high-energy ion bombardment on the deposition surface during the hot-wire chemical vapor deposition process. The density of defect states in the prepared amorphous silicon material should be relatively low while the test did reveal the presence of residual tungsten in the film, of which the concentration may be higher than 10^{18} atoms·cm^{-3}. Generally, the content of residual tungsten is related to the old or new hot wire. The content of tungsten in the material deposited by the new hot wire is relatively high while the content of tungsten in the material deposited by the old hot wire is low. However, it is now possible to heat the hot wire to a temperature higher than the temperature of the hot wire at the time of material deposition after the new hot wire is installed; therefore, the tungsten oxide on the surface of the hot wire can be evaporated, resulting in reducing the content of tungsten in the amorphous silicon. After such treatment, the content of tungsten in the amorphous silicon can be lowered to a level below the measurement limit of the secondary particle spectrum. Meanwhile, residual tungsten has no influence on the characteristics of amorphous silicon. Second, the lifetime of the hot wire is not preferable. Specifically, metal silicide is easily formed on the surface of the hot wire when the temperature of the hot wire is not particularly high. Since the metal silicide easily breaks the hot wire, the lifetime of the tungsten wire depends on the formation of the metal silicide. Generally, metal silicide is formed at both ends of the filament because the temperature there is lower than the temperature in the middle of the filament. Therefore, many different engineering designs are conducted to increase the lifetime of the hot wire. For example, an additional outlet of hydrogen or inert gas is installed at both ends of the hot wire to reduce the concentration of silane there, contributing to reducing the formation of metal silicide and increasing the lifetime of the hot wire. Finally, the uniformity of the film prepared by the hot-wire chemical vapor deposition process is worse than that of the RF plasma chemical vapor deposition process, and the performance of the prepared amorphous silicon cells is not as high as that of the latter. To sum up, in the hot-wire chemical vapor deposition process, much work in device design and interface control needs to complete to obtain a place in industrial production.

In addition to the two main technologies of the PECVD process and the HWCVD process, many other deposition processes of a-Si thin-film have been developed. The following methods have not been widely used for various reasons. Firstly, the reactive sputtering deposition technology. It is used to bombard the target silicon with a mixed gas of hydrogen and argon and deposit the material of the target onto the substrate to form amorphous silicon or microcrystalline silicon. Secondly, the electron beam evaporation technology. The deposited material should be hydrotreated to reduce the density of defect states in the material because there is no hydrogen in the electron beam evaporation process. Thirdly, spontaneous chemical vapor deposition technology. Fourth, ultraviolet excitation and mercury sensitization can be used to perform photoinduced chemical vapor deposition. Fifth, the pulsed laser deposition technology. However, most of these deposition processes can only obtain poor-quality a-Si films compared to films and cells prepared by the PECVD process; besides, the cell efficiency is much lower compared to the latter; therefore, it cannot be applied to large-scale manufacturing of a-Si cells.

5.3.5 Deposition Technology of Microcrystalline Silicon

As mentioned above, microcrystalline silicon and nanocrystalline silicon essentially refer to the same material, suggesting that the silicon thin-film material contains a certain number of small crystal grains. Although the size of the grains is on the order of nanometers, it was called microcrystalline silicon in earlier studies. Later, people began to call it nanocrystalline silicon with the flourishing development of nanoscience. Besides, it is also referred to a material with small crystal grains as nanocrystalline silicon and a material with slightly larger crystal grains as microcrystalline silicon. However, there is no uniform regulation at present; microcrystalline silicon and nano silicon are mixed in the literature. The term microcrystalline silicon will be used in the following sections.

The rapid growth of high-quality microcrystalline silicon films is of great significance both in laboratory research and in the production process of the company. Since the absorption coefficient of microcrystalline silicon is not very high, it is necessary to use a microcrystalline silicon-absorbing layer with a thickness of more than 1 μm to effectively absorb the red portion of sunlight. However, the amorphous silicon layer in a-Si/uc-Si multi-junction amorphous silicon solar cells produced in large-scale industrial manufacturing is generally only 0.3 μm or even thinner. Therefore, the output of solar cells is mainly limited by the production of microcrystalline silicon in the pipeline operation, and the high growth rate becomes more and more essential. With the increase in the growth rate, the output will increase and the cost will fall; therefore, it can be further industrialized. Gradually, various solutions and growing processes have been proposed to solve this problem. The two main technologies are the HW CVD process and the PECVD process mentioned earlier. In the PECVD process, different plasma sources are used to prepare amorphous silicon films. Commonly used ones are the high-frequency/ultra-high-frequency capacitively coupled plasma, the thermal plasma, the inductively coupled plasma with high density and low gas pressure, the surface microwave plasma, and the electron cyclotron resonance plasma. Different plasma sources are applied in these techniques; there is a significant difference in the growth rate and film quality. These plasma sources are described in detail as follows.

Radio-Frequency Capacitively Coupled Plasmas (RF-CCP) is one of the simplest plasma-generating devices, a capacitor consisting of two parallel plate electrodes. Plasma can be generated by adding an RF voltage (typically 13.56 MHz). However, the simple device has a relatively limited number of plasmas generated at normal pressure (several hundred mTorr or less), and the deposition rate of a-Si films is only 1–2 Å/s, which cannot meet the needs of mass production.

Therefore, it is necessary to increase the power of the RF source to promote the decomposition of the silane gas in order to increase the deposition rate. However, the voltage on the capacitor plate increases and the energy of the plasma also increases as the power of the RF source increases. The high-energy ions bombard the surface of the growing film causing the defects of the film to increase and the quality of the film to deteriorate; thus, the photoelectric conversion efficiency of the cells is lowered causing more serious photodegradation.

Hence, the method of increasing the pressure of the reaction chamber to reduce the bombardment energy of the plasma is used to solve this problem. The gas pressure is typically 1–10 Torr, and the plasma energy is reduced after collision with the molecules of source gas, resulting in avoiding the problem of poor film quality caused by excessive plasma energy. The microcrystalline silicon film is deposited at a high gas pressure, a high density of power, a spray gas flow, and a small electrode spacing. Deposition in this mode not only increases the growth rate of the microcrystalline silicon film but also increases the density and quality of the microcrystalline silicon film. Besides, the source gas of silane or disilane is also fully decomposed and utilized in a near-depletion mode. Such a high-pressure zone is called the high-pressure depletion region (HPD) technology.

The deposition of amorphous silicon films using the high-pressure depletion region technology has been widely accepted. Increasing the pressure can significantly improve the film quality whether it is a small amount of research in the laboratory or mass production in the factory.

The relationship between process parameters and cell efficiency in different periods of the high-pressure depletion deposition technology is displayed in Table 5.6. It can be revealed from the table that the higher the growth rate, the worse the film quality, and the lower the photoelectric conversion efficiency of the produced cells. With the advancement of technology, the growth rate of the films and the photoelectric conversion efficiency of cells are getting higher and higher;

Table 5.6 Relationship between process parameters and cell efficiency in different periods of the high-pressure depletion deposition technology

Time	1998	2001	2001	2003	2003	2006	2006	2006
Gas pressure (torr)	4	1–8	1–8					
Growth rate (Å/s)	9.3	9	6	12	60	10	20	5
Temperature (°C)	350	<200						
Cell efficiency		6.2%	7.1%	8.1%		9.2%	7.9%	10%
Literature	[37]	[38]	[38]	[39]	[39]	[40]	[40]	[40]

however, the contradiction among the growth rate, film quality, and cell efficiency still exist. Under the same process conditions, the cells with a higher growth rate are relatively inefficient while the cells with a lower growth rate are relatively more efficient.

The basic structure of the UHF capacitively coupled plasma (VHF-CCP) process is identical to the RF capacitively coupled plasma (RF-CCP) process, except that the excitation frequency is higher, typically 27–300 MHz. Then, a more intense decomposition of the silane gas, a higher plasma density, and an increased growth rate are caused by the higher frequency RF excitation. The growth rate of VHF-CCP is higher than that of the microwave plasma. The microwave plasma has a higher excitation frequency with a standard value of 2.45 GHz. The growth rate of the microwave plasma process is extremely fast.

The thermal expansion plasma, also known as the plasma jet stream, was first developed by the University of Eindhoven in the Netherlands. The basic structure of the generating device is a low-pressure (typically several hundred mTorr) deposition chamber and a plasma-producing high-pressure (tens to thousands of Torr) small tubes or nozzles. The plasma is derived from a radio-frequency glow discharge or a direct current series arc; then, the high-pressure plasma generated is injected into the low-pressure chamber to promote the reaction of the silane. In the process, the plasma-generating region and the thin-film deposition region are separated, and the generated plasma is substantially free ions when it reaches the surface of the substrate; therefore, the grown amorphous silicon film has a good quality.

The generating device of the electron cyclotron resonance plasma (ECR) technology is an electron cyclotron resonance chamber surrounded by an electromagnet. There is a dielectric window on the chamber, through which the microwaves are transmitted by the waveguide. Besides, a plasma of cyclotron resonance is generated in the chamber by appropriately adjusting the magnetic strength of the electromagnet and the frequency of the incident microwave. It follows the principle of the electron cyclotron. After the electron is accelerated to a certain energy, the silane in the cyclotron resonance chamber can be decomposed to generate plasma and then diffused to the surface of the substrate to form an amorphous silicon film. Moreover, the energy of the plasma can be controlled by adjusting the voltage of the substrate because the substrate is not directly bombarded by the plasma to form a film, contributing to controlling the growth rate and quality of the film. A diagram of a common ECR generator is illustrated in Fig. 5.13.

5.3.6 Optimization of Thin-Film Materials Based on Silicon

The quality and characteristics of thin-film materials based on silicon are closely related to the preparation conditions. Increasing the quality of the film by optimizing the process conditions to improve the photoelectric conversion efficiency of the solar cell while increasing the growth rate and the yield to reduce the cost is an eternal goal of all amorphous silicon cells. However, various vapor deposition processes include

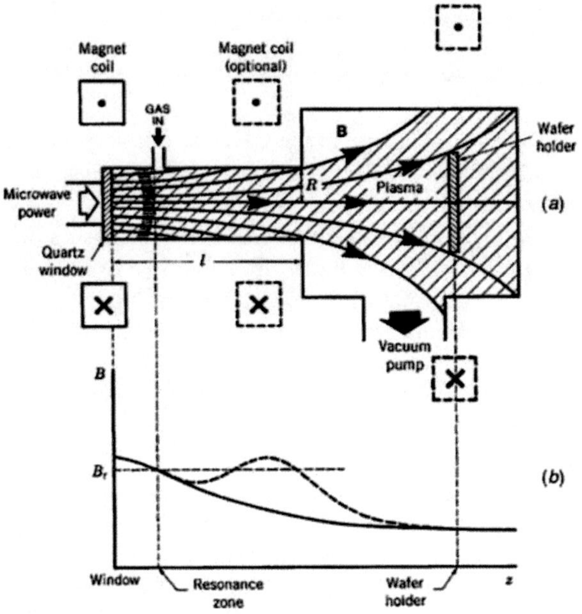

Fig. 5.13 Diagram of ECR plasma generation device [41]

many parameters such as the gas pressure of the reaction chamber, the power of the glow, the temperature of the substrate, and the flow of the gases. These parameters affect each other. Moreover, the structure of each reaction system is different, and a set of optimization parameters obtained from a specific system under specific conditions maybe not applicable to other systems. Therefore, the optimization of the preparation conditions of the film based on silicon still has achieved many improvements after several decades of research.

First, the optimization of the substrate temperature is discussed. The substrate temperature is an essential parameter determining the quality of the amorphous silicon film in various deposition methods. The temperature of the substrate can be considered in two aspects. From one perspective, increasing the substrate temperature helps to increase the diffusion coefficient of the particles and ions reaching the surface of the substrate thus, the particles and ions can diffuse a sufficient distance on the growth surface to find a position with lower energy. Therefore, raising the temperature of the substrate helps to improve the quality of the material (reduce the density of defects and micro-voids). From another perspective, the hydrogen content in the amorphous silicon could be lowered by a too high temperature of the substrate, resulting in increasing the density of the defect states in the material. These two opposite effects result in an optimum temperature of the substrate for the deposition of amorphous silicon, typically between 200 and 300 °C. The optimum temperature of the substrate also depends on other parameters. Firstly, the optimum temperature of the substrate is related to the deposition rate, and high-speed deposition requires a higher temperature of the substrate. During the high-speed deposition, a large surface

Table 5.7 Pressure of best reaction chambers in common growth processes

Name of processes	RF glow plasma chemical vapor deposition	Deposition process of microcrystalline silicon	Hot-wire chemical vapor deposition	Microwave plasma chemical vapor deposition	Electron resonance plasma chemical vapor deposition
Pressure of best reaction chamber	1 Torr	5–10 Torr	A few mTorr to tens of mTorr	A few mTorr to tens of mTorr	A few mTorr

diffusion rate is required for the particles and ions on the growth surface to find locations with lower energy in a shorter period. Secondly, a large number of hydrogen atoms cover the growth surface under the condition of higher hydrogen dilution, leading to effectively increasing the surface diffusion coefficient of particles and ions. The optimum temperature of the substrate can be relatively low under these conditions.

The optimum temperature of the substrate for amorphous germanium and amorphous germanium silicon is higher than that of amorphous silicon because the germanium hydride particles are heavier than the corresponding silicon hydride particles. Especially, the optimization of substrate temperature in the high-speed deposition process of the silicon-germanium alloy is particularly significant. Both heavier germanium hydride particles and high-speed deposition require higher substrate temperatures while germanium-hydrogen bonds are weaker than silicon-hydrogen bonds; thus, many dangling bonds of germanium in the material could be caused by a too high substrate temperature. Therefore, the density of defect states in amorphous silicon-germanium alloys is higher than that in amorphous silicon, and increases as the germanium content increases. The pressures of the best reaction chambers vary greatly in different deposition methods, as exhibited in Table 5.7.

The optimal pressure during the glow plasma deposition process is mainly discussed in this section since the current methods commonly used in the deposition of thin-film cells based on amorphous silicon are the radio-frequency and ultra-high-frequency glow plasma, and the two methods are essentially the same. The stability of the plasma should be considered in the choice of pressure. The lowest power capable of maintaining a stable plasma is related to the product of the pressure in the reaction chamber and the distance between the cathode and the substrate. Typically, a shorter distance between the cathode and the substrate requires higher pressure while a longer distance between the cathode and the substrate requires lower pressure. Another factor determining the optimal pressure is the dilution of hydrogen. A pure silane plasma requires low pressure while a high-hydrogen-diluted silane plasma requires high pressure. A simple guiding idea is to increase the dilution of hydrogen and keep the partial pressure of the silane in the reaction chamber constant simultaneously.

5.3.7 Production Process and Industrialization of Solar Cells Based on Amorphous Silicon

The production process of photovoltaic cells based on amorphous silicon is: washing of glass substrate → laser scribing → ultrasonic cleaning → PECVD → laser scribing → vacuum sputtering → laser scribing → connection → pre-test → encapsulation → junction box installation → test → processing → packing/storing.

The complete production line of thin-film photovoltaic cells based on amorphous silicon consists of a material handling system and synthesis equipment. Besides, an analytical inspection center and a series of quality monitoring equipment are required in the line to ensure the production of high-quality and low-cost products. In addition to the main production line, it is equipped with supply facilities of safe and efficient gas, as well as a complete gas monitoring and decontamination system to meet the requirements of safe operation and environmental protection.

The raw materials required for thin-film photovoltaic cells based on amorphous silicon are mainly substrate glass, TCO conductive glass, and special gases such as SiH_4, PH_3, and TMB. Other gases used majorly include N_2, O_2, He, H_2, Ar, NH_3, CH_4, as well as compressed air, and natural gas. The targets used for physical vapor deposition (PVD) mainly include ZnO, Al, and Ag.

The assembly line is also called the packaging line. The packaging is a key step in the production of photovoltaic cells; a good component board cannot be produced without a good packaging process. The package of cells ensures the lifetime of cells and enhances the anti-shock strength of cells. Customer satisfaction is based on the high quality and long lifetime of the product; therefore, the package quality of the component board is significant. The raw materials for the cell component mainly include backplane glass, PVB film, and junction box.

The manufacturing equipment of thin-film solar cell based on amorphous silicon accounts for more than 70% of the total cost of a production line of thin-film solar cells; the most critical equipment in the entire production line of thin-film solar cells is the deposition equipment of amorphous silicon film, accounting for the main part of the total cost of equipment. The simplest of these is the single-chamber device, that is, the p-i-n layer of the amorphous silicon cell is deposited in the same reaction chamber. The device has the advantages of low cost and stable operation and the disadvantage of the cross-contamination of the gas. Due to the low cost of the equipment, the cost of the corresponding solar cell is low, and the investor can recover the investment in a short period. The biggest complication of single-chamber equipment is the cross-contamination of reactive gases. The simple structure of the cell is p-i-n, in which a boron-containing gas is required for the growth of the p-layer; the usual gas is borane, trimethyl-boron, or boron trifluoride. After the p-layer is deposited, some residual boron-containing gas would always exist in the reaction chamber, resulting in affecting the quality of the intrinsic layer. Similarly, phosphorus-containing gas such as phosphine (PH_3) is required for the deposition of the n-layer. After the n-layer is deposited, the residual phosphorus-containing gas also has a certain influence on the p-layer of the bottom cell. Therefore, the reaction chamber is flushed with hydrogen after the deposition of each layer to reduce the effects of cross-contamination.

The single-chamber equipment has attracted the attention of many companies due to its low cost and stable operation, regardless of the cross-contamination of reactive gases. A single-chamber reaction system can effectively reduce the production costs if it is technically effective to reduce cross-contamination of dopant gases.

Besides, the multi-chamber reaction system is an essential means to produce high-efficiency thin-film solar cells based on silicon. The multi-chamber system can effectively avoid cross-contamination of reactive gases, reduce the content of impurities in the intrinsic layer, and improve the photoelectric conversion efficiency of solar cells. Simultaneously, different layers of the cell can be deposited. However, it has the disadvantage of high equipment costs and the high maintenance cost caused by maintaining many parts, which is not affordable for small companies. For large-scale production enterprises, the multi-chamber separation deposition system is still an essential deposition device for thin-film solar cells based on silicon and with glass as the substrate. Some major semiconductor equipment companies, including Applied Materials in the United States, have begun to investigate and develop large-scale plasma glow discharge deposition equipment for thin-film solar cells based on silicon.

The technical characteristics of thin-film photovoltaic cells based on amorphous silicon contribute to broad market prospects. However, the large-scale conductive glass, raw materials of the main gases, and the production equipment of thin-film cells required for the development of the thin-film photovoltaic industry are all dependent on imports and expensive, making the total investment of the project huge and the financing pressure great. Particularly, there are few technologies with Chinese independent intellectual property since the core technologies are monopolized by foreign countries, resulting in excessive dependence on foreign countries and then making the industry face bottlenecks. The cost is high, and a large amount of value-added profits are earned by countries with technological advantages. Therefore, it is necessary to master core technology, core equipment technology, and core management mode to form core competitiveness, improve the dominant role and the market voice of thin-film photovoltaic cell technology based on amorphous silicon, enhance the long-term competitiveness of the entire industry, reduce medium and long-term risks, and occupy a larger market of technology transfer and equipment manufacturing [42].

5.4 Common Structure and Working Principle of Solar Cells Based on Amorphous Silicon

Thin-film solar cells based on amorphous silicon have been developed for nearly 40 years since their birth. During this period, people designed cells of various structures to improve energy conversion efficiency. However, its basic structure determined by the basic characteristics of amorphous silicon still cannot be separated from the basic p-i-n structure, no matter how the structure changes. Multi-junction cells based on amorphous silicon are also obtained by connecting two or more p-i-n cells. In this chapter, the common structure and working principle of amorphous silicon solar cells in the sections of single-junction thin-film solar cells are introduced according to amorphous silicon-based amorphous silicon and multi-junction thin-film solar cells.

5.4.1 Structure and Working Principle of Single-Junction Thin-Film Solar Cell Based on Amorphous Silicon

1. **p-i-n and n-i-p structures and their working principles**

The mobility and lifetime of carriers in amorphous silicon materials are much lower than those in corresponding crystalline materials. The diffusion length of the carriers is also relatively short. With the usual PN junction cell structure, photogenerated carriers are recombined before they diffuse into the junction region. The absorption rate of light will be low and the corresponding photogenerated current will be small if a very thin material is used. Therefore, a thin-film cell based on silicon employs a p-i-n structure (for a transparent substrate) or an n-i-p structure (for an opaque substrate) to solve this problem.

The structure of the p-i-n cell is illustrated in Fig. 5.14b. First, a transparent conductive film (TCO) is deposited on a glass substrate. The transparent conductive film has two functions, that is, letting light enter the solar cell through the substrate and providing an electrode for collecting current (called a top electrode). Then, p-type, i-type, and n-type films of amorphous silicon are sequentially deposited, followed by the vapor deposition of the metal electrode, such as aluminum/titanium (A1/Ti). It has the following advantages. First, the p-i-n structure is a body junction formed using the p-layer and the i-layer, contributing to avoiding the influence of the interface state between the metal and the amorphous silicon on the cell characteristics; therefore, the cells prepared have preferable repeatability and stable performance. Second, the layers of the cells are all made of amorphous silicon with cheap material, simple process, and continuous production ability. Third, the flexibility of the design is excellent. Recently, people have mainly focused on the research of p-i-n cells because of these advantages of the p-i-n structure. Various forms of p-i-n structure have emerged, and the photoelectric conversion efficiency of cells has been greatly improved.

The structure corresponding to the p-i-n structure is an n-i-p structure, as illustrated in Fig. 5.14c. Such structures are typically deposited on opaque substrates such as stainless steel and plastic. Besides, the p-region of the thin-film cell based on silicon should be grown on the side close to the light-receiving surface since the mobility of holes in the film based on silicon is nearly two orders of magnitude smaller than that of electrons. Taking the substrate of opaque stainless steel as an

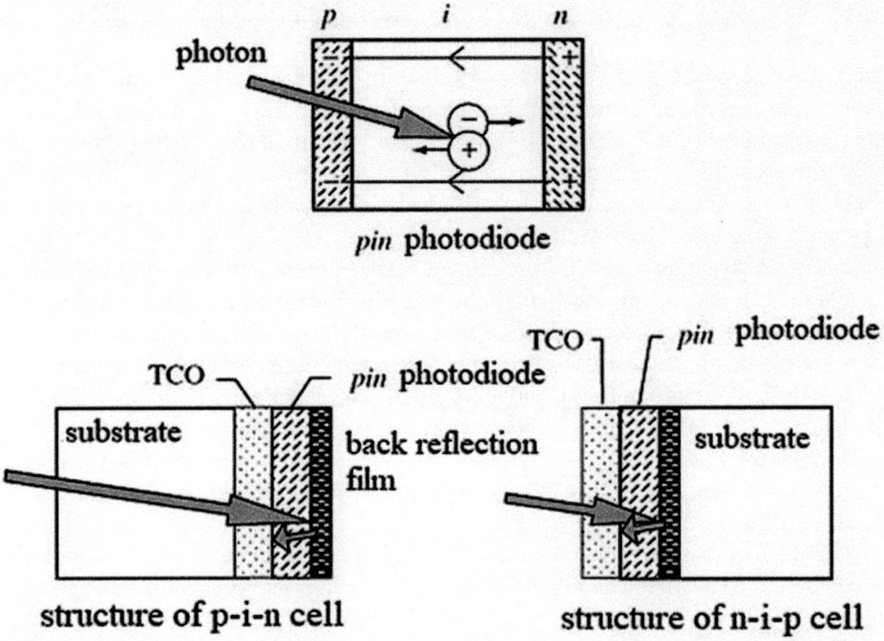

Fig. 5.14 Structures of p-i-n cells with a transparent substrate and n-i-p cells with an opaque substrate [22].

example, the best way to prepare the cell structure should be the n-i-p structure. Particularly, a back-reflection film is first deposited on the substrate; commonly used back-reflection films include silver/zinc oxide (Ag/ZnO) and aluminum/zinc oxide (Al/ZnO). Considering the factor of cost, silver/zinc oxide is commonly used in laboratories while aluminum/zinc oxide is generally used in the industrial production of large quantities of solar cells. An n-type, i-type, and p-type amorphous silicon or microcrystalline silicon material is sequentially deposited on the back-reflection film; then, a transparent conductive film TCO is deposited on the p-layer. Besides, the transparent conductive film may also be made of indium tin oxide (ITO) material. However, the thickness is made very thin, generally, only 70 nm to reduce the reflection loss of incident light since the surface conductivity of the ITO film is not high enough. The conductivity is further reduced; metallic gate lines are generally added to the ITO surface to increase the collection ratio of photocurrent.

Compared with the p-i-n structure, the n-i-p structure has the following characteristics. First, the n-layer is deposited on the back-reflection film. Particularly, zinc oxide is relatively stable and is not easily etched by hydrogen ions in the plasma because the usual back-reflection film is metal/zinc oxide; therefore, the n-layer may be amorphous silicon or microcrystalline silicon. Besides, the range of deposition parameters of the n-layer is relatively wide because the mobility of electrons is much higher than the mobility of holes. Second, the p-layer is deposited on the intrinsic layer; thus, the p-layer can be made of microcrystalline silicon. There are many advantages to using a p-layer made of microcrystalline silicon. For example, the short-wave response of the cell is good since the microwave silicon has a smaller short-wave absorption coefficient than amorphous silicon; the doping efficiency of the p-layer made of microcrystalline silicon is higher than that of the p-layer made of amorphous silicon, and the corresponding electrical conductivity is high. Moreover, the open-circuit voltage of the cell can be effectively increased through using a p-layer made of microcrystalline silicon.

The n-i-p structure also has some drawbacks. First, the effective light-receiving area of the cell is reduced since a metallic gate electrode is added to the top electrode ITO to increase the collection ratio of the current. Second, the requirements for the back-reflection film are relatively high because the thickness of ITO is very thin, ITO itself is difficult to have a rough and textured structure, and the light scattering effect of such a cell mainly depends on the textured structure of the back-reflection film.

Its working body is the p-i-n photodiode in the middle whether it is a cell with p-i-n or n-i-p structure. The working principle of p-i-n photodiode is discussed in detail as follows.

The p-i-n-type structure made of a-Si: H is a cell structure produced with the development of a-Si:H. The main difference between it and crystalline silicon solar cells is the addition of an intrinsic layer, namely, the i-layer, which is also an inevitable result caused by the properties of amorphous silicon. In amorphous silicon materials, the diffusion length is short due to the low mobility of carriers. The photogenerated carriers will quickly recombine due to the limitation of the diffusion length and cannot be collected when the photogenerated carriers are generated and there is no electric field at or near the vicinity. As mentioned earlier, doping introduces defects into amorphous silicon. p-type and n-type amorphous silicon has a high density of defects and a high recombination rate of photogenerated carriers; they can only be used in amorphous silicon cells to establish built-in potential and ohmic contact instead of as a light-absorbing layer. Therefore, amorphous silicon solar cells rely on the intrinsic layer to absorb sunlight and must adopt a p-i-n structure. Then, the following features are required to effectively collect photogenerated carriers: first, the intrinsic i-layer must be used to absorb photons; second, the entire range of light injection in the amorphous silicon solar cell should be filled with electric fields. In the cell of the p-i-n structure, the theory of the PN junction indicates that the space charge region is mainly distributed on the side with the lower doping concentration. Therefore, the space charge region of both the PI junction and the IN junction is mainly on the side of the i-region. Besides, the i-layer can be made very thin; the built-in electric field formed by the PI junction and the IN junction almost spans the entire intrinsic layer. The photogenerated carriers in the i-layer are completely placed in the electric field and can be collected once generated. Thus, the photoelectric conversion efficiency of cells can be significantly improved by the p-i-n structure. Apparently, this also requires that the intrinsic layer have a higher generation rate of photogenerated carriers, a low density of defect states, and suitable thickness. Thus, it is critical to prepare a high-quality i-layer and to explore a suitable thickness of i-layer. The thickness of the cell with the p-i-n structure is approximately within 1 μm if the glass substrate is not considered.

An ideal band diagram of a p-i-n cell based on a-Si:H in dark and light conditions is illustrated in Fig. 5.15. The value of E_{Fh} at $x = 0$ is the open-circuit voltage, and the built-in potential V_B is also given in the figure. The bandgap of the p-type layer is 2.0 eV, and the bandgap of the intrinsic i-type layer in the middle and the bandgap of the n-type layer are both 1.8 eV. The Fermi level E_F in the p-type layer is 1.7 eV lower than the bottom of conduction band E_C while the Fermi level E_F in the n-type layer is 0.05 eV lower than E_C. Generally, it is expected that the bandgap of the p-type layer is larger enough to increase the built-in potential and then increase the electric field in the i-layer, contributing to the collection of carriers. In a generally designed cell structure based on amorphous silicon, light is incident through a p-type layer (generally referred to as a p-type layer as a window layer). Besides, the

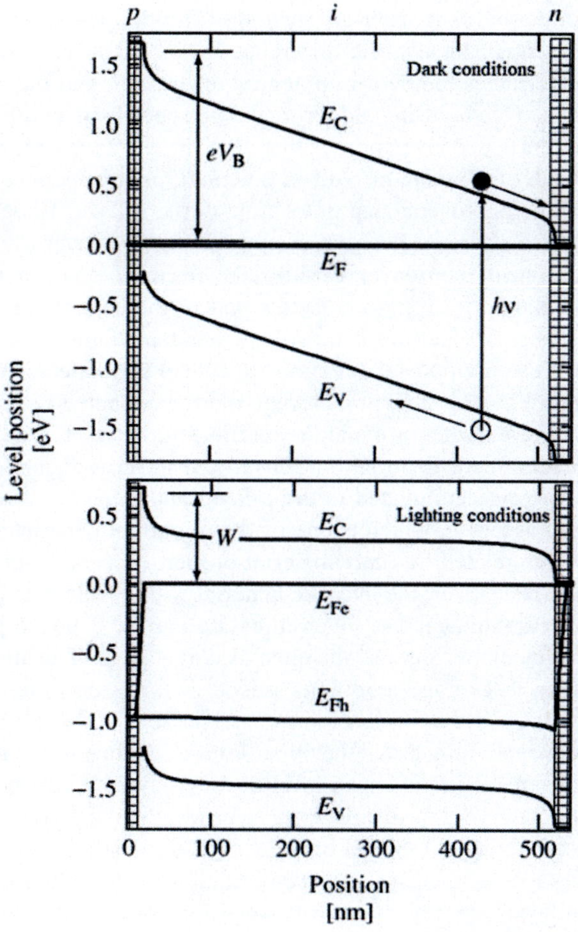

Fig. 5.15 Band diagram of the p-i-n cell under open-circuit conditions [22]

mobility of holes is greatly reduced because the negatively charged dangling bonds introduced by doping are very effective hole traps. However, the transmission of electrons and holes is an essential element for photovoltaic devices. The photons absorbed in the doped a-Si layer contribute little to the output power of the solar cell, and a larger bandgap can reduce the absorption of light by the p-type layer, allowing more light to reach the i-layer. Moreover, these Fermi levels must reach a level to achieve thermal equilibrium, and excess electrons in the n-type layer diffuse into the p-type layer when the p-i-n junction is formed, creating a built-in electric field. Thus, the positions of the E_C and E_V vary with the position within the device while the Fermi level remains unchanged. The value of the built-in potential V_B is the difference of the initial Fermi level, and the electrons and holes generated by the device absorbing photons drift in the direction suggested by the built-in electric field.

In the thermal equilibrium state without illumination, the three layers in the p-i-n structure have the same Fermi level, and the conduction band and the valence band in the intrinsic layer are inclined from the p-layer to the n-layer to form a built-in potential. Ideally, the difference in the Fermi level of the p-layer and the n-layer determines the built-in potential of the cell, and the corresponding electric field is called the built-in field. From the perspective of a high concentration of defect states in the doped layer, photogenerated carriers are mainly produced in the intrinsic layer. Under the action of built-in potential, photogenerated electrons flow to the n-layer while photogenerated holes flow to the p-layer. Under the open-circuit conditions, photogenerated electrons accumulate in the n-layer while photogenerated holes accumulate in the p-layer. Besides, the electric field generated in the p-layer and n-layer of the intrinsic layer offsets a part of the built-in field. Simultaneously, the photogenerated electrons in the n-layer and the photogenerated holes in the p-layer tend to diffuse in opposite directions to cancel the collected current of the photogenerated carriers. The diffusion current and the collected current under the built-in field are achieved when a dynamic equilibrium between the two opposite currents; meanwhile, there is no net current in the intrinsic layer. The voltage generated by the charges accumulated in the p-layer and the n-layer at this time is the open-circuit voltage V_{OC}, which is one of the important parameters of solar cells [43] with the value related to many material properties. First, it depends on the bandgap of the intrinsic layer; the intrinsic material with a wide bandgap can produce a large open-circuit voltage. The material with a narrow bandgap produces a small open-circuit voltage; for example, the open-circuit voltage of an amorphous silicon cell is smaller than that of an amorphous silicon cell. Besides, the magnitude of the open-circuit voltage also depends on the characteristics of the doped layer, especially the doping concentration. Since the upper limit of the open-circuit voltage is determined by the difference in the Fermi level of the n-layer and the p-layer, the optimization of the doped layer is also quite critical, especially the p-layer. Therefore, amorphous silicon carbide alloy a-SiC:H or microcrystalline silicon is usually used as the material of p-layer to increase the open-circuit voltage. Although amorphous silicon carbide alloys generally have a relatively high density of defect states, their relatively wide bandgap allows their Fermi level to be relatively low. Moreover, its wide bandgap

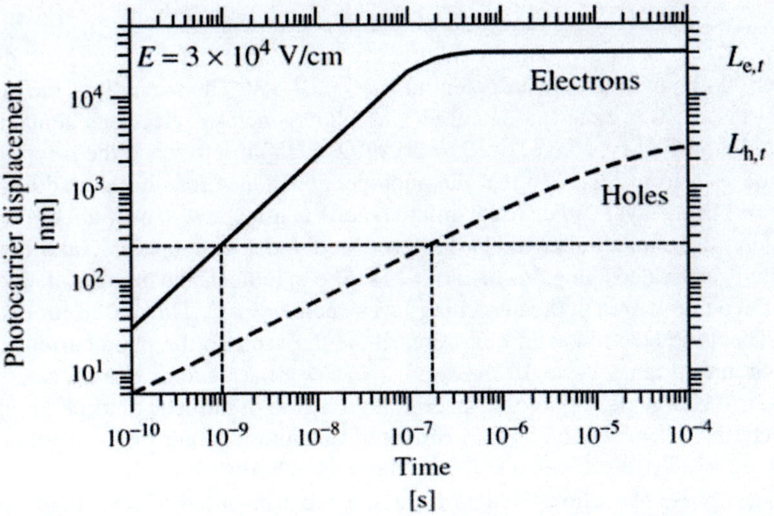

Fig. 5.16 Drift length of photogenerated electrons and holes in a-Si:H at room temperature with time going by [44]

can reduce the absorption in the p-layer. The microcrystalline silicon has a small width of band tail and high doping efficiency; besides, the Fermi level can approach the top of the valence band; thus, the microcrystalline silicon can also increase the magnitude of the open-circuit voltage. Finally, the magnitude of the open-circuit voltage also depends on the quality of the intrinsic layer, namely, the magnitude of the reverse leakage current determined by the width of the band tail and the density of the defect states.

Next, the behavior of electrons and holes in the p-i-n junction will be further discussed. The relationship between the drift length of photogenerated electrons and holes in a-Si:H at room temperature with time is illustrated in Fig. 5.16. The strength of the built-in electric field is $E = 3 \times 10^4$ V/cm, which is a typical value of the built-in electric field of the p-i-n cell with an i-type layer and a thickness of 500 nm under short-circuit conditions [44].

The behavior of electrons is first considered. The drift distance of the electron is proportional to the time (in logarithmic coordinates) at the first 10^{-10} to 10^{-7} s of photogenerated electron generation, $L(t) = \mu_e Et$. Among them, μ_e is the mobility of the electron, with a typical value of 1 $cm^2/(V{\cdot}s)$, which is much smaller than the mobility of electrons in crystalline silicon (about 1000 $cm^2/(V{\cdot}s)$ at room temperature). After 10^{-7} s, the drift distance no longer changes with time, reaching a saturation value of $L_{e,t} = 3 \times 10^{-3}$ cm. This effect is caused by deep traps and defects of trapping electrons.

How the behavior of electrons affects the operation of a-Si:H cells under short-circuit conditions is discussed as follows. In this section, the charge generation inside the cell under illumination conditions is mainly described, considering the movement time of electrons in the cell under short-circuit conditions. The built-in electric field is $E \approx V_B/d = 3\times10^4$ V/cm if the thickness of the i-type absorbing layer (undoped) is $d = 500$ nm and the built-in potential is $V_B = 1.5$ V. The strength of the electric field in Fig. 5.16 is exactly this value. The photogenerated electrons generated in the middle of the i-type layer need to drift $d/2 = 250$ nm to reach the n-type layer. It can be seen from Fig. 5.16 that the photogenerated electrons in the middle of the i-type absorbing layer can drift 250 nm to reach the n-layer with only about $t = 1$ ns. Therefore, the total number of charges produced under illumination conditions can be roughly estimated using this result. If ζ is used to indicate the number of electrons produced per unit area in the absorbing layer, then $\zeta = jt/2$. The reason for dividing by 2 is because electrons and holes contribute the same to the photocurrent. If the short-circuit current $J_{SC} = 10$ mA/cm^2, $\zeta = 5 \times 10^{-12}$ C/cm^2 can be calculated. The typical charge density of the space charge region in intrinsic amorphous silicon is about 10^{-8} C/cm^2, which is four orders of magnitude larger than the value of ζ; therefore, drifting electrons have little effect on the built-in electric field.

Next, the case of the holes will be discussed. It can be observed from Fig. 5.16 that the migration speed of holes is much smaller (a few orders of magnitude) than that of electrons. Compared with electrons, it takes about 180 ns for the photogenerated holes in the middle of the i-layer to reach the p-layer. The total number of holes per unit area with the same short-circuit current density is about 9×10^{-10} C/cm^2, which shifts a certain amount from the density of the space charge region of silicon; however, it would not have much impact. To sum up, the effect of photogenerated carriers on the built-in electric field is very small.

As the absorbing layer of light, the design optimization of the i-type absorbing layer has a great influence on the absorption efficiency of light and even the photoelectric conversion efficiency of the p-i-n cell. First, the effect of the thickness of the absorbing layer on the absorption of light is analyzed. In Fig. 5.17, the computerized efficiency of an a-Si:H p-i-n cell as a function of the thickness of the absorbing layer is illustrated. The different curves in the figure represent the results of illumination with monochromatic light of different wavelengths at the same illumination intensity.

The different curves in Fig. 5.17 are the result of different absorption coefficients (5000–100,000 cm^{-1}) for different kinds of monochromatic light. For a typical a-Si:H material, the photon energy corresponding to this range of absorption coefficients ranges from 1.8 to 2.5 eV. The solid dots correspond to the incident light passing through the p-type layer, and the hollow dots are the incident light passing through the n-type layer.

First, the incident light passing through the p-t[ype layer (window layer) is considered (the solid point in Fig. 5.17). Since the intrinsic absorbing layer is very thin, the absorbed energy is generally proportional to the number of absorbed photons (the product of the absorption coefficient α and the thickness of the absorbing layer d). Besides, the absorbed energy gradually becomes saturated as the thickness of the absorbing layer increases. The highest absorption coefficient can be observed in the figure ($\alpha = 100{,}000$ cm^{-1} corresponds to a photon with an energy of 2.3 eV). The photon absorption would reach saturation when the thickness of the absorbing layer is $d > 100$ nm (the typical thickness value when the incident light can be completely

Fig. 5.17 Absorption of light by the a-Si:H p-i-n cell as a function of the thickness of the absorbing layer [22]

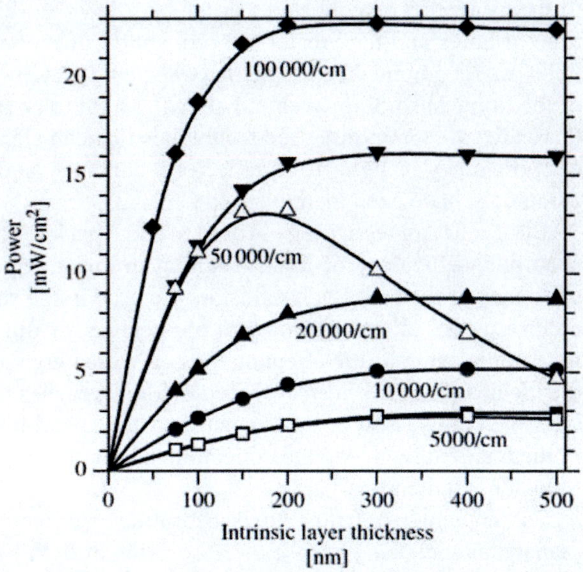

absorbed). Since no more photons can be absorbed by increasing the thickness, the electric photoconversion efficiency of the cell no longer changes with thickness.

Besides, the calculation of the photoelectric conversion efficiency of the cell when the incident light is incident from the n-type layer (hollow point) is also illustrated in Fig. 5.17. For the case of incident light with a small absorption coefficient (5,000 cm^{-1}), the photogenerated carriers in the range of the entire absorption layer are substantially the same as the case where the incident light passes through the p-type layer, and the photoelectric conversion efficiency of the cell is also substantially the same. Then, the case where incident light with a large absorption coefficient (50,000 cm^{-1}) passes through the n-type layer is considered. Moreover, the photogenerated carriers are still substantially the same as the case where incident light passes through the p-type layer when the thickness of the absorbing layer is less than the saturated thickness (here, 200 nm); meanwhile, the photoelectric conversion efficiency of the cell is also substantially the same. However, the photoelectric conversion efficiency of the cell in which the n-type layer is incident drops significantly when the thickness of the absorbing layer is further increased; meanwhile, the photoelectric conversion efficiency of the cell in which the p-type layer is incident remains unchanged. The reasons are as follows: (1) increasing the thickness cannot increase the absorption of light when the thickness of the absorbing layer is sufficient to absorb most of the incident light; (2) generally, the holes generated in the absorption layer need to drift longer distances to reach the p-type layer compared to the case where light is incident from the p-type layer when incident light passes through the n-type layer. As mentioned earlier, the mobility of holes is several orders of magnitude lower than that of electrons. The time of holes generated in the i-layer drifting to the p-layer to be collected is much longer than the time of electrons drifting to the n-layer to be collected. The thicker the absorbing layer, the longer the drift time of holes and the greater the probability of recombination. A large number of holes are recombined before being collected. In addition to optimizing the thickness of the i-type absorbing layer and allowing light to be incident from the p-type layer, there are two commonly used techniques that can significantly increase the absorption efficiency of light: back reflector technology and texturing technology. These techniques have been mentioned in Chap. 3.

Back reflector technology is used to add a back reflector to the cell to increase the absorption efficiency of light, resulting in improving the photoelectric conversion efficiency of the cell. Back reflectors are used in the structural diagrams of the p-i-n and n-i-p cells at the beginning of this section, as illustrated in Fig. 5.14. The back reflector has no or little effect on the cell with a high absorption coefficient because the incident light has been absorbed before it reaches the reflector. It can be revealed after the comparison of the thickness of the absorbing layer, the absorption length of the incident light, and the collection length of the holes that the effect of the back reflector is quite remarkable.

For incident light with a small absorption coefficient, the back reflector in a simple planar solar cell can increase the absorption of light by a factor of two, contributing to increasing the output power by a factor of two and greatly improving the device performance. The basic principle is the total reflection of light, and the working principle is the same as that of fiber. A beam of light entering the fiber from one end can travel for several kilometers without loss. For a solar cell, the light energy can be completely absorbed by using a good back reflector, especially when the thickness of the absorbing layer is less than $1/\alpha$ of the absorption length of the light with a wavelength of α. The reflection and diffraction of light are employed in the texture technique. Light trapping of A-Si: H and some other solar cells can be achieved by comparing the roughness of the texture with the wavelength of the light under sunlight. The textured topography can induce random reflection and diffraction, resulting in prolonging the optical path and increases the absorption of light, as illustrated in Fig. 5.18. Yablonovitch's [45] research indicates that the simultaneous use of these two techniques, that is, the deposition of a layer of texture film on an ideal reflector, can achieve a theoretical maximum gain of $4n^2$, where n is the refractive index of the texture. The refractive index n of a silicon film is about 3.5 and the expected gain is 50 (for light with a very small absorption coefficient).

Generally, the main function of back reflector technology and the texturing technology is used to reduce the threshold energy of the photon absorbed by the cell for incident light with lower energy (approximately 0.2 eV can be reduced, as illustrated in Fig. 5.17). The light intensity absorbed by a cell with a thickness of 0.5 μm can be increased from 420 to 520 W/m^2 by using the reduced threshold energy of 0.2 eV. The effect of the reflectivity of the back reflector on the output power of the a-Si:H p-i-n solar cell and the effect of the texture technique on the quantum efficiency of the a-Si:H p-i-n cell are illustrated in Fig. 5.19.

In both p-i-n and n-i-p cells, the implementations of back reflector technology and texture technology are different. For n-i-p cells, a layer of texturing TCO film is typically spin-coated on a transparent substrate (typically glass). There are

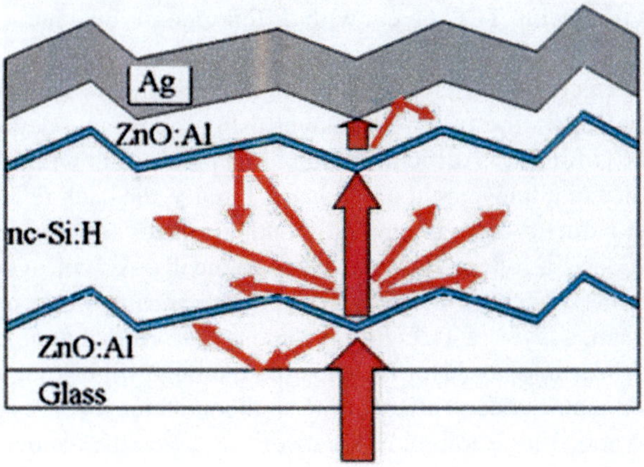

Fig. 5.18 Propagation path of light in p-i-n cells using texturing ZnO:Al reflector technology [46]

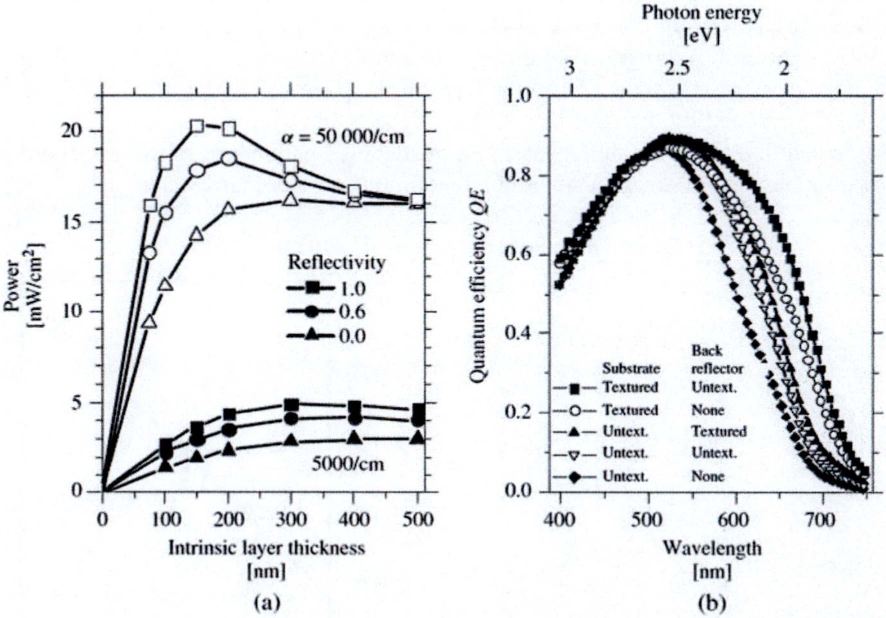

Fig. 5.19 Function of back reflector technology and texture technology. **a** Effect of reflectivity of the back reflector on the output power of a-Si:H p-i-n solar cells. **b** Effect of texture technology on the quantum efficiency of a-Si:H p-i-n solar cells [47]

many processes producing TCO layers with different texturing characteristics using various materials (typically SnO_2 and ZnO for a-Si:H cells). After the TCO layer is completed, a semiconductor layer is deposited thereon. However, some problems are caused by the deposition of a p-type layer using the plasma deposition method on a texturing TCO film; it is difficult to prepare a p-type film with good properties due to the presence of a transparent oxide layer. Finally, the back reflector deposited on the top semiconductor layer is typically a double-layer structure: a thin layer of TCO, below which is a reflective metal layer (typically Ag with good reflectivity or Al with low cost). In the p-i-n cell, a layer of texturing silver or aluminum is first deposited; then, a layer of TCO film is deposited to form a double-layered back reflector and a semiconductor layer is deposited thereon [48]. Moreover, a top TCO layer is deposited after the deposition of the semiconductor layer is completed.

In addition to absorbing sunlight as an absorbing layer, the bandgap of the material of the i-type layer also affects the open-circuit voltage of the cell [49]. The relationship between the open-circuit voltage of the solar cell based on a-Si:H and the optical bandgap of the intrinsic absorption layer [50] are illustrated in Fig. 5.20. The empirical formula for the open-circuit voltage is $V_{OC} = (E_g/e) - 0.80$.

The relationship between V_{OC} and E_g is so simple because the open-circuit voltage is mainly determined by the band structure of the p-i-n cell under illumination and is hardly affected by the thickness of the i-type absorption layer and the intensity of incident light. The result suggests that the detailed parameters of the cell and the measurement environment have little effect on the open-circuit voltage.

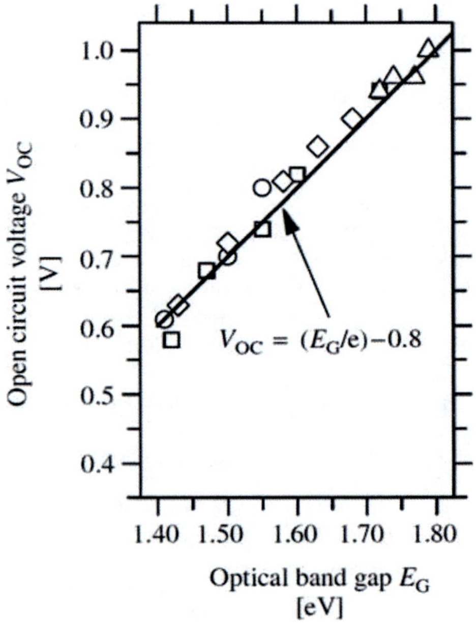

Fig. 5.20 Relationship between the bandgap of i-layer and the open-circuit voltage of the cell

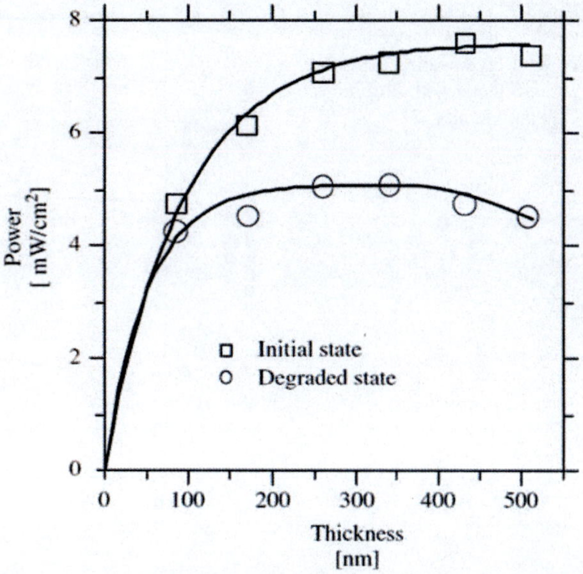

Fig. 5.21 Initial power and power after 25,000 h of illumination of n-i-p solar cells in which the intrinsic absorption layer has different thicknesses [51]

Finally, the effect of the thickness of the absorbing layer on the photodegradation effect is discussed. In Fig. 5.21, the relationship between the output power and different thicknesses of the absorption layer in a batch of n-i-p solar cells produced by United Solar Corporation under standard illumination is illustrated. These cells are n-i-p solar cells grown on stainless steel. One of the two curves is the initial output power; the other is the output power after 25,000 h of operation. It can be observed from the initial output power curve of the battery that the output power increases as the thickness increases when the thickness is small (<400 nm) while it saturates when the thickness is greater than 400 nm; meanwhile, the output power remains almost unchanged. In the output characteristic curve of the degraded cell, the cell has reached its maximum output power at a thickness of 200–300 nm. Afterward, the output power decreases to some extent as the thickness increases. The specific cause of this phenomenon is still unclear at present; it may be related to the back reflector or to the characteristics of the i-type layer itself. Besides, the greater the thickness of the absorbing layer, the greater the degree of degradation; this should be considered when optimizing the thickness of the absorbing layer.

2. **Single-junction solar cell based on a-SiGe:H**

Since the bandgap of amorphous silicon is 1.7–1.8 eV, the corresponding long-wave absorption is relatively small. Therefore, the a-SiGe:H alloy is the first choice for intrinsic materials with a narrow bandgap to improve the long-wave response of the cell. Specifically, the bandgap of the material can be adjusted accordingly by controlling the ratio of germanium to silicon in the material. Besides, the bandgap of the material decreases as the content of germanium increases. The long-wave response of the cell is increased and the corresponding short-circuit current is increased. This

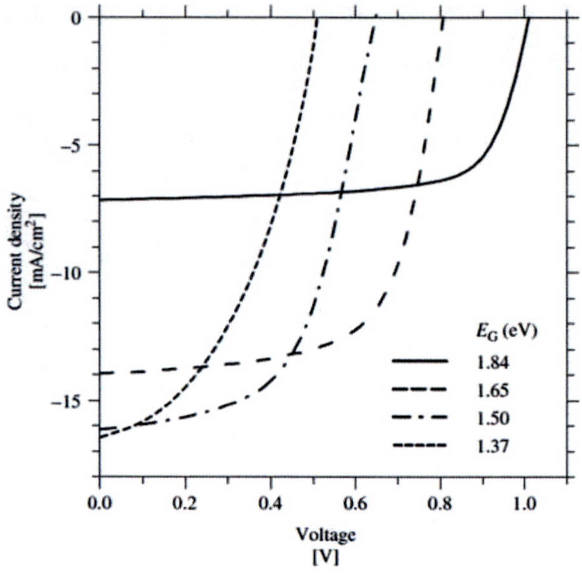

Fig. 5.22 Characteristic curve of p-i-n solar cells in which i-layers being a-Si:H and a-SiGe:H with different Ge component [52]

trend is illustrated in Fig. 5.22. At the cost, however, the open-circuit voltage of the cell is reduced and the fill factor of the cell is also reduced. Moreover, the density of defects in the a-SiGe:H alloy also increases as the Ge composition increases. With the increase of the density of defects in the i-type intrinsic absorption layer, the effect of the recombination center of defects begins to exceed that of the band-tail states and becomes the main factor affecting the performance of the cell.

It can be observed that with the increase in the Ge composition, the fill factor of the cell drops rapidly when the bandgap of the a-SiGe alloy falls below 1.4 eV. This is because the Ge component is too large, introducing too many defects. Although more photons are absorbed, a too high density of defects could cause the recombination of the carrier to occupy a more essential position because the energy lost by the recombination has exceeded the excess energy absorbed due to the reduction of the bandgap.

High-hydrogen doped a-SiGe:H films and devices have higher quality and stability [53]. Meanwhile, the bandgap grading technique can improve the collection ability of holes [54, 55], leading to an increase in the fill factor of the a-SiGe cell. Device performance can be improved by adjusting the Ge composition in the i-type layer to form an asymmetric V-shaped bandgap profile. Near the n-type layer and the p-type layer, the a-SiGe:H material with a wider bandgap is used, the bandgap from the two ends to the middle is narrower and narrower, and the narrowest part is near the p-type layer. Such a gradual design allows incident light to be absorbed near the p-type layer, and holes with less mobility can be collected over a shorter distance. Simultaneously, the inclination of the valence band also makes it easier for the holes generated in the intermediate region of the i-type layer or near the N-region to move toward the p-type layer. During the growth of a-SiGe:H, the output current of the a-SiGe:H cell can be optimized to 24.4 mA/cm^2 if the hydrogen-doping technology, the bandgap grading technique, and optical enhancement techniques such as back reflectors are adopted [56].

3. **Single-junction solar cell based on microcrystalline silicon**

Recently, microcrystalline silicon cells have been deeply investigated and applied as a bottom or intermediate cell for multi-junction cells. Compared with amorphous silicon, the diffusion length of carriers in the material is still small, even though the transport characteristics of carriers in intrinsic microcrystalline silicon are significantly improved. For example, the diffusion length of holes in higher quality microcrystalline silicon is still less than 1 um. Therefore, the microcrystalline silicon cell still adopts a p-i-n or n-i-p structure similar to amorphous silicon in order to improve the collection efficiency of photogenerated carriers by the cell.

Although microcrystalline silicon absorbs light at a higher frequency compared to amorphous silicon, its absorption coefficient is much smaller than that of amorphous silicon. The absorption spectra of amorphous and microcrystalline silicon are illustrated in Fig. 5.23. Besides, the intrinsic layer of the microcrystalline silicon cell is required to be much thicker than that of the amorphous silicon cell to increase the short-circuit current of the cell. Generally, the thickness of the intrinsic layer of microcrystalline silicon is 1–2 μm. Since the microcrystalline silicon cell can absorb the light with a wider frequency band, the short-circuit current of the single-junction cell based on microcrystalline silicon is larger than that of the single-junction cell based on amorphous silicon in the case where the thickness is sufficient.

The optical bandgap of microcrystalline silicon is between amorphous silicon and single crystalline silicon; the long-wave response of the cell is determined by the amount of nano-grain composition in the material. Besides, the electrical bandgap of microcrystalline silicon is close to that of single crystalline silicon. By measuring the

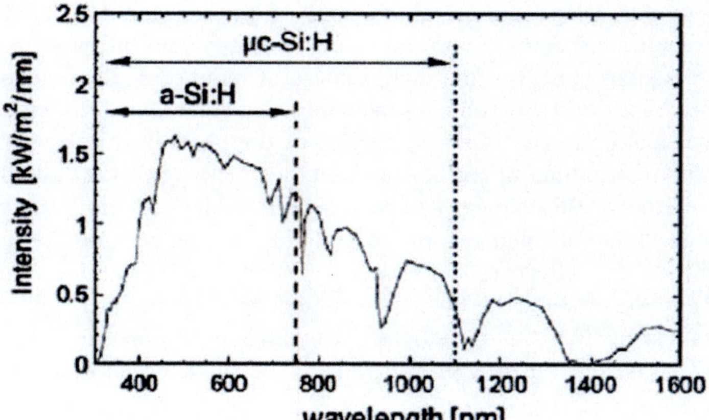

Fig. 5.23 Spectrum distribution of AM1.5 sunlight and absorption spectrum of a-Si:H and uc-Si:H [57]

relationship between temperature and the reverse saturation current of a microcrystalline silicon cell, B. Yan et al. revealed that the electrical bandgap of microcrystalline silicon is 1.2–1.3 eV. The smaller bandgap determines that the open-circuit voltage of the microcrystalline silicon cell is smaller than that of the amorphous silicon cell; meanwhile, the open-circuit voltage of the microcrystalline silicon cell is generally about 0.5 V.

Compared with amorphous silicon cells, microcrystalline silicon cells are more sensitive to impurities. The sensitivity of microcrystalline silicon cells to impurities mainly comes from the sensitivity to oxygen. Oxygen atoms could form a weak donor doping in the microcrystalline silicon, causing the Fermi level in the intrinsic layer of the microcrystalline silicon cell to move upward. Simultaneously, the built-in field in the intrinsic layer is concentrated on the p-i interface, resulting in reducing the strength of the built-in field between most regions in the intrinsic layer and the N-region; then, the long-wave response of the cell is deteriorated.

Non-optimized microcrystalline silicon materials are porous. The prepared material is placed in a normal room temperature environment; its resistivity increases with time, leading to a decrease in efficiency. The diffusion of impurities has a potential impact on the performance of microcrystalline silicon solar cells while the efficiency of p-i-n microcrystalline silicon cells generally does not decline. This is because this cell is deposited on a glass substrate and the back of the cell is a metal electrode. First, impurities cannot diffuse into the microcrystalline silicon through the glass substrate. Second, the metal back electrode is a good impurity barrier. Therefore, the performance of this p-i-n cell will not significantly decline with time, even if microcrystalline silicon has a certain degree of porosity.

For microcrystalline silicon cells with n-i-p structure, the diffusion of impurities is a more serious problem. Since the upper electrode of the n-i-p cell is ITO that is not a good impurity (mainly water and oxygen) isolation layer, the problem of impurity diffusion of n-i-p microcrystalline cells is more prominent. Experiments have revealed that the photoelectric conversion efficiency of non-optimized microcrystalline silicon cells decreases significantly with time; the decrease in current is most pronounced mainly due to the decrease of quantum response in the long-wavelength region. However, the compactness of microcrystalline silicon can be significantly improved by optimizing the deposition conditions of microcrystalline silicon, contributing to reducing the decline of the photoelectric conversion efficiency of microcrystalline silicon cells caused by impurity diffusion. Another major feature of microcrystalline silicon cell is great light-induced stability. Many research institutions have investigated that microcrystalline silicon cells do not have light-induced degradation. However, not all microcrystalline silicon cells are stable after prolonged exposure to light due to the structural diversity of microcrystalline silicon, especially those microcrystalline silicon cells with a large proportion of amorphous components. It is not surprising that degradation occurs in bright light.

5.4.2 *Structure and Working Principle of Multi-junction Thin-Film Solar Cells Based on Amorphous Silicon*

1. **Advantages of multi-junction solar cells**

Since sunlight has a very wide spectrum, useful spectral regions for solar cells cover ultraviolet light, visible light, and infrared light. Apparently, the energy of all solar photons cannot be effectively used through a semiconductor material with a single bandgap. From one perspective, the absorption coefficient for light with photon energy smaller than the bandgap of semiconductor in the semiconductor is very small; it does not contribute to the photoelectric conversion efficiency of the solar cell. From another perspective, the effective energy for light with photon energy much larger than the bandgap of semiconductor is only the part of the bandgap while the part of the energy larger than the bandgap is lost through the formation of thermal electrons. Based on this principle, the use of multi-junction cells can effectively apply photons with different energies, contributing to the significant improvement of the photoelectric conversion efficiency of solar cells. The efficiency chart in the past few decades of various types of solar cells published in 2012 by the National Renewable Energy Laboratory (NREL) of the United States is illustrated in Fig. 5.24. They all represent the highest efficiency achieved in the world at that time and predict the future growth trend of the photoelectric conversion efficiency of solar cells. It can be observed from the figure that the multi-junction solar cell has the highest efficiency.

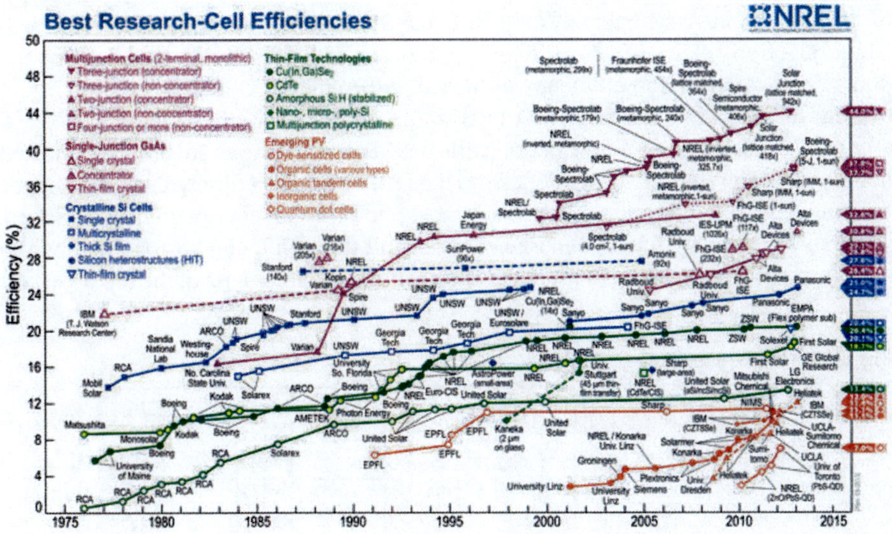

Fig. 5.24 Highest photoelectric conversion efficiency of different types of solar cells in different periods [58]

Apparently, the multi-junction solar cell may not be an amorphous silicon multi-junction cell while the multi-junction structure can greatly improve the photoelectric conversion efficiency of the cell and be applied to various types of solar cells.

In solar cells using amorphous silicon, amorphous germanium-silicon alloy, and microcrystalline silicon as absorbing materials, a double-junction or triple-junction cell structure is mostly used. The use of multi-junction cells can improve not only the effective utilization of photons in different spectral regions but also the stability of solar cells. As mentioned earlier, amorphous silicon and amorphous silicon-germanium produce light-induced defects under long-term light conditions. Besides, the effect of the same density of light-induced defect states on the photoelectric conversion efficiency of a solar cell with a thin intrinsic layer is smaller than that on a solar cell with a thick intrinsic layer. In a multi-junction cell, the thickness of each junction can be relatively thin, contributing to increasing the built-in field (assuming that the p-doped and n-doped layers of each sub-cell are the same as those of the single-junction cell). Therefore, multi-junction thin-film cells based on silicon are not only more efficient than single-junction cells but also more stable than single-junction cells.

Amorphous silicon solar cells can be prepared into a series structure through a controlled process to form a multi-junction solar cell. This multi-junction structure is particularly effective for improving the photoelectric conversion efficiency of amorphous silicon solar cells. There are two reasons: first, this multi-junction structure does not affect its single-node performance because the amorphous silicon material does not require lattice matching like crystalline silicon PN junctions; second, the optical bandgap of amorphous silicon can be well adjusted by forming an alloy. Multi-junction solar cells based on a-Si:H have a higher conversion efficiency than single-junction solar cells. Specifically, multi-junction solar cells can improve the photoelectric conversion efficiency of the cell due to the segmental absorption effect of light; besides, this has been involved in the introduction of III–V solar cells. A common structure of double-junction cells based on amorphous silicon is illustrated in Fig. 5.25. The sunlight is incident from the left. The a-Si:H film with a thickness of 500 nm in the figure is enough to absorb most photons with energy greater than 2 eV while the photons with lower energy can transmit through it and be absorbed by the a-SiGe:H absorption layer. In practical applications, the thickness of the top p-i-n cell

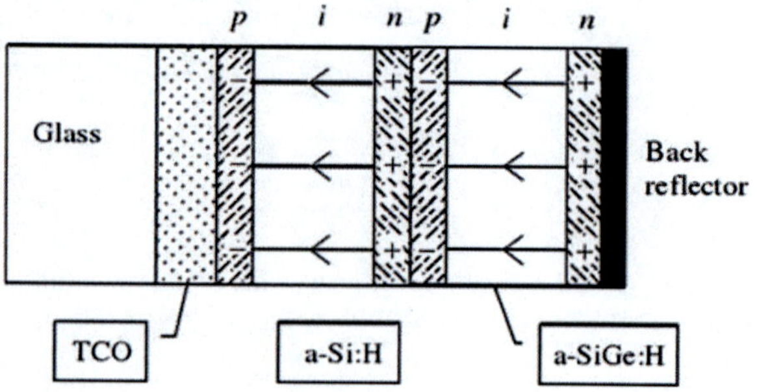

Fig. 5.25 Double-junction solar cell with two p-i-n junctions in series [22]

is adjusted to absorb about half of the photons; meanwhile, the rest photons transmit through the top layer and are absorbed by the bottom p-i-n junction. Moreover, a material with a relatively large bandgap can be used as the absorption layer of the top cell because the photons absorbed on the top layer have relatively high energy. In this way, a relatively high open-circuit voltage can be obtained on the top p-i-n structure. The bottom layer uses a material with a narrower bandgap to absorb photons with less energy that are not absorbed by the top layer. Ideally, the photovoltage of the entire device is equal to the sum of the photovoltages of the two sub-cells; the photocurrent is equal to the smaller one of the photocurrents of the two sub-cells. The fill factor of the entire device is determined by the fill factors of the two sub-cells and the difference between the photocurrents of the two sub-cells. Another important link is the connection of the two sub-cells. Specifically, the n-layer of the top cell and the p-layer of the bottom cell at the junction of the two cells are connected. This is a reverse PN junction. Assuming that the photocurrent flows in a tunnel recombination manner. In short, the electrons in the n-layer of the top cell enter the p-layer of the bottom cell through the tunneling effect and recombine with the holes therein, or the holes in the p-layer of the bottom cell enter the n-layer of the top cell through the tunneling effect and recombine with the electrons therein. Therefore, increasing the mobility of carriers is the most effective method to improve the tunneling effect. Consequently, a microcrystalline silicon p-layer or a microcrystalline silicon n-layer is usually used in practical devices.

Such a typical a-Si/a-SiGe double-junction tandem solar cell is illustrated in Fig. 5.25. The a-SiGe:H material of the bottom p-i-n junction has a bandgap of 1.55 eV while the a-Si:H material of the top layer has a bandgap of 1.80 eV. Before depositing the a-Si:H p-i-n junction with a bandgap of 1.80 eV on the top layer, the a-Si:H cell with a bandgap of 1.55 eV on the bottom layer can obtain a J_{SC} of 20 mA/cm^2 and a V_{OC} of 0.65 V under standard illumination. Besides, the output power can be calculated as $P = 9.1$ W/m^2 if its fill factor FF is 0.7. Moreover, the current density through each junction is about half of this value when two p-i-n cells are connected in series; meanwhile, the open-circuit voltage is greater than twice the open-circuit voltage of the bottom layer ($V_{OC} = 0.65 + 0.90 = 1.55$ V). The output power is increased to 11.2 W/m^2, which is 19% higher than that of a single-junction cell.

In the ideal case, the greater photoelectric conversion efficiency of from single-junction to tandem double-junction and triple-junction solar cells under the sunlight can be achieved through the proper matching of semiconductors with different bandgaps [59]. Computer simulations indicate that the photoelectric conversion efficiency of the dual-junction tandem solar cell can reach 20% using a-Si:H with a bandgap of 1.8 eV as the top absorption layer and a-Si:H with a bandgap of 1.2 eV as the bottom absorption layer. Apparently, such a high photoelectric conversion efficiency cannot be achieved in practice at present.

There are three important reasons why the photoelectric conversion efficiency of a multi-junction solar cell is better than that of a single-junction solar cell. First, the multi-junction solar cell can absorb sunlight in sections, with high absorption efficiency and low energy loss. Second, the thickness of the i-type intrinsic absorption layer in an optimized multi-junction cell is thinner than that of a single-junction cell [60]. The thinner i-type layer suggests that the fill factor of every single junction in a multi-junction cell is higher than that of a single-junction cell, and the stable efficiency of the cell is not much different from the initial efficiency (the degree of degradation is lighter). Finally, multi-junction solar cells have higher operating voltages and lower operating currents compared with single-junction cells. The lower operating current can reduce the excess loss during the current flowing from the p-i-n junction to the load. From one perspective, the manufacturing process of a multi-junction solar cell based on a-Si:H is more difficult than that of a single-junction solar cell, corresponding to these advantages. Besides, the performance of multi-junction solar cells is more sensitive to the spectrum of the incident light due to the need to absorb the incident light in sections, making it critical to accurately control the bandgap and thickness of each film. Moreover, most multi-junction cells based on a-Si:H use a-SiGe:H alloy. The Ge source in this alloy is germane (GeH_4) gas. The price of germane gas is several times that of silane and it is highly toxic. Extremely stringent safety measures must be established to handle these gases in the manufacturing process.

To sum up, the advantages of multi-junction solar cells based on a-Si:H, especially the advantages that can reduce the degradation degree of the cell and increase the stable output power of the cell, are still the important direction of the development of a-Si:H cell in the future, even though there are some difficulties in the production process.

2. **Several Common Multi-junction Solar Cells Based on Amorphous Silicon**

(1) **Double-junction solar cell based on a-Si/a-Si**
Double-junction solar cells based on a-Si/a-Si is not only the simplest multi-junction cell but also a structure of device that is currently widely used in mass production. Although the top cell and the bottom cell are both amorphous silicon, the bandgap can be made different by adjusting the deposition parameters of the intrinsic layer in the top cell and the bottom cell. Generally, the intrinsic layer of the top cell is deposited at a lower temperature of the substrate. Since the hydrogen content in the material is higher at low temperatures, the bandgap is wider. However, the intrinsic layer of the bottom cell can be deposited at a relatively high temperature of the substrate. The hydrogen content in high-temperature materials is relatively low, and the bandgap of the materials is small. Meanwhile, the adjustable range of the bandgap of amorphous silicon is very small. Therefore, the intrinsic layer of the bottom cell is required to be much thicker than the intrinsic layer of the top cell in order to make the bottom cell have sufficient current. Furthermore, the photoelectric conversion efficiency of the cell can be effectively improved by optimizing the tunnel combination junction.

(2) **Double-junction solar cell based on a-Si/a-SiGe**
The advantages of double-junction solar cells based on a-Si/a-Si are low material cost and cheap product price. However, the use of only one material of amorphous silicon results in a single bandgap, which is not good for light absorption

Table 5.8 Energy conversion efficiency of small-area solar cells prepared in different laboratories

Structure	Initial efficiency η (%)	Stable efficiency η (%)	Manufacturer
a-Si/a-SiGe/a-SiGe	15.2	13.0	United Solar
	11.7	11.0	Fuji
	12.5	10.7	U.Toledo
		10.2	Sharp
a-Si/a-SiGe	11.6	10.6	BP Solar
		10.6	Sanyo
		12.0	United Solar
a-Si/μc-Si		12.0	U.Neuchatel
	13.0	11.5	Canon
a-Si/poly-Si/poly-Si	12.3	11.5	Kaneka
a-Si/a-SiGe/uc-Si	11.4	10.7	ECD

in long-wavelength bands; thus, the short-circuit current is small, limiting its energy conversion efficiency. Then, different intrinsic materials can be used to improve the long-wave response of the bottom cell; amorphous silicon-germanium alloy is an ideal choice. As mentioned earlier, the ratio of silane (or disilane) and germane in the plasma can adjust the ratio of silicon and germanium in the material to adjust the bandgap of the material. For the bottom cell of a double-junction solar cell based on a-Si/a-SiGe, the optimal germanium composition is from 15 to 20%, and the corresponding bandgap is about 1.6 eV. Besides, a total current of about 22–23 mA/cm^2 can be obtained using this double-junction cell.

(3) **Triple-junction solar cell based on a-Si/a-SiGe/a-SiGe**
The energy conversion efficiency of small-area solar cells with common structures prepared in different laboratories is listed in Table 5.8.

Among these cells in Table 5.8, the three-junction cell with a structure of a-Si(1.8 eV)/a-SiGe(1.6 eV)/a-SiGe(1.4 eV) has the highest efficiency [13]. A typical three-junction solar cell with an n-i-p structure grown on a substrate of stainless steel is illustrated in Fig. 5.26. Figure 5.25 corresponds to a p-i-n series cell (glass substrate). In both cell structures, the incident light passes through the p-type layer; therefore, holes need to travel a shorter distance than electrons to be absorbed by the electrodes. Next, the design of these two types of cells and the most widely used deposition processes will be briefly outlines.

Two steps are required to prepare an n-i-p cell on a substrate of stainless steel: first, a reflective metal layer is deposited on the substrate by sputtering or thermal evaporation; then, a ZnO buffer layer is deposited by sputtering. Silver is usually used as a reflective metal layer in scientific research because it has a very high reflectance. However, aluminum is generally used due to cost issues in actual production. The reflective metal layer is deposited at a temperature of 300–400 °C. The self-segregation of the metal thin-film during the process of growth forms a texturing

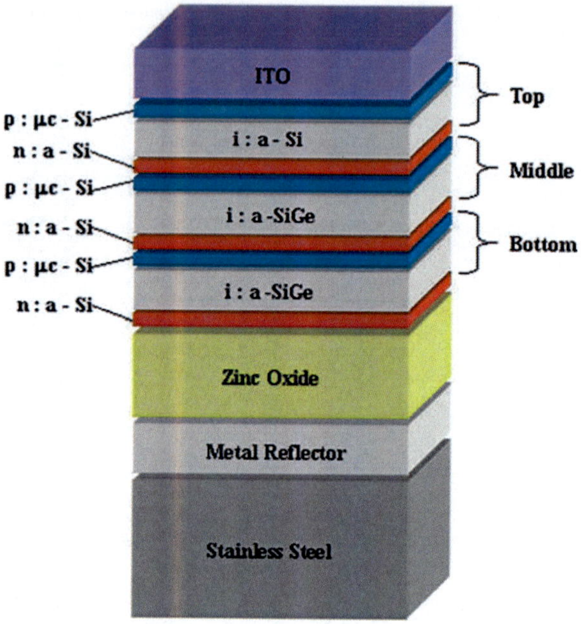

Fig. 5.26 Structure of a three-junction n-i-p solar cell

topography that can capture photons. Then, the sample is transferred to an RF PECVD deposition system to continue the next process for depositing the semiconductor layer. First, the bottom n-i-p cell, whose intrinsic absorption layer is a-SiGe with a bandgap of 1.4–1.5 eV, is deposited; then, the middle n-i-p cell based on a-SiGe, whose intrinsic absorption layer has a bandgap of 1.6–1.65 eV, is deposited; finally, the top n-i-p cell, whose intrinsic absorption layer is a-Si with a bandgap of 1.8 ~ 1.85 eV, is added and prepared by a high hydrogen-doping process at a lower temperature. After the growth of the three-junction structure is completed, an ITO layer with a thickness of 70 nm is deposited on the top by thermal evaporation or sputtering, serving as a top electrode and an anti-reflective film. A layer of the metal grid is deposited on top of ITO by thermal evaporation or sputtering to further reduce the contact resistance.

In a multi-junction cell with a p-i-n structure, a transparent and conductive textured oxide layer (typically SnO_2 or ZnO) is first spin-coated on a glass substrate. Atmospheric pressure chemical vapor deposition (APCVD) is commonly used [61, 62].

Then, a top-level p-i-n cell using a-Si:H as the i-type layer is deposited, followed by the middle a-SiGe cell. Finally, there is the bottom a-SiGe cell with a relatively narrow bandgap. The vertical structure ends by depositing a ZnO buffer layer and a metal reflective layer on the back surface.

5.5 Development History and Future Prospects of Amorphous Silicon Solar Cells

5.5.1 Development Process of Amorphous Silicon Solar Cells

1. **Initial development of amorphous silicon solar cells**

A-Si:H solar cells caused a wave of research frenzy in developed countries and even the world when they first came out; this was inseparable from the background of the times. Since oil played a leading role in the energy structure around the world, it has become a key factor determining the economy, the survival, development, and decline of a country. The Arab-Israeli Conflict broke out in October 1973. The Organization of Petroleum Exporting Countries has adopted methods such as reducing oil production and raising prices to support the struggle of the people in the Middle East and safeguard their own interests. Consequently, those countries that rely on large quantities of cheap oil imported from the Middle East are hit hard economically. Some Westerners exclaimed that "the energy crisis" (also known as the "oil crisis") occurred in the world! The "crisis" objectively made people realize in advance that the existing energy structure must be completely changed, and the transition to the future energy structure should be accelerated. Therefore, many countries, especially industrially developed countries, have re-intensified their supports for the development of solar energy and other renewable energy technologies. Thus, a wave of development and utilization of solar energy has emerged in the world.

In 1976, D. E. Carlson of the RCA company in America developed an amorphous silicon solar cell with p-i-n structure with a photoelectric conversion efficiency of 2.4%. The photoelectric conversion efficiency continued to be increased by 8% in 1980, making it possible to industrialize. Amorphous silicon solar cells have stimulated researchers and research units around the world to invest in this field. Besides, it has also attracted the attention of the business community and many national governments, contributing to promoting the development of amorphous silicon solar cells. Amorphous silicon solar cells quickly walked out of the laboratory and into factories and relatively large production lines. From a technical perspective, the progress of amorphous silicon solar cells at this stage is mainly manifested in the following aspects. (1) A more complex and practical structure is indicated from simple ITO/p-i-n(a-Si)/Al to SnO_2(F)/p-a-SiC/i-a-Si/n-a-Si/Al. SnO_2 transparent conductive film is more stable than ITO; it has a lower cost and is easy to realize texture, leading to increasing the absorption of light by solar cells. Using a-SiC:H as the p-type window layer makes the bandgap wider, reduces the absorption loss of light by the p-layer, and utilizes incident solar energy better. (2) The laser scribing segmentation of the a-Si layer, the two thin-layers of electrodes, and the production of integrated components are realized. (3) There are two methods of single-chamber mass production and multi-chamber flowing production of amorphous silicon thin films. Besides, there are also two production methods of modules in production, including the production method of modules with transparent conductive glass as the substrate and the production method of modules with flexible materials (such as stainless steel) as the substrate. Many companies or business branches using a-Si solar cells as their main products had appeared in the world. For example, Chronar, Solarex, and ECD in the United States, Sanyo, Fuji, and Sharp in Japan. Some companies in America and Japan had also installed test stations with their own products to generate power outdoors; the largest test station had a capacity of 100 kW. In the mid-1980s, amorphous silicon accounted for 40% of the total sales of solar cells in the world. A tripartite confrontation of amorphous silicon, polysilicon, and single crystalline silicon had emerged.

2. **Development of amorphous silicon solar cell slowing down**

Although amorphous silicon solar cells have many advantages, their disadvantages are also apparent, such as the low initial photoelectric conversion efficiency and poor stability. The initial solar cell product has an initial photoelectric conversion efficiency of 5–6%. After being irradiated under standard solar light for one year, the stable efficiency of the solar cell is 3–4%. Apparently, application in low light is not a problem; however, stability under strong light outdoors becomes a more serious problem when the solar cell is used to generate power. The performance of test power stations for power generation is severely degraded, with a short life span, which seriously affects the confidence of consumers and causes difficulties in market development. Consequently, some production lines such as Chronar closed.

In the first stage, the problem of performance degradation of a-Si solar cell product actually has two aspects, namely, the packaging problem and the instability problem of a-Si material constituting the cell. The packaging problems mainly include the aging of packaging materials and the defects in packaging. The harmful atmosphere in the environment causes damage to the electrode material and electrode contact of cells, leading to reducing or even disabling the cell performance. This problem is mainly solved by improving packaging technology. Gradually, the packaging problem is basically solved after adopting a glass laminate package (referring to a cell with a glass substrate) and a hot-pressed package with multiple protective layers (for a cell with a substrate of stainless steel). At present, the service life of solar cells has reached more than 10 years.

3. **Perfection and improvement of technology for amorphous silicon solar cells**

Due to setbacks in development momentum, the development of amorphous silicon solar cells experienced a period of adjustment, perfection, and improvement in the late 1980s and early 1990s. From one perspective, exploration and research have been strengthened. From another perspective, a larger-scale industrialization development at a higher technological level is in preparation. Thus, many new device structures, new materials, new processes, and new technologies have been explored. Based on successful exploration, a larger-scale industrialization climax appeared in the mid-1990s, and several high-level production lines of cell modules ranging from several megawatts to ten megawatts have been established. The area of modules is on the order of square meters, and the production process is fully automatic. Besides, the life span of cell modules can be prolonged to more than 10 years by using new packaging technology. The production of modules is based on the perfect junction technology and laminated cell technology. The efficiency of the product module reaches 6–8%, the efficiency of the pilot test module (the area is about 900 cm^2) reaches 9–11%, and the maximum efficiency of the small-area cell reaches 14.6%. At present, dozens of universities, national laboratories, and companies in the world are engaged in the research of thin-film solar cells based on silicon; besides, their industrialization technology is becoming increasingly mature.

Thin-film cells based on amorphous silicon have a history of more than 30 years from invention to true industrial production. At present, thin-film cells based on silicon have been developed into an important branch of the solar cell industry. Moreover, the market size of solar energy will continue to expand as the price of crude oil continues to rise and the price of solar energy continues to decrease. Due to different requirements for solar cells in different environments, different structures of cells have their own unique development spaces. Thin-film cells based on silicon not only have many advantages but also have some technical problems. We believe that thin-film cells based on silicon will become an important part of the solar industry with continuous research and development in the field of science and technology.

5.5.2 *Prospect and Challenges in the Future*

Among all renewable energy, the photovoltaic industry with the fastest growth will play an essential role in the future energy landscape. Governments around the world are promulgating active policies to help the development of the photovoltaic industry because is it clean and pollution-free. According to statistics from the European Photovoltaic Industry Association (EPIA), the total global photovoltaic industry has increased from 1.8 GW in 2000 to 97.4 GW in 2011. It has increased 37 times in just 10 years, with an average annual growth rate of 44% [63]. The expansion of the production scale indicates a reduction in cost. According to statistics, the cost of photovoltaic modules can be reduced by 20–22% each time the industrial size is doubled.

The silicon-based thin-film solar cells is a significant member of the thin-film solar cell family. The development speed of conventional solar cells based on single crystalline silicon and polycrystalline silicon has been limited due to the shortage of raw materials. Therefore, the development of new thin-film solar cells is particularly rapid. For example, the production of thin-film solar cells in the United States has exceeded the production of solar cells based on polycrystalline and single crystalline silicon in 2007. The main thin-film cells are amorphous silicon cells and cadmium telluride cells.

Despite the considerable development, amorphous silicon solar cells still have a long way to go before they can enter the market for large-scale applications. Specifically, the following aspects can be considered.

(1) **Further explore the principle of light-induced degradation**
Photo-induced degradation has always been a major weakness of solar cells based on amorphous silicon. The decline rate of single-junction cells can reach 30% while the decline rate of multi-junction cells can even reach 10–15%. Although the photo-induced degradation of amorphous silicon cells has been investigated intensively in the past few decades, and attempts have been made to reduce the magnitude of photo-induced degradation, the photo-induced degradation of amorphous silicon cells has not been completely resolved. Particularly, some aspects of the device have to be sacrificed to limit the optical saturation effect; for example, a relatively thin i-type absorption layer might be used, even though this will reduce the absorption efficiency of light. The stable efficiency of the cell will be further improved if the problem of photo-induced degradation can be solved.

(2) **Further reduce costs**
In the current situation, the production cost of solar cells based on amorphous silicon is still higher than that of thin-film solar cells based on cadmium telluride, even though the cost of amorphous silicon cells is far lower than that of crystalline silicon cells. Therefore, the cost of amorphous silicon cells needs to be further reduced to make the electricity price of solar cells the same as that of power grids. From one perspective, the process is continuously improved to increase the growth rate and the quality of materials, resulting in increasing the efficiency and yield of cells and reducing the production cost of the amorphous silicon cells. From another perspective, optimizing the design of devices and improving the life and reliability of cells can also reduce costs. It is believed that the price of thin-film solar cells based on silicon can be reduced to the same level as the electricity price of power grids in the near future; In addition to the existing photovoltaic market, it is also possible to explore new application fields of a-Si, such as space energy and large-scale photovoltaic power stations.

References

1. S. Xiong, M. Zhu et al., *Solar Cell Foundation and Application* (Science Press, Beijing, 2009), p. 10 (in Chinese)
2. T. Bradford, *Solar Revolution* (MIT Press, 2006)
3. X. Zhang et al., Application and prospect of amorphous silicon thin film cell. Lamps Light. **1**, 39–41 (2010)
4. W. Spear et al., Solid State Commun. **17**, 1193 (1975)
5. D.E. Carlson et al., Appl. Phys. Lett. **28**, 671 (1976)
6. J. Yang et al., Appl. Phys. Lett. **70**, 2975 (1997)
7. C. Wronski et al., *Electron Devices Meeting*, vol. 22 (International., 1976), pp. 75–78
8. S. Fonash, *Solar Cell Device Physics* (Access Online via Elsevier, 1981)
9. J. Tauc, *Optical Properties of Solids* (Academic Press, 1987)
10. T. Hama et al., J. Non-Cryst. Solids **59**, 333 (1983)
11. A. Middya et al., J. Appl. Phys. **78**, 4966 (1995)

12. Y. Hamakawa et al., in *Proceedings of the 16th Photovoltaic Specialists Conference* (1982), p. 679
13. J. Yang et al., Appl. Phys. Lett. **70**, 2977 (1997)
14. M. Vanecek et al., J Non Cryst Solids. **227**, 967 (1998)
15. S. Guha et al., Non-Cryst. Solids **97–98**, 1455 (1987)
16. S. D, in *Proceedings of the 24th Photovoltaic Specialists Conference* (1994), p. 670
17. K. Emery et al., in *Proceedings of the 25th Photovoltaic Specialists Conference* (1996), p. 1275
18. D. Carlson et al., in *Proceedings of the 28th Photovoltaic Specialists Conference* (2000), p. 707
19. J. del Cueto et al., Prog. Photovoltaics **7**, 101 (1999)
20. R.A. Street, *Technology and Applications of Amorphous Silicon* (Springer, 2000)
21. S. Guha et al., Technical digest, in *International Photovoltaic Science and Engineering Conference (PVSEC-7)* (1993), p. 43
22. A. Luque et al., *Handbook of Photovoltaic Science and Engineering* (Wiley, 2003)
23. R. Chittick et al., J. Electrochem. Soc. **116**, 77 (1969)
24. W. Spear, J. Non-Cryst, Solids **8–10**, 727 (1972)
25. A. Matsuda et al., Sol. Energy Mater. Sol. Cells **78**, 3 (2003)
26. X. Deng et al., in *Proceedings of the 1st World Conference on Photovoltaic Energy Conversion* (1994), p. 678
27. J. Yang et al., in *Proceedings of the 25th Photovoltaic Specialists Conference* (1996), p. 1041
28. A. Shah et al., Mater. Res. Soc. Symp. Proc. **258**, 15 (1992)
29. X. Deng et al., in *Proceedings of the 26th Photovoltaic Specialists Conference* (1997), p. 591
30. N. Ito et al., in *Proceedings of the 28th Photovoltaic Specialists Conference* (2000), p. 900
31. T. Watanabe et al., Jpn. J. Appl. Phys. **25**, 1805 (1986)
32. S.-S. Lee et al., J. Crystal Growth **310**, 3659 (2008)
33. A. Mahan, J. Appl. Phys. **69**, 6728 (1991)
34. Q. Wang et al., Phys. Rev. B **47**, 9435 (1993)
35. Q. Wang et al., J. Non-Cryst. Solids **299**, 2 (2002)
36. A. Mahan et al., Appl. Phys. Lett. **78**, 3788 (2001)
37. L. Guo et al., Jpn. J. Appl. Phys. **37**, L1116 (1998)
38. B. Rech et al., Sol. Energy Mater. Sol. Cells **66**, 267 (2001)
39. A. Shah et al., Sol. Energy Mater. Sol. Cells **78**, 469 (2003)
40. B. Rech et al., Thin Solid Films **511**, 548 (2006)
41. M.A. Lieberman, A.J. Lichtenberg, *Principles of Plasma Discharges and Materials Processing* (Wiley, 2005)
42. 赵志明, *Technical Development Status and Trends of Amorphous Silicon Thin-film Solar Cell and Production Equipment, Sci-Tech Information Development & Economy*, vol 19 (2009), pp. 136–140
43. J. Nelson, *The Physics of Solar Cells* (Properties of semiconductor materials) (Imperial College, 2003)
44. Q. Gu et al., J. Appl. Phys. **76**, 2310 (1994)
45. Y. E, J. Opt. Soc. Am., **72**, 899 (1982)
46. Y. Liu, Very high frequency plasma deposited amorphous/nanocrystalline silicon tandem solar cells on flexible substrates. Ph.D thesis, Utrecht University, 2010
47. S. Hegedus et al., in *Proceedings of the 25th Photovoltaic Specialists Conference* (1996), pp. 1061–1064
48. A. Banerjee, J. Appl. Phys. **69**, 1030 (1991)
49. L. Jiang et al., Mat. Res. Soc. Symp. Proc. **609**, A18.3.1 (2000)
50. R. Crandall et al., in Conf. Proc. 13th NREL Photovoltaics Program Review, eds. by H. Ullal, C. Witt, vol. 353 (1996), p. 101
51. S. Guha et al., IEEE Tran. Electron Devices **46**, 2080 (1999)
52. P. Agarwal et al., J. Non-Cryst. Solids **299**, 1213 (2002)
53. L. Yang et al., Mater. Res. Soc. Symp. Proc. **219**, 259 (1991)
54. S. Guha et al., Appl. Phys. Lett. **54**, 2330 (1989)
55. J. Zimmer et al., J. Appl. Phys. **84**, 611 (1998)

56. J. Yang et al., in *The 26th Photovoltaic Specialists Conference* (1997), p. 563
57. M. Kabir et al., Recent Patents Electr. Eng. **4**, 50 (2011)
58. Solar Cell efficiency, http://en.wikipedia.org/wiki/Solar_cell
59. S. Sze, *Physics of Semiconductor Devices* (Wiley, 1981)
60. A. Madan et al., *The Physics and Applications of Amorphous Semiconductors* (Academic Press, 1988)
61. R. Gordon et al., Sol. Energy Mater. **18**, 263 (1989)
62. H. Iida et al., IEEE Electron Device Lett. **4**, 157 (1983)
63. G. Masson et al., *Global Market Outlook for Photovoltaics Until 2016* (European Photovoltaic Industry Association, 2012)

Chapter 6
Cu(InGa)Se_2 Solar Cell

6.1 Properties of the Cu(InGa)Se_2 Material

6.1.1 Structure and Composition

$CuInSe_2$ and $CuInS_2$, ternary compounds consisting of I B, III A, and VI A elements, are adopted as absorption layers of copper indium selenium (CIS) thin-film solar cells. Copper indium gallium selenium (CIGS) solar cells are developed when Ga is doped into the CIS thin-film solar cells to replace part of the elements of the cells. The crystal structures with similar component ratios (e.g., $CuInSe_2$ (ABC_2) solar cells) exhibit high similarity to chalcopyrite structure. Chalcopyrite structure is largely identical to the diamond structure; it is an ordered structure with I and III elements replacing II elements. A and B pertain to cations, while C belongs to anion. Chalcopyrite structure originates from the composition of $CuFeS_2$, exhibiting high similarity to sphalerite structure, which is a body-centered tetrahedral structure. The sphalerite structure has a high-temperature phase and pertains to the cubic crystal system. Besides, the chalcopyrite structure belongs to cubic crystal system, whereas it exhibits a low-temperature phase with a lattice constant slightly less than that of the former [1]. In chalcopyrite structure, four atoms are coordinated in each cell. The nearest neighbor of each A or B atom is four C-type atoms. The C-type atoms are orderly surrounded by two A atoms and two B atoms. Since A–C bonds and B–C bonds exhibit different lengths and ionic properties, the tetrahedral structure composed of C atoms as the center is not overall symmetrical, as presented in Fig. 6.1. A stronger bond connection is identified between C-type atom and nearest neighbor A-type atom, which results from the contribution of d-electrons in A-type atom to bonding. As impacted by the difference in bonding between different atoms, the lattice constant ratio c/a of the chalcopyrite structure reaches approximately 2 [2].

As revealed from the results of thermodynamic analysis, $CuInSe_2$ exhibits the solid-state phase transition temperatures of 665 °C and 810 °C, respectively, and its melting point reaches 987 °C. At the temperature below 665 °C, the crystals exhibit a chalcopyrite structure, while at the temperature 810 °C, they show a sphalerite

C. Zhang et al., *Semiconductor Photovoltaic Cells*,
https://doi.org/10.1007/978-981-15-9480-9_6

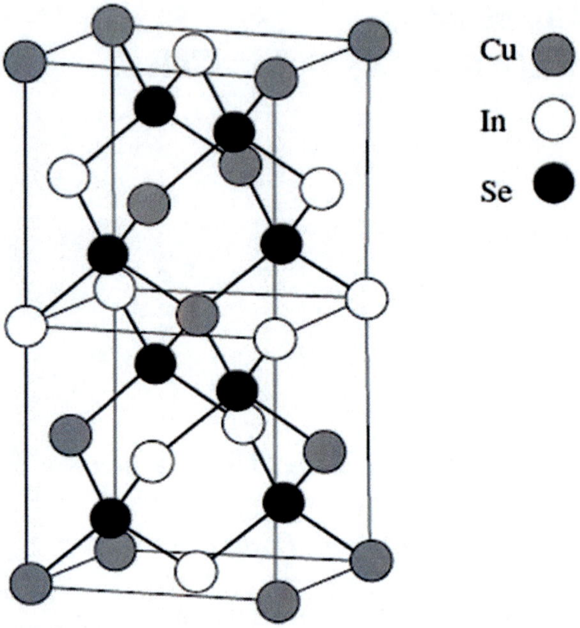

Fig. 6.1 CIS chalcopyrite structure [2]

structure. By adding Ga element to CIS materials, Ga will partially replace In in the material to synthesize $CuIn_xGa_{1-x}Se_2$. Since the atomic radius of Ga is smaller than that of In, the lattice constant of chalcopyrite structure will decrease with the increase in Ga content. If Cu and In atoms are arranged randomly on the sublattice position, such structure complies with the sphalerite structure [2, 3].

In the analysis of crystal phases, phase diagrams can act as a vital tool to intuitively describe the state of multicomponent systems with the variations of temperature, pressure, and composition. CIS refers to a ternary compound material, and the phase diagram of the Cu–In–Se system is illustrated in Fig. 6.2. Overall, the amount of Se element is excessive for growing Cu (InGa) Se_2 thin-film materials. The excess of Se in the Cu–In–Se system is close to the direct connection between Cu_2Se and In_2Se_3 (the dotted line illustrated in Fig. 6.2). The chalcopyrite structure of $CuInSe_2$ and many other phases on this line are termed as Ordered Defect Compounds (ODC) [2], which is attributed to the regularly arranged intrinsic defects in their chalcopyrite crystal structure. Figure 6.3 presents the pseudobinary phase diagram of $CuInSe_2$ represented by the direct connection between Cu_2Se and In_2Se_3 in Fig. 6.3. In the phase diagram, the α phase represents the morphological structure of chalcopyrite $CuInSe_2$, the δ phase refers to the sphalerite structure (HT) at high temperature, and the β phase refers to the phase state of ODC structure. The CIS materials applied for solar cells are chalcopyrite structure (α phase), pertaining to the cubic system. At the growth temperature over 500 °C, the material will pertain to pure α phase, and the composition of Cu is 22–24% on average. Thus, a Cu-poor/In-rich CIS thin film

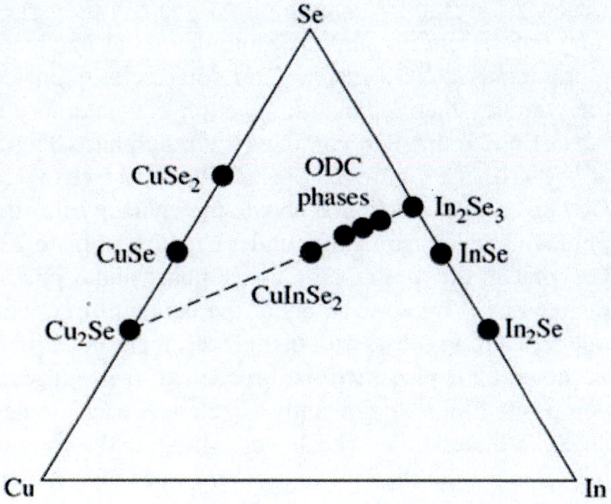

Fig. 6.2 Phase diagrams of the Cu–In–Se system [3]

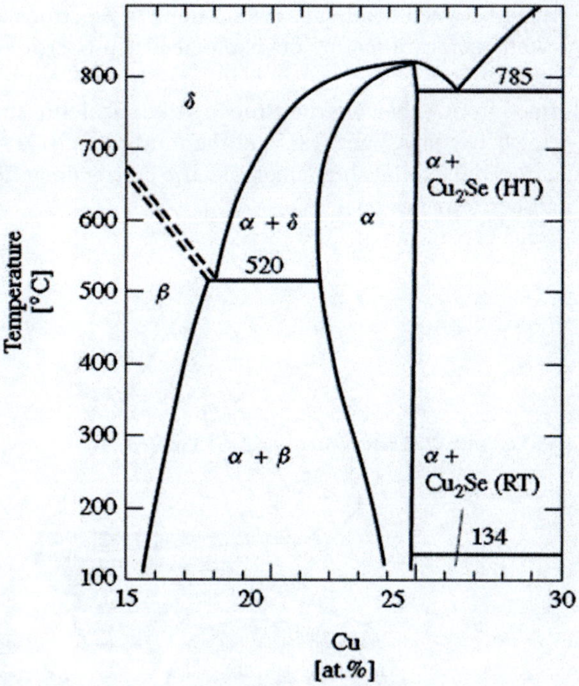

Fig. 6.3 $CuInSe_2$ pseudo binary phase diagram [2]

exhibiting superior properties will be formed, which acts as the absorption layer of solar cells, facilitating the fabrication of high-performance solar cell.

$CuInSe_2$ is alloyed with any proportion of $CuGaSe_2$ to synthesize $Cu(InGa)Se_2$ (GICS). CIGS is a type of quaternary composite crystal structure. It has a wide variety of crystal systems (e.g., Al and S systems) except for chalcopyrite structure, thus exhibiting high freedom degree of bandgap engineering. At present, the representative CIGS material for solar cells is chalcopyrite crystal structure. For the complex thermodynamic reaction of quaternary compound CIGS, the insight into its phase diagram complies with the phase diagram of Cu_2Se–In_2Se_3–Ga_2Se_3 at 550–810 °C. Figure 6.3 presents the phase change of CIGS with the variation of Cu content. As mentioned above, the chalcopyrite structure (α phase) employed in photovoltaic industry exists under Cu content from 22 to 24%. With the increase in Cu content, the film mixes Cu-rich phase and α phase, while the other phase states denoted by β phase exist under the content of copper less than 20%, i.e., the Cu-poor side. In Fig. 6.4, with the increasement in Ga/In ratio in Cu-poor film, the area occupied by α phase will be broadened. It is suggested that the phase state of the composite film is significantly effected by each elemental constituent [3, 4].

CIGS materials are capable of regulating the composition variation in a relatively large range, and the optical properties of CIGS materials do not vary significantly with the variation in the composition. As impacted by different stoichiometric compositions and different distribution of intrinsic defects, a range of band gaps will be formed between band gaps, while traps of electrons and holes in transport will be created, thereby affecting the photoelectric properties of materials and changing the structure. For instance, when Ga/(Ga + In) = 0.6, the CIGS material will be transformed from a chalcopyrite structure to a sphalerite structure. At the ratio of Cu/(Ga + In) between 0.7 and 1.0, and the ratio of Ga/(Ga + In) ranges from 0.2 to 0.3, the material exhibits high photoelectric conversion efficiency, making it suitable for high-performance devices.

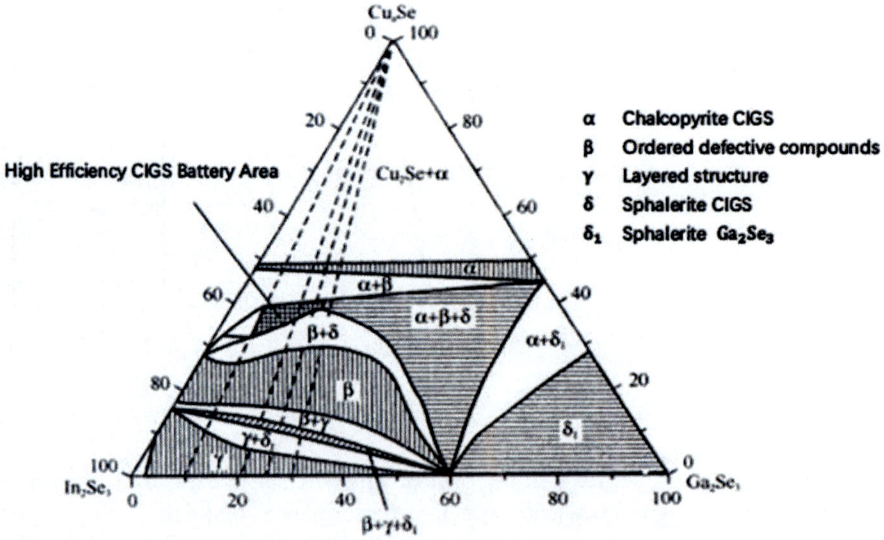

Fig. 6.4 Phase diagram of Cu_2Se–In_2Se_3–Ga_2Se_3 system [4]

During the growth of CIGS material, the content of Cu can alter the phase state of the material surface. Copper-rich matrix can make the surface of materials to reach a secondary Cu_ySe ($y < 2$) phase state. In the film growth, the grain size can exceed 1 mm under the copper-rich condition. Set $x = In/(In + Cu)$, the materials are transformed from p-type to n-type with the increase in x value. The proportion of Ga element in the material affects the band gap [4].

The ratio of (In + Ga)/(In + Ga + Cu) is altered by substituting Ga for In and adding Na in the substrate. Adding a small amount of Na from the substrate can achieve good morphology and conductivity of the material during the growth of the film. Moreover, since Na may substitute In or Ga, the distribution of defects in the film will be affected. Meantime, the contents of Na and Ga also affect the phase field of chalcopyrite, since Ga substitutes for Cu to form ordered defect compounds, and Na substitutes for InCu defect with the decrease in compensating donor density. The ratio of Ga content in total Ga and In content will affect the band gap of the material [5–8]. Set

$$x = \frac{Ga}{Ga + In} \tag{6.1}$$

In CIGS solar cells exhibiting 17.7% efficiency, the Ga content takes up 10–30% of the total Ga and In. Then, the band gap reaches 1.1–1.2 eV [9]. Such ratio will also affect the distribution of defects in the films. The properties of CIGS materials with different components are listed in Table 6.1.

Table 6.1 Energy characteristic analysis of CIGS materials with different components [61]

Materials	Preparation methods	Analysis methods	E_A/meV	Notes
$CuInSe_2$	Co-evap	σ(T)	250 170	The initial stage is the annealing in air (Na-free)
$Cu(In_{1-x}Ga_x)Se_2$ (x = 0–0.22)	Co-evap	σ(T)	100 30	$x = 0$ $x = 0.22$
$Cu(In_{1-x}Ga_x)Se_2$	Two-step	σ(T)	40 350	Na-doped Na-free
$Cu(In_{1-x}Ga_x)Se_2$ (x = 0.2)	Co-evap	σ(T)	60–120 250	Na-doped Na-free
$CuInS_2$	Co-evap	Hall measurement	90 50–60	n-type p-type
$CuInGa_2$	Co-evap	Hall measurement	60–135	Cu-rich
$CuInGa_2$	Co-evap	Hall measurement	<20	Cu-rich
$CuInGa_2$	Single-epi-GB	Hall measurement	32	Cu-rich
$Cu(In_{1-x}Ga_x)Se_2$	Two-step	Hall measurement	50	Na-doped

6.1.2 Optical Properties

Absorption coefficient refers to one of the critical parameters to characterize the optical properties of semiconductor materials. Light absorption process represents the process of electron transition. As suggested from the results of experiments, the CIGS material acts as a direct band gap semiconductor, and its absorption coefficient can reach 10^5 cm^{-1}. In the thin-film semiconductor materials, it reaches a higher level, and only 1–2 μm thickness can achieve more than 99% solar absorption [2]. Figure 6.5 presents the absorption coefficients of a range of semiconductor materials.

Existing studies reported that absorption coefficient α is a function of photon energy hv and optical band gap E_g, that is,

$$\alpha = \frac{A(E - E_g)^2}{E} \tag{6.2}$$

where A denotes constant related to the energy density of light absorption.

The absorption coefficients of semiconductor films can be determined by measuring reflectivity, transmittance, and thickness under different polarization states. If interference is not considered, the relationship between absorption coefficient and transmittance T and reflectance R is expressed as

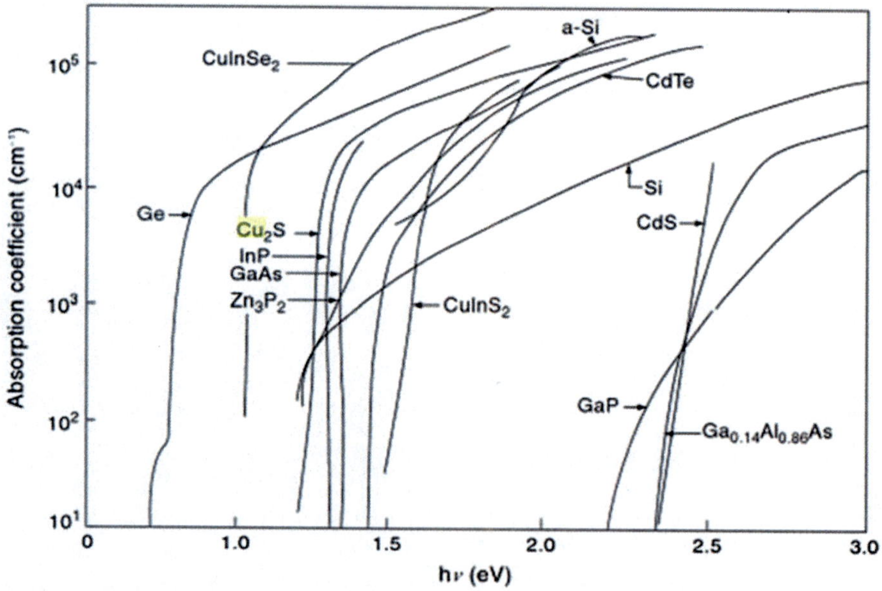

Fig. 6.5 Optical absorption coefficient of different semiconductor materials [5]

$$T = (1 - R)^2 \exp\frac{-\alpha d}{1} - R^2 \exp(-2\alpha d) \tag{6.3}$$

where d denotes the thickness of the film. Under a small reflection and a large absorption, it can be simplified into

$$T = (1 - R)^2 \exp(-\alpha d) \tag{6.4}$$

Accordingly, absorption coefficient is expressed as

$$\alpha = \frac{\ln \frac{(1-R)^2}{T}}{d} \tag{6.5}$$

Subsequently, the size of optical band gap E_g can be calculated.

The optical band gap of GIGS material is associated with temperature, and the relationship is defined as

$$E_g(T) = E_g(0) - \frac{\alpha T^2}{T + b} \tag{6.6}$$

where a and b are constants, and gaps will exist in different measurement environments.

The optical band gap of CIGS films is also associated with the ratio of Ga/(Ga + In) and the content of Cu. Under the Ga content of s zero, the optical band gap is 1.02 eV. Under the Ga content of 100%, the optical band gap is 1.67 eV. If Ga in the film is uniformly distributed, and x = Ga/(Ga + In), the relationship between E_g and x is defined as

$$E_{gCIGS} = (1 - x)E_{gCIS} + xE_{gCGS} - bx(1 - x) \tag{6.7}$$

where b denotes the bending coefficient (different from b in Eq. 6.6), and the range of values determined by various experiments is nearly $0.11 < b < 0.26$ eV, and the empirical value of 0.21 is usually used [10–12].

6.1.3 Electrical Properties

$CuGaSe_2$ materials commonly exist as p-type conductivity. The conduction type of CIGS films can be varied by regulating the atomic ratio of cations or different doping forms. For $CuInSe_2$ materials, the material is p-type when the amount of Cu is excessive, while the In-rich materials can be p-type or n-type. Under the over-pressure annealing of Se, n-type can be transformed into p-type, and p-type can be converted into n-type material when Se is annealed at low pressure. For CIGS materials, the conductive type of the thin film is directly associated with its composition. Under the small content of Se in the material, the material lacks Se to form vacancies, and the external electrons of the nearest neighbor Cu and In atoms in the chalcopyrite structure lose covalent electrons. Then, the Se vacancies correspond to donor impurities and provide electrons for the conduction band. When Ga doping makes part of Ga atom replace In atom, for the difference of electron affinity, the external electrons of Cu and Ga form electron pairs, making the Se vacancy unable to provide free electrons. Thus, the n-type conductivity of the material decreases with the increase in Ga content. The Cu(InGa)Se$_2$ thin films applied for devices are commonly p-type materials in Se-rich atmosphere, and the carrier concentration reaches about 10^{16}/cm^3 [13, 14].

Chalcopyrite CIGS materials exhibit numerous inherent defects. According to the calculation of CIS materials, there are five types of donor defect levels (D1–D5) and six types of acceptor defect levels (A1–A6), as illustrated in Fig. 6.6. To be specific, A1 is formed by Cu vacancies and pertains to shallow acceptor level that can be formed at ambient temperature. In vacancies (A3/A5) and substitution defects (A4) created by Cu substituting for In both pertain to acceptor-type point defects, whereas the substitution defect induced by replacing Cu with In is the donor-type point defect. When the overall number of donor-type point defects reaches over that of acceptor-type point defects, the material is n-type; otherwise, the material pertains to p-type.

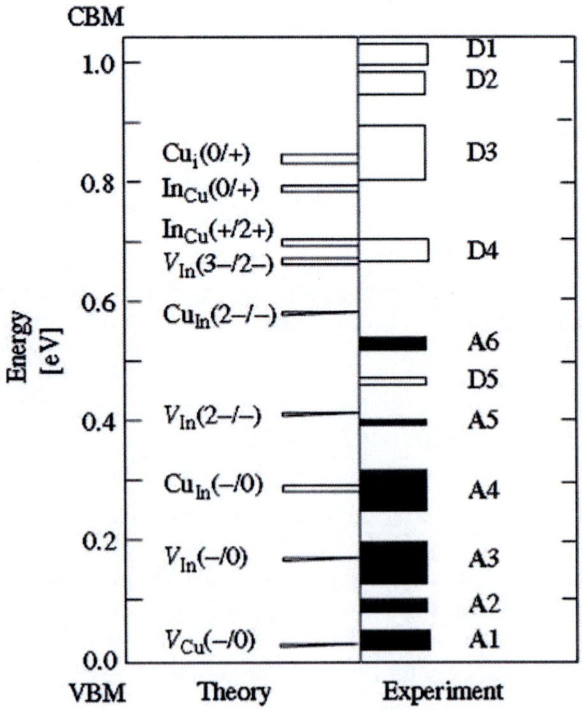

Fig. 6.6 Schematic diagram of donor/acceptor point defect energy levels in CIS [2]

Accordingly, the constituent element ratio of CIGS material is a critical mechanism to regulate the conductive type [15]. A range of electron transfer processes can be observed under photoluminescence, photoconductivity, photovoltage, optical absorption, and electrical measurements. Moreover, the theoretical calculation and comparison of formation and conversion energies can lay a basis for distinguishing different intrinsic defects. The specific data are presented in Fig. 6.6.

In terms of mobility, the mobility of CIGS materials exhibits a larger range of values. The hole mobility of $Cu(InGa)Se_2$ ranges from 15 to 150 $cm^2/V{\cdot}s$ in single crystal state, peaking at 200 $cm^2/V{\cdot}s$. The electron mobility of single crystal falls in the range of 90–900 $cm^2/V{\cdot}s$. As revealed from Hall measurements of p-type polycrystalline materials, the mobility of p-type polycrystalline materials can reach 5–50 $cm^2/V{\cdot}s$, which may probably be limited and impacted by the correlation between grains and transport between grain boundaries [16].

With the increase in Ga content percentage, the open-circuit voltage V_{OC} of CIGS-based solar cells increases, and the short-circuit current J_{SC} is down-regulated.

6.1.4 Surface, Grain Boundary, and Substrate

Given the experience of manufacturing, crystalline thin films with a grain size of about 1 μm can be exploited to fabricate devices. The grain morphology and grain size vary with the fabrication method and environment. Large grain structure or columnar structure perpendicular to the surface of the substrate are commonly employed in the film-making process to reduce the probability of carrier reflection or capture by grain boundaries, thereby enhancing the carrier transport characteristics. Scanning electron microscopy (SEM) and transmission electron microscopy (TEM) are vital methods of characterizing the surface morphology and grain structure of thin films. Figure 6.7 presents the SEM image of CIGS thin films obtained by electron co-evaporation on Mo substrates, from which the existence of displacement, stacking fault, and other defects are suggested [2].

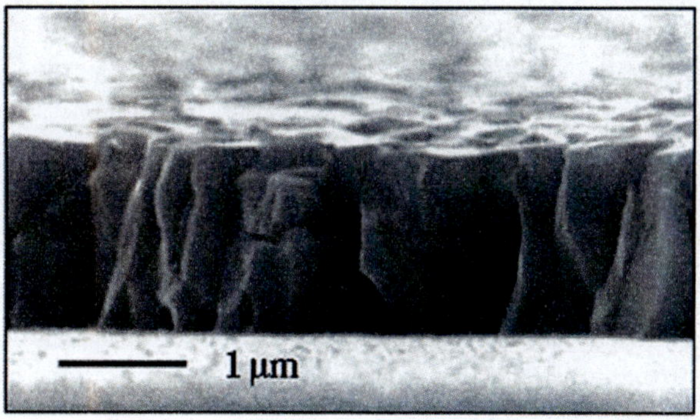

Fig. 6.7 SEM image of CIGS thin film

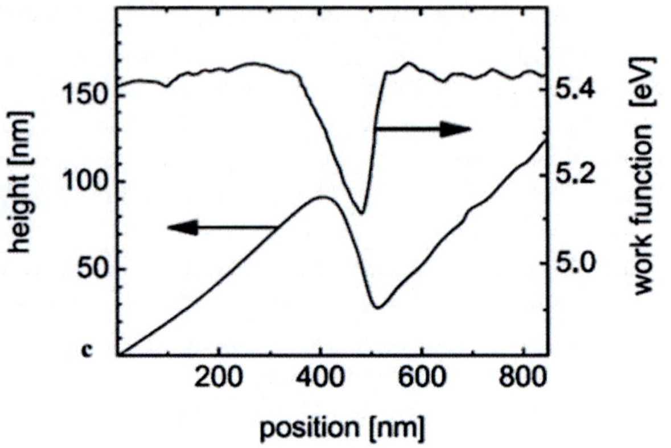

Fig. 6.8 KPFM image of CIGS surface in any direction

KPFM is capable of measuring the height variation of the sample surface. Meantime, the work function of a specific point on the surface can be determined by the contact potential generated by the electrostatic force between the sample and the tip, Fig. 6.8 illustrates the variation of work function in random directions on KPFM surface. As suggested from the observation of the curve trend, the work function around the grain boundary decreases obviously, similar to the band bending when not exposed to air, while it disappears when the band is exposed to the atmosphere [17].

The band gap near the surface of CIGS absorption layer differs from that in body. A typical example is the downward shift of the valence band of the copper-poor surface phase (e.g., Cu(In, $Ga)_3Se_5$). For the electric migration of copper from the surface to the body, the energy band is bent. When CIGS is composed by $CuIn_3Se_5$, the electromigration terminates since it cannot satisfy the requirements of the novel structure. The electromigration of copper in $CuInSe_2$ is also associated with the transformation type of chalcopyrite structure.

When the surface is poor in copper, the phase shift of Cu(In, $Ga)_3Se_5$ appears on the surface. However, such offset cannot be determined with structural detection methods (e.g., TEM), whereas it can be measured by X-ray photoelectron spectroscopy (XPS). As revealed from the results, the free surface of $CuInSe_2$ is similar to that of $CuIn_3Se_5$ in the copper-poor case. In the presence of the $CuIn_3Se_5$ component, the loss of Cu attributed to the electromigration of Cu will be reduced. As indicated from the analysis of conductive type, the conductive type of Cu(In, $Ga)_3Se_5$ itself is n-type. In its presence in CIGS material, the conductive type of the whole material may be altered [17]. When CIGS materials are exposed to the atmosphere, the surface band bending will disappear, and oxide will be formed. In the presence of Na, the oxidation will be strengthened, and surface components (e.g., In_2O_3, Ga_2O_3, SeO_x and Na_2CO_3) will appear.

From the microcosmic aspect, polycrystalline semiconductors are composed of grains, and each grain pertains to a complete crystal; however, different grains follow different orientations, and randomly arranged in the crystal. The grain size of CIGS polycrystals is large, and the band structure and absorption coefficient do not significantly differ from those of single crystals, whereas the carrier transport and recombination are affected by the grain boundaries. As impacted by the different orientations of adjacent grains, numerous defects are created at grain boundaries (e.g., lattice dislocations, vacancies, interstitial atoms, and impurities). Defects at grain boundaries will introduce trap states into the energy gap, capture the charge carriers, and then form barriers to alter the energy structure. Interfacial defects in doped-semiconductors also form interface states, which capture the majority of the charge carriers and hindering the charge carrier transport. The effect of grain boundaries on carrier transport is dependent on the direction of grain boundaries. For the grain boundary and the space charge region of PN junction paralleling and perpendicular to the current direction, the majority of carriers are blocked by the barrier, the mobility decreases, and the minority of charge carriers are recombined at the grain boundary, which will exert the greatest impact [18].

For grain size and grain boundary, oxygen atoms will passivate the Se vacancies on the grain surface. The vacancies will act as recombination centers at the grain boundary, thereby down-regulating the number of effective vacancies and hindering carriers transport. Oxygen atoms can replace vacancies and eliminate trapping effects. The mechanism of Na element action remains controversial. Na is considered a catalyst for oxidation, which can maximally transform oxygen molecules into oxygen atoms and enhance the passivation of Se vacancies. One view is that a small amount of Na is introduced to the material to generate a point defect rather than a secondary phase of the similar material. Na may induce substitution defects of Cu or In in CIGS materials. The former is inactive and generally does not introduce impurity energy level, whereas the latter commonly induces shallow acceptor levels to increase effective hole density. Another explanation is that when Na is doped in large quantity, it will substitute Cu and form $NaInSe_2$, which is more stable and exhibits a larger energy gap. The increase in energy gap will help increase the open-circuit voltage. As presented in Fig. 6.9, the grains and grain boundaries of the Na-samples and Na-free samples, were measured by TEM, it suggests that the two images in Fig. 6.9a are homogeneous, and the grain sizes are large; While in Fig. 6.9b the two images display smaller grains and wider CL contrast region at the grain boundary [1, 6–8]. Thus, the addition of Na will significantly increase the tolerance of the film to component mismatch [19].

During the growth process of Cu (InGa) Se_2, glass, metal, and flexible polymer are commonly adopted as substrates. Various substrates exhibit diverse characteristics, and they have their own unique advantages as growth substrates of CIGS materials. Substrates should satisfy the physical and chemical requirements below [3]:

(1) Vacuum compatibility. It is not easy to decline in various vacuum deposition processes and can resist heat.
(2) Thermal stability. On the whole, the growth temperatures of high-performance CIGS materials range from 500 to 600 °C, so the substrates can definitely withstand temperatures above 350 °C at least.

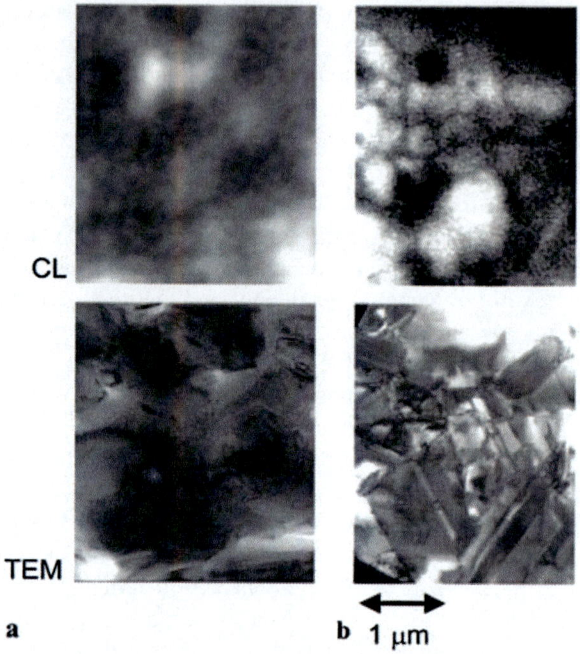

Fig. 6.9 TEM images of **a** Na and **b** Na-free samples [67]

(3) Proper thermal expansion. Thermal expansion coefficients of substrates should fall in the range of thermal expansion coefficients of CIGS materials.
(4) Chemical inertia. In theory, the substrate should not be corroded in the growth and should not cover any impurities that may diffuse into the film.
(5) Sufficient humidity barrier. Substrates should ensure that the film is not affected by the environment in the long-term operation, i.e., water vapor is essentially prevented from penetrating the film.
(6) Surface smoothness. Substrate with smooth surface is conducive to film growth and good contact with other parts.
(7) Cost, energy consumption, availability, weight, and other factors. Ideal substrates are optimally appropriate for their low cost, low consumption, abundant available materials, and lightweight, displaying associations with the thickness of substrates and indirectly affecting the efficiency of CIGS absorption layer in solar cells.

The effects of substrates on the preparation of thin films can be classified into three types (i.e., thermal expansion, chemical effects, and surface nucleation effects). When the growth of film is terminated, the substrate and film remain at the growth temperature, and the $Cu(InGa)Se_2$ film bears less pressure. In the cooling process, if the thermal expansion of the substrate does not fit that of the film, the stress will act on the film. For the thermal expansion coefficient of the substrate lower than 9×10^{-6}/K, little crevasse and vacancies may exist in the films. For the coefficient higher than this value, pressure will act on the film, causing adhesion failure.

Na ions diffuse into the chalcopyrite structure of CIGS materials when soda-lime glass acts as the substrates. The impact of Na on the microstructure of CIGS is reflected by the growth of grains. Besides, under such impact, the material displays a higher degree of preferred orientation to the <112> crystal orientation that is perpendicular to the substrate.

The different preferential orientation is primarily attributed to the nucleation of chalcopyrite materials. Compared with two CIGS films on soda-lime glass substrates and Mo substrates, the latter more fits <112> crystal orientation if the effect of Na is not considered.

When metals act as flexible substrates, the reaction of metals (e.g., Cu, Al, and Se) is noteworthy.

6.2 Device Properties

6.2.1 Photocurrent Generation

In this section, the photocurrent generation of $Cu(InGa)Se_2$ solar cells is discussed [1–3, 19–24].

$Cu(InGa)Se_2$ solar cell exhibits broad promising applications. Its material exhibits high advantages in optical and electrical properties; to be specific, the carrier lifetime in CIGS can reach hundreds of nanoseconds, the diffusion length can reach several microns, and the absorption length reaches below 1 μm. All the mentioned data lay a basis for increasing the efficiency of photovoltaic cells. It has been reported that the photocurrent J_{SC} of CIGS cell can reach 35.2 mA/cm^2, close to the maximum value of 42.8 mA/cm^2 of J_{SC} under AM1.5 illumination and 1.12 eV gap. Quantum efficiency (e.g., external quantum efficiency) is a vital parameter to characterize the loss mechanism of current, which can be referenced to analyze the photocurrent generation of CIGS cells. There are numerous factors affecting the external quantum effect, primarily the band structure of absorption layer, window layer, and a range of loss mechanisms. The existing photoelectric conversion efficiency of devices has reached 20%.

CIGS thin-film acts as absorption layer in CIGS cells, absorbing light to generate photogenerated carriers and provide photocurrent. The magnitude of photocurrent generation is determined by the absorption coefficient α (λ) and the mechanism of the electron absorption transition. As illustrated in Fig. 6.10, the absorption coefficient is ascribed to a range of the transition processes [16].

Set $\Phi(z)$ as the luminous flux of incident light (unit is cm^{-2} s^{-1}), α as Optical absorption coefficient, and $G(z)$ as defined as photogenerated carrier generation rate. Luminous flux decays exponentially with depth, which is expressed as follows:

$$\Phi(z) = \Phi(z_0)\exp[-\alpha(z - z_0)] \tag{6.8}$$

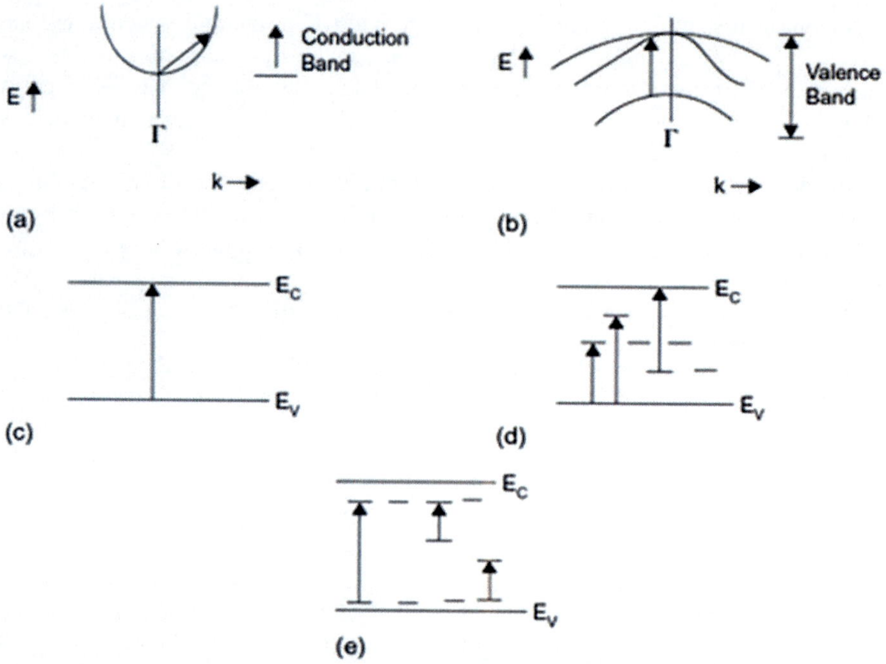

Fig. 6.10 Schematic diagram of optical absorption transition mechanism

$$G(z) = -\frac{d\Phi}{dz} = \alpha\Phi(z_0)\exp[-\alpha(z - z_0)] \tag{6.9}$$

If the variations of energy gap are considered, the situation will become more sophisticated and should be elucidated. Only the reflection of the front and back surfaces of the absorption layer independent of the wavelength is considered, and the reflection ratios are set as R_F and R_B, respectively, both of which are less than 1. If the thickness of the CIGS absorbing layer is W, the intensity of light reflected by the surface is $\Phi_1 = R_B \cdot \Phi(W)$. Under AM = 1.5 long and short wavelength illumination, the charge generation rate in CIGS varies with depth as presented in Fig. 6.11. Figure 6.11a suggests that CIGS is more significantly absorbed at short wavelength (0.56 μm), whereas no significant absorption at a depth greater than 0.5 μm. At long wavelength (1.06 μm), absorption occurs primarily in the deeper region, and the back reflection fraction is larger, and the spectral response is poor. Figure 6.11b characterizes the full charge generation rate, suggesting that the absorption largely occurs at the depth of the first micron level of CIGS material. Accordingly, a theoretical basis is laid for enhancing the thickness of absorption layers in the design of photovoltaic cells.

In practical CIGS photovoltaic devices, photoelectric current consumption is impacted by various factors, as characterized by quantum efficiency (QE). In the previous chapters, the concept of QE is presented. The definition of QE is expressed

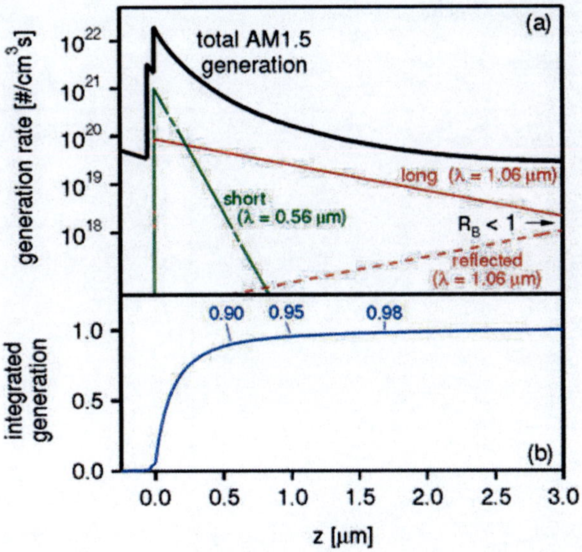

Fig. 6.11 The relationship between charge generation rate and depth in CIGS [22]

as follows:

$$\mathrm{QE}(\lambda) = \frac{\Delta J/q}{\Phi} = \frac{\text{Number of hole-electron pairs collected}}{\text{Number of incident photons}}$$

$$\mathrm{QE}_{\mathrm{ext}}(\lambda, V) = [1 - R(\lambda)][1 - \mathrm{AZnO}(\lambda)][1 - \mathrm{ACdS}(\lambda)]\mathrm{QE}_{\mathrm{int}}(\lambda, V)$$

$$\mathrm{QE}_{\mathrm{int}}(\lambda, V) = \frac{1 - \exp[-\alpha(\lambda)W(V)]}{\alpha L + 1}$$

The quantum efficiency $\mathrm{QE}_{\mathrm{int}}$ reveals the ability of thin-film materials exhibiting different band gaps to absorb sunlight in different bands and generate photocurrent; it is associated with the quality of materials in different layers and PN junction characteristics [25]. Where α denotes the optical absorption coefficient of Cu(InGa)Se_2, W is the width of Cu(InGa)Se_2 space charge region, and L is the diffusion length of minority carriers. AZnO(λ) and ACdS(λ) are the optical absorptivity of CdS and ZnO window layers, respectively. Photocurrent refers to the integral of the external quantum efficiency and spectrum. The external quantum efficiency is affected by the energy gap of the Cu(InGa)Se_2 absorption layer, CdS and ZnO window layer that constitute the CIGS photovoltaic cell; it is also susceptible to various photocurrent losses. Under the bias voltage of 0 and −1 V, the simulated quantum efficiency curve of CIGS thin-film solar cell is plotted in Fig. 6.12 (where the solid line denotes the bias voltage 0 V, and the dotted line is the bias voltage −1 V). The analysis of the photocurrent loss in the respective region is illustrated as follows [19].

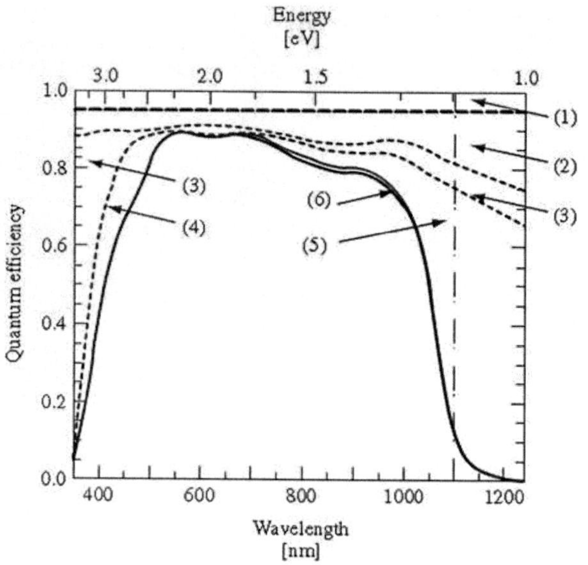

Fig. 6.12 Quantum efficiency curve of CIGS thin-film cell

Zone 1: Current collection gate electrode blocks the surface of the cell and reduces the light loss attributed to the illumination area. In the interconnection module, such loss will be determined by the area of the interconnection part.

Zone 2: Reflection loss between air and Zno/CdS/CIGS interface, i.e., frontal contact interface. Such loss can be down-regulated by evaporating an anti-reflective layer of MgF_2 that exhibits a thickness of about 100 nm, whereas this anti-reflective layer is impractical in glass-covered devices on the whole.

Zone 3: Light loss attributed to ZnO window layer absorption. The loss is split into two parts. Some electrons with energy greater than the band gap are absorbed to form an electron–hole pair, whereas they cannot form photocurrent. The infrared light with small energy is absorbed by free electrons to generate heat energy. The absorption wavelength covers visible and infrared lights. The infrared light is smoothly absorbed by free carriers when $\lambda > 900$ nm. The electron energy of $\lambda < 400$ nm is similar to that of the ZnO band gap and gets easily absorbed.

Zone 4: Light loss attributed to absorption in CdS zone. The energy gap of CdS is 2.42 eV, corresponding to the photon energy of 520 nm. Accordingly, photons with $\lambda < 500$ nm are smoothly absorbed by CdS, and the absorption intensity is proportional to the thickness of CdS layer. It is generally considered that the electron-hole pairs generated by the CIS layer can be hardly collected. In practical product manufacturing processes, the absorption layer is commonly thicker, and the absorption loss is larger than expected.

Zone 5: The absorption of $Cu(InGa)Se_2$ layer is incomplete. The gradient of $Cu(InGa)Se_2$ energy gap affects the steepness of the quantum effect curve, making the curve vary with a certain slope. When the photon energy approaches to the band gap of $Cu(InGa)Se_2$, the photon cannot be absorbed overall by the absorption layer.

Table 6.2 Photoelectric current loss of CIGS photovoltaic cells [2]

The corresponding zone in Fig. 6.12	Optical loss mechanism	$\Delta J/(mA/cm^2)$	$(\Delta J/J_{SC})/\%$
1	Loss of gate electrode shielding area	1.7	4
2	Reflection of ZnO/CdS/CIGS	3.8	8.9
3	Absorption of ZnO	1.9	4.5
4	Absorption of Cds	1.1	2.5
5	Incomplete absorption of CIGS	1.9	4.4
6	Incomplete collection of CIGS	1.0	2.3

Under the thickness of the absorption layer less than the reciprocal of the absorption coefficient, the material itself cannot fully absorb long-wavelength photons.

Zone 6: Incomplete collection of photogenerated carrier pairs by $Cu(InGa)Se_2$. The loss is associated with the thickness of $Cu(InGa)Se_2$ layer. It has been demonstrated experimentally that the performance is better when the thickness ranges from 0.1 to 0.5 μm under zero bias. Moreover, the loss is also a function of bias voltage as well as an electrical loss.

From the analysis of Table 6.2, the total loss current can reach 11.4 mA/cm^2, taking up 26.6% of the total current. According to Fig. 6.12 and Table 6.2, the basic way causing current loss can be determined, which provides a possible method for the performance optimization of cells. Indeed, compared with the simulation results, in actual situations, the loss mechanism and results will change correspondingly as impacted by the differences in device structure design, optical and electrical properties of the materials applied, which should be elucidated.

The quantum effect also considers the diode characteristics of the PN junction itself. The PN junction in CIGS is buried junction, and the junction is located 30–80 nm below the CIGS surface layer. Under long-wavelength light irradiation, the defects in the space charge region of PN junction are one of the causes of the light absorption loss. The better the diode characteristics, the higher the cell efficiency and the quantum efficiency will be. Moreover, the resistance also significantly impacts the photoelectric conversion efficiency. Thus, elevating the quality of CIGS absorption layer can noticeably improve quantum efficiency QE and enhance absorption [26].

For the larger surface band gap of CIGS thin films, low-energy photons cannot be absorbed. Considerable experimental results suggest that the high content of Ga on the film surface will increase the band gap, defects, and carrier recombination centers, causing surface recombination, shorten minority carrier lifetime, and diffusion length, as well as cause more loss of photogenerated carriers. To improve this situation, the doping method for Ga element gradient distribution is developed. As suggested from the experimental results, the parabolic gradient distribution curve can relatively facilitate the absorption of low-energy photons by CIGS and enhance the absorption efficiency.

6.2.2 Recombination

Free electrons in conduction bands and free holes in valence bands in semiconductors are the basis of solar cells. When the electron is excited to transit to the conduction band, a hole will appear in the valence band, or the free electron-hole pair may be generated by the exciton separation. Free electrons are recombined in a variety of forms during their motion [1–3, 19–22, 27]. The recombination mechanism can be split into radiation recombination and non-radiation recombination: with electron transited from the conduction band to the valence band, energy is converted to photon in the process of emission; the recombination between the bands is achieved; energy is transferred to other carriers. In practice, defects in CIGS materials commonly form energy levels or tails in valence and conduction bands, instead of as discrete energy. Accordingly, the defect-related recombination mechanism should be considered in the operation of CIGS cells, as described by SRH model based on carrier transition probability statistics. It is assumed that there are only two possible states for the defect, the donor defect level and the acceptor defect level capture the hole and the electron, respectively; the recombination process is illustrated in Fig. 6.13. In equilibrium state, the probability of generation and recombination is identical. During

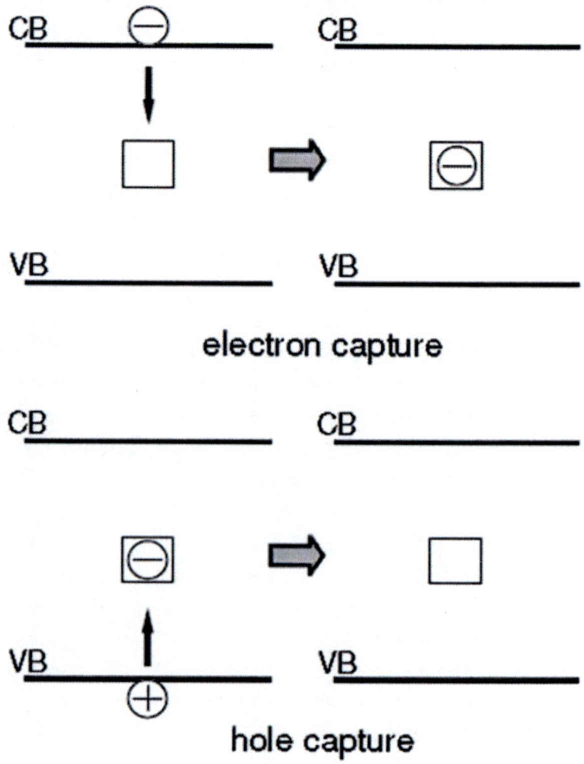

Fig. 6.13 Recombination mechanism in SRH model

light or electrical injection, the recombination rate increases linearly with the increase in free carrier concentration, which may be accompanied by the emergence of other recombination mechanisms as well [22].

The recombination rate can be expressed by

$$R = \frac{\sigma_e \sigma_h v_{th} N_t (np - n_i^2)}{\sigma_e (n + n_i \exp[(E_t - E_i)/kT]) + \sigma_h (p + n_i \exp[(E_i - E_t)/kT])} \tag{6.10}$$

where σ_e and σ_h are electrons and hole capture cross sections; v_{th} is carrier velocity; N_t is recombination center concentration; n and p denote electron and hole concentration, respectively; n_i is intrinsic carrier concentration at equilibrium; E_i is the intrinsic Fermi level; E_t is the recombination center energy level.

To be specific, the electronic lifetime is

$$\tau_e = \frac{1}{(N_t v_{th} \sigma_e)} \tag{6.11}$$

The hole lifetime is

$$\tau_h = \frac{1}{(N_t v_{th} \sigma_h)} \tag{6.12}$$

When $n \gg p$ or $p \gg n$, the equation can be simplified as follows:

$$R_n = \frac{\Delta p}{\tau_h} \quad R_p = \frac{\Delta n}{\tau_e} \tag{6.13}$$

where Δp, Δn is the excess carrier concentration.

Moreover, surface recombination is critical to CIGS operation. It primarily occurs at the material interface or grain boundary, i.e., the front and back surfaces of the photovoltaic cell. The recombination rate is expressed by

$$R = \frac{n + n_1}{S_h} + \frac{p + p_1}{S_e} \tag{6.14}$$

where S_h and S_e denote the surface recombination rate of holes and electrons in unit time and unit length.

Other composite mechanisms (e.g., direct inter-band and Auger recombination) are slight in SRH-type materials, which cannot be considered under appropriate conditions.

The composite principle can be adopted to explain the variations of J-V at different temperatures and wavelengths. The J-V relationship of CIGS photovoltaic cells is expressed as

$$\mathrm{J} = J_0\left(\exp\left(\frac{q(V - R_s J)}{nk_B T}\right) - 1\right) + \frac{V - R_s J}{R_{sh}} - J_{ph} \tag{6.15}$$

This equation is covered in previous chapters. Where $J_0 = J_{00} \exp(-\varphi/nk_B T)$, φ is the barrier height, and J_{00} is determined by the recombination mechanism. The recombination of interface, space charge region and body will affect the performance of the diode, parasitic resistance, and *J-V* characteristics of the cell. The effect of the open-circuit voltage V_{OC} can be analyzed.

Different recombination paths are parallel, so the open-circuit voltage is commonly affected by a single recombination mechanism under larger current. The *J-V* curves of $x = \text{Ga}/(\text{Ga} + \text{In})$ with values of 0, 0.24 and 0.61 are determined, as illustrated in Fig. 6.14. The optical band gaps are 1.02, 1.16 and 1.40 eV, respectively. The curve trend complies with Eq. (6.15), and the value of n is 1.5 ± 0.3. For general CIGS cells, n ranges from 1 to 2. It is therefore suggested that the recombination mechanism of $Cu(InGa)Se_2/CdS$ photovoltaic cells is primarily SRH mechanism, achieved by deep trap states in space charge region of $Cu(InGa)Se_2$. The energy level can provide both electrons and holes. When n varies from 1 to 2, defects play the role of deep trap level. When the change of n approaches to 1, the recombination mechanism begins to fall in inter-band recombination, mostly in the neutral region. Surface recombination will make the change of n value more than 2. The recombination at grain boundaries should also be considered, causing the loss of carrier collection and reducing the voltage value. When the doping type is p-type, the grain boundary recombination effect can significantly decrease.

In practical $Cu(InGa)Se_2$ materials, trap defects commonly occur in the form of energy band or the tail of valence band and conduction band, instead of in the form of discrete energy. The energy level range of the defect will act as a vital factor to determine the recombination current. The recombination with tail will make n altered with temperature, thereby generating the tunnel recombination current. Admittance spectroscopy can effectively characterize the distribution of electron defects in solar cells. The lifetime of minority carriers is also a critical factor affecting V_{OC}. The

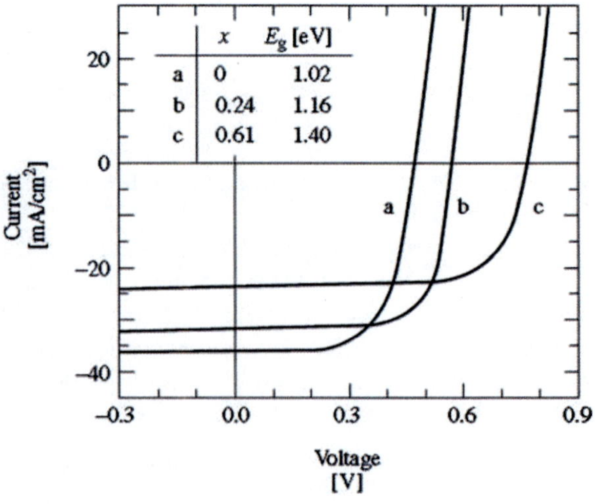

Fig. 6.14 *J-V* curves of $Cu(InGa)Se_2$ photocells doped with different Ga doping [18]

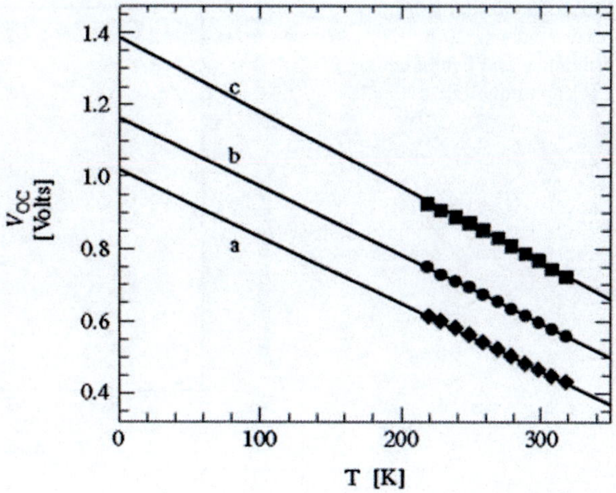

Fig. 6.15 The relationship between V_{OC} and temperature (light intensity from a to c decreases in turn) [2]

lifetime of minority carriers in high-speed devices is measured at 10–100 ns. When the carrier diffusion length approaches to the thickness of $Cu(InGa)Se_2$ absorption layer, the recombination in back contact cannot be ignored; when the back contact is Schottky contact, Schottky barrier will cause resistance loss. With the front contact interface recombined between CIGS absorption layer and buffer layer, the open-circuit voltage decreases with the increase in temperature at a certain light intensity. For samples with $Ga/(Ga + In) = 0.24$, the open-circuit voltage is determined at the light intensity of 1, 10, and 100 mW/cm^2, and the curve is illustrated in Fig. 6.15.

Various discussions about carrier transport and recombination mechanisms of CIGS cells do not consider grain boundary effects, i.e., the grain is assumed to be columnar and all transport processes are not required to cross the grain boundary, whereas this assumption is difficult to achieve practically. The grain boundary effect will impact the carrier diffusion length, causing micro grain boundary recombination to reduce. If the grain boundary is heavily doped with p-type, the electrons are prevented from reaching the grain boundary, and the defect recombination at the grain boundary will be adversely affected.

Series resistance and parallel resistance are also critical to the photoelectric conversion efficiency. Figure 6.16 presents the relationship between the photoelectric conversion efficiency and the incident light intensity of a CIGS thin-film solar cell on a glass substrate. The photoelectric conversion efficiency reaches 14–15% under standard light intensity. When the light intensity decreases to 50 mW/cm^2, the efficiency remains 10%. However, when the light intensity decreases to 10 mW/cm^2, the efficiency is only 3%. As revealed from the results, the major reason for the decrease in conversion efficiency of cells under low light intensity is that the series and shunt resistance of cells increase, while the short-circuit current displays a linear decrease. The increase in series resistance reduces the filling factor of the cell, while rising shunt resistance cannot prevent the decrease in cell's open-circuit voltage attributed to the decrease in short-circuit current. The result is that the decrease in filling factor

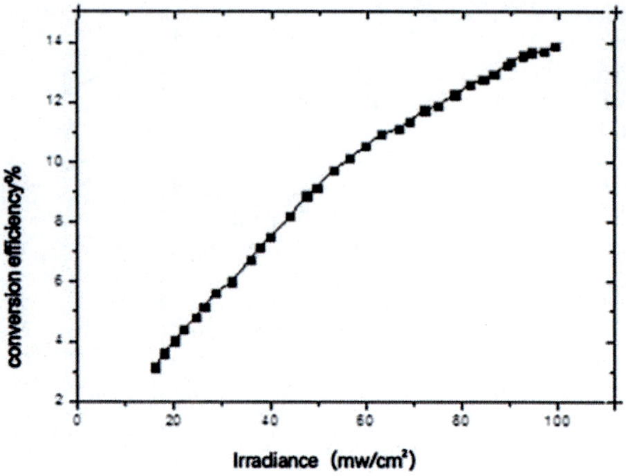

Fig. 6.16 Relationship between conversion efficiency and irradiance of CIGS photovoltaic cells [19]

and open-circuit voltage noticeably down-regulates the efficiency of the cell at a low light intensity. For CIGS thin-film solar cells, such a situation can be optimized by regulating copper content. Table 6.3 lists the parameters of CIGS solar cells with different copper contents under 0.1 mW/cm^2 light intensity. It is suggested that for CIGS solar cells covering 18% copper, the shunt resistance can reach 142 kΩ cm^{-2} at a low light intensity. The photoelectric conversion efficiency is 5.45% at 0.1 mW/cm^2 light intensity, 10% at 5 mW/cm^2 light intensity and 12% at standard light intensity. The prominent low-light performance of CIGS thin-film solar cells is dependent on the tolerance of copper content in the absorbing materials. It allows the use of a lower copper content deviating from the stoichiometric ratio to enhance the resistivity of CIGS thin-film solar cells and to prepare solar cells exhibiting high performance [19].

Moreover, temperature acts as a critical factor affecting the performance of solar cells. Since temperature impacts parameters (e.g., carrier concentration, carrier mobility, and band gap), resistivity and band gap are different at different temperatures. Furthermore, the defect excited states at each interface are a function of temperature as well, so temperature impacts solar cell parameters. The *J-V* curve and temperature relationship of CIGS cells with 16.4% efficiency between 350 and 200 K are simulated, well complying with the measured curve. From the results of the simulation curve illustrated in Fig. 6.17, the temperature coefficient of the short-circuit current reaches 0.0046 mA/cm^2/K. If dE_g/dT is ignored, the short-circuit

Table 6.3 Cell parameters of CIGS photovoltaic cells at 0.1 mW/cm^{-2} light intensity with different Cu content

Content of Cu/(%)	Rsh (kΩ cm^{-2})	FF/(%)	V_{OC}/(mV)	J_{SC}/(μA cm^{-2})	η/(%)
23.3 (Cu-rich)	3.5	25.4	93.9	27.8	0.66
21.5 (Standard)	11.8	31.6	256.0	25.4	2.05
18 (Standard)	142.0	57.1	405.0	23.6	5.45

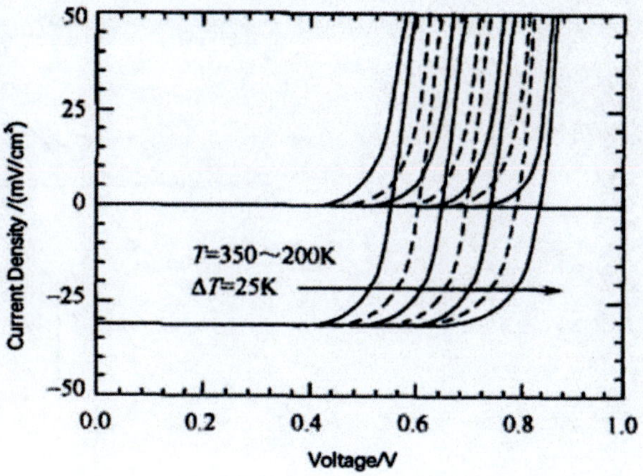

Fig. 6.17 Relationship between light-dark *J-V* curve and temperature of ZnO/CdS/CIGs/Mo cells under AM 1.5 [19]

current will be altered to −0.007%/K, which can be overlooked. Thus, short-circuit current is generally considered to remain unchanged with the variation of temperature. However, the temperature coefficient of the open-circuit voltage is negative, and the filling factor increases slowly with the decrease in temperature.

6.2.3 *Interface Characteristics of Cu(InGa)Se_2/CdS*

In the majority of heterojunction devices, the characteristics of PN junction interface will be critical to the performance of solar cells [26, 28–33]. In CIGS solar cells, by depositing CdS onto the absorption layer of CIGS, the possibility of bad shunt paths is down-regulated, and the sputtering damage attributed to ZnO deposited onto the absorption layer of CIGS solar cells is prevented [24]. The CIGS absorber layer and the CdS buffer layer are lattice-matched for generating a heterojunction, in which the CIGS and the CdS are p-type and n-type, respectively. The characteristics of its contact surface will noticeably impact the electrical characteristics of the device.

The band diagram of Cu(InGa)Se_2/CdS is illustrated in Fig. 6.18. Discontinuous mutation ΔEc is identified in conduction band between Cu(InGa)Se_2 and CdS, underpinning the inversion of Cu(InGa)Se_2. It is also considered that Cd doping, as a well-known donor impurity, causes n-type doping on the surface of the absorption layer, which is one of the reasons for Cu(InGa)Se_2 to form an inversion layer, whereas there has been no specific model to prove such perspective. For the doping of Ga in p-type Cu(InGa)Se_2, the band gap E_g is determined by the doping ratio. The band gap of n-type CdS is 2.4 eV. In the window layer, there exists an N^+ type transparent ZnO layer exhibiting a band gap of 3.2 eV. If ZnO exploits an intrinsic layer, the interface recombination will be promoted. ΔE_C denotes the difference between the minimum value of the conduction band of CdS and the minimum value of the conduction band of Cu(InGa)Se_2, which is a vital factor affecting carrier transport and recombination.

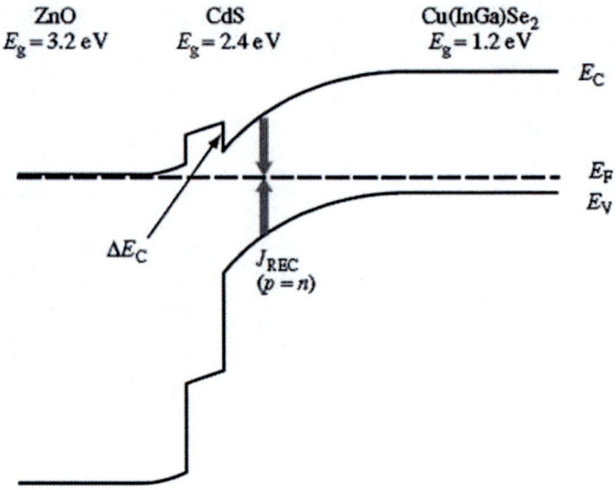

Fig. 6.18 Energy band diagram of ZnO/CdS/Cu(InGa)Se_2 device under zero bias and zero illumination [23]

Such effect will be revealed in device parameters (e.g., short-circuit current J_{SC}, open-circuit voltage V_{OC}, filling factor FF, and conversion efficiency), as presented in Fig. 6.19. When $\triangle E_C$ varies from −0.7 to 0.4 eV, J_{SC} is altered slowly. When $\triangle E_C > 0.4$ eV, a significant barrier will be identified at the interface, preventing the transport of photogenerated carriers; thus, J_{SC} shows a sharp decline trend.

When Cu(InGa)Se_2 exhibits a smaller band gap and $\triangle E_C > 0$, open-circuit voltage can increase, and device performance can be improved. When $\triangle E_C > 0.4$ eV, the conduction band of CdS is lower than that of CIGS, thereby forming a peak barrier. The collection of photogenerated carriers in Cu(InGa)Se_2 will be blocked, and the short-circuit current and filling factor will decrease dramatically. Since electrons can pass through the contact surface via thermionic emission, the open-circuit voltage V_{OC} can rise to 0.8 V. At the large ratio of Cu(InGa)Se_2 doped Ga, $\triangle E_C < 0$ varies from −0.7 to 0 eV. With the rise in the absolute value, the inversion of Cu(InGa)Se_2 near the contact interface disappears, the surface state recombination is initiated, the carrier lifetime at the interface decreases, and the open-circuit voltage declines. The filling factor FF also decreases with the increase in the absolute value of the band-side offset value, which is the identical trend to the open-circuit voltage V_{OC}. In this scenario, the open-circuit voltage is dependent on the quality of the contact surface and no longer limited by the energy gap at the absorption layer. When $\triangle E_C > 0$ and the value ranges from 0 to 0.4 eV, FF is basically a constant. When such value exceeds 0.4 eV, FF will decrease dramatically, which is identical to that of short-circuit current J_{sc}. The mentioned theoretical deduction can confirm that the optimal range of $\triangle E_C$ falls into 0–0.4 eV.

Recombination phenomena may occur on the contact surfaces of the two materials, not affecting the open-circuit voltage V_{OC}. The Fermi level on the contact surface is close to the conduction band, so the electrons near the surface of Cu(InGa)Se_2 act as the majority carriers. The SRH model can be adopted to calculate the total recombination current at the contact surface. To simplify the calculation, the surface

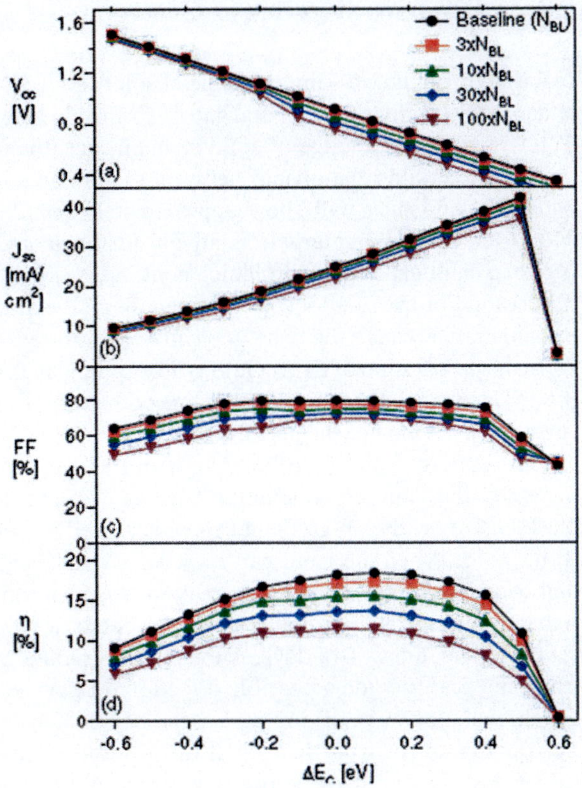

Fig. 6.19 Schematic diagram of the relationship between parameters of CIGS photocell and ΔE_c [33]

recombination speed of electrons and holes can be considered $S_e = S_p = S$, an essential parameter to assess the quality of the contact surface. Carriers involved in the recombination at the interface may occur between the electrons and holes of Cu(InGa) Se_2 and CdS, or between those of the two materials. On the whole, the CIGS surface is set as a copper-deficient layer, thereby reducing the valence band near the absorption layer to form a wider hole transport barrier, the hole concentration at the interface, as well as the interface recombination.

The quality of the contact surface is associated with the lattice characteristics of the two materials in the depositing process. From the microscopic crystal orientation analysis, an epitaxy relationship between the contact surface of CdS and Cu(InGa)Se_2 with chalcopyrite structure in [112] crystal orientation is observed under a transmission electron microscopy. The contact surface parallels to the [111] crystal orientation of cubic CdS or the [002] crystal orientation of hexagonal CdS. The lattice mismatch size is significantly small. The distance between CdS and the chalcopyrite Cu(InGa)Se_2 in the [112] crystallographic direction is nearly 0.334 nm. With the rise in the content of Ga in Cu(InGa)Se_2, the mismatch degree varies. For instance, when Ga/(Ga + In) = 0.3, the spacing is 0.331 nm; when the proportion rises to 0.5, the spacing is 0.328 nm [34].

6.2.4 Gradient Bandwidth Device

In the study on the working principle of solar cells, the efficiency of solar cells can be enhanced by regulating the band gap of absorbing layer semiconductor [1–7, 19, 35–37]. Given the knowledge of solid-state physics, the identical crystal structure type displays a certain relationship between the lattice constant and the semiconductor band gap, and they will show approximately parallelogram shapes on the graph. Moreover, such characteristic relatively fits chalcopyrite crystals. The relationship between band gap width and lattice constant is illustrated in Fig. 6.20. Through the observation of the relationship between them, the area enclosed by the curve is the possible range where the band gap can be controlled.

In the preparation of Cu(InGa)Se_2 absorption layer, CIGS by incorporating appropriate Ga to regulate the band gap is one of the ways to optimize the matching of solar spectrum and enhance the performance of solar cells. With the variation of Ga doping ratio, the energy gap of CIGS ranges from 1.04 to 1.67 eV, and the valence band is generally flat, and the conduction band is elevated. In the thickness range, p-type Cu(InGa)Se_2 with gap gradient can be achieved by setting the non-uniform Ga/(Ga + In) ratio. When the band gap $E_g < 1.3$ eV or Ga/(Ga + In) < 0.5, the efficiency and band gap of the cell are basically independent of each other. With the increase in band gap, the open-circuit voltage V_{OC} increases to 0.8 eV or higher, whereas the efficiency declines. The devices exhibiting gradual bandwidth can achieve higher efficiency and provide a possible direction for devices exhibiting smaller thickness.

Three major ways of doping Ga can be adopted for gradual bandwidth devices, i.e., doping on the front surface of the absorption layer (front gradient), doping on the back surface of the absorption layer (back gradient), as well as doping on both sides (double gradient). The energy band variations are illustrated in Fig. 6.21, Fig. (a) is the back gradient, Fig. (b) is the double gradient, Fig. (c) represents the front gradient, ΔE_{Fr} indicates the front gradient barrier energy, and ΔE_{Ba} denotes the back gradient barrier energy difference. The back gradient shows a slightly improved

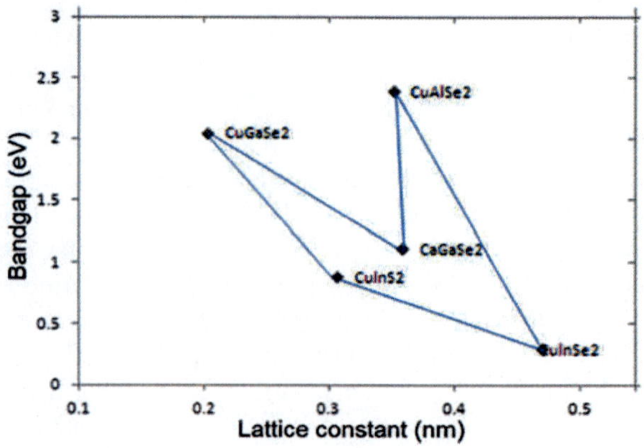

Fig. 6.20 The relationship between lattice constants (*x*-axis) and band gap (*y*-axis) of chalcopyrite crystals

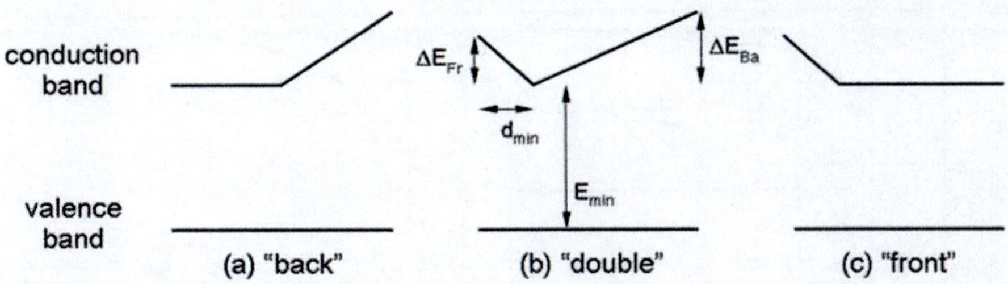

Fig. 6.21 Schematic diagram of energy band variation corresponding to different gap gradient implementations [2]

efficiency by about 1%. The front gradient enhances the efficiency by approximately 2% or more, while the double gradient enhances the efficiency by 3% or more.

The backward gradient method is adopted to build an additional electric field for minority carriers, so the collection of carriers can be promoted, and the recombination of the back field can be reduced. Compared with the CIGS photovoltaic cell exhibiting the minimum energy gap (1.04 eV), the open-circuit voltage and short circuit current of the solar cell under the back gradient increase, the maximum value of short circuit current is 39.5 mA/cm^2, and the overall efficiency is up-regulated by approximately 1%. The factors affecting the parameters consist of minimum thickness $d_{\min}$ and energy band difference $\triangle E_{\text{Ba}}$. With the increase in the energy band difference, the efficiency increases, and the filling factor shows a slight decrease. The most effective back gradient occurs in areas that exhibit high photogenic density and unsatisfactory collection efficiency.

Front gradient mode is not commonly employed in high-efficiency devices, and it is primarily adopted to achieve double gradient mode. The front gradient method is capable of up-regulating the open-circuit voltage more than the back gradient method, and the short-circuit current declines. Under the minimum thickness of about 0.3 μm, the filling factor decreases evidently; this finding is because a barrier will appear at this time, which hinders the collection of carriers and, therefore, causes FF to decrease.

In theory, the double gradient mode rises the short circuit current by minimizing the energy gap, while the local wide band gap up-regulates the open-circuit voltage; thus, it can highly likely to enhance the device performance. Under the minimum thickness less than 0.3 μm, the performance of the device is basically the identical to that of the back gradient, whereas for the larger thickness, the short-circuit current will decrease. Accordingly, surface vulcanization will act as a follow-up process to achieve a double gradient band gap structure. The double gradient can facilitate the absorption of photons and photogenerated carriers with the energy between the minimum energy gap and the CdS energy gap, while regulating $\triangle E_B$ to minimize the interface recombination of PN junctions, as an attempt to obtain the optimal short-circuit current and open-circuit voltage and optimize the performance.

The variation of photogenerated current density of the gradual band device with thickness is illustrated in Fig. 6.22. The photogenerated current density varies with

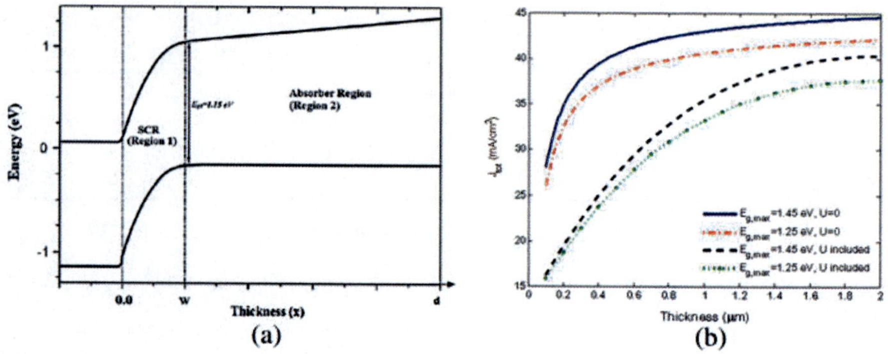

Fig. 6.22 The relationship between photogenerated current density and thickness of gradual bandwidth devices [44]

the energy gap, since the energy gap is associated with the absorption coefficient. When recombination is considered, the recombination current down-regulates the photogenerated current, whereas it does not decrease uniformly with thickness. The recombination current will be different in a range of regions [38–42].

The spectral energy of visible light is primarily concentrated at 400–700 nm, and the maximum energy is 500 nm. When the band gap of CIGS film is calculated by the overall Ga content exceeding 1.3 eV, the surface band gap will reach over 2.5 eV due to gradient distribution, and the absorbable photons are near ultraviolet light, thereby making the quantum efficiency of the cell decrease significantly; thus, the Ga content cannot be excessively high. As suggested from the experimental results, the quantum efficiency is better when the Ga content ranges from 0.2 to 0.3.

The most effective way to achieve the surface gap gradient is to dope S near the positive surface. The major effect of S doping is to reduce the energy level of valence band, rather than the mechanism of Ga doping, which is achieved by primarily elevating the energy value of conduction band. Accordingly, the impact on the collection of generated electrons can be minimized, and a more optimized gradual bandwidth device can be achieved.

6.3 Manufacturing of $Cu(InGa)Se_2$ Solar Cells

6.3.1 Material Deposition Technology

$Cu(InGa)Se_2$ should achieve a thickness of at least 1 μm in the device, and the composition is relatively maintained below a certain ratio. Overall, a $Cu(InGa)Se_2$ layer is deposited on a substrate made of glass, metal and other materials. There are numerous deposition techniques for $Cu(InGa)Se_2$ [1–4, 19, 34, 38, 43–55]. When taking a certain technology as the optimal commercial application, we should not only highlight high deposition rate and deposition quality, while considering cost economic factors.

1. **$Cu(InGa)Se_2$ substrate**

There are many types of $Cu(InGa)Se_2$ substrates with different characteristics, corresponding to different performance parameters.

Soda-lime glass (SLG) is the most common substrate, which more applies to the requirements of industrial large-scale production. SLG exhibits the advantages of low cost, high surface smoothness, and high temperature resistance, and the thermal expansion coefficient matches the thermal expansion coefficient of $Cu(InGa)Se_2$. SLG contains alkaline elements, e.g., Na and Ca, and diffuses into the $Cu(InGa)Se_2$ layer during the heat treatment to optimize the electrical characteristics. However, SLG also has defects. The high brittleness and non-flexibility of glass significantly limit the scope of SLG applications. Secondly, SLG may have considerable defects, which impacts the film formation quality.

Metal exhibits greater flexibility, better mechanical stability and thermal stability. Under the requirements of a range of $Cu(InGa)Se_2$ growth substrates, Cr-steel and Ti have been considered the ideal substrates. The former shows a cost advantage, and the latter exhibits lightweight.

Polyimide (PI) in polymers is currently the only commercially available polymer substrate. The PI layer is capable of withstanding a high temperature of more than 400 °C, lightweight, good insulation, and smooth surface, whereas the thermal stability is slightly worse and the thermal expansion coefficient is higher than the previous two substrates.

When metal and polymer act as substrates, Na is added externally to achieve high performance of the absorbing layer, which can be added during growth or after growth.

2. **Film growth mode [32]**

The growth process of the film directly affects the final structure and properties of the film. In the film deposition process, particles reaching the substrate interact with other particles and with the substrate to form an ordered or disorderly arranged film. There are three major modes of film formation:

(1) Island growth mode. In the initial stage of film formation, no favorable nucleation sites exist on the substrate. However, with the growing number of deposited atoms, multiple three-dimensional cores are present on the substrate. Firstly, they grow into individual islands, and subsequently the islands are expanded into a film. In the island growth mode, a poor wettability is between the deposit and the substrate, and rarely bonds with the substrate atoms, whereas it tends to bond between the material atoms. The growth of most metals on non-metal substrates complies with this mode.

(2) Layered growth mode. If a high wettability is identified between the material atom and the substrate, and the material atom and the substrate atom are easily

bonded, the film adopts a two-dimensional expanded layered growth mode from the nucleation stage. The bond between the deposited atoms remains greater than the tendency to form the external surface, and the layered growth mode can be maintained.

(3) Compound growth mode. During the two-dimensional mode of layered growth, stress appeared in the film, and the growth mode turned to island mode.

3. **$Cu(InGa)Se_2$ film formation method**

1) Co-evaporation method

Co-evaporation is the most extensive and successful method for the deposition of $Cu(InGa)Se_2$. Currently, CIGS/CdS solar cells with efficiency greater than 15% use such method. Co-evaporation is a multiple evaporation process that transfers Cu, In, Ga, and Se to a thermal substrate in an environment where the amount of Se is excessive. Evaporation temperature varies with metal. Overall, 1300–1400 °C is used for the evaporation of Cu, In is 1000–1100 °C, Ga is 1150–1250 °C, and Se is 300–350 °C. During evaporation, the evaporation rate is dependent on the flux distribution and gas flow rate of each source. Ga/(Ga + In) ratio slightly impacts film growth, Cu is a more significant factor. The equipment of the co-evaporation method is illustrated in Fig. 6.23. The Cu, In, Ga, and Se evaporation sources provide the elements required for film formation. Parameters (e.g., composition and evaporation rate) of the thin film are determined with instruments (e.g., atomic absorption spectroscopy (AAS) and electron impact scattering spectroscopy (EEIS)).

According to the different evaporation methods of Cu, the co-evaporation method can fall into one-step method, two-step method, and three-step method, and others,

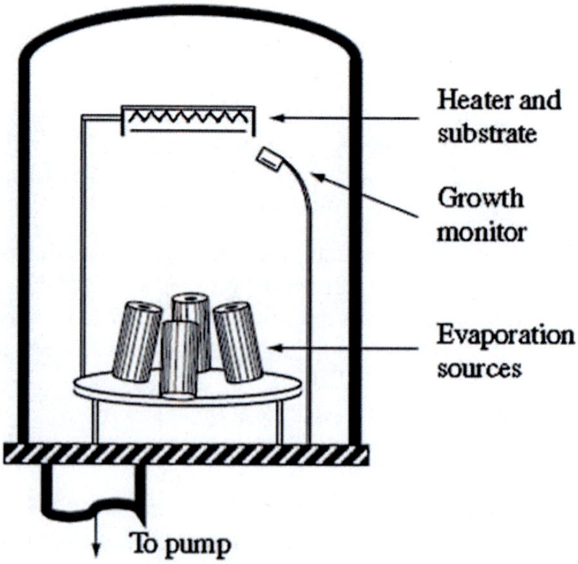

Fig. 6.23 Schematic diagram of co-evaporation equipment [2]

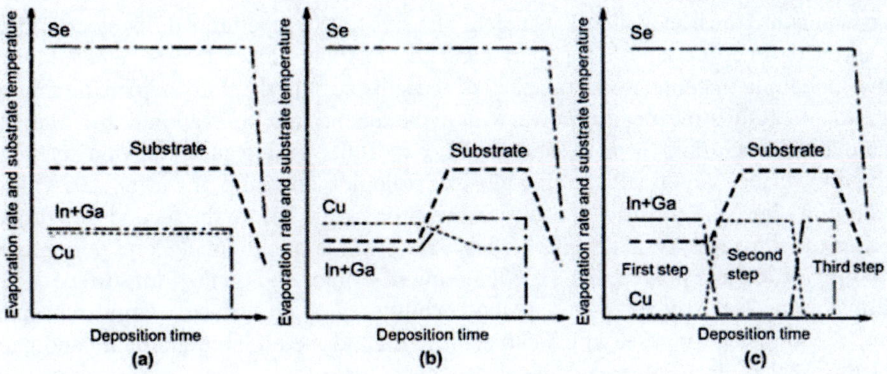

Fig. 6.24 Relationship between deposition time, substrate temperature and element flux of co-evaporation method

as illustrated in Fig. 6.24. As impacted by the advantage of diffusion speed, Cu can achieve a uniform distribution in the thickness of the film. However, the diffusion of In and Ga is slow, and the variation of the flow rate will cause the element to exhibit a gradient thickness distribution. The co-evaporation method is performed in an excessive Se atmosphere to ensure that the Se content in the film is sufficient, whereas the excessive Se cannot be combined in the absorption layer, and will be re-evaporated from the surface.

The one-step method is to maintain the flow rates of the four evaporation sources Cu, In, Ga, and Se constant in the depositing process. The changes in substrate temperature and evaporation source flow rate during deposition are illustrated in Fig. 6.24. The one-step method has simple process control and is suitable for large-area production, whereas the prepared film has relatively small crystal grains and cannot form a gradient band gap.

The two-step method is termed as the Boeing double-layer process. The specific steps of the two-step process are elucidated below. First, the substrate is heated to up-regulate the substrate temperature to 400–450 °C, and a first layer of Cu-rich CIGS film is deposited. The so-called Cu-rich, i.e., the ratio of Cu/(In + Ga) reaches over 1, such films exhibit small grain size and low resistivity. The second thin film refers to a Cu-poor film deposited when the substrate temperature is high at 500–550 °C, exhibiting a larger grain size and a higher resistivity. After two thin-film depositions, a Cu-poor CIGS film is obtained. Compared with the one-step process, the double-layer process can achieve a larger grain size. As demonstrated in existing studies, when the film is Cu-rich, Cu_xSe will be present on the film surface. At high substrate temperatures, Cu_xSe will liquefy, thereby enhancing the atomic mobility and increasing the grain size compared with the case when there is no liquid phase.

The steps of the three-step method are elucidated below. In the first step, 90% of the In, Ga, and Se elements are evaporated at a substrate temperature of 300–400 °C to form an $(In_{0.7}Ga_{0.3})_2Se_3$ preset layer. The second step aims to evaporate Cu and Se at the substrate temperature of 550–580 °C. Then, the heating power of the substrate is constant until the film is slightly Cu-rich. The substrate temperature drops since liquid phase Cu–Se compounds are generated. The substrate temperature is detected by a thermocouple in contact with the back of the substrate. In the third step, maintaining the identical substrate heating power to as in the second step, co-evaporation of a small amount of In, Ga, and Se on the slightly Cu-rich film, forming an In-rich thin layer on the film surface, and finally getting close to stoichiometry ratio of Cu($In_{0.7}Ga_{0.3}$)Se_2 thin film; then, the substrate temperature returns to that before the second step drop. After the third step is achieved, the flow rate of Se is continuously kept during the cooling process to prevent the Se from being re-evaporated on the film surface. The film obtained with the three-step method exhibits a smooth surface, a large grain size and a complete structure, which can achieve a Ga-doped double-gradient band gap [19].

The existing three-step method is the most effective method to prepare high-efficiency devices. In the three-step co-evaporation process, there are a range of phase transitions. After the first step is completed, a precursors layer is formed. In the second step, the precursors' layer reacts with Cu and Se to form a quaternary compound, and the film thickness is up-regulated. At the beginning of the second step, since less Cu is evaporated, a Cu-poor ordered vacancy compound (OVC) is formed, and there are Cu vacancy defects and substitution defects in which III group elements substitute for Cu. With the increase in Cu, Cu diffuses to the defect lattice, and In/Ga diffuses outward. It reacts with evaporated Cu and Se on the film surface to form new unit cells, and more new phases are generated (e.g., Cu(In, $Ga)_5Se_8$, Cu(In, $Ga)_3Se_5$, and Cu(In, Ga)Se_2), the film thickness increases, and the film will vary from Cu-poor to Cu-rich. Then, the film preferentially complies with the [112] orientations. In Cu-rich samples, cubic and tetragonal phases coexist with CIGS chalcopyrite phase. The former exists above the Cu(InGa)Se_2 grains, and the latter primarily exists at the grain boundaries; besides, the former is formed before the latter. The film displays a layered structure, and the grain size can reach 1–2 μm. The TEM image of Cu-rich CIGS is illustrated in Fig. 6.25.

In the third step, In, Ga, Se evaporated to the surface of Cu(InGa)Se_2 and Cu_xSe react with Cu_xSe, Cu diffuses outward, and In, Ga, Se enter the crystal lattice, leading to the formation of a Cu(InGa)Se_2 phase. After Cu_xSe is reacted, the grains remain significantly large, and considerable domain boundaries are present. With the addition of In and Ga In large quantities, the thin film is transformed into a Cu-poor phase, and the number of Cu vacancies and substitution defects that replace Cu is up-regulated, causing lattice mismatch at the domain boundary, and the stress decompositions the large grains and shrinks the lattice; as a result, the grains become smaller. Cu in the film surface obtained in the third step is depleted, and In(Ga)-rich crystal phase exists. Three-step process is capable of forming thin films with a grain size of 3–5 μm, and the grains grow in columns exhibiting high density [5, 46].

When using a three-step method to achieve vacuum deposition on a polymer network, if the rate is too fast, source elements will concentrated, especially Cu element, which will cause a short circuit of the device. The substrate temperature during deposition can be appropriately reduced [44]. Low-temperature deposition can be performed with PVD method at 300 °C. Low-temperature co-evaporation

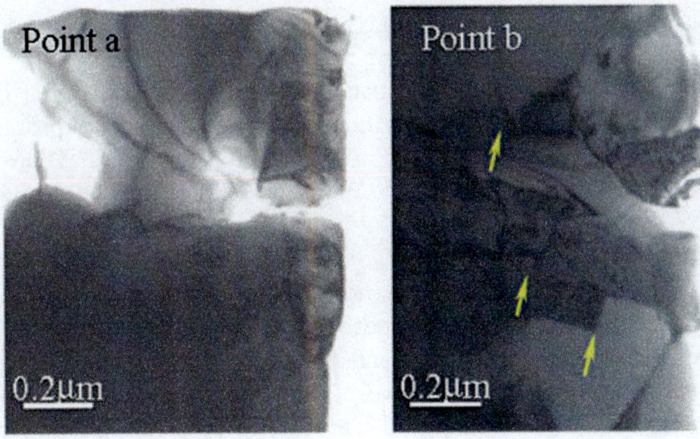

Fig. 6.25 TEM image of Cu-rich CIGS [5]

deposition is capable of declining the mobility of constituent atoms, thereby causing smaller grain size and better film uniformity.

2) Post-selenization of metallic precursors

The post-selenization process has many types of deposition and Se reaction methods. It exhibits the advantages of easy and precise control the stoichiometry of each element in the film, uniform film thickness, and uniform composition distribution. It also raises low requirements on equipment, and it contributes to large-area film growth. It is the preferred process for industrial production. CIGS solar cells currently implemented using post-selenization processes can achieve efficiencies over 16%.

The post-selenization process is to first deposit metallic precursors on a substrate covered with a back electrode; it is mainly composed of Cu, In, Ga, as well as metallic precursors in a Se atmosphere to obtain a CIGS film at an ideal stoichiometric ratio. Compared with the co-evaporation process, the post-selenization process is difficult to regulate the content and distribution of Ga, and a double-gradient energy gap is difficult to achieve. Overall, after the selenization is completed, a sulfur treatment is performed, a part of S atoms is substituted, instead of Se atoms, to form a $Cu(InGa)S_2$ layer on the surface of the film. The material exhibits a wide band gap, thereby hindering interfacial recombination and elevating open-circuit voltage.

Metallic precursors can be deposited by vacuum or non-vacuum processes. Under vacuum conditions, evaporation and sputtering processes can be adopted. Under non-vacuum, electrodeposition, spray pyrolysis, and chemical spraying can be exploited. Sputtering process has become the main process method because of its equipment economy, high deposition rate, good uniformity, and applicability to large-area films. Overall, a DC magnetron sputtering method at ambient temperature is used to prepare a Cu–In–Ga precursors. Because the sputtering sequence has a large effect on the composition of the thin film, the sputtering of the three elements should be performed in a certain order at ambient temperature. In the sputtering process, the sputtering order of the elements, the thickness of the sputtering, and the ratio of the elements significantly impact the degree of alloying and surface morphology of the film and, therefore, directly affect the adhesion between the film and the back electrode. Selenization process is considered the difficulty of post-selenization process. In the selenization process, the working temperature is about 450–600 °C. There are three main selenium sources, i.e., gaseous H_2Se, solid particulate Se, and organometallic selenium source ((C_2H_5)Se_2: DESe).

Selenization process with selenium source H_2Se: Gaseous H_2Se is generally diluted with nitrogen or argon, and the flow rate is rigorously controlled. In the selenization process, H_2Se gas is decomposed into Se atoms and reacts with precursors. Meantime, the precursors are annealed to make a high-quality Cu(InGa)Se_2 film. The reaction device is illustrated in Fig. 6.26. The control of heating annealing is a major factor affecting the quality of the film. To achieve higher quality, rapid thermal processing (RTP) has been commonly carried out. In the RTP process, the precursors covered with a thin layer of Se is rapidly heated to 500 °C by a heating source directly facing the substrate in 1–2 min, and then annealed. The rapid heating and annealing treatment can significantly shorten the heating time, lower the cost, and avoid the diffusion of harmful gases, saving materials without affecting the uniformity of film formation. However, H_2Se gas requires high-pressure storage for highly toxic and volatile [19].

The solid selenium source is to place Se particles on an evaporation boat and evaporate to produce selenium vapor to achieve selenization. The selenization of solid selenium sources commonly uses a closed process, operating in a closed graphite box or other closed containers to obtain a stronger selenium pressure. Such a process can selenize considerable proportioned precursors at one time with high production efficiency. The solid selenium source process has high safety and low cost, whereas selenium vapor is more difficult to control, which easily causes the loss of group III elements in precursors, affecting the stoichiometric ratio, film quality, and uniformity.

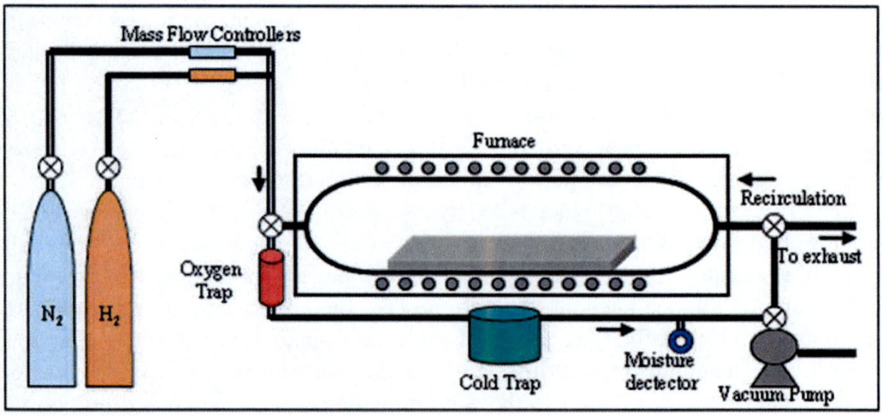

Fig. 6.26 Schematic diagram of H_2Se selenization device [2]

The use of organometallic selenium sources is based on the improvement of the acute toxicity of H_2Se gas. DESe is liquid at ambient temperature and can be stored at atmospheric pressure. Compared with the other two methods, the amount of selenization is less and the cost is lower. CIGS thin films currently implemented with organometallic selenium have good structural characteristics and optical properties, good adhesion between the substrate and the film, and great potential for development [19].

In the operation of the precursors selenization method, a high-concentration Se reaction at a low temperature stage is commonly used, so the precursors can be completely selenized without harmful binary phases. Pure inert annealing is used at high temperature, so it will not cause damage to the back electrode, and it can effectively eliminate lattice defects and obtain a uniform CIGS film [48, 50, 51].

3) Other deposition methods

(1) Electrodeposition method. The environment of electrodeposition refers to an acidic solution, primarily chloride or sulfate. The solution is composed of a conductive salt and an acidic complexing agent in a certain ratio. The solar cell prepared under the chloride solution exhibits higher efficiency. The reaction temperature is ambient temperature, and the thickness of the deposited film is nearly 2 μm. The composition range is $CuIn_{0.32}Ga_{0.01}Se_{0.93}$–$CuIn_{0.35}Ga_{0.01}Se_{0.99}$. To improve the problem of large stoichiometric deviation, a vacuum vapor phase method should additionally deposit some In, Ga, and Se for adjustment. The existing devices prepared with electrodeposition methods can exhibit an efficiency of 15.5% [34, 56, 57].

(2) Particle deposition method. Particle deposition method refers to a non-vacuum process. The main process is to make high-purity Cu and In metal powders into a liquid alloy in a hydrogen atmosphere at a certain ratio at a high temperature, and then convert it into a powder under argon. Mix powder with size less than 20 μm with wetting agent and dispersant, and then grind them into ink. The ink is sprayed on the Mo-coated substrate and dried to prepare precursors. The precursors is annealed at 440 °C in an atmosphere of nitrogen and hydrogen to form a CIGS film at an ideal stoichiometric ratio.

(3) Spray pyrolysis method. This method aims to dissolve a metal salt or an organic metal containing Ga and In into a solution in a certain proportion, and subsequently spray it on a high-temperature substrate to prepare a CIGS film via a decomposition reaction. The film formation quality is dependent on the solution ratio, substrate temperature, and spray rate. The thin film made by spray pyrolysis can achieve a thickness of 2 μm and suppress the generation of secondary phases. However, this deposition method exhibits the defects of loose structure and pinholes.

(4) Liquid phase method. The liquid phase method for preparing CIGS thin film is to arrange the four elements into a precursor solution according to a planned stoichiometric ratio at a certain temperature, and then uniformly

coat the substrate. Finally, under the protection of nitrogen, it was placed in a tube furnace and heated to 500 °C for a certain period, followed by natural cooling to develop a CIGS polycrystalline thin-film method [46]. The proportion of elements, the heating rate, and Se content critically impact the final film formation quality. The method is characterized by simple operation and high safety, and can be used for large-area film formation, and is not restricted by the shape of the substrate, and the cost is low.

(5) Nanoparticle ink printing and spin coating method. The characteristic of such a process is to first prepare CIGS nanoparticles, and then select a suitable dispersant to make a uniform CIGS Nano ink. The substrate can be spray-coated, drip-coated, spin-coated, and blade-coated; then, it can be used to form CIGS film by heat treatment. Such process has many advantages: low cost, high raw material utilization rate, unlimited choice of substrates, large-area thin films, and easy control of film thickness. However, since nanoparticles or quantum dots are adopted, various parameters (e.g., surface morphology, particle size and distribution and stoichiometric ratio) should be controlled, thereby increasing the difficulty and complexity of the process. Meantime, the synthesis of several high-quality CIGS nanoparticles is one of the key points of such a method.

(6) Screen printing method. Screen printing is a non-vacuum low-cost CIGS thin-film preparation process. A precursor solution mixed with a liquid binder is prepared according to the expected stoichiometric ratio of the elements. The deposited precursor is sintered into a CIGS film under the control of the atmosphere. Screen printing method is easy to control the growth and film components, and can better control the uniformity and thickness of the film. Moreover, the screen printing effectively exploits materials, lowering the production cost. However, to grow thin films with good uniformity and crystallinity, we should try to prevent the formation of heterogeneous phases.

(7) Mixing method process. Mixing method process is an improvement on the co-evaporation process. Such process integrates the advantages of evaporation and sputtering processes, which is capable of achieving the coordinated high film quality and large area. The specific steps of the mixing method include: firstly, a linear evaporation source is exploited to evaporate the prefabricated layer to enhance the adhesion of the film and easily regulate the distribution of Ga; secondly, Cu is sputtered to accurately regulate the Cu content and reduce heat loss; thirdly, the In–Ga–Se layer is co-evaporated to form a thin film with a certain stoichiometry. The crystal quality of the thin film made by the mixing method depends primarily on the substrate and the annealing temperature. At the overly low temperature, the atom mobility is poor, which adversely affects crystal growth. If the annealing temperature is overly high, the film will crack for surface tension.

Moreover, the methods to fabricate CIGS films consist of electrophoretic deposition method, gas phase transport method, mechanochemical method, as well as laser-induced synthesis method [19, 34].

6.3.2 *Junction and Device Formation*

1. **Formation of heterojunctions in CIGS thin-film solar cells**

On the whole, solar cells refer to devices with PN junction structures composed of two materials of opposite conductivity types with thin barriers between them; CIGS thin-film solar cells are no exception [39–42, 45, 56–60]. The realization of the initial CIS cell is a PN heterojunction consisting of n-type single-crystal $CuInSe_2$ and p-type CdS. Subsequently, the device draws upon vacuum evaporation to deposit the intrinsic CdS onto the absorption layer, and then deposits a doped CdS layer as a window layer. With the development of materials science, doped ZnO gradually replaces the doped CdS layer to increase the energy gap and obtain the gain of photocurrent. The intrinsic CdS layer is deposited on the $Cu(InGa)Se_2$ absorption layer via a chemical bath deposition (CBD) process. The CBD method shows high similarity to chemical vapor deposition (CVD), except that the vapor phase is replaced by the liquid phase of the deposit. The growth of CdS with the CBD method is achieved via the reaction between ions and ions or by collecting colloidal particles. The lattice structure of the generated CdS is cubic, hexagonal, or a mixture of the two. CdS employed in CIGS thin-film solar cells is generated via the reaction between ions, and the lattice structure exhibits a hexagonal structure. Deviations in the stoichiometric ratios of oxygen, hydrogen, nitrogen, and carbon atoms in the depositing process overall affect the quality of CdS [5]. The inverse heterojunction currently employed in CIGS solar cells is illustrated in Fig. 6.27. The p-type region is composed of a CIGS thin film, while the n-type region is more complicated, not only having n^+-ZnO, i-ZnO, and CdS, but also having an inverse CIGS layer. Practice has shown that CIGS solar cells that can achieve high efficiency generally use CIGS absorption layers with copper-poor surfaces. As demonstrated in the existing studies, for differences in chemical ratios, ordered vacancy compound (OVC) can be formed. As impacted by the existence of this type of n-type OVC layer, the CIS or CIGS homojunction penetrates deep into the absorption layer, hinders the interface recombination, and exhibits higher resistivity. However, some controversies remain in this inference.

In early research, a PN heterojunction has been formed by infiltration of Cd or Zn into p-type $CuInSe_2$ at 200–450 °C. Above 150 °C, S and Se will penetrate each other on the $CuInSe_2$/CdS contact surface, and promote the diffusion of Cd to $CuInSe_2$ at temperatures above 350 °C. As impacted by the diffusion of Cd, the presence of mixed components can be observed in the $Cu(InGa)Se_2$/CdS/ZnO heterojunction, i.e., the heterojunction is formed deep into $Cu(InGa)Se_2$. The mixed component can be CdSe obtained by reacting the Cd element with Se in $Cu(InGa)Se_2$. The Cd on the $Cu(InGa)Se_2$ surface began to accumulate from the early growth stage. As shown in

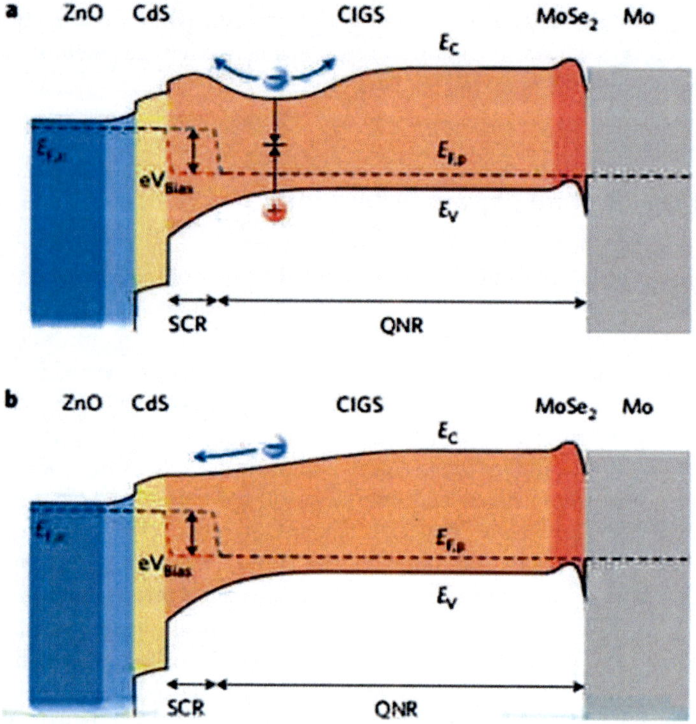

Fig. 6.27 Heterojunction energy band diagram of CIGS solar cell [46]

TEM images, there is a Cd accumulation layer over 10 nm on the Cu(InGa)Se_2 Cu-poor surface. Since the ionic radii of Cd^{2+} and Cu^+ are similar, the phenomenon that Cu^+ in Cu(InGa)Se_2 is replaced by Cd^{2+} will be termed as the electromigration of Cd. The existence of CdS buffer layer can enhance the device electrical properties. An electrical heterojunction is formed on Cu (InGa) Se_2, and there are nearly no macro defects at the interface except for the columnar grain boundaries. Such a phenomenon is obtained by SEM image analysis of Cu(InGa)Se_2 solar cell with an efficiency of 19.3% by NREL laboratory. The effect of the CdS/Cu(InGa)Se_2 contact surface on the characteristics of the cell has been explained in the foregoing.

The existence of intrinsically high-resistance ZnO will impact the band structure, causing the energy difference between the conduction band bottom and the equilibrium Fermi level at the interface to increase, and the interface recombination to increase. However, experimental parameters show that the existence of intrinsic ZnO is conducive increasing open-circuit voltage. As suggested from the experimental results, if the intrinsic ZnO is removed, the open-circuit voltage will decrease by at least 20–40 mV, and the efficiency will decrease as well.

For Cu(InGa)Se_2/CdS/ZnO three layers, there is usually a separate donor trap in CdS, and a deep acceptor trap in CIGS layer. The high defect density in the CdS layer is equivalent to the shallow-level donor density, which is critical to generate photoconductivity. The electron mobility in the CdS layer is more than 50% lower than the normal value in the crystal. For the diode performance of the formed PN junction, the parameters of ZnO have nearly no effect on it, and the change of CdS parameters only exert a small effect. However, the characteristics of the $Cu(InGa)Se_2$ absorption layer critically impact the performance of the diode.

2. **Device structure**

The CIGS cell exhibits the classic structure of a substrate, back contact, absorption layer, buffer layer, window layer, and electrode from bottom to top. Its classic structure and material characteristics of each part are illustrated in Fig. 6.28. The substrate of CIGS solar cells is usually composed of soda-lime glass, stainless steel metal, or polyimide. The back contact commonly uses metal Mo to cover the substrate to form a non-blocking contact with the absorption layer. On the whole, the thickness of the Cu(In, Ga)Se_2 absorbing layer is set to 2–3 μm, which can be achieved with a range of deposition methods. Under certain conditions, a dual-gradient energy gap can be achieved, and Cu-poor surface structures are often employed in devices, so high device performance can be more effectively achieved. The CdS buffer layer and the ZnO window layer are generally deposited onto the absorption layer using a chemical bath process (CBD) or an electronic sputtering process to form a heterojunction structure, which is a vital part of solar cell operation. Electrodes commonly use metals with excellent conductivity (e.g., Al or Ti) to reduce parasitic resistance and enhance the electrical characteristics of the cell [19].

According to the different light paths, CIGS solar cells have two structures. When light is incident from the substrate, the cell structure is identical to the conventional structure. When light is incident from the top of the cell, a front contact layer made of TCO is added to the cell. The specific differences are illustrated in Fig. 6.29.

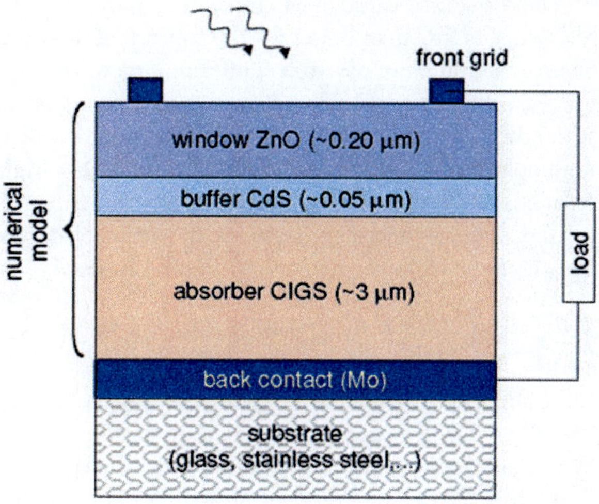

Fig. 6.28 The classic structure of a solar cell

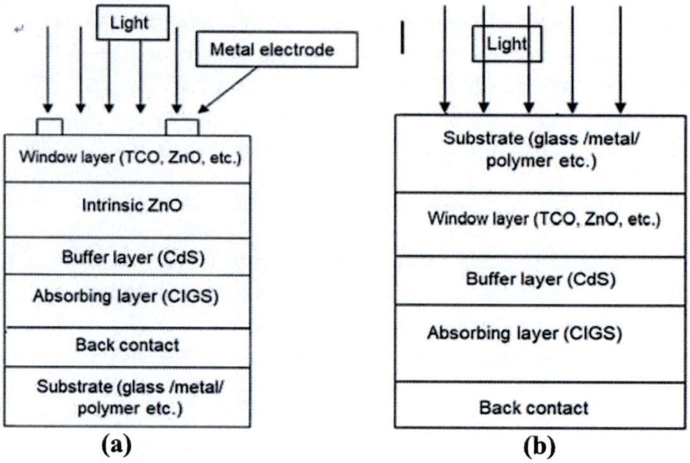

Fig. 6.29 Two structures of CIGS solar cells

The substrate is the basic part of the cell, and it has several characteristics, e.g., thermal expansion and transparency. For details, please see Sect. 6.1.4.

1) Back contact

In CIGS solar cells, Mo is the most frequently applied material for back contact; it acts as an optical reflector. By developing CIS or CIGS solar cells, Pt, Au, Cu, and Mo have been adopted as back-contact materials. However, Mo has been extensively employed in GIGS solar cells for its relative stability at processing temperature, resistance to alloying of materials containing Cu and In, low contact resistance with CIGS materials, and good substrate adhesion.

The resistance value of Mo reaches usually 5×10^{-5} Ω cm or less and the contact resistance is less than 0.3 Ω. Its deposition method is usually carried out by electron beam evaporation or electron sputtering process below 500–600 °C. In the sputtering process, the electrical characteristics and adhesion of Mo are associated with the sputtering Ar gas pressure. The current process is to first deposit a Mo layer with high adhesion and high resistivity at a relatively high pressure. On the whole, this step exploits a pressure of 10 mTorr; the thickness of the deposited layer is nearly 0.1 μm. The second step is a 0.9 μm Mo layer deposited at a pressure of 1 mTorr. This layer has a lower resistivity, which can reach 1×10^{-5} Ω cm.

The mechanical and electrical properties of Mo impact the performance of photovoltaic devices [60]. The internal stress of the Mo layer will cause the structure and electrical properties to vary. Under tensile stress, the Mo layer exhibits a porous and loose structure. When the CIGS absorption layer grows on the substrate covered with Mo, the CIGS first forms isolated and randomly distributed islands on the surface. With the extension of the deposition time, the islands and islands communicate with each other to achieve epitaxial growth. Meantime, the crystal orientation of the thin film began to be determined. The growth of GIGS film exhibiting a thickness of 1 μm shows that the growth process of the film on Mo is a process from non-fixed crystal growth to mostly <112>. When the film growth encounters the Mo <110> grain boundary, the direction will be altered to <220/204>. A diffusion effect is identified on the contact surface between Mo and CIGS. Se atoms and Mo atoms will form $MoSe_2$. The CIGS/Mo heterojunction with $MoSe_2$ will not be a Schottky junction, whereas a good ohmic contact will be achieved. In the presence of oxygen atoms, Mo–O or MO–O–Se components occur at a deposition temperature of 600 °C, and a Schottky barrier is formed at the CIGS-Mo/MoO_2 interface. The generation period of the $MoSe_2$ layer on the CIGS/Mo contact surface is formed in the second step of the three-step deposition process, and the presence of Na element stimulates the production of $MoSe_2$ layer. The diffusion of Na atoms into the absorption layer is dependent on the deposition state of the back contact layer. Thus, to achieve the contact resistance that satisfies the requirements of the device, there is a compromise between the thickness of Mo and the fill factor. The optimal thickness of the Mo layer is 0.2 μm, and the desired morphology is a columnar layer structure, thereby expediting the diffusion of Na and growth on the CIGS film.

To improve the solar cell's absorption of light, TCO is used the back contact layer, which can transmit light through the entire cell, whereas the material will deteriorate at the higher deposition temperature of the absorption layer.

2) Buffer layer

A semiconductor suitable for serving as a buffer layer pertain to n-type semiconductors and exhibit the band gap of 2.0–3.6 eV. Though many materials satisfy the requirements, CdS remains the most widely used buffer layer material. CdS is a direct band gap semiconductor material exhibiting a band gap of 2.4 eV. It has a cubic sphalerite structure and a hexagonal wurtzite structure, and has a small lattice mismatch coefficient with CIGS. Growing CdS in high-efficiency CIGS cells widely uses a large-area, low-cost chemical bath deposition (CBD), whereas it is not compatible with the in-line vacuum process. The electron sputtering deposition PVD can also be used, whereas there are common problems (e.g., poor film uniformity). At present, the improvement of the buffer layer complies with two directions, i.e., to find alternative materials without Cd and wide band gap; to facilitate the deposition process.

The function of the buffer layer is to positively affect the electrical properties of the formed heterojunction, and at the identical time protect the heterojunction from chemical corrosion and mechanical damage. For the effect of electrical properties, the existence of buffer layer optimizes the band correction; a wide depletion layer is built to reduce the tunnel current; a higher contact potential energy is established to allow the open-circuit voltage to reach a higher value. For protection function, the buffer layer can hinder mechanical and electrical damages attributed to oxygen deposition. The intrinsic ZnO in the buffer layer can also prevent defects in the CIGS film from dominating the open-circuit voltage of the entire device.

For device performance, the thickness of the buffer layer acts a critical factor. In the early study on CIS solar cells, the thickness of CdS was about 1–3 μm, and the evaporation process at the substrate temperature or the sputtering process under certain conditions was used [61]. CBD has now become the mainstream process. The CBD process will bring a large amount of oxygen atoms into the buffer layer, the proportion can even reach 10–15%, the Cd(S, O, OH) component exists in the buffer layer, and the Cd atoms are also introduced with the deposition solution, and its proportion can affect conversion efficiency of the device. According to some data, the use of the CBD method up-regulates the conversion efficiency from 17.6% to 18.5%. Based on the CBD process, the thickness of the CdS buffer layer achieving complete conformal coverage of the CIGS surface can reach 10 nm. Such coverage is dependent on the deposition conditions, especially the pre-deposition fluid with a high S/Cd ratio. For the obvious green light absorption of CdS, the unsafe state of Cd in the manufacturing process, and environmental pollution, the study on Cd-free buffer layer materials has gradually been essential.

In the study on materials that replace CdS as a buffer layer, Cd-free devices can be implemented in two ways. The first is to find a buffer layer material capable of replacing CdS. Plenty of materials in the experiment show promising applications. The second is to omit CdS and directly deposit ZnO. In fact, there is a tendency for the two approaches to merge. In numerous experiments, a range of methods and materials have been tried to bring some considerable results, as listed in Table 6.4.

Among the many choices, existing data show that devices made of materials, e.g., CBD-ZnS, MOCVD-ZnSe, ALD-ZnSe, CBD-ZnSe, PVD-$ZnIn_2Se_4$, and ALD In_2S_3 can achieve conversion efficiency more than 11%. CBD-ZnS with a gap of 3.8 eV can reach even higher efficiency. Aoki University in Japan used a solution

Table 6.4 Performance parameters of Cu(InGa)Se_2 solar cells with different deposition methods and buffer layer materials [2]

Buffer layer material	Deposition process	Efficiency/%	Voc/mV	Jsc/(mA/cm^2)	FF/%
None		10.5	398	39.0	68
None		15.0	604	36.2	69
ZnO	MOCVD	13.9	581	34.5	69
ZnO	ALCVD	11.7	512	32.6	70
Zn treatment	$ZnCl_2$ Solution method	14.2	558	36.3	70
Zn(O,S,OH)x	CBD	14.2	567	36.6	68
ZnS	CBD	16.9	647	35.2	74
ZnS + Zn treatment	CBD	14.2	559	35.9	71
Zn(Se,OH)	CBD	13.7	535	36.1	71
ZnSe	ALCVD	11.6	502	35.2	65
ZnSe	MOCVD	11.6	469	35.8	69
In_xSe_y	Co-evaporation method	13.0	595	30.4	72
$ZnIn_xSe_y$	Co-evaporation method	15.1	652	30.4	76
$In_x(OH,S)_y$	CBD	15.7	594	35.5	75
In_2S_3	ALCVD	13.5	604	30.6	73

made of zinc salt, ammonia, and thiourea to deposit ZnS in three consecutive CBDs, and the efficiency of the CIGS cell reached 18.6%. The cell made by ZnS (O, OH) with a slightly optimized process can exhibit an efficiency of 18.5%. In the analysis and comparison of Table 6.4, it should be stated that the quality of the $Cu(InGa)Se_2$ material employed in each experiment is different, and the quality of the heterojunction formed varies with the use of process method. For several processes not suitable for junction formation, the defects introduced in the process will primarily affect the efficiency.

However, Zn-based compounds may form blocking barriers with $Cu(InGa)Se_2$ for band alignment. Under the layer thickness of the compound below 50 nm, the barrier is easily tunneled by the carrier and a tunnel current is generated. The realization of this situation is dependent on the quality and surface uniformity of $Cu(InGa)Se_2$. If the quality is great and the contact surface is covered uniformly, the probability of this tunneling is relatively large. In the case of the introduction of hydroxides, this barrier will be offset under certain conditions. Chemical bath deposition commonly introduces this hydroxide [59].

For the buffer layers of ZnSe and $ZnIn_2Se_4$, prepared via a dry process, the efficiency is nearly 15%. ZnSe is prepared with the MOCVD method and exhibits a band gap of 2.67 eV, which will generate a photocurrent higher than that of the reference structure CIGS solar cell, whereas the open circuit voltage will become lower. $ZnIn_2Se_4$ exhibits a low band gap of 2.0 eV, whereas it has good lattice matching with CIGS materials. Meantime, the material preparation process is compatible with the co-evaporation process, which is suitable for large-scale production.

A buffer layer of Zn compound (e.g., ZnO) deposited by the CBD method, due to the characteristics of the process itself and the atmosphere of ammonia, a variety of Zn compounds will exist in it. Overall, when CBD-ZnO acts as a buffer layer, it will undergo an annealing process at 200 °C for 15 min, so the Zn-O-Zn bond in the buffer layer can be firmly established. As impacted by the wide band gap characteristics of ZnO compared with CdS, devices using CBD-ZnO buffer layers can absorb shorter wavelengths and elevate collection efficiency. A comparison of absorption wavelength and quantum efficiency can be observed in Fig. 6.30. Another more promising material is (Zn, Mg)O. By regulating the content of Mg, the film band gap can be adjusted between 3.3 and 7.7 eV. Compared with ZnO, this material exhibits the advantages of completely dry growth environment, improved energy band alignment to reduce interfacial recombination of heterojunction, and so on. However, for the sputtering loss, the efficiency has not reached a high degree, and the implementation is mostly through the buffer layer with CdS.

3) Window layer

For the window layer of CIGS solar cells, two points should be met. The first is sufficient transparency to transmit light to the bottom of the device. The second is a sufficiently high electrical conductivity, and the parasitic resistance is minimized in the process of the photo-generated current being transported to the external circuit. Transparent metal oxide TCO is a common material for front contact. A narrow

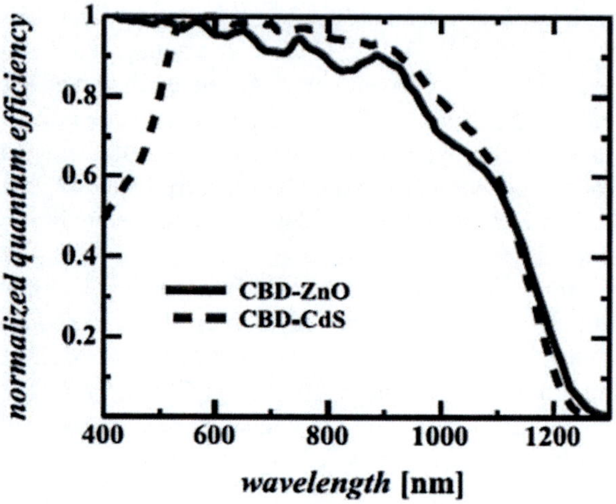

Fig. 6.30 Comparison of CBD-ZnO and CBD-CdS collection efficiency and absorption wavelength [42]

metal grid (Ni–Al) is deposited on the TCO, so series resistance can be reduced. The quality assessment of the front contact is dependent on several factors: the surface resistance, absorptivity, and reflectivity of the TCO, and the pitch of the metal grid.

In the early manufacture of CIS or CIGS solar cells, doped CdS acts as the window layer. Doped CdS achieve high conductivity through the density of donor defects generated by doping Al or In. Spectral absorption loss is achieved by forming an alloy with ZnO to broaden the band gap. With the application of TCO materials with band gap 3 eV, the doped CdS is gradually replaced. The existing window layer is either doped with tin in In_2O_3 (In_2O_3: Sn, ITO), or doped with Al in ZnO. The latter is implemented by RF sputtering and more frequently adopted. The window layer is not only the core of the built-in electric field in the solar cell heterojunction, but also the upper surface of the cell together with the electrode to form a channel for power output. Moreover, its series resistance with the buffer layer can protect the device from the electrical loss attributed to the unevenness of the surface of the absorption layer.

To achieve high cell efficiency, the TCO should be deposited at a low temperature below 150 °C to prevent harmful diffusion between the contact surface with the CdS/CIGS. A commonly used deposition process is a sputtering process. In industrial production, DC sputtering is commonly used. The sputtering gas environment commonly uses a mixed gas atmosphere of Ar: O_2, and the sputtering rate is commonly controlled between 0.1 and 10 nm/s.

Intrinsic ZnO is a direct band gap metal oxide semiconductor exhibiting a band gap of 3.4 eV at ambient temperature. Under natural growth, ZnO is n-type, hexagonal wurtzite structure, and has good lattice matching with CdS. Depending on the ZnO doped elements, there are ZnO:Al (AZO), ZnO:Ga (GZO), and ZnO: B (BZO). The deposition method includes a CVD process and a sputtering process. For BZO, the MOCVD deposition method is usually used, because the material doped with B element has high carrier mobility and high absorption rate for long-wavelength spectrum. It is extensively employed in current CIGS solar cells and has achieved higher efficiency. For GZO and AZO, DC RF sputtering method and Zn/Al alloy targets can be exploited, and sputtering rate can reach 4–5 nm/s. However, under the hysteresis effect, the optimal photovoltaic effect can only be achieved in a narrow window. In comparison with AZO, GZO exhibits lower resistivity and better application advantages [39]. Compared with MOCVD, the DC RF sputtering process has stronger damage to the junction. Accordingly, the filling factor (FF) and short-circuit current of CIGS solar cells at GZO are not as good as those of BZO solar cells. However, GZO is characterized by low equipment requirement and low production cost, which satisfies the requirements of mass production, so it has theoretically the value of further optimization. LPCVD deposition of ZNO is also being studied broadly. The atomic layer exhibits a lower deposition rate, whereas the control of the gas flow increases the uniformity, and the high throughput provides a prerequisite for mass production.

4) Metal contact electrode of the device

After the layers of the device are deposited, a metal contact consisting of Al–Ni is deposited on the window layer exhibiting a thickness of 1–2 μm. To allow light to penetrate the device as much as possible, the metal contacts are commonly made in a grid type. Before the deposition of metal Al, tens of nanometers of Ni can be deposited to prevent the formation of a high-resistance oxide layer and enhance ohmic contact. Meantime, the presence of Ni can also prevent Al from diffusing to ZnO and improve cell stability. At present, the metal electrode can be prepared by a deposition method that is evaporated through an aperture mask.

5) Packaging and interconnection of small-area CIGS solar cells [39]

For CIGS solar cells based on small-area flexible substrates, considerable ways can be adopted to achieve interconnection and packaging to make them module components. To be specific, monolithic circuit integration is a more successful and ideal method for glass substrate cells, which has been adopted for industrial production. Continuous scribing of thin layers of different components allows direct separation of the cell, so the back contact layer of a CIGS solar cell can be directly deposited on the front contact layer of the next cell. Such a process can be achieved with mechanical or laser scribing technology.

Wire drawing line refers to another silicon wafer technology based interconnection method. Such a method can be used for CIGS solar cells with a stainless steel substrate. The covering method is formed for a conductive substrate or a substrate where a conductor and an insulator are mixed, and it is formed when the front contact of one cell is in physical contact with the substrate of the next cell. The cells connected into a module allow for differences in efficiency, elevating the efficiency of the entire unit.

Proper packaging can promote the preservation and efficiency stability of CIGS solar cells. Practice has proven that packaging can make performance stable for more than 20 years. The package should meet the following requirements. The first is to exhibit sufficient transparency. The second is to effectively prevent water vapor infiltration. To ensure the 20-year life of CIGS solar cell modules, the daily water vapor transport rate should not exceed 10^{-4}~10^{-5}g/(m^2 day). Third, it exhibits high resistance and stability under ultraviolet light. Fourth, it is best to achieve low cost. Transparent polymers (e.g., ethylene-vinyl acetate (EVA)) can be considered a suitable choice. The glass plate can effectively protect the CIGS solar cell module, whereas it exhibits the defects of weight increase and loss of flexibility.

6.4 Development of Cu(InGa)Se_2 Solar Cells

6.4.1 The Development Process of Cu(InGa)Se_2 Solar Cells

CIGS solar cells have undergone the development for half a century [1–9, 19, 62–65]. In the 1960s, the studies on ternary chalcopyrite semiconductor materials of group I-III-VI were initiated. In 1974, Bell Labs in the United States developed the initial PN heterojunction photodetector displaying a CIS/CdS structure by evaporating n-type CdS onto p-type $CuInSe_2$; besides, a 5% photoelectric conversion efficiency was initially reported for single crystal $CuInSe_2$ solar cell.

In 1976, Maine University developed the first polycrystalline CIS/CdS heterojunction thin-film solar cell. The CIS thin-film material of the cell was prepared by co-evaporation of single-crystal $CuInSe_2$ and Se, exhibiting a thickness of 5–6 μm and a p-type deposition on the glass substrate covered with gold film. Next, a CdS of 6 μm thick was deposited on the film material as a window layer to form heterojunction exhibiting an efficiency of 4–5%, which set a research precedent for CIS thin-film solar cell.

In the 1980s, Boeing and ARCO were committed to tackling down a range of problems in CIS cell production and throughput. In 1981, the Boeing Company produced a CIS thin-film solar cell exhibiting an efficiency of 9.4%, so the development potential of CIS thin-film solar cells have been fully exploited in the photovoltaic field. In this period, CIGS thin-film cells employed ordinary glass or alumina as the substrate, and the back electrode adopted a sputtered Mo layer. The absorption layer $CuInSe_2$ is synthesized complying with a two-step process. First, a Cu-rich film is deposited; subsequently, a high-resistance Cu-poor film is grown. The intrinsic CdS and the low-impedance CdS film doped with In were used as the n-type window layer, and Al was selected as the electrode. This structure underpins the device structure of CIGS thin-film cells. The specific structure is illustrated in Fig. 6.31. For a long time since then, Boeing Company has taken the lead in this area. In 1982, a novel structure with $Cd_{1-x}Zn_xS$ instead of CdS as a buffer layer was achieved, making the solar cell's efficiency reach 11.9%. After that, the efficiency of CIS solar cells has been elevated rapidly, and the process has also been significantly optimized.

In 1988, ARCO adopted a sputtered Cu and In precursors and H_2Se post-selenization to achieve the short-circuit current of CIS thin-film cells as 41 mA/cm^2,

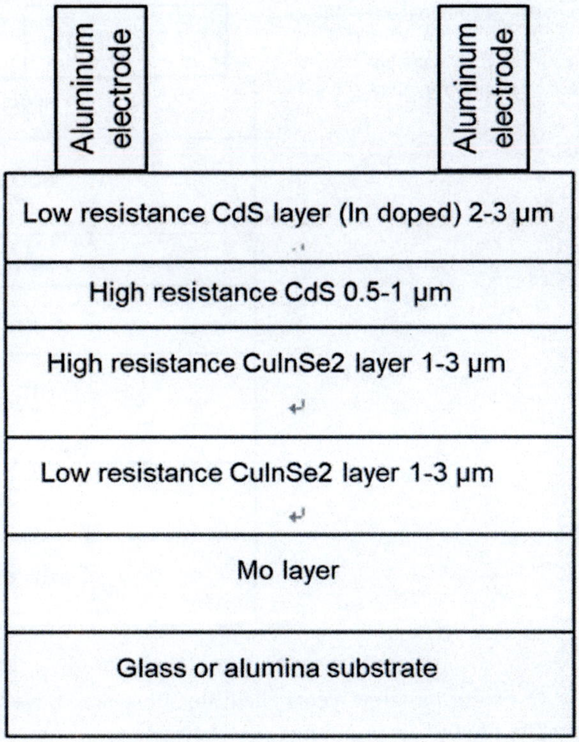

Fig. 6.31 Typical structure of CIS solar cell made by Boeing Company

which has made a major breakthrough in CIS thin-film solar cell research. The cell uses a glass substrate/Mo layer/CIS/CdS/ZnO/electrode structure, with the thickness of CdS less than 50 nm, the light transmittance significantly increased, and the spectral absorption of the absorption layer broadened; the presence of ZnO impacts the inhibition of the optical reflection. Under the effort of ARCO, sputtering precursors post-selenation method and co-evaporation method successfully became the mainstream technologies to fabricate high-efficiency CIGS thin-film solar cells.

In the research, the band gap of $CuGaSe_2$ was reported to be wider than that of $CuInSe_2$. The incorporation of elements Ga and S can broaden the band gap and elevate the matching degree of the solar spectrum. As impacted by the requirements of improving the band gap of the absorbing film material of CIS thin-film cells, CIGS cells have been implemented. In 1989, the Boeing Company prepared Cu $(In_{0.7}Ga_{0.3})Se_2$/CdZnS solar cells by doping Ga. The open-circuit voltage and the conversion efficiency reached 555 mA and 12.9%, as unmatched by CIS cells without Ga. In 1994, a three-step co-evaporation process was developed, and small-area CIGS cells achieved a breakthrough exhibiting an efficiency of 15.9%. Its typical structure is illustrated in Fig. 6.32. In 2008, the efficiency reached 19.9% [2, 19].

Fig. 6.32 Classic structure of CIGS thin-film solar cell achieved by three-step co-evaporation method

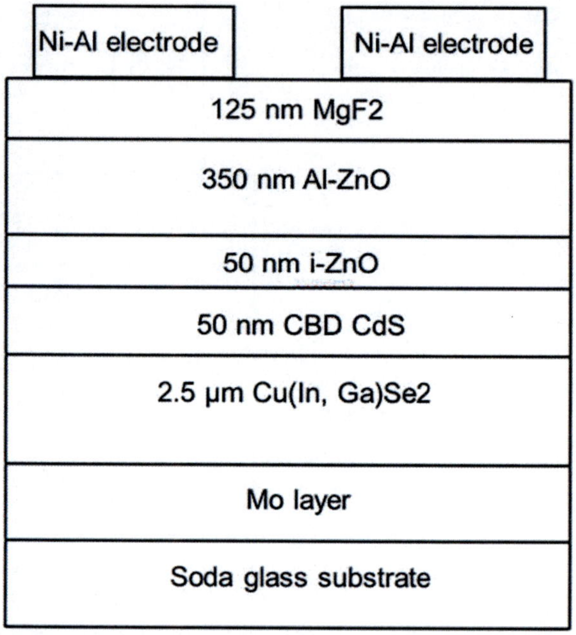

Over the past few years, flexible substrates have become a novel hotspot. The results of studies on a range of substrates are even more encouraging. Swiss scientists have developed a CIGS thin-film solar cell exhibiting an efficiency of 18.7% on a polyimide substrate via an optimized co-evaporation three-step process. In the specific processing procedure, the In/Ga flux gradient configuration replaces the process conditions where the flux is constant throughout the process, and the temperature in the second step drops below 500 °C, thereby elevating device efficiency. A research team from Japan adopted a three-step co-evaporation process to achieve 17.4% efficiency on a Ti substrate and 17.7% efficiency on a substrate made of ZrO material; higher efficiency values were achieved than these efficiency values by another Japan research team in 2009. In 2009, Aoyama University reported a CIGS thin-film solar cell based on a Ti substrate exhibiting an efficiency of 17.9%. The team used CBD-Zn(S, O, OH) instead of the traditional CdS buffer layer. Meantime, it has been proved that proper substrate temperature is critical to form an ideal dual-gradient band gap structure and high device performance. The roughness of the substrate is crucial to the final performance of the CIGS thin-film solar cell with Ti substrate. In 2011, ZSW created an efficiency of 20.3%, a data comparable to that of polysilicon solar cells. The manufacture of high-efficiency CIGS thin-film solar cells provides a rich experience for the proportion of Cu and Ga elements at the absorption layer, so CIGS can achieve industrialization [65].

As the study on CIGS thin-film solar cells has been gradually deepened, and the efficiency is continuously elevated, and these solar cells exhibit robust stability and high radiation resistance in outdoor testing, which lay a solid foundation for their widespread applications in the future. Numerous companies have begun to focus on the industrial development of CIGS thin-film solar cells, and have achieved considerable progress in the development of modules. In 2007, the first CIGS-based solar cell printed on aluminum foil was implemented by Nanosolar Inc. [26].

In industrial production, glass, stainless steel, PI, Ti are primarily adopted as substrates. For glass substrates, a three-step scribing method has been commonly exploited to achieve a monolithically integrated CIGS thin-film solar cell. For CIGS thin-film solar cells with metal foil as the substrate, an insulating layer should be laid in advance, so the technology applicable to the glass substrate can also be applied to the metal foil substrate. However, the mechanical scribing process is hard to control since it is not underpinned by a rigid substrate, and it can easily damage the insulating layer. In the process of finding a method to replace the monolithic integrated CIGS thin-film solar cell, the GSE method is proposed, and a single CIGS thin-film solar cell is connected in series. In the manufacturing process of the absorbent layer, the post-selenization method of metallic precursors more applies to large-scale and large-area film preparation. In practical industrial production, the co-evaporation and the post-selenization are considered common preparing processes. Nanoparticle printing process refers to an effective production with low cost and suitable for large-scale production, whereas the current stage of technology cannot make the film exhibit good performance. At present, the representative companies in the industrial production of CIGS thin-film solar cells are primarily located in the United States, Japan, and Germany.

Global Solar Energy in the United States refers to a company established in 1996 to produce CIGS solar cells and photovoltaic systems. The company's products use stainless steel foil as the substrate, and the entire production uses a rolling bar process. The co-evaporation process acts as the CIGS absorption layer deposition process, while the other layers are deposited by electronic sputtering. In 2009, the company achieved a production of 20 MW. Miasolë is an American company committed to developing flexible CIGS thin-film solar cells, and was set up in 2001. The product feature is that CBD-Zn (S, O, OH) is acted as a buffer layer, the absorbent layer is deposited with post-selenization method of metallic precursors and then vulcanized. In 2010, the company's output exceeded 43 MW. Wrath Solar, a German company founded in 1999, uses glass-based CIGS thin-film solar cells as its major product and exploits co-evaporation to deposit absorbent layers. In 2009, the output reached 30 MW.

6.4.2 Challenges

The existing efficiency of CIGS thin-film solar cells is constantly improved and has great potential for development. However, in the development process, some technical challenges are inevitable [1, 4, 19, 65].

The first refers to the device's requirements for progressively thinner absorbing layers. Though some calculations show that there is no direct relationship between the thinning of the absorption layer and the device efficiency, the thinning of the absorption layer will cause a range of problems for material preparation and growth. The second problem indicates the replacement of buffers and window areas. Novel materials should be discovered and applied to the buffer and window areas of the device, reducing absorption to minimize current loss. The novel material should be able to establish a better band offset with CIGS material, and it can be exploited for transparent front contact connection in the series solar cell. The third problem refers to the content of Ga in Cu(In, Ga)Se_2. In the present study, devices with high-content Ga absorption layers are characterized by high voltage and low current, which is capable of reducing the current loss in module integration. An excessively high content of Ga is conducive to tandem arrangement, whereas it will cause more defects in the absorption layer, which adversely affects efficiency improvement. Finally, the realization of the series thin-film cell. The realization of this configuration requires a single CIGS to achieve the goal of stability and efficiency. Moreover, no comprehensive model and theory has been proposed to explain the mechanism of Na doping.

In terms of production costs, how to reduce costs by selecting an appropriate production mode and optimizing the production is an important issue to be considered in the industrial development of CIGS thin-film solar cells. There are two main aspects of cost savings in the production, one is energy consumption and the other is mechanical damage rate.

Environmentally, because Cu(InGa)Se_2 contains rare elements, its storage capacity may be a limiting condition that affects the large-scale application of CIGS solar cells. In the production, some of the reagents and products used are toxic, which may jeopardize the environment and human health, and should be set up for subsequent processing.

For stability, under humid and hot conditions, the cell fill factor will be reduced by 20–50%, the open-circuit voltage will be relatively reduced, the efficiency will be affected, and the stability will decrease. The reasons for such finding are that the density of defect states of the CIGS thin film are increased in heat and humid conditions, and thus the resistivity of the ZnO layer is reduced, causing more photo-generated carrier collection barrier and promoting the recombination; as a result, the cell performance is reduced. Thus, the packaging process and materials should be improved, and the internal connection structure of the cell should be optimized, thereby enhancing the stability of the cell.

6.4.3 Prospect Forecast

As promoted by numerous research institutions and organizations, the efficiency of Cu(InGa)Se_2 solar cells has been significantly enhanced. Besides, with the increase in material science and engineering experience, the material deposition range and equipment selection have also been optimized. It can be optimistically estimated that the efficiency of CIGS solar cells will increase continuously under the current 20%

efficiency. However, we also have to face up to the current lack of semiconductor knowledge and should invest time and energy to address problems (e.g., equipment and manufacturing scale).

In the preliminary development of CIS and CIGS solar cells, their future economic efficiency was predicted to be much higher than that of Si solar cells, whereas thin-film solar cells have not exhibited any cost advantages thus far. In the future for a long time, considerable work should be conducted to make CIGS thin-film solar cells display their special advantages [22].

On the one hand, novel manufacturing technologies should be developed based on existing advanced engineering models, which starts from improving manufacturing processes and deposition instruments. In the cell manufacturing process, it is required to determine more measurable attributes, develop and improve the measurement and diagnosis methods of indicators, optimize the control of the entire production, enhance the reliability of the manufactured devices, and underpin performance improvement. Indeed, to achieving this goal, a comprehensive and in-depth knowledge of materials and equipment is required.

On the other hand, the core knowledge of materials and devices should be continuously enriched and advanced. The effective enhancement of efficiency can only be achieved by increasing the open-circuit voltage V_{OC}, and the defects in chemical and electrical properties are factors affecting V_{OC}. Accordingly, the knowledge concerning the origin of defects should be more specifically studied, which significantly impacts the establishment of a complete growth model of $Cu(InGa)Se_2$ and more effectively determines the process parameters during defect and PN junction formation. For basic theory, the action mechanism of Na element, the properties of the grain boundary, and the free surface will be worth researching. In the manufacturing process, the buffer layer CdS deposited by the chemical bath may be substituted by some other materials, which will be Cd-free and exhibit a wide band gap. The search and application of the mentioned materials will also become a hotspot in improving cell efficiency.

The subsequent development of CIGS thin-film solar cells should make breakthroughs in the following aspects [2]:

(1) The alloy exhibiting a band gap $E_g > 1.5$ eV in cell will not adversely affect efficiency, and it facilitates cell manufacturing. Polycrystalline thin-film tandem cells exhibiting a band gap of nearly 1.7 eV are suggested to achieve significant development, and the conversion efficiency is expected to reach 25% or more. Such a tandem structure requires a transparent top and bottom cell connection, absorbing blue and red photons, respectively; thus, the bottom structure can be modified without affecting the top structure.

(2) A low-temperature process is achieved during the $Cu(InGa)Se_2$ deposition. The low-temperature process requires the substrate temperature to be lower than the conventional requirements, which can broaden the selectable range of the substrate; for instance, a flexible polymer network can be adopted. At low substrate temperature, the thermal stress on the substrate can be reduced, and the heating and annealing rates can be up-regulated; as a result, losses can be

reduced, and a thinner Cu(InGa)Se_2 film exhibiting a thickness less than 1 μm is obtained.

Furthermore, in Cu(InGa)Se_2 thin-film solar cells development, the substrate materials tend to be diversified and flexible, and the manufacturing steps will be streamlined. As fueled by the enhancement of efficiency, the applications of the cells will also be diversified and extensive [63, 66].

In brief, Cu(InGa)Se_2 thin-film solar cells are considered to effectively improve efficiency and lower cost. In the future, more efforts will be required to optimize their performance and industrialization.

References

1. T. Nakada, Electron. Mater. Lett. **8**, 179–185 (2012)
2. A. Luque, S. Hegedus, *Handbook of Photovoltaic Science and Engineering* (Wiley, Com, 2011)
3. Y. Zhong, *Study on the application of surface copper indium gallium selenium thin films in solar cells,* the Master Degree Dissertation of National Cheng Kung University, 2009
4. S. Kodigala, *Cu($In_{1-x}Ga_x$)Se_2 Based Thin Film Solar Cells* (Academic Press, 2010)
5. Y. Xue, B. Yang, Evolution of (In, Ga) 2Se3-Curich-In (Ga) rich of CIGS thin film. J. Optoelectron. Laser **19**, 348–351 (2008) (in Chinese)
6. K. Chopra, Prog. Photovolt. Res. Appl. **12**, 69–92 (2004)
7. S.H. Wei, Appl. Phys. Lett. **72**, 3199 (1998)
8. W. Li, J. Synth. Crys. **35**, 131–134 (2006)
9. B. Stanberya, Crit. Rev. Solid State Mater. Sci. **27**, 73–117 (2002
10. P. Reinhard et al., IEEE J. Phothovoltaics **3**, 572–580 (2013)
11. A. Hultqvist et al., Sol. Energy Mater. Sol. Cells **95**, 497–503 (2011)
12. H. Neumann et al., Solar Cells **16**, 317–333 (1986)
13. L. Gütay et al., Thin Solid Films **517**, 2222–2225 (2009)
14. L. Gutay et al., in *34th IEEE Photovoltaic Specialists Conference (PVSC)* (2009), pp. 874–877
15. A. Shimizu, Jan. J. Appl. Phys. **39**, 109 (2000)
16. U. Rau, Appl. Phys. A **69**, 131–147 (1999)
17. I.L. Repins et al., in *34th IEEE Photovoltaic Specialists Conference (PVSC)* (2009), pp. 978–983
18. W.C. Lim et al., Surf. Interface Anal. **44**, 724–728 (2012)
19. S. Xiong, M. Zhu et al., *Solar Cell Foundation and Application,* vol. 10 (Science Press, Beijing, 2009) (in Chinese)
20. T. Schlenker et al., Thin Solid Films **480**, 29–32 (2005)
21. J. Nelson, *The physics of solar cells* , trans. Y. Gao (Shanghai Jiaotong University Press, Shanghai, 2011). (in Chinese)
22. S. Fonash, *Solar Cell Device Physics* (Elsevier, 1981)
23. M. Gloeckler, *Device Physics of Cu (In, Ga) Se2 Thin-Film Solar Cells* (Colorado State University, 2005)
24. A. Rockett et al., Thin Solid Films **237**, 1–11 (1994)
25. M. Igalson, Optpelectron. Rev. **4**, 261–268 (2003)
26. F. Liu, Study on the diode characteristics of Cu(In, Ga)Se2 thin film solar cells. J Synth. Cryst. **38**(2), 455–459 (2009)
27. A. Niemegeers et al., Prog. Photovolt. Res. Appl. **6**, 407–421 (1998)
28. K. Hamakawa, *Solar Photovoltaic Cell and Its Application,* trans. H. Zhang and X. Cui (Science Press, 2008) (in Chinese)
29. R.N. Bhattacharya et al., Sol. Energy **77**, 679–683 (2004)

30. T. Nakada et al., in *The 24th Photovoltaic Energy Conversion*, vol. 1 (1994), pp. 95–98
31. A. Yamada et al., Thin Solid Films **480**, 503–508 (2005)
32. T. Nakada et al., Appl. Phys. Lett. **74**, 2444–2446 (1999)
33. M. Topic et al., in *14th EU-PVSEC,* Barcelona (1997), pp. 2139–2142
34. W. Yang, *Preparation and Properties of CIGS Thin Film Solar Cells.* The Master Degree Dissertation of Shanghai Jiaotong University, 2007 (in Chinese)
35. A.O. Pudov et al., J. Appl. Phys. **97**, 064901-064901-6 (2005)
36. M. Contreras et al., Appl. Phys. Lett. **63**, 1824–1826 (1993)
37. V. Gremenok et al., Phys. Stat. Sol. (c) **6**, 1237–1240 (2009)
38. R. Wuerz et al., Thin Solid Films **517**, 2415–2418 (2009)
39. F. Kessler et al., Sol. Energy **77**, 685–695 (2004)
40. T. Sugiyama et al., Jan. J. Appl. Phys. **39**, 4816 (2000)
41. A. Yamada et al., in *The 28th Photovoltaic Specialists Conference* (2000), pp. 462–465
42. B. Sang et al., Sol. Energy Mater. Sol. Cells **67**, 237–245 (2001)
43. N.E. Gorji et al., Sol. Energy **86**, 920–925 (2012)
44. S. Yu, J. Zhong et al., *Solar Energy Photovoltaic Device Technology* (University of Electronic Science and Technology Press, 2011) (in Chinese)
45. B. Dai, H. Zheng, *Solar Cells Technical Manuals, Chapter 7, p. 153* (Posts & Telecom Press, Beijing, 2012). (in Chinese)
46. L. Pengwei, *Preparation and Characterization of CIGS Thin Film Solar Cell Materials* (Henan University, 2012)
47. Hernández J. Sastré, Rev. Mex. Fis. **57**, 441–445 (2011)
48. P.K. Johnson et al., Prog. Photovolt. Res. Appl. **13**, 579–586 (2005)
49. S. Wu, L. Xu, Research on preparing technology of CIGS thin film by electrodeposition method. Anhui Chem. Ind. **33**(6), 32–33 (2007)
50. C. Liao, J. Han et al., Effect of Se vapor concentration on CIGS film preparation. Acta Phys. Chim. Sin. **27**(2), 432–436 (2011)
51. M. Venkatachalam, J. Instrum. Soc. India **38** (2008)
52. E. Eser et al., in *The 23th Photovoltaic Specialists Conference* (2005), pp. 515–518
53. K. Mukati et al., in *IEEE 4th World Conference on Photovoltaic Energy Conversion*, vol. 2 (2006), pp. 1842–1845
54. C. Li, *Research on CIGS Thin-Film Solar Cell Made by Magnetron Sputtering.* The Master Degree Dissertation of Henan Normal university, 2011 (in Chinese)
55. E. Eser et al., in *The 35th Photovoltaic Specialists Conference (PVSC)* (2010), pp. 661–666
56. A. Grimm et al., Thin Solid Films **515**, 6073–6075 (2007)
57. R.N. Bhattacharya et al., Appl. Phys. Lett. **89**, 253503 (2006)
58. S. Chaisitsak et al., Jan. J. Appl. Phys. **38**, 4989 (1999)
59. R.N. Bhattacharya et al., J. Phys. Chem. Sol. **66**, 1862–1864 (2005)
60. F. Kessler et al., Thin Solid Films **480**, 491–498 (2005)
61. U. Rau, Appl. Phys. A **96**, 221–234 (2009)
62. M. Pagliaro et al., Chemsuschem **1**, 880–891 (2008)
63. W.K. Batchelor et al., in *The 29th Photovoltaic Specialists Conference* (2002), pp. 716–719
64. X. Zhang et al., Int. Soc. Opt. Photon. 83120H-83120H-10 (2011)
65. C. Li, Research progress of Cu(In, Ga)Se2 thin film solar cells. Chin. J. Power Sources **33**(2), 77–80 (2009). (in Chinese)
66. V. K. Kapur et al., Lab to large scale transition for non-vacuum thin film CIGS solar cells: Phase I. Ann. Tech. Rep. **1** (2003)
67. U.P. Singh et al., Int. J. Photoenergy **2010**, 468174 (2010)

Chapter 7
CdTe Solar Cells

7.1 Introduction

CdTe was fabricated in 1947 by Frerichs [1] based on the direct reaction Cd vapor and Te vapor in a hydrogen environment. The CdS/CdTe thin-film solar cell was initially trial-produced by Bonnet and Rabenhorst [2]. The n-CdS/p-CdTe thin-film heterojunction cells fabricated with the gas-carrying and vacuum evaporation methods exhibits photoelectric conversion efficiency of 5.4%. In 1956, Loferski of RCA Lab initially proposed the application of CdTe material in solar photovoltaic [3]. In 1959, Rappaport, also in the RCA Lab, developed CdTe single crystalline homojunction solar cells through the diffusion of In element onto p-type single crystalline CdTe [4]. The cells exhibited a photoelectric conversion efficiency of 2% (the structure was not established under AM1.5, standard photovoltaic metering was not employed, and a 73 mW/cm^2 light source was exploited), which are considered the first CdTe solar cells born. In 1979, the CNRS research team in French produced CdTe solar cells exhibiting photoelectric conversion efficiency of 7% [5]. They employed the close-space vapor transport to deposit As-doped p-type CdTe thin films onto n-type single crystalline CdTe. Besides, they initially applied CdTe thin films in solar photovoltaic. Not before long, they considerably enhanced the photoelectric conversion efficiency of solar cells to a 10.5%, and then achieved an open-circuit voltage of 820 mV as well as a short-circuit current of 21 mA/cm^2 [6]. Since then, however, a decreased number of research on CdTe homojunction solar cells has been conducted subsequently. Moreover, studies on CdTe heterojunction solar cells were conducted in the 1960s. In the early 1960s, researchers delved into heterojunction solar cells consisting of n-type CdTe and p-type Cu_2Te [7]. To be specific, n-type single crystalline or polycrystalline thin-film of CdTe was employed to fabricate a p-type Cu_2Te layer on the CdTe surface by performing a surface reaction in an acid solution supplemented by Cu salt; thus, a $CdTe/Cu_2Te$ heterojunction solar cell was prepared. After a decade, $CdTe/Cu_2Te$ heterojunction solar cells have exhibited a photoelectric conversion efficiency over 7% (the intensity of the test light source reached 60 mW/cm^2) [8]. Subsequently, more stable and effective materials of window layer (In_2O_3:Sn,

C. Zhang et al., *Semiconductor Photovoltaic Cells*,
https://doi.org/10.1007/978-981-15-9480-9_7

of which the abbreviated forms include ITO, ZnO, SnO_2, CdS, etc.) and single crystalline CdTe heterojunction solar cells have been extensively studied. Besides, the short-wavelength response of solar cells is restrained by the transmittance of the mentioned materials of window layer. In 1977, a research group in Stanford employed ITO and p-type single crystalline CdTe fabricated via electron beam evaporation and then successfully developed heterojunction solar cells exhibiting the photoelectric conversion efficiency of 10.5% [9]. In 1987, a solar cell fabricated by reactive deposition of In_2O_3 onto p-type single crystalline CdTe exhibited a conversion efficiency of up to 13.4% [10]. In the mid-1960s, Mulle et al. [11, 12] initially introduced n-type CdS thin films into CdTe solar cells. They adopted the evaporation to form n-type CdS thin films onto single crystalline CdTe, as an attempt to develop CdS/CdTe heterojunction solar cells; the fabricated cells exhibited a photoelectric conversion efficiency of nearly 5%. The introduction of CdS thin film is considered a milestone in CdTe solar cells development.

Adrivich et al. [13] developed the initial CdS/CdTe thin-film heterojunction solar cell in 1969. They deposited a TCO film, a CdS film, and a CdTe film (termed as a "superstrate" structure) on a glass substrate in order. Then, on the ninth European Photovoltaic Conference in 1972, Bonnet and Rabenhorst reported another structure of CdS/CdTe thin-film solar cells (termed as "substrate" structure) [2]. By chemically vapor-depositing (CVD) CdTe thin films on Mo substrates and then vacuum-evaporating the CdS thin films, they fabricated CdS/CdTe thin-film solar cells; the obtained solar cells exhibited photoelectric conversion efficiency of 5–6%. Furthermore, they proposed the factors critically affecting the high-energy conversion efficiency of CdS/CdTe thin-film solar cells:

(1) p-type doping of Cu in CdTe;
(2) Effective control of the doping efficiency of CdTe;
(3) Element distribution at CdTe-CdS interface;
(4) Activation and passivation of grain boundaries;
(5) Acquisition of the back contact barrier of p-type CdTe.

In the 1980s and 1990s, studies on CdS/CdTe thin-film solar cells placed an emphasis on device design, post-processing technology, as well as low-resistance contact. For the stability and adaptability of CdTe materials, a wide variety of methods of preparing thin film have been adopted to develop CdTe thin-film solar cells. It is noteworthy that the CdTe thin-film solar cells fabricated with the mentioned methods all exhibited high photoelectric conversion efficiency. Such result was achieved since the "Shockley-Read-Hall" combination of the space charge region is the factor primarily limiting the photoelectric conversion efficiency of CdTe thin-film solar cells [14], which is not directly associated with the specific preparation method of the material. In the following years, several research groups have attempted to enhance the performance of CdTe thin-film solar cells, and they have achieved considerable progress. In the 1980s, Tyan and Aluerne enhanced the photoelectric conversion efficiency of the solar cell to 10% [15]. Subsequently, Ferekides et al. successfully produced CdTe solar cells exhibiting photoelectric conversion efficiency of 15.8%

[16]. In 2001, Professor Wu Xuanzhi, a Chinese scientist at the National Renewable Energy Laboratory (NREL) in the United States, successfully fabricated a CdTe thin-film solar cell exhibiting the photoelectric conversion efficiency of 16.5% by optimizing the front-electrode and the back-electrode process of the CdTe thin-film solar cells [17].

Though CdTe thin-film solar cells fabricated with different methods have exhibited high-energy conversion efficiency, the CdS/CdTe solar cells generally consist of the two structures as mentioned above. For example, in the "superstrate" structure, CdTe is firstly deposited on CdS, and then the CdTe and CdS layer will undergo post-heat treatment with Cl and O in later process, respectively. In the 1980s, the solar cell performance was primarily enhanced by regulating and optimizing device process parameters (e.g., the deposition temperature of CdTe, post-annealing, growth, as well as the composition of chemical atmosphere) to treat and prepare CdTe ohmic contacts. For instance, Panasonic Company has reported the screen printing technology for the fabrication of CdTe solar cells. The components of $CdCl_2$, O, and Cu were rigorously controlled by regulating the slurry, temperature, and time in the sintering process [18]. In the preparation of CdTe through the electrodeposition, adding Cl to the plating bath and the post-processing process led to 10% enhanced performance of the device [19]. Researchers at Kodak also achieved 10% enhanced device performance by optimizing the deposition temperature of CdTe with the close-space sublimation method and up-regulating the oxygen content in the surrounding atmosphere [20]. A turning point in enhancing the performance of CdTe solar cells refers to covering the CdS/CdTe structure with a layer of $CdCl_2$ and performing thermal annealing in the air [21, 22]. In 1993, combining post-annealing treatment with $CdCl_2$ and low-resistance ohmic contact led to more than 15% photoelectric conversion efficiency of CdTe solar cells produced with the close-space sublimation [23]. By optimizing the window layer process [24] and the vapor treatment with $CdCl_2$ [25], the cell performance was enhanced again. Thus far, the maximum energy conversion efficiency of CdTe solar cells reaches 17.3% [9], $V_{OC} = 845$ mV, $J_{SC} = 25.9$ mA/cm^2, FF = 75.5%. Moreover, the efficiency of its commercial modules has reached nearly 10%.

Studies on solar cells primarily aim to enhance conversion efficiency, lower costs, and facilitate stability. Polycrystalline CdTe thin-film solar cells are considered the easiest to manufacture, and the most cost-efficient and long-term operation life among thin-film solar cells that are represented by CdTe, $CuInSe_2$, and polycrystalline silicon thin films. Accordingly, it has been most rapidly commercialized. To enhance the photoelectric conversion efficiency, the cell structure and the material process of the respective layer should be optimized. Appropriately decreasing the thickness of the CdS layer of the window layer can limit the loss of incident light; as a result, the short-wave response of the cell increases, and the short-circuit current density is up-regulated. Moreover, it critically impacts the device characteristics. To enhance the power conversion efficiency, a buffer layer is usually introduced between the CdTe and the metal electrode in order to enhance ohmic contact. To lower the cost, the deposition temperature of CdTe should be down-regulated to below 550 °C to apply to cheap glass as a substrate. Existing research and industrialization of CdTe thin-film solar cells are being progressively deepened. To commercialize the cells, the preparation optimization of CdTe thin film is prioritized, so comprehensive comparison should be drawn to find a process solution suitable for industrialization. Second, component stability should be ensured. In the next two decades, CdS/CdTe solar cells are estimated to dominate the market of solar cells.

7.2 Material Properties

CdTe refers to a p-type compound semiconductor of group II-VI; its crystal exhibits a sphalerite structure. Figure 7.1 presents the lattice structure of (111) and (100) directions of CdTe crystals exhibiting a sphalerite structure. The arrangement of atoms in different directions is observed in the figure, and the face-centered cubic arrangement of atoms is clearly identified. The space group of CdTe is F-43 m. The lattice constant is 6.48 Å [26]. As a semiconductor material for thin-film solar cells showing promising applications, CdTe has aroused extensive attention over the past few years for the three major characteristics of CdTe materials, which are critical to thin-film solar cells:

Firstly, thin-film solar cells exhibit the bandgap of 1.45 eV, and the absorption edge is nearly 855 nm. The response to the solar spectrum appears in the most ideal solar spectral band. Single-junction thin-film solar cells adopting CdTe as the absorbing layer can exhibit relatively high energy conversion efficiency; its theoretical conversion efficiency reaches 30%.

Secondly, CdTe refers to a direct-gap semiconductor, with the absorption coefficient in the range of visible light reaching 10^5 cm^{-1}. Besides, 99% of the photons

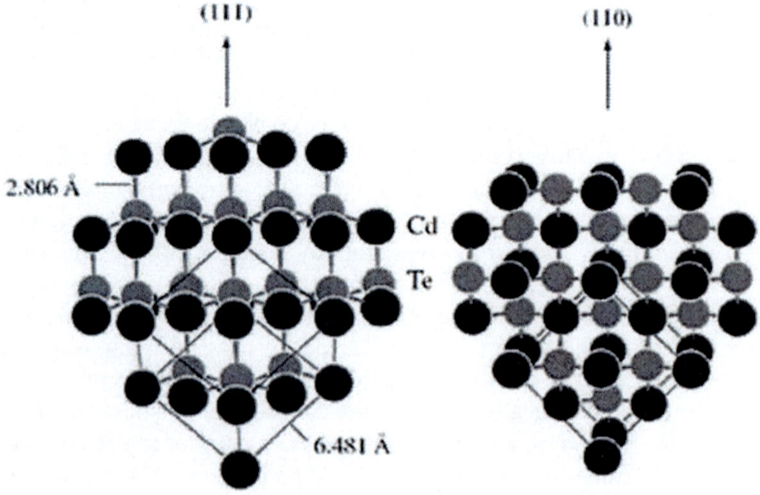

Fig. 7.1 Lattice structure of CdTe exhibiting a sphalerite structure

in the sunlight with higher energy than the bandgap of CdTe can be absorbed in a 2 micron-thick absorption layer. Solar cells employ CdTe as an absorption layer. In theory, the required thickness of its absorption layer is approximately several microns, the material consumption is relatively small, and the device cost is low.

Moreover, CdTe exhibits a crucial chemical property, i.e., sublimation. The sublimation reaction of CdTe occurs at high temperatures (commonly over 460 °C). The specific reaction is expressed as follows:

$$CdTe \rightarrow Cd + 1/2\, Te_2$$

At low temperatures, the sublimation can be reversed, i.e., the sublimated steam is re-condensed into CdTe. As a result, CdTe thin-film materials exhibit unique advantages. Since the sublimation temperature is significantly lower than the melting point of the CdTe material (1092 °C), CdTe should not be heated to the melting point to be further sublimated and deposited on the substrate surface for the formation of a high-quality film. Meantime, in comparison with other semiconductor materials (e.g., $CuIn_xGa_{1-x}Se_2$ (CIGS) thin film), the method of preparing polycrystalline CdTe thin film is highly tolerant. Moreover, multiple preparation approaches can be adopted to form high-quality polycrystalline CdTe thin films. The existing methods to fabricate CdTe in the laboratory include close-space sublimation, electrochemical deposition, vapor transport deposition, screen printing, and magnetron sputtering. CdTe cells fabricated with the mentioned methods have exhibited high-energy conversion efficiency [27, 28].

However, as a solar cell material, CdTe also has defects in physical properties:

Firstly, CdTe has a strong self-compensating effect. The so-termed as self-compensation effect is the spontaneous compensation effect of considerable intrinsic defects on the donor impurity or the acceptor impurity. For CdTe, its intrinsic defect is mainly Cd vacancies that play a critical role in compensating for defects attributed to other doping. Thus, CdTe is difficultly similar to semiconductors (e.g., silicon) to enhance the electrical performance by doping impurity elements. CdTe thin film exhibits a low carrier concentration (commonly 10^{14} cm^{-3}), and mobilities of electrons and holes inside the film are 1100 and 80 cm^2 V^{-1} s^{-1} [29];

Secondly, the work function of CdTe reaches 5.7 eV, and a material of back electrode with extremely high work function is required to form a good ohmic contact with it. However, the mentioned materials are rare and ordinary metal electrode materials cannot establish good ohmic contact with them.

At the present stage, many key problems remain, which are concerned with material and manufacturing process in the preparation and research of CdTe thin-film solar cells. Accordingly, though CdTe thin-film solar cells have aroused widespread attention and achieved large-scale production, the photoelectric conversion efficiency of their laboratory cells and commercial cell templates remains low. The maximum efficiency of 17.3% of the cells is far from 30% in theory since CdTe still has considerable key problems, which requires further investigations.

7.3 Structure and Process Implementation of CdTe Solar Cells

As mentioned above, CdTe thin-film solar cells consist of two main structures, but to obtain higher energy conversion efficiency, over the past few years, more "superstrate" structures have been adopted, as shown in Fig. 7.2. Besides the simple manufacturing process of the cell, another crucial advantage of this structure is that it facilitates the preparation of the back electrode of the cell. The surface of the CdTe film can be treated by etching, doping, epitaxy, and others to establish effective ohmic contact. Then, the basic PN junction is formed between the p-type CdTe absorption layer and the n-type CdS window layer. Moreover, there are other factors affecting the performance of the device. For instance, since the thickness of CdS is significantly thin, a high-resistance oxide layer is required. Besides, interdiffusions are identified between CdS and CdTe, and potential barriers may exist in the preparation of back contact electrodes. In the following, each of the mentioned factors is analyzed.

7.3.1 *Substrate*

When sunlight enters a solar cell, it firstly penetrates the substrate, the substrate is the supporting structure of the cell as well. This requires the substrate to exhibit two characteristics (i.e., high transmittance of light and a certain intensity). On the whole, soda-lime glass acts as the glass substrate of the cell. It is noteworthy here that to reduce the reflection of light on the solar cell surface, a layer of antireflective film can be deposited onto the glass substrate surface. The optical thickness of this film is 1/4 times that of the peak wavelength of solar radiation intensity. This can make

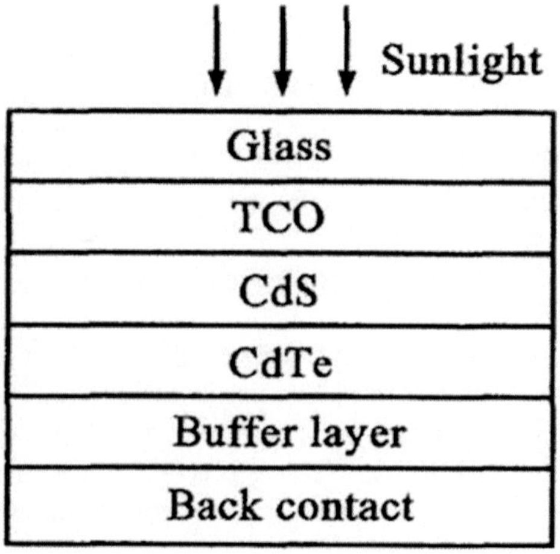

Fig. 7.2 Structure of superstrate in CdTe solar cell

the incident light and the reflected light interfere with each other on the glass surface for the phase difference, the reflection of light on the surface can be reduced, and the utilization efficiency of sunlight is effectively enhanced.

7.3.2 *Front Electrode*

In the CdTe thin-film solar cell structure, the front electrode is a window via which light passes, and it also acts as a collection layer for photo-generated electrons. This makes the material of the front electrode should meet the conditions below:

Firstly, a high transmittance in the visible and near-infrared bands is required, which is appropriate to approach or exceed 90%;

Secondly, the material should have high n-type conductivity;

Thirdly, as a vital component of solar cells, the material should exhibit high stability, which can ensure that the material properties remain stable during the subsequent preparation and working life of the solar cell.

The existing materials of the front electrode that people choose are primarily transparent conductive oxides (referred to as TCO). The mentioned oxides cover SnO_2:F (FTO), ZnO:Al (AZO), In_2O_3:Sn (ITO), Cd_2SnO_4 (CTO), etc. When the preparation is optimized, the transmittance of the mentioned oxides is greater than 90%, the concentration of carriers reaches 10^{18}–10^{21} cm^{-3}, and the stability is excellent, which fully satisfies the requirements of CdTe thin-film solar cells. The mentioned oxides are the most common materials of front electrode. Over the past few years, to further enhance the photoelectric conversion efficiency of CdTe thin-film solar cells and effectively avoid the short-circuit phenomenon attributed to thinning the CdS layer, a novel system of front electrode has been developed, i.e., a front electrode made of Cd_2SnO_4/Zn_2SnO_4 composite layer. In such system, Cd_2SnO_4 is a low-resistance front-electrode layer, and Zn_2SnO_4 refers to a high-resistance buffer layer. Compared with the conventional system of single-layer oxide, the Cd_2SnO_4/Zn_2SnO_4 composite layer exhibits lower resistivity and better transmittance of light, and it is easy to prepare in a large area.

7.3.3 *Window Layer*

For CdTe solar cells exhibiting high photoelectric conversion efficiency, the acquisition of high-quality window film CdS is another critical link. CdS is also a direct-gap compound semiconductor of group II-VI exhibiting a bandgap of 2.2 eV and an absorption edge wavelength of 512 nm. It allows most photons in sunlight to pass through without being absorbed. Thus, CdS has been extensively employed as a window layer in CIGS and CdTe thin-film solar cells. Since the defect of CdS is primarily sulfur vacancy, it is a donor defect. Accordingly, CdS exhibits n-type characteristics, and the concentration of carriers in it reaches 10^{16}–10^{17} cm^{-3}. Thus, in a CdTe film, CdS acts as an n-type film in the PN junction. CdS exhibits two different crystal structures, i.e., a hexagonal wurtzite structure and a cubic sphalerite structure. Between them, the lattice mismatch between the hexagonal wurtzite structure and CdTe is smaller, which is most appropriate to CdTe thin-film solar cells. The existing methods to fabricate CdS films are: magnetron sputtering, close-space sublimation, vacuum thermal evaporation, and chemical water bath deposition, etc. [30–33]. Preparation of high-quality CdS thin films is a key to obtain CdTe thin-film solar cells exhibiting high energy conversion efficiency. From the theoretical study on the bandgap of CdTe, the short-circuit current of CdTe thin-film solar cells can theoretically reach 30.5 mA/cm^2 [34]. About 7 mA of this current is contributed by photons (wavelengths less than 510 nm) with higher energy than the bandgap of CdS. If this part of the photon is absorbed by the CdS layer, there is no contribution to the current of the CdTe thin-film solar cell. This part of the photons should pass through the CdS layer and reach the CdTe absorption layer smoothly to further enhance the device performance. Accordingly, it is important to reduce the thickness of the CdS layer, reduce the effective photon loss, and improve the short-wavelength response of the cell. At present, the thickness of CdS is generally regulated at nearly 100 nm; however, if the thickness of the polycrystalline CdS film is excessively reduced, micro pinholes are easily generated in the film. The existence of the mentioned pinholes noticeably increases the leakage current of the cell, and the open-circuit voltage and fill factor of the cell will be affected [27]. Meantime, in the heat treatment of device preparation, the interdiffusion between CdS and CdTe will also lead to the occurrence of CdS pinholes. The appearance of pinholes causes CdTe to come into contact with TCO, forming a short-circuit and current shunt, which adversely affects the preparation of large-area cell. To avoid the mentioned negative effects attributed to thinning CdS, an ultra-thin, high-resistance oxide film is added between TCO and CdS. The thickness is commonly regulated at about tens of nanometers. This thin film acts as a buffer layer in the cell. For the tunneling effect, electrons can pass through the ultra-thin layer almost unhindered, and meantime, it can effectively avoid the short circuit attributed to direct contact between CdTe and TCO, and effectively enhance the device performance [34].

There are numerous methods to fabricate polycrystalline CdS thin films (e.g., electrodeposition (ED), close-space sublimation (CSS), screen printing (SP), molecular beam epitaxy (MBE), physical vapor deposition (PVD), vacuum evaporation, chemical spraying, sputtering, high-temperature spraying, and chemical bath deposition (CBD)). To be specific, the CBD method is simple in process, low in cost, and easy to achieve large-scale production; it has been broadly valued by people. Moreover, it exhibits the following advantages:

(1) Vacuum system is not required, the equipment requirement is simple;
(2) The purity of reaction materials is relatively low (analytical grade), and the selectivity of reaction materials is large. It is easy to buy, and the price is low;
(3) The reaction temperature is low, and the selectivity of the substrate is high.

Thus, the chemical bath deposition is considered a low-cost and low-temperature deposition, and the film deposited with such method is uniform and dense and exhibits good film properties. Preparation of polycrystalline CdS thin film as a window layer with a chemical bath deposition method can extend the service life of minority carriers in the cell and optimize the bandgap of the cell.

There are two major competing reactions in the growth of CdS with the CBD, i.e., the arrangement of homogeneous particles in solution and heterogeneous surface reactions on the substrate. Heterogeneous surface reactions consist of two processes:

One is the process of adsorbing CdS particles on the substrate, termed as the cluster–cluster mechanism, which causes a rough and loose morphology of the film. The specific process of the reaction is that free cadmium ions react with free sulfur ions to form CdS particles, which are deposited on the substrate surface under the relatively weak inter-particle force; as a result, the surface of the film becomes rough, loose, and poor in denseness;

The other is termed as an ion-ion mechanism, which first adsorbs the complex of Cd^{2+} on the substrate and then adsorbs a sulfur source to form a mesophase; lastly, the mesophase is decomposed to produce CdS. The specific process is that ammonia is provided by an alkaline solution composed of ammonium hydroxide, and a cadmium salt (e.g., $CdCl_2$) generates a free cadmium particle (Cd^{2+}) via a dissociation reaction. The cadmium particle and the ammonia molecule compound to form a dominant complex particle: $Cd(NH_3)_4{}^{2+}$. On the surface of substrate, $Cd(NH_3)_4{}^{2+}$ complex particles react with OH^- to produce $Cd(OH)_2(NH_3)_2$ complex particles adsorbed on the surface of substrate. Subsequently, such a complex particle reacts with thiourea following the Rideal-Eley mechanism to form the metastable complex $Cd(OH)_2(NH_3)_2SC(NH_2)_2$ adsorbed on the surface. Moreover, as the metastable complex adsorbed on the surface breaks down, a novel surface is formed, and lastly CdS is fabricated [35]. The CdS film formed by this mechanism is dense and flat.

During the reaction of depositing CdS film, the mentioned two reactions coexist. In the initial stage of the reaction, for the high concentration of ions in the solution, the ionic mechanism becomes the major process. With the progress of the reaction, the CdS particles increase for the arrangement of similar particles, and the cluster mechanism becomes the major process. Next, CdS particles are adsorbed on the surface of the substrate, causing the CdS film to be loosened. Accordingly, the deposition processes dominated by various mechanisms compete with each other.

The chemical reaction is expressed as follows:

$$NH_3 + H_2O \rightarrow NH_4^+ + OH^-$$

$$Cd^{2+} + 4NH_3 \rightarrow Cd(NH_3)_4^{2+}$$

$$(NH_2)_2CS + OH^- \rightarrow CH_2N_2 + H_2O + HS^-$$

$$HS^- + OH^- \rightarrow S^{2-} + H_2O$$

$$Cd(NH_3)_4^{2+} + S^{2-} \rightarrow CdS \downarrow + 4NH_3 \uparrow$$

The overall chemical reaction is

$$Cd(NH_3)_4^{2+} + SC(NH_2)_2 + 2OH^- \rightarrow CdS + CH_2N_2 + 2H_2O + 4NH_3$$

It can also be seen from the above equation that the growth process of CdS is primarily the following 4 processes:

(1) Free cadmium ions in the solution are combined with ammonia to generate $Cd(NH_3)_4^{2+}$ complex ions, which gather near the substrate;
(2) A reversible adsorption of $Cd(OH)_2$ takes place on the substrate surface;
(3) Thiourea and $Cd(OH)_2$ adsorb with each other again, thereby forming a metastable $Cd(OH)_2SC(NH_2)_2$ complex;
(4) $Cd(OH)_2SC(NH_2)_2$ is further decomposed to produce CdS, while a thin film is being formed, an adsorption vacancy is regenerated on the outer surface.

Since the solubility product of CdS ($K_{sp} = 1.4 \times 10^{-29}$) is obviously small in size, CdS precipitates easily. Thus, only by regulating the decomposition rate of Cd^{2+} and S^{2-}, can a uniform and dense CdS film be formed.

As the window layer of solar cells, CdS films exhibit their unique and important functions. On the one hand, a high transmittance is required to ensure that CdTe can absorb maximal photons. On the other hand, a large photoconductivity is required to reduce the internal resistance of the cell. The Fermi level also should be in appropriate place. Newly deposited CdS thin films are often not ideal in the mentioned parameters, and subsequent annealing significantly impacts the structure, grain size, and optical energy gap of the thin film, which adversely affects the performance of the entire cell [36–38].

7.3.4 Absorptive Layer

The absorption layer of a CdS/CdTe thin-film solar cell refers to a CdTe film; its thickness is commonly regulated at 4–7 μm. For the preparation of the absorption layer, there are commonly the following ways:

1. **Physical vapor deposition method (PVD)**

The basis of preparing CdTe thin films by vapor deposition is the balance between Cd, Te_2 vapor, and the formation of CdTe solid phase, $Cd + 1/2Te_2 <=> CdTe$. Accordingly, CdTe can be fabricated by simultaneously evaporating Cd and Te sources or by sublimation using CdTe sources, or can be obtained by carrying and transporting Cd and Te_2 or depositing CdTe vapor on a substrate. The CdTe thin film deposited by the CdTe sublimation method makes the composition of the compound constant, while requiring requires a vapor pressure lower than the vapor pressures of the elements Cd and Te, so a single-phase thin film can be formed over a wide range of substrate temperatures. Likewise, the film of multicomponent compound (e.g., $CdZn_{1-x}Te_x$ and $CdTe_{1-x}S_x$) can be formed by simultaneously evaporating multiple sets of II-VI binary compound sources.

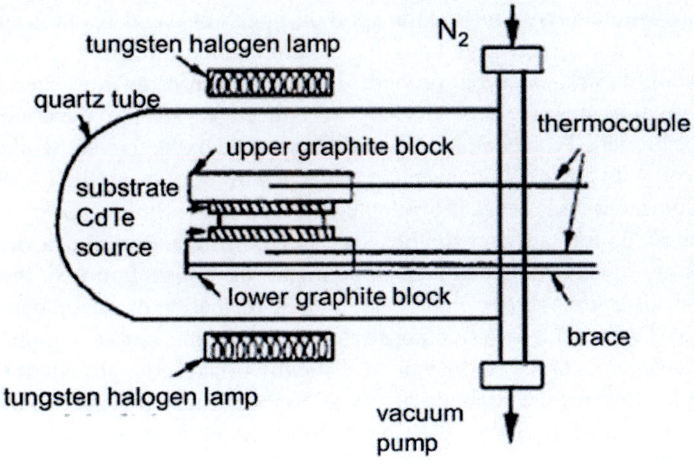

Fig. 7.3 Diagram of CSS structure

2. **Close-space sublimation**

The close-space sublimation method refers to a method of preparing CdTe thin film developed using the sublimation characteristics of CdTe materials. In 1982, Tyan et al. [15] first employed the close-space sublimation method to prepare CdTe thin films and CdTe thin-film solar cells with an energy conversion efficiency of 10%. The deposition device is shown in Fig. 7.3. CSS uses the flakes or powder of high-purity CdTe as the source, and the distance between the two graphite blocks is about 1–30 mm, which acts as the heater of substrate and source, respectively. Under the irradiation of a tungsten halogen lamp, the upper and lower pieces of graphite absorb light to heat up rapidly, heating the source and the substrate. The substrate temperature ranges from 550 to 650 °C, and the source temperature is 80–100 °C higher than the substrate temperature. The reaction chamber is filled with N_2 and the vacuum degree is 7.5×10^2–7.5×10^3 Pa. The CdTe powder is sublimated and decomposed into Cd atoms and Te_2 molecules under the heating of a source graphite heater. The sublimated Cd and Te collided with Ar gas molecule in the growth chamber, and then diffused and transported into the substrate at a lower temperature Subsequently, a reverse reaction of the sublimation reaction took place, and condensation was deposited on the substrate surface. The deposition rate is primarily dependent on the temperature of source and the pressure of reaction chamber. The general deposition rate is 1.6–160 nm/s, and the maximum can reach 750 nm/s. The microstructure of the material is dependent on the temperature of substrate, the temperature gradient between the source and the substrate, and the crystallization status of the substrate. The grain size is $(2–5) \times 10^3$ nm, increasing with the rise in temperature of substrate and the thickness of film on the whole. In the depositing process, inert gases, e.g., Ar and He are commonly used. They do not react with CdTe, Cd, and Te_2, but exist as a transitional environment. Cd and Te_2 molecules collide with the mentioned inert gases to prevent the Cd and Te_2 molecules from sublimating and directly splashing onto the surface of substrate, increasing the transport distance of Cd and Te_2 molecules in the growth process, and effectively improving the uniformity of the film. To enhance the nucleation ability and electrical properties of CdTe thin films fabricated by close-space sublimation, many researchers have introduced oxygen into the atmosphere of close-space growth. Tyan et al. [15] initially employed oxygen in the working gas to prepare a cell with an energy conversion efficiency of more than 10% (AM2) and an area of 0.1 cm^2. Ferekides et al. [16] adopted the identical technique to deposit a CdTe thin film on a CdS film grown with the CBD, and the efficiency of the device reached 15.8%. If oxygen is introduced to the ambient gas, the concentration of the acceptor in the CdTe absorption layer can increase, and the formation of a deep buried homo-junction can be prevented. As demonstrated by numerous studies, doping a small amount of oxygen in an inert atmosphere critically impacts the growth of CdTe thin films, which enhances the conductivity of CdTe thin films, up-regulates the number of nucleation in the initial growth stage, reduces the grain size of CdTe thin films, increases the density of CdTe thin films, and effectively decreases holes in films. At the temperature of the substrate over 600 °C, the interdiffusion of S and Te will occur at the CdS/CdTe interface, thereby forming the $CdTe_{1\text{-}y}S_y$ ternary phase; as a result, the PN junction will be deviated from the CdS/CdTe interface, the service life of minorities is extended, and the electrical performance of the junction is enhanced. The presence of oxygen is conducive to the formation of compounds in the transition layer between CdS/CdTe and hinders the lattice mismatch; Meantime, the CdTe source is oxidized unevenly, and a thin oxide layer of 5×10^2–5×10^3 nm is formed on the surface of the CdTe source, which decreases the sublimation effect of the source, down-regulates the efficiency of the device, impacts the repeatability of the process, and adversely affects large-scale production. Rose et al. [39] generated the cell without oxygen, with the efficiency of 13%. Existing studies reported that, for the self-compensating effect of CdTe, high-efficiency cells can also be fabricated without intentionally doping the CdTe absorption layer.

Moreover, the CSS method applies to the preparation of CdS films. Oxygen is doped in the deposition chamber, and a high-temperature nucleation and low-temperature growth process is exploited. The oxygen-doped CdS film facilitates the formation of $CdS_{1\text{-}y}Te_y$ ternary phase, elevates the open-circuit voltage and fill factor, and down-regulates the reverse saturation current.

3. **Sputtering method**

The magnetron sputtering technology refers to exploiting an orthogonal magnetic field and the electric field between the cathode and the target (anode) being sputtered; the directions of the electric and magnetic fields are perpendicular to each other. When the coating chamber is evacuated to the set condition, a suitable amount of argon will be filled, and several hundred volts will be applied between the cathode (e.g., column target or flat target) and the anode (e.g., coating chamber wall). Subsequently, the glow discharge will be generated in the coating chamber, and the argon gas will be ionized. As impacted by the orthogonal electromagnetic field, the electrons advance along the target surface in a cycloidal manner, and the movement of the electrons is limited in a certain space, which rises the probability of collision with the working gas molecules and enhances the ionization efficiency of the electrons. After multiple collisions, the electrons losing energy and termed as "final electrons" enter into the weak electric field. When they lastly reach the anode, they are already low-energy electrons, which will no longer overheat the substrate. Meantime, high-density plasma is bound near the target surface and establishes no contact with the substrate. The atoms on the surface of the target are sputtered and then deposited on the work piece surface to form a thin film. For the CdTe thin film, the high-energy ion Ar^+ accelerates against the surface of the CdTe target material as impacted by magnetic field lines, and via the momentum conversion, the surface material of the CdTe is splashed out, and then deposited on the substrate to form a thin film. The deposition reaction commonly takes places on a substrate below 300 °C, and the pressure of the reaction chamber reaches nearly 10 mtorr.

4. **Vapor-phase transport deposition method**

The vapor-phase Cd and Te_2 are transported to a hot substrate by a suitable gas to directly deposit and grow CdTe. Common transport gases are H_2 and He. This technology enables precise control of the composition of the film and the concentration and distribution of the dopant. The deposition rate of the film is dependent on the partial pressure of Cd, Te_2 and the temperature of the substrate. At the molar ratio of Te/Cd slightly higher than the chemical ratio, the film is p-type conductivity [40]. At the molar ratio of slightly lower than the chemical ratio, the film is n-type conductivity. The resistivity of n- and p-type films at an ambient temperature close to chemical ratio is significantly high, about $1 \times 10^4\ \Omega$ cm. Doping PH_3 or AsH_3 in the reaction gas can reduce the resistivity of the p-CdTe film to 200 Ω cm.

5. **Screen printing method**

A sintering paste composed of a mixture of CdS, $CdCl_2$ flux, and a propylene glycol inhibitor is applied to the substrate after passing through a screen; then, it is sintered in a nitrogen atmosphere at 700 °C. The CdS film is grown by recrystallization [41]. A mixture of Cd, Te mixture (or CdTe), $CdCl_2$ flux, and propylene glycol inhibitor is adopted at an equimolar ratio, sintered at 590–620 °C, Cd, Te react (or CdTe recrystallize) to form CdTe film. Rising the sintering temperature and exploiting $CdCl_2$ flux can facilitate the mutual diffusion of S and Te across the interface and form $CdTe_{1\text{-}y}S_y$ and $CdS_{1\text{-}y}Te_y$ at the CdS/CdTe interface. The transmittance of short wave in the window layer is reduced, and the long-wave cutoff wavelength of the spectral response is extended to more than 850 nm.

6. **Spray thermal decomposition method**

The aqueous solution of Cd salt and Te salt is sprayed on a hot substrate to form a CdTe film, exhibiting the minimum cost. Using CdTe spray slurry instead of the solution of Cd salt and Te salt can significantly enhance the efficiency of device. The aqueous solution of $CdCl_2$ and thiourea act as reactants, and a CdS film of 8 nm × 10 nm and a CdTe film of about 6 nm × 10 nm are sequentially spray-deposited at 375–400 °C; the efficiency of the device reaches 12.7% [42].

7. **Metal–organic chemical vapor deposition (MOCVD)**

CdTe film is grown by reacting dimethyl cadmium (DMCD) and isopropyl telluride (DIPTE) in hydrogen [43]. The deposition rate is about 0.1 nm/s. The film is densely packed with columnar grains. The conductivity type can be regulated by the molar ratio of DMCD/DIPTE. At the ratio lower than 0.5, the film is p-type conductivity; at the ratio higher than 0.5, the film is n-type conductivity.

With triethylgallium (TEG) and AsH_3 as n-type and p-type dopants, respectively, the resistivity of the film can be reduced. On the CdS film grown with the solution method, the efficiency of the cell made by growing the CdTe film by the MOCVD technology reaches 9.87%.

8. **Electrochemical deposition method**

The electrochemical deposition method can be adopted to prepare CdTe films with the optimal composition ratio. In an acid solution supplemented by Cd salt and TeO_2 (the main component in the solution is $HTeO^{2+}$), CdS/TCO/Glass acts as the cathode, and the potential of the cathode is −0.2 to −0.65 V (compared with a standard calomel electrode). Reaction on the cathode:

$$HTeO^{2+} + 3H^+ + 4e^- \rightarrow Te^+ + H_2O$$

$$Cd^{2+} + Te + 2e^- \rightarrow CdTe$$

Limited by the solubility of TeO_2, the deposition rate is remarkably low, about 0.27–0.55 nm/s. Switching to Te anode can supplement $HTeO^{2+}$. The newly deposited film pertains to n-type conductivity, and p-CdTe can be obtained after proper heat treatment [44]. Doping can be achieved by electrochemical co-deposition or electromigration. In 1978, Panicker et al. [45] initially employed the aqueous solution system composed of $CdSO_4$ and TeO_2 to fabricate CdTe amorphous films exhibiting a grain size of about 50–100 nm with the cathodic DC deposition method. The conversion efficiency of CdTe thin-film solar cells made by Morris et al. [46] in Australia and Song et al. [47] of the Colorado School of Mines in the United States based on the electrochemical deposition reached 13.1 and 13.4%. The efficiency of solar modules produced by the PB Solar company in British reached over 10%.

The heat treatment of CdTe thin film is a critical part of preparing high-efficiency polycrystalline CdTe solar cells. Overall, the CdTe films obtained are all n-type semiconductors. In CdTe thin-film solar cells, the CdTe thin film should be p-type to form a PN junction with the n-type CdS window layer. Thus, CdTe should be heat-treated by $CdCl_2$ to transform it from an n-type material to a p-type semiconductor material. CdTe films can be heat-treated in multiple ways. For instance, a CdTe film is immersed in a $CdCl_2$:CH_3OH or $CdCl_2$:H_2O solution, and then dried in a $CdCl_2$ atmosphere, a HCl atmosphere, and a Cl_2 atmosphere to form a $CdCl_2$ film. Furthermore, Cl can be introduced in the formation of Cl^- or as a component (screen printing) in the reaction chamber (electrochemical deposition) in the depositing process of the CdTe thin film.

$CdCl_2$ heat-treating CdTe and CdS films are the most common in preparing devices. It directly affects the properties of thin films and devices, which includes boosting the growth of grains and the passivation of grain boundary defects. This heat treatment enhances efficiency primarily because: (1) facilitate the growth and recrystallization of CdTe grains and increase the grain size in CdTe films [48]; (2) passivate the interface of grains, thereby reducing the density of defect states and the recombination of carriers [49]; (3) facilitate the interdiffusion at the interface between the CdTe and CdS, and effectively reduce the lattice mismatch between CdTe and CdS [50, 51]; (4) for the self-compensation effect of CdTe, high-concentration doping is difficult to obtain. However, Cl^- ions in $CdCl_2$ participate in the formation of Cd vacancies, thereby doping CdTe crystals, hindering the recombination of carriers in and between grains, thereby obtaining good transport properties and improving p-type doping, extending the lifetime of minorities [44]. Overall, the heat treatment of the CdTe film is conducted directly in air or vacuum, or it may be performed in a non-oxidizing atmosphere (e.g., nitrogen).

In the heat treatment of $CdCl_2$, the small grains in the CdTe thin film begin to recrystallize and present a more ordered lattice structure. The reaction principle is shown in the following formula [52]:

$$\mathrm{CdTe(s) + CdCl_2(s) \rightarrow 2Cd(g) + 1/2Te_2(g) + Cl_2(g) \rightarrow CdCl_2(s) + CdTe(s)}$$

As expressed in the above reaction formula, the solid CdTe film and $CdCl_2$ are first decomposed into gaseous Cd, Te_2, and Cl_2 under heat treatment, and then recrystallization takes place. The gaseous material is recrystallized to form CdTe and $CdCl_2$ solids. This reaction indicates that the Cl_2 atmosphere facilitates the regrowth of CdTe film grains through a local gas phase transport reaction. As impacted by this process, the small grains in the CdTe film gradually disappear, and the interface between CdS and CdTe begins to be reorganized to a certain degree. The reaction is remarkably sensitive to temperature. Overall, the temperature of heat treatment is regulated at about 400 °C. Meantime, there is a certain degree of dependence on the amount of $CdCl_2$ and the reaction time. Adequate $CdCl_2$ can make the reaction proceed more efficiently, whereas it will negatively affect the adhesion of the film and the interface of CdS and CdTe if the amount of $CdCl_2$ is excessive or the processing time is excessively long. If the heat treatment is performed in air, the effect of oxygen in the air on the sample cannot be overlooked. In the heat treatment, oxygen in the air will undergo some oxidation reactions with the surface of the CdTe film and $CdCl_2$. The oxides produced by the mentioned oxidation reactions should be primarily removed before the back electrode is fabricated. The common removal method is to use a chemical solution to etch the surface of the heat-treated CdTe film to remove oxides generated during the heat treatment.

The effect of $CdCl_2$ heat treatment and subsequent annealing on the material composition distribution is very significant. The entire film becomes Cd rich, and the closer it is to the surface, the higher the concentration of Cd and Cl. The S in CdS will diffuse into CdTe and cause the redistribution of impurities, and the higher concentration of oxygen will be distributed at the CdTe/CdS interface as a substitute impurity [50, 53]. Another effect after heat treatment with $CdCl_2$ refers to the randomness of the direction of each grain in the CdTe crystal. The native CdTe grows preferentially in the (111) direction. At the relatively low deposition temperature, CdTe grows in the (111) direction. As calculated by Kim et al. [54] with XRD, the thermal activation energy of recrystallization reaches 2.5 ± 0.3 eV, equivalent to the diffusion activation energy of Cd from 2.44 eV to 2.57 eV. When the time of the heat treatment is extended, the grains of CdTe film will have a certain direction. Moutinbo et al. [50] used AFM to study the recrystallization process at different temperatures. It is considered that fine grains will appear between the grains in the initial stage of heat treatment, and then the growth of the mentioned fine grains will lead to random direction of the grains. At 500 °C, there is a eutectic line in the binary phase diagram of CdTe and $CdCl_2$, and the eutectic point is at 77% $CdCl_2$. When the CdTe crystal surface is covered with $CdCl_2$, the equilibrium state is severely deviated. Some existing studies reported that $CdCl_2$ is enriched on the CdTe grain boundaries, probably because the solubility of $CdCl_2$ in CdTe is significantly small, causing $CdCl_2$ to segregate at the grain boundaries of CdTe; as a result, electrical active regions are formed at the grain boundaries. After treatment with $CdCl_2$, the CdTe surface becomes smoother, and large cracks appear at the grain boundaries. The I-V curve with temperature reveals that the current transport mechanism of the primary film and the film after heat treatment has varied. In the native or air-annealed state, the current transport is regulated by the tunnel or interface recombination mechanism; after treatment with $CdCl_2$, the current transport mechanism is converted into the junction recombination mechanism. The density of interface states decreases significantly, and the interface recombination rate is down-regulated. Recent studies revealed that the density of dislocations and stacking faults in the depletion region is noticeably reduced, which significantly elevates the quality of the crystal in the depletion region.

7.3.5 Back Contact

To prepare a high-efficiency and stable polycrystalline CdTe thin-film solar cell, a stable low-resistance ohmic contact between the back electrode and p-CdTe of the CdTe solar cell is essential. Accordingly, there should be no barrier to block hole transport between the CdTe absorption layer and the back contact, as well as no contact resistance below 0.1 Ω cm^2. However, several difficulties remain in forming an ohmic contact with p-CdTe:

(1) In accordance with the Schottky theory, to form an ohmic contact between a p-type semiconductor and a metal, the work function of the semiconductor should be lower than the work function of the metal; otherwise, the interface barrier will generate high contact resistance. Since the work function of CdTe is about 5.5 eV, it is difficult to find a metal or alloy with a high work function to form an ohmic contact with it;

(2) For the influence of the surface state, the pinning effect of the Fermi level on the CdTe surface deviates from the Mott-Schottky theory;
(3) Being difficult to obtain low-resistance CdTe is another limitation to establish ohmic contact.

CdTe is a defective semiconductor, and the concentration of carriers inside is largely dependent on its intrinsic defect. The mentioned intrinsic defects included Te vacancies and Cd interstices, both of which are electron donors, and also Cd vacancies and Te interstices, both of which are electron acceptors. External doping, if it is ionized alone or constitutes a compound defect with the intrinsic defect, impacts the electrical properties of CdTe. The most typical example refers to the compound defect of V_{Cd}–Cl_{Te}, considered to act as an acceptor. In single crystalline CdTe, a doping concentration of 10^{17}–10^{18} cm^{-3} is achieved by the diffusion of impurities into the crystal. In the growth of CdTe single crystals, with the addition of Group III elements Al, Ga, and In achieves a donor concentration of 2×10^{18} cm^{-3} [55], and the addition of Group V elements also achieves an acceptor concentration of 6×10^{17} cm^{-3}. The I A element introduces a shallow acceptor level in the single crystalline CdTe, which covers the addition of Na and Li obtaining a carrier concentration of 10^{15} cm^{-3}. Moreover, the I B element (e.g., Au, Ag, and Cu) has also been successfully used to dope CdTe. For a single crystalline CdTe solar cell, the concentration of carriers above 10^{17} cm^{-3} can achieve a significantly low series resistance and ohmic contact. For polycrystalline materials, however, the situation is significantly more complicated. Since in polycrystalline materials, the additional series resistance contributes to back contact barrier as well as the grain boundary barrier. The height of the mentioned barriers is determined by the state of impurities and dangling bonds, as well as the concentration of carriers inside adjacent grain boundaries. Moreover, for the self-compensation effect of II-VI compound semiconductors in CdTe, it is difficult to achieve heavy doping with conventional technology. On the whole, the concentration of carriers reaches the order of 10^{15} cm^{-3}, which cannot satisfy the order of magnitude required to achieve ohmic contact, and ohmic contact cannot be achieved under the quantum tunneling effect.

One way to achieve better back ohmic contact performance of CdTe is to deposit a highly doped p^+ layer on the CdTe surface for the reduction of the effect of back contact barrier. The higher the doping level of the p^+ layer, the thinner the barrier region will be; as a result, the original thermal electron emission transport mode is converted to a tunnel or heat-assisted tunnel transport mode. Another method refers to depositing a layer between p-CdTe and the metal back electrode to achieve a heavily doped back contact layer, so the Fermi level of CdTe matches the metal back electrode. Low-resistance contact is achieved by the quantum tunnel transport mechanism between the back contact layer and the metal, which requires that the top of the valence band of the back contact material is lower than or substantially the same as that of CdTe with respect to the vacuum level, so the interface region does not have a valence band spike that hinders hole transport to the back electrode.

Currently used back contact materials are:

(1) Gold

Gold refers to one of the most back contact materials studied. It can be used alone as back contact material or as the last layer of back contact. Existing studies reported that when Au is used alone as back contact material, while Cd diffuses out of CdTe, Au diffuses to Cd vacancies in CdTe, so an additional p-type layer is formed. Another possibility is the formation of a thin layer of $AuTe_X$, which helps to reduce the contact barrier. At present, the most common method is to first chemically etch the CdTe film, and then evaporate Au or Au–Cu alloy. Te-rich surfaces formed by chemical etching noticeably reduce contact resistance. Though the Cu/Au back contact layer has exhibited a lower contact resistance, it impacts the efficiency of the cell, and the effect is unstable.

(2) HgTe

HgTe refers to a semiconductor with a narrow bandgap, and its work function is 5.9 eV [56]. Its advantage is that the lattice mismatch with CdTe is small (nearly 0.3%) [57], and it can form solid solution $Hg_{1-x}Cd_xTe$ with CdTe in the entire range of composition. Accordingly, the potential barrier between the valence bands of CdTe and HgTe can be eliminated via interdiffusion. Such material is significantly suitable as back contact material of CdTe. On the whole, such material has been used for the contact of single crystalline CdTe, However, Chu et al. [58] applied it to CdTe solar cells to achieve a series resistance of 0.4–0.8 Ω cm, whereas the repeatability is poor, and the device tends to cause short circuit during the preparation.

(3) Doped graphite paste

At present, the maximal photoelectric conversion efficiency of CdTe solar cells is achieved with doped graphite paste as the back contact [59]. Graphite paste doped by HgTe:Cu is coated on the CdTe surface, and then drying and annealing are conducted in a He atmosphere to achieve back contact. In the preparation process, the p-type highly doped regions of Cu_2Te and $Hg_{1-x}Cd_xTe$ are formed at the interface between CdTe and doped graphite paste, which facilitates the tunneling of carriers. Such back contact exhibits the advantage that a fill factor over 70% can be obtained steadily. The performance of back contact is closely associated with the corrosion process of CdTe and the annealing temperature and time of back contact [60]. The annealing of the back contact will impact the diffusion degree of the contact material in CdTe. Such back contact exhibits the defect that the preparation is relatively complicated and is not suitable for large-scale production.

(4) Copper

The back contact of copper is achieved by forming a heavily doped layer at the interface between CdTe and Cu. The preparation is as follows [61]: First, the Cu deposited on the CdTe surface is annealed (to diffuse copper), and then the elemental copper on the surface is removed with an etching solution, and then the electrode is evaporated. The merits of this back contact are relatively low contact resistance, relatively high surface conductivity, and insensitivity to the size of the work function, so cheap metals can be taken as electrodes. The defect of this back contact is that the introduction of considerable Cu will affect the stability of device performance.

(5) Cu_2Te

The essence of adopting elemental Cu as the back contact is to form a layer of Cu_2Te between CdTe and metal. Thus, the University of South Florida conducted a study on this layer. A Cu_2Te film was deposited on the CdTe surface by sputtering to compare the Cu_2Te layer formed by heat treatment. Such a method is capable of regulating the amount of elemental Cu and reducing the decay of junction characteristics over time. The composition analysis by EDS also demonstrated that the chemical composition of Cu_2Te films deposited at 250 °C conforms to the molecular formula, and films deposited at higher temperatures cover more copper. The characteristic of the back contact is associated with the deposition temperature, thickness, and annealing conditions of Cu_2Te [62]. Cu_2Te exhibits two functions as back contact, i.e., the back contact and the diffusion source of Cu. When Cu_2Te film is taken as back contact layer, Cu will enter the acceptor impurity of p-CdTe, thereby increasing the doping concentration of CdTe. Moderate Cu diffusing into CdTe facilitates the enhancement of device performance, whereas excessive Cu can cause device performance to degrade. Ferekidss et al. adopted Cu_2Te as back contact layer to prepare a polycrystalline CdTe thin-film solar cell with a fill factor of 69.5%.

(6) ZnTe

ZnTe is an excellent II-VI compound semiconductor. Its direct bandgap at ambient temperature is 2.26 eV, and commonly exhibits p-type conductivity for the lack of Zn in the structure. The concentration of Zn vacancies is dependent on the partial pressures of Zn and Te. The higher the partial pressure of Te, the higher the concentration of Zn vacancies. In the structure of the ZnTe thin film fabricated by the vacuum evaporation method, the number of Zn vacancies is strongly dependent on the substrate temperature, and the films deposited at lower temperatures exhibit a higher Te/Zn ratio. The resistivity of undoped ZnTe films reaches 10^5–10^6 $\Omega\,cm^2$ [63]. To achieve p-type ZnTe, Cu, Ag, Au, Li, P, N, Sb, and Al have been studied as dopants. To be specific, Cu is the most effective dopant for ZnTe, which can enter the Zn vacancies to form the acceptor center, replacing the original ${V_{Zn}}^{++}$. The ionic radius of Cu^+ (0.96 Å) is significantly close to the ionic radius of Cd^{2+} (0.95 Å), and it is easy to replace Cd. The ionization energy of Cu is about 0.15 eV. The resistivity of ZnTe:Cu films is associated with the doping concentration of Cu. The more Cu doped, the higher the conductivity of the film. To obtain a carrier concentration over 10^{18} cm^{-3}, a Cu-doped concentration of 2–6% is required [64].

Theories and experiments demonstrated that the difference of valence band between ZnTe and CdTe is only 0.05 eV [65]. Such a small shift of valence band causes a significantly low or zero barrier between the ZnTe and CdTe interface, and carriers easily pass through, which is the primary reason why p-ZnTe is suitable for acting as a back contact layer of CdTe solar cells.

Meyers [65, 66] firstly used heavily doped ZnTe as an intermediate transition layer between the high-resistance CdTe and the metal back contact, and achieved a thin-film solar cell exhibiting the photoelectric conversion efficiency of 11.2% by thermally evaporating ZnTe:Cu. Gerssert et al. [67] fabricated ZnTe:Cu thin films by sputtering; they also delved into the relationship between electrical conductivity and substrate temperature. It was demonstrated that at the substrate temperature over 260 °C, the conductivity is noticeably reduced. Tang et al. [68] analyzed the polycrystalline ZnTe:Cu film fabricated with the co-evaporation; they found that with the rise in the post-treatment temperature, there is an abnormal phenomenon of conductivity temperature.

There is a potential barrier of 1.1 eV between the conduction band of CdTe and ZnTe:Cu, which can reflect the electrons drifting towards the back electrode; as a result, the collection efficiency is effectively enhanced, especially the long-wave collection efficiency. However, a reverse junction may be formed between ZnTe:Cu and the metal, so the ZnTe:Cu layer should be relatively thin. Moreover, the diffusion of Cu in ZnTe:Cu into CdTe will bring adverse effects. Accordingly, an intrinsic ZnTe layer should be introduced between the ZnTe:Cu and CdTe layers. The ZnTe layer can improve the lattice mismatch, prevent the doped atoms in the back contact layer or the metal atoms of the back electrode from diffusing into CdTe, and block the leakage channel of CdTe; as a result, the photoelectric conversion efficiency and stability of CdTe solar cells are enhanced.

(7) Ni–P

Since Cu back contact impacts the stability of CdTe cells, researchers have begun to seek materials and methods without Cu back contact. The electrodeless deposition method is common to evaporate Ni onto metal surfaces. B. Ghosh et al. used such method to prepare the Ni–P back contact layer of CdTe [69]; they found that the growth of Ni–P layer on CdTe is dependent on the concentration, PH value and temperature of solution. XRD analysis of the microstructure on the contact interface reveals the presence of NiP, $NiTe_2$, NiP_2, and P. The annealing temperature also significantly impacts the contact characteristics. The optimal annealing temperature is 250 °C. Then, the XRD diffraction peak of NiP_2 is the most robust, since NiP_2 has a high work function, which is equivalent to Au, so it is considered conducive to enhancing the contact characteristics. Moreover, annealing helps the diffusion of p at the contact interface into CdTe, doping it into a p-type, and assisting in the formation of a tunnel junction. However, the annealing temperature is overly high, and the series resistance is enhanced for the formation of the Ni_3P layer. The role of P in it is not significantly clear, and further systematic research is required. At present, the open-circuit voltage and fill factor of CdTe cells contacted with Ni–P remain relatively small [70].

7.4 Characteristics of CdS/CdTe Junction

There are three main interfaces for CdTe thin-film solar cells: TCO/CdS interface, CdS/CdTe interface, and CdTe/back-electrode interface. As the main PN junction of CdTe thin-film solar cells, the CdS/CdTe interface depletion region is the site for photoelectric conversion of CdTe thin-film cells, and is the core of the entire cell. At the interface, there is a 9.7% lattice mismatch between the cubic CdS and the cubic CdTe. Such a high lattice mismatch causes considerable defects at the CdS/CdTe interface. Accompanying the mentioned interface defects is considerable defect energy levels. The existence of the mentioned defect energy levels causes photo-generated carriers to be captured during the interface transmission, thereby seriously reducing device performance. As mentioned earlier, to reduce the impact of the mentioned defect levels, people use the interdiffusion characteristics between CdS and CdTe to make CdS/CdTe devices form an intermediate product CdS_xTe_{1-x} during heat treatment [48–71]. The intermediate product can effectively reduce considerable defect energy levels attributed to lattice mismatch at the interface, and the intermediate product also exerts the following effects on the device:

(1) CdTe surface can be passivated;
(2) It can act as an interface buffer layer, so the electrical interface between CdTe and CdS is separated from the alloy interface between them, reducing the negative impact of interface defect energy levels;
(3) It can effectively enhance transmission characteristics of carriers and photoelectric conversion efficiency of devices.

Moderate CdS/CdTe interface diffusion is critical to obtain high-efficiency CdS/CdTe solar cells. Interdiffusion is a two-way process. Te diffusing into CdS forms the $CdS_{1-x}Te_x$ ternary phase, and S diffusing into CdTe forms the CdS_yTe_{1-y} ternary phase. The degree of interdiffusion and the depth of diffusion depend on the deposition temperature and annealing temperature of the CdTe film. Moreover, the thickness of the CdS layer will become thinner for interdiffusion. Though the thinner window layer can increase the transmittance and facilitate cell efficiency enhancement, the diffusion consumption of CdS films in CdTe deposition and its post-processing is not always uniform, and pinholes may appear in local areas, making CdTe and TCO form a junction by direct contact. Since its reverse saturation current is higher than Cds/CdTe, the parallel resistance of the cell will be reduced, and the performance of the device will be adversely affected.

7.5 Device Characteristics of CdTe Solar Cells

A solar cell is a large-area PN junction. Thus, the current–voltage equation of a solar cell is the current–voltage equation of the PN junction under light:

$$J = J_0 \exp[(V - JR)/AkT] - J_{sc} + V/r \tag{7.1}$$

where A is the quality factor of the diode, J_0 is the reverse saturation current of the diode, the series resistance is R, and the shunt resistance is r [72].

Figure 7.4 shows the forward characteristics of CdTe and GaAs solar cells [73]. Since the bandgap of CdTe and GaAs are almost the same, if they are both limited by direct recombination, then they should exhibit the identical open-circuit voltage(V_{OC}) and current–voltage characteristic curve. But from the Fig. 7.4 we can see that the V_{OC} difference between the two is close to 200 mV, and the voltage difference at the maximum power approaches to 300 mV. This is because the quality factor of CdTe is 1.9, while the quality factor of GaAs is only 1.0. The physical difference is that CdTe solar cells have other current recombination mechanisms. Under normal operating conditions, this mechanism causes the forward current of CdTe solar cells to be almost two orders of magnitude greater than GaAs. Such a scenario also reveals that by down-regulating the recombination in CdTe, the V_{OC} of the device can further increase. Corresponding to the energy band diagram, we can clearly see that for the existence of considerable forbidden band traps, low-concentration doping of CdTe will cause a large increase in recombination current. On the other hand, low-concentration doping and the resulting electric field distribution will cause a decrease

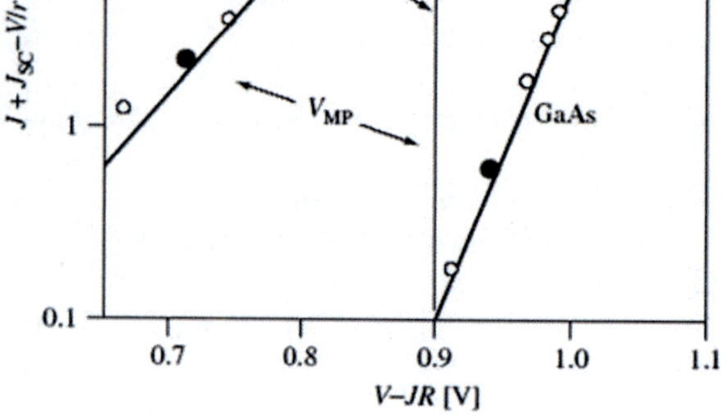

Fig. 7.4 Comparison of current–voltage characteristics of GaAs and CdTe solar cells [73]

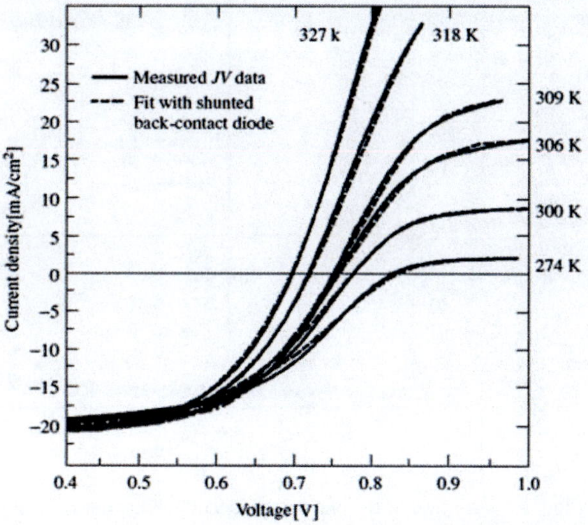

Fig. 7.5 Effect of back-electrode contact barrier on CdTe characteristics [74]

in minority carrier transport, which will cause a more severe change in the collection efficiency associated with voltage, resulting in a decrease in J_{mp}.

Figure 7.5 illustrates the effect of back-electrode contact barrier on the energy band and the J-V curve of the solar cell. Assuming that the two depletion regions do not overlap, the mentioned two diodes can be considered two independent circuit elements. Under the barrier height of the back barrier diode set to 0.3 eV, the theoretical curve well complies with the experimental curve. With the decrease in the temperature, the effect of the back barrier becomes increasingly serious. This effect is most severe in the first quadrant. Then, the shape of the J-V curve is termed as "flip" [74–76]. The magnitude of the "flip" is associated with the doping amount of Cu when preparing the back contact electrode. If the doping amount of Cu is small, the "flip" phenomenon is only identified at high temperatures, the larger back contact barrier is also suggested to most severely impact device performance [77].

Doping considerable Cu to reduce the back contact barrier can suppress the "flip" phenomenon, while causing CdTe solar cells to be less stable at high temperatures. It has been reported so far that when CdTe solar cells are operated at high temperatures for long periods, significant performance variations have been observed. The mentioned studies are often referred to as "stress" experiments, in which a decrease in fill factor is first identified, followed by a decrease in open-circuit voltage [74–78]. Variations to the short-circuit current can only occur in extreme cases. Figure 7.6 presents the performance test results of CdTe solar cells fabricated by NREL at different time intervals before stress and at 100 °C under open-circuit conditions. As shown in the dark current curve of the device, with the increase in the temperature stress, the "flip" phenomenon becomes increasingly serious. With the decrease in temperature or the extension of the stress time in the open-circuit state, the

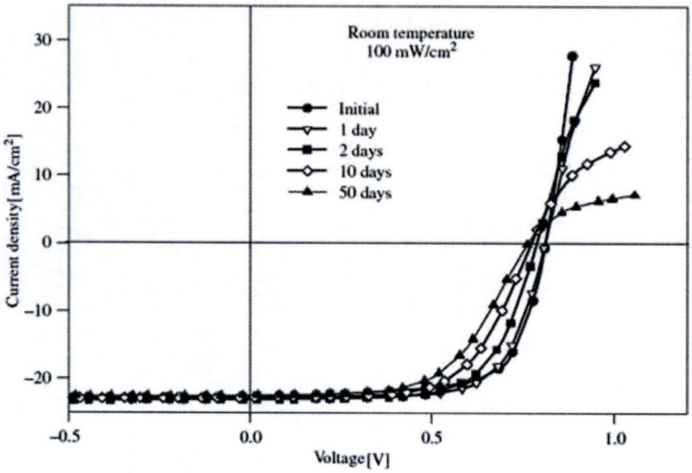

Fig. 7.6 Variation of the characteristics of CdTe solar cells at high temperatures of 100 °C [78]

phenomenon of "flip" tends to increase. This finding is explained as the concentration of acceptor decreases when the test temperature is lowered, and the effect of the back barrier becomes more serious, as presented in the Fig. 7.6. The variation in device performance under stress is similar to the decrease in the concentration of carriers in the CdTe absorption layer. However, the study on the stress of the device in the short-circuit state or the maximum power state indicates that the degradation of the device performance is significantly smaller than the open-circuit state. This finding is explained as the reduction of the electric field inside the cell causes Cu atoms to overflow from the back contact area when the device is in forward bias, which has two effects on the device performance: first, the height of the back barrier increases, similar to the case without Cu; On the other hand, the performance of the device decreases due to the movement of Cu toward the PN junction. However, whether Cu will increase the combination centers in CdTe remains unclear.

Considerable loss exists before the photons reach the absorption layer of the CdTe solar cell, and the incomplete absorption of the photons causes the photons to pass through the absorption layer and cause the loss of the photons again. The quantum efficiency can be calculated by measuring the reflection and absorption of the respective layer of the solar cell separately. Figure 7.7 shows several major mechanisms that cause J_{SC} reduction in CdTe solar cells.

It is suggested that the measured external quantum efficiency is multiplied by the spectrum (unit is photon/cm^2/nm), integrated over the entire wavelength range, and lastly multiplied by the electron power to get J_{SC}. Under the bandgap $E_g = 1.5$ eV, the maximum current density reaches nearly 30.5 mA/cm^2. The optical region presented in the figure reflects solar cells on light, the absorption of the glass substrate, the absorption of the SnO_2 front electrode, as well as the loss attributed to the absorption of light below 500 nm by CdS window layers. The loss near the bandgap is attributed

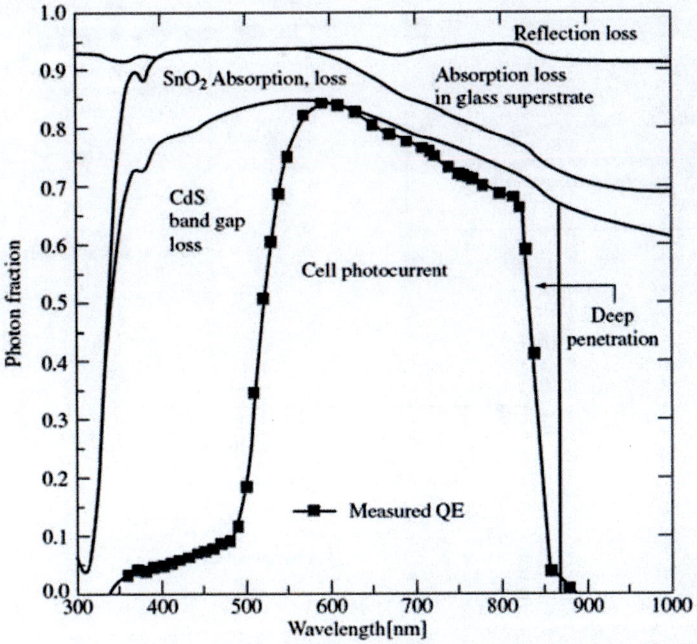

Fig. 7.7 Photon loss and quantum efficiency of CdTe solar cells as a function of wavelength [78]

to photons passing via the entire solar cell. Integrating the mentioned loss curves gives the current density of the losses attributed to each mechanism. The mentioned losses are marked in the blank area of the figure, and the sum of all the loss parts is equal to the difference between the actual measured value of the current density and the maximum value [79]. As suggested from the mentioned calculation results, the current density can be noticeably increased by down-regulating the thickness of the CdS window layer, and the current density can be enhanced by altering the glass substrate type or optimizing the SnO_2 process flow. Only by introducing an antireflective film or more effectively regulating deep-penetrated photons slightly impacts current density.

Information regarding trap states in the forbidden band can be obtained by determining the capacitance of solar cells, and it also shows the distribution of carrier density in the absorption layer [80]. Figure 7.8a illustrates the capacitance test results of CdTe solar cells under the biases of 0, −1, and −3 V. The capacitance is relatively small, attributed to the low carrier concentration and the large depletion width. According to the large depletion width, most photons will be absorbed in this region, so the photocurrent does not vary evidently with voltage variations [81]. When the frequency variations in the range of three orders of magnitude, the capacitance remains almost the same, indicating that the effect of the trap on the device is relatively small, and the upswing of the curve at high frequencies is for the self-induction phenomenon of the circuit.

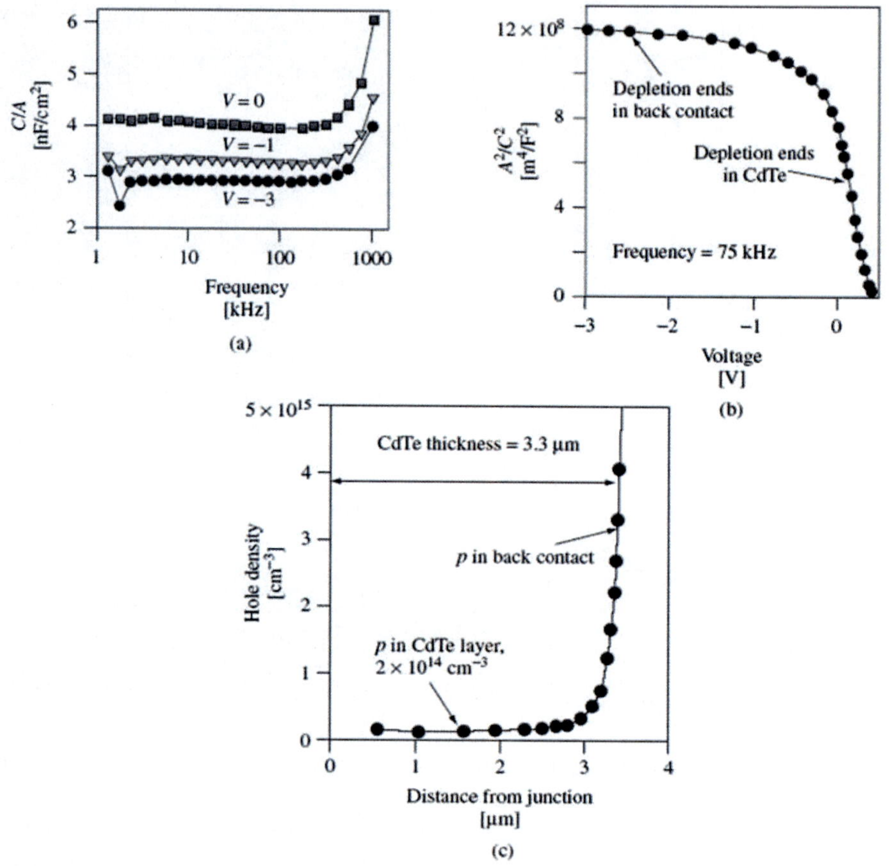

Fig. 7.8 C-V test results of CdTe solar cells [81]

Figure 7.8b is the capacitance–voltage test result obtained at 75 Hz. Where the vertical axis (C^{-2}) is proportional to the square of the depletion width w (at a given voltage, $C/A = \varepsilon/w$). The slope of the curve is inversely proportional to the carrier density at the boundary of the depletion region. In this scenario, there are two different regions: C^{-2} and the width of the depletion region vary slightly with the voltage when the bias is reversed; under the voltage close to zero or positive, the width of the depletion region decreases evidently.

Figure 7.8c shows the trend of hole density as a function of the depletion width (length from the junction). The variations in the two areas are now clearer. Located at a distance of 3 μm from the CdTe surface, the concentration of holes is significantly low (approximately 10^{14} orders) and then rapidly increases. This is because the CdTe layer exhibits the thickness of exactly 3 μm; when the depletion region enters the back contact region, the concentration of holes there rapidly rises.

7.6 Development Prospects of CdTe Solar Cells

Compared with other solar cells, CdTe thin-film solar cells exhibit lower manufacturing costs, depending on their cell structure, raw materials, and manufacturing processes. Firstly, CdTe thin-film cells are photovoltaic devices formed by sequentially depositing multiple thin films on glass or other flexible substrates. In comparison with other solar cells, CdTe solar cells exhibit a relatively simple structure. The simple structure noticeably shortens the production time and significantly reduces the manufacturing cost. As revealed by data from First Solar in the United States, the full-process production time of cadmium telluride thin-film solar cell modules is less than 2.5 h. Secondly, the absorption coefficient of cadmium telluride reaches 10^5 cm^{-1} in the range of visible light (100 times higher than that of silicon materials). In the sunlight, 99% of the photons with higher energy than the bandgap of cadmium telluride can be absorbed in a 2 μm thick absorption layer. Accordingly, in theory, the thickness of the absorbing layer in the cadmium telluride thin-film solar cell is about several micrometers, and the consumption of raw materials is significantly small, so the manufacturing cost of the cadmium telluride solar cell is relatively low. Lastly, cadmium telluride pertains to a simple binary compound system, which is easy to produce single-phase materials; besides, the preparation method is easy to implement, which effectively lowers the manufacturing cost. More than 10 existing technologies can be adopted to fabricate small-area cadmium telluride solar cells with a conversion efficiency of more than 10%. To be specific, 5 technologies have been used for industrial production, especially the near-sublimation and vapor-phase transport deposition technologies. They exhibit the merits of high deposition rate (3–10 nm/min), high film quality, large crystal grains, and effective utilization of raw material (>85%), which especially applies to large-scale production.

Compared with other types of solar cells, e.g., crystalline silicon and gallium arsenide, cadmium telluride thin-film solar cells exhibit the maximum theoretical photoelectric conversion efficiency of nearly 30%. From the perspective of laboratory conversion efficiency, the laboratory conversion efficiency of cadmium telluride thin-film solar cells has been approaching that of crystalline silicon solar cells. On September 9, 2013, the photoelectric conversion efficiency of the cadmium telluride thin-film solar cells developed by First Solar company reached 18.7%, and the component efficiency was 16.1%, as verified by the National Renewable Energy Laboratory (NREL) of the United States Department of Energy. Both set a new world record. From the perspective of photoelectric conversion efficiency of commercial products, the maximal photoelectric conversion efficiency of mass-produced cadmium telluride thin-film solar cells is second only to copper-indium-gallium-selenium thin-film solar cells among various types of thin-film solar cells. At present, the average conversion efficiency of global thin-film solar cells is about 10%. To be specific, the maximum photoelectric conversion efficiency of mass-produced CIGS, CdTe, and amorphous/microcrystalline silicon thin-film solar cell modules were 15.5%, 13.1%, and 9.8%, respectively.

Some experiments revealed that with the rise in the temperature of the module, the performance of all solar cells will be degraded, primarily because of the decrease in the open-circuit voltage of the solar cell. The temperature coefficient indicates the rate at which the output power of a solar cell module varies with the increase in the operating temperature. Overall, the temperature coefficient of crystalline silicon solar cell modules is −0.45%/°C to −0.5%/°C, i.e., for every 1 °C rise in the module temperature, the output power of solar cell modules decreases by 0.45–0.5%. The temperature coefficient of CdTe thin-film solar cell modules is approximately − 0.25%/°C, nearly half lower than that of crystalline silicon solar cells; thus, it more applies to harsh environments (e.g., high temperatures, deserts, and humid regions). A lower temperature coefficient reveals that the output power of the CdTe thin-film solar cell module exhibits less susceptibility to temperature, demonstrating that the CdTe thin-film solar cell can provide more energy at higher temperatures. As suggested from the experimental data of First Solar company in America, at the temperature of the solar cell module lower than 25 °C, the performance of the polycrystalline silicon solar cell module (expressed by the ratio of the output power of the DC power supply and the rated power under standard test conditions) is better than that of the CdTe solar cell module. At the temperature of the solar cell module higher than 25 °C, the performance of the CdTe solar cell module is superior over that of the polycrystalline silicon solar cell module. For instance, at the temperature of the solar cell module of 65 °C (40 °C higher than the standard temperature), the output power of the conventional crystalline silicon solar cell module decreases by 20%, while the CdTe thin-film solar cell module developed by First Solar company only decreases about 10%. This reveals that in hot summer or high-temperature areas, the actual power generation of CdTe solar cells is more robust than that of crystalline silicon solar cells.

The weak-light effect of CdTe thin-film solar cells is another noticeable advantage over crystalline silicon solar cells. For the wide spectral response range of CdTe thin-film solar cells, it exhibits a high susceptibility to weak-light and exerts a good weak-light effect, enabling it to generate electricity in weak-light environments (e.g., early morning, evening, or cloudy and rainy days). Thus, the CdTe thin-film solar cell has a significantly longer time to generate power per day than crystalline silicon, and its actual power generation is higher than that of the crystalline silicon solar cell, which remedies its relatively low luminous efficiency.

Considerable studies suggested that cadmium telluride thin-film solar cell modules are environmental-friendly products. CdTe, unlike the toxic element cadmium, refers to a compound exhibiting stability and safe utilization. The amount of CdTe in the CdTe thin-film module is significantly small, and only about 250 kg of cadmium telluride is required for 1 MW cadmium telluride module. CdTe is sealed between two pieces of glass, and no cadmium is released at ambient temperature. Even at a high temperature of 1100 °C, 99.96% of CdTe was sealed by two pieces of molten glass without leaking, as reported by the Brookhaven National Laboratory (BNL) in the United States. Comparing several other solar cells and other energy sources, the overall emissions of cadmium are the minimum during the entire life cycle of CdTe solar cell modules for their preparation and use. As revealed from the European PV Accept Project report, the energy recovery period of cadmium telluride thin-film solar cell only takes 10.8 months. Besides, the practice made by the First Solar company in the United States has also proven the period as 10 months, while that of crystalline silicon cells is 2.5–3 years. Manufacturers of CdTe thin-film solar cell modules recycle waste CdTe modules and exploit the reuse technology of waste modules to reuse the major raw materials, which can strengthen environmental protection while gradually achieving the development model of circular economy. In 2011, the European Union has exempted the requirements of RoHS ("Directive on Restricting the Use of Certain Hazardous Components in Electronic and Electrical Equipment"). Furthermore, considerable CdTe solar cell modules have been extensively employed in several EU countries (e.g., Germany, Spain, and Italy).

The CdTe thin-film solar cells exhibit the merits of simple process, easily large-scale production, stable performance, low cost, and high efficiency. However, they also face prominent problems in the following aspects:

First, the back electrode of the solar cell should be optimized. How to select a suitable material or structure to prepare a stable back junction exhibiting low series resistance is a research hotspot of CdTe solar cells. Since the electronic affinity of CdTe is 4.3 eV, and the bandgap is 1.5 eV at ambient temperature, the common conductive materials exhibiting high work functions (e.g., Au, Ni, and C), remain difficult to form a good ohmic contact with p-CdTe. This finding has become one of the main reasons for the major performance parameters of the currently fabricated CdTe solar cells (e.g., short-circuit current, open-circuit voltage, fill factor, as well as photoelectric conversion efficiency), which are significantly different from theoretical expectations. Cu is capable of replacing Cd atoms in CdTe to form CuCd substitution defects, which can increase the doping concentration of p-CdTe as an acceptor impurity, and it can effectively improve the contact between the CdTe layer and the back electrode to form a quasi-ohmic contact. Accordingly, it is extensively employed in the preparation of high-efficiency CdTe solar cells. Most of the existing back contact materials of CdTe solar cells are Cu or Cu-containing compounds (e.g., Cu, Cu_2Te, ZnTe:Cu). However, the study reports that excessive Cu content in the back contact layer will cause the overall device performance to decrease. Thus, the effective control of Cu content is critical to prepare such back contact materials.

The second refers to the mechanism of CdS/CdTe interface. Whether the mechanism is expounded is of vital significance to efficiency enhancement. The quality of the CdS/CdTe interface directly affects the performance parameters of solar cells. The diffusion at CdS/CdTe interface is associated with the substrate temperature and the treatment of $CdCl_2$. As reported by H. R. Moutinho et al. of NREL, CdTe deposited with the PVD method can be adopted to achieve effective recrystallization under the treatment of $CdCl_2$ at appropriate temperatures and periods. After the optimization, the polycrystalline thin film is significantly recrystallized and shows grain growth, and the stress of the material is further reduced. Meantime, the treatment of $CdCl_2$ further down-regulates the number of impurity energy levels in the forbidden band, and also plays a critical role in improving the performance of solar cells. Moreover, the heat treatment of $CdCl_2$ can enhance the crystallinity of CdS, while preventing the interface from oxidizing, and facilitate the diffusion at CdS/CdTe interface as well as the generation of CdS_xTe_{1-x}. The diffusion at CdS/CdTe interface in the heat treatment of $CdCl_2$ and its effect on the efficiency of the solar cell should be studied in depth.

Lastly, given the scarcity of raw materials, the limited natural reserves of tellurium (Te) will hinder the long-term development of CdTe thin-film solar cells. Tellurium refers to a silvery rare metal element, primarily associated with copper, lead, zinc, and other metal minerals; it has been extensively applied in metallurgy, electronics, chemistry, glass, and other industrial fields. In industry, tellurium is primarily recycled from the waste of electrolytic copper or smelted zinc. The existing geological reserves of global tellurium are inconsistent globally. Given data from the United States Geological Survey (USGS), global reserves of tellurium reach only about 24,000 tons, primarily distributed in the United States, Peru, Canada, and other countries and regions. As suggested from relevant domestic reports in China, the global geological reserve of tellurium is 149,000 tons. If it is estimated that 130–140 kg of tellurium is required to manufacture 1 MW thin-film solar cell modules, the resources of tellurium on the earth, based on the reserves of tellurium (24,000–149,000 tons), can be exploited for 100 production lines exhibiting an annual production capacity of 100 MW for 17–115 years. Accordingly, it is optimistic that global reserves of tellurium can satisfy production needs. However, from an economic perspective, the rising price of tellurium raw materials has significantly hindered the development of cadmium telluride solar cell industry. In 2000, the price of the tellurium raw material was only $34.4 per kilogram. However, as fueled by the advancement of the cadmium telluride thin-film solar cell industry, the price of the tellurium raw material has surged continuously. In 2011, the average global market price of tellurium raw materials reached $349 per kilogram. With the continuous enhancement of the capacity of CdTe solar cells, the market price of this scarce raw material tellurium will be inevitably up-regulated; as a result, the production cost of cadmium telluride thin-film solar cells will continuously increase, and the economics of the solar cell will also decrease.

References

1. R. Frerichs, Rev. **72**, 594–601 (1947)
2. D. Bonnet et al., *Conference Record 9th IEEE Photovoltaic Specialist Conference* (1972), pp. 129–132
3. J. Loferski, J. Appl. Phys. **27**, 777–784 (1956)
4. P. Rappaport, RCA Rev. **20**, 373–397 (1959)
5. J. Mimilya-Arroyo et al., Sol. Energy Mater. **1**, 171 (1979)
6. G. Cohen-Solal et al., *Conference Record 4th ECPVSC* (1982), pp. 621–626
7. A. Fahrenbruch, *Fundamentals of Solar Cells* (Academic Press, 1983)
8. J. Ponpon et al., Rev. Phys. Appl. **12**, 427–431 (1977)
9. K. Mitchell et al., J. Appl. Phys. **48**, 829–830 (1977)
10. T. Nakazawa et al., Appl. Phys. Lett. **50**, 279–280 (1987)
11. R. Muller et al., J. Appl. Phys. **35**, 1550–1556 (1964)
12. R. Dutton, Phys. Rev. **112**, 785–792 (1958)

13. E. Adirovich et al., Sov. Phys. Semicond. **3**, 61–65 (1969)
14. J. Phillips et al., Phys. Status. Solidi. B **194**, 31–39 (1996)
15. Y.S. Tyan et al., *16th IEEE Photovoltaic Specialist Conference* (1982), pp. 794–798
16. C. Ferekides et al., *23rd IEEE Photovoltaic Specialist Conference* (1993), pp. 389–392
17. X. Wu et al., *17th European Photovoltaic Solar Energy Conference*, vol. II (2001), pp. 995–999
18. N. Suyama et al., *Conference Record 21st IEEE Photovoltaic Specialist Conference* (1990), pp. 498–503
19. B. Basol, *Conference Record 21st IEEE Photovoltaic Specialist Conference* (1990), pp. 588–594
20. Y. Tyan et al., *Proceedings of 16th IEEE Photovoltaic Specialist Conference* (1982), pp. 794–800
21. P. Meyers et al., U.S. Patent, 4, 710, 589 (1987)
22. R. Birkmire, *Proceeding Polycrystalline Thin Film Program Meeting* (1989), pp. 77–80
23. X. Wu, J. Appl. Phys. **89**, 4564–4569 (2001)
24. B. Mccandless et al., *Conference Record 25th IEEE Photovoltaic Specialist Conference* (1996), pp. 781–785
25. X. Wu, *Conference Record 17th European Photovoltaic Solar Energy Conversion* (2001), pp. 995–1000
26. C.S. Ferekides et al., Sol. Energy Mater. Sol. Cells **35**, 255–260 (1994)
27. B.E. McCandless et al., *Proceedings of 22nd IEEE PVSC* (1991), pp. 967–971
28. J. Zhou et al., Mater Res. Soc. Sympos. Proc. **1012**, Y13–Y23 (2007)
29. K.M. Dugan, Masters thesis, University of South Florida, 1995
30. N. Romeo et al., *Proceedings of Third World Conference on Photovoltaic Energy Conversion*, vol. 1 (2003), pp. 469–472
31. C. Ferekids et al., Proc. AIP Conf. **353**, 39–44 (1995)
32. A. Romeo et al., Sol. Energy Mater. Sol. Cells **67**, 311–316 (2001)
33. J. Barker et al., Int J. Sol. Energy **12**, 79–94 (1992)
34. C.S. Ferekides et al., Sol. Energy **77**, 823–827 (2004)
35. Y.J. Chang et al., Surf. Interface Anal. **37**, 398–405 (2005)
36. P.J. George et al., Appl. Phys. Lett. **66**, 3624–3626 (1995)
37. C.J. Santos et al., Thin Solid Films **518**, 1791–1795 (2010)
38. H. Matsumoto et al., Jan. J. Appl. Phys. **22**, 1832–1836 (1983)
39. D.H. Rose, *The 25th Photovoltaic Specialists Conference* (1996), pp. 777–780
40. T.L. Chu et al., Appl. Phys **58**, 1349–1355 (1985)
41. Y.K. Jun et al., J. Electrochem. Soc. **135**, 1658–1661 (1988)
42. S.P. Albright et al., IEEE Trans. Electron Devices **37**, 434–437 (1990)
43. T.L. Chu et al., J. Appl. Phys. **71**, 3870–3876 (1992)
44. B.M. Basol et al., J. Appl. Phys. **58**, 3809–3813 (1985)
45. M. Panicker et al., J. Electrochem. Soc. **125**, 566–572 (1978)
46. G.C. Morris et al., *23rd IEEE Photovoltaic Specialists Conference* (1993), pp. 469–472
47. W. Song et al., *26th Photovoltaic Specialists Conference* (1996), pp. 873–876
48. B.E. Mccandless et al., Prog. Photovoltaics Res. Appl. **5**, 249–260 (1997)
49. P.D. Paulson et al., Thin Solid Films **37**, 299–303 (2000)
50. H. Moutinho et al., J. Vac. Sci. Technol. **16**, 1251–1255 (1998)
51. U. Jahn et al., J. Appl. Phys. **90**, 2553–2558 (2001)
52. A. Bosio, Prog. Cryst. Growth Charact. Mater. **52**, 247–279 (2006)
53. N. Romeo et al., *Proceedings of 3rd World Conference on Photovoltaic Energy Conversion* (2003), pp. 469–470
54. H. Kim et al., J. Appl. Phys. **25**, 2673–2679 (1993)
55. A.L. Fahrenbruch, Solar Cells **21**, 399–412 (1987)
56. E. Janik et al., J. Phys. D: Appl. Phys **16**, 2333–2337 (1983)
57. G. Asa et al., J. Appl. Phys. **77**, 4417–4424 (1995)
58. T.L. Chu et al., *20th IEEE Photovoltaic Specialists Conference* (1988), p. 1422
59. X. Wu, Sol. Energy **77**, 803–814 (2004)

60. J. Britt et al., Appl. Phy. Lett. **62**, 285–291 (1983)
61. B.E. McCandless et al., *IEEE First World Conference Photovoltaic Energy Conference* (1994), p. 107
62. V. Viswanathan, RF sputtered back contacts for CdTe/CdS thin film solar cells. Master thesis, University of South Florida, 1997
63. J. Tang, Study of back contact formation on CdTe/CdS thin film solar cells. Ph.D. thesis, Colorado School of Mines, 2000
64. T.A. Gessert et al., J. Vac. Sci. Technol. **14**, 806–812 (1996)
65. P.V. Mayers, Polycrystalline cadmium telluride n-i-p solar cell. Final report to SERI under subcontract no. ZL-7-06031-2 (1990)
66. P.V. Meyers, Polycrystalline thin-film, Cadmium Telluride solar cells fabricated by electrodeposition cells. Final report to SERI under subcontract no. ZL-7-06031-2 (1988)
67. T.A. Gessert et al., *26th IEEE Photovoltaic Specialists Conference* (1997), p. 419
68. J. Tang et al., *25th IEEE Photovoltaic Specialists Conference* (1996), p. 925
69. B. Ghosh et al., Semicond. Sci. Tech. **10**, 71–76 (1995)
70. R.W. Miles et al., J. Cry. Growth **161**, 148 (1996)
71. M. Herndon et al., Appl. Phys. Lett **75**, 3503 (1999)
72. J. Sites et al., Sol. Cells **27**, 411–417 (1989)
73. S. Kurtz et al., *23rd IEEE Photovoltaic Specialist Conference* (1990), pp. 138–140
74. G. Stollwerck, et al., *Conference Record 13th European Photovoltaic Solar Energy Conversion* (1995), pp. 2020–2022
75. A. Niemegeers et al., J. Appl. Phys. **81**, 1886–2881 (1997)
76. B. McCandless et al., *Conference Record 2nd WCPVSEC* (1998), pp. 448–452
77. A. Fahrenbruch, Sol. Cells **21**, 399–412 (1987)
78. S. Hegedus, *28th IEEE Photovoltaic Specialist Conference* (2000), pp. 535–538
79. R. Hulstrom et al., Sol. Cell **15**, 365 (1985)
80. P. Mauk et al., IEEE Trans. Electron Dev. **37**, 1065–1068 (1990)
81. X. Liu et al., J. Appl. Phys. **75**, 577–581 (1995)

Chapter 8
Dye-Sensitized Solar Cell

8.1 Introduction

The major components of the dye-sensitized solar cell (DSSC) consist of a dye, a semiconductor electrode, an electrolyte layer, as well as a counter electrode layer. The dye molecules in the dye-sensitized solar cell are adsorbed on the nanostructured anode as a photosensitive layer, and the photosensitive layer is capable of fully absorbing sunlight and works like chlorophyll, which are different from those of the solar photovoltaic cell based on the semiconductor PN junction introduced earlier. The electrons generated by the dye molecules after light absorption can be efficiently harvested by the nanostructured electrodes and then transmitted to the external circuit. The electrolyte layer facilitates electrons conduction between the photosensitive layer and the counter electrode layer, so the reaction can be performed smoothly.

Since DSSC technology complies with an overall different structure and photoelectric conversion principle from that of semiconductor PN junction solar photovoltaic cells, its raw material cost and the manufacturing process cost are significantly down-regulated, only one-tenth or lower of silicon cells. Besides, it has a relatively simple process, low energy consumption, low pollution, as well as environmental-friendly characteristic. Moreover, dye-sensitized cells exhibit other advantages (e.g., low requirements for lighting conditions and availability even in indoor spaces with insufficient sunlight). Furthermore, DSSC performance exhibits extreme insensitivity to temperature variations. For instance, in natural sunlight, the cell temperature can easily reach 60 °C. At the temperature increasing from 20 to 60 °C, DSSC exhibits basically unchanged photoelectric conversion efficiency, whereas the conventional silicon cell will drop by 20%. Moreover, if a flexible substrate (e.g., plastic or metal plate) is applied instead of glass, a flexible cell can be generated. The flexible fiber dye-sensitized solar photovoltaic cells reported over the past few years have further expanded the flexible space of the cell as impacted by the use of the fibrous substrate. Meantime, since the mentioned cells exhibit low-cost, non-toxicity, and environmental protection, they have been progressively practical. Moreover, when the cell acts as a display, it is capable of emitting light while generating electricity,

C. Zhang et al., *Semiconductor Photovoltaic Cells*,
https://doi.org/10.1007/978-981-15-9480-9_8

so energy self-sufficiency can be achieved. For the mentioned characteristics, dye-sensitized solar cells exhibit promising and broad commercial applications and potential competitiveness, and they have become one of the research hotspots in the field of solar cell technology.

The discovery of the dye photovoltaic effect dated back to the nineteenth century, which was in 1839, half a century after the photovoltaic phenomenon of electrodes in solution was discovered by Becquerel [1]. Dr. Moser from the Physical Chemistry Laboratory of the Vienna University initially reported the photoelectric effect of dye sensitization [2]. This result was efficiently applied by scientists studying photography, and eventually, color photography was achieved. Until a century later, people have begun to explore the exploitation of dye-sensitized photoelectric effect in solar energy conversion. In the 1960s, people began to delve into the photoelectric effect of single crystal semiconductors in dye solutions. The research studies showed that only dye molecules directly adsorbed on the semiconductor surface can exert a photovoltaic effect. The reason is that a thick film can hinder the transfer of electrons from the excited state of dyes to the semiconductor and thus a very thin film, such as a single layer of molecules densely packed on the semiconductor surface, can maximally facilitates photoelectric generation. However, since the single-layer molecular dye on the single crystal semiconductor surface exhibits significantly low light absorption efficiency, such photovoltaic device has remarkably low conversion efficiency (less than 0.5%) and poor light stability. The low photoelectric conversion efficiency of dye-sensitized single crystal semiconductors has been troubling and limiting the application of dye-sensitized semiconductors in solar energy conversion. The first breakthrough in this field was made in 1976, when Tshubomura et al. substituted single crystal semiconductors with porous polycrystalline ZnO, and the dye sensitizer was Rose Bengal [3]. Compared with the single crystal, the surface area of the polycrystalline ZnO film has been evidently expanded, and the absorption of light by monolayer dyes adsorbed on the surface has been significantly promoted; thus, its photoelectric conversion efficiency reaches 1.5%, an order of magnitude higher than that of electrodes based on single crystal ZnO. Moreover, according to Tshubomura et al., I^-/I_3^- outperforms other redox electrolytes in dye-sensitized solar cells. In the mid-1980s, the Gralzel research group had acted as a major force in the study on dye-sensitized solar cells. They have generated inexpensive and safe TiO_2 nanocrystalline film electrodes on conductive glass. The electrode exhibits a significantly large specific surface area and a roughness reaching 1000. If the dye is adsorbed on the film surface, the incident light can be absorbed efficiently. In 1991, new dye-sensitized solar cells made of TiO_2 by Professor Graetzel achieved the conversion efficiency of 8% [4]. In 1997, the Graetzel team adopted optimized Ru dyes to enhance the conversion efficiency to 10% [5]. Since the breakthrough of the DSSC was made in the Swiss Federal Institute of Technology, DSSC research has aroused rising attention. When the patent announcement came into effect, seven companies including those in Australia, Switzerland, and Germany purchased the right to use the patent and invested manpower and material resources for practical and industrial studies. From 1992 to 1999, the research institute represented by the German Photovoltaic Research Institute (INAP) and the Australian STA Company had conducted an exploratory study in the preliminary stage of industrialization. To be specific, the Australian STA company established a unique DSSC display roof with an area of 200 m^2 in October 2002, indicating the prospect of future industrialization. Swiss EPFL, European Union ECN Research Institute, Sharp Corporation of Japan, and Arakawa of Tokyo University of Science and Technology have achieved the efficiency of the DSSC comparable to that of small cells on stripe cells larger than 1 cm^2. Sharp Corporation of Japan, reported the photoelectric conversion efficiency of 6.13% achieved by the DSSC module.

The plastic color cell based on dihydroindole organic dye-sensitized electrodeposited nano-zinc oxide films developed by Gifu University in Japan exhibited an efficiency of 5.16% [6]. Fully flexible DSSC based on low-temperature TiO_2 electrode preparation technology developed by the Tongyin University of Yokohama, Japan, exhibited an efficiency of over 6% [7]. Japan's Peccell Technologies Company successfully developed a polymer flexible substrate-based DSSC exhibiting an output voltage of 4 V or more, an output current of 0.11 A or more, as well as an efficiency of 4.13–5.12%. Swiss Leclanche S1A (Swiss), Solaronix (Swiss), and Greatcell have been developing practical indoor and outdoor products. The US Konarka High-Tech Company invested millions of dollars in 2002, for practical and industrial studies on dye cells with flexible films (e.g., transparent conductive polymers as substrates and electrodes). The research complies with the requirements of the US military for flexible and mobile energy conversion applications and has been sufficiently supported by the US military. The US GE Company also developed and implemented the DSSC project in 2000, and organized a research team at a certain scale. As suggested from the results of the domestic and international authorized patent enquiries, the number of DSSC invention patents currently authorized worldwide has reached 1,659, primarily distributed in Japan's Sharp, Fuji, Hitachi, South Korea's Samsung, and some European and American companies, covering engineering, Chemistry, and other fields. However, as impacted by the internal resistance of the cell and the interface between materials, the efficiency of DSSC currently reaches only 5–6%.

8.2 DSSC Device Structure

The structure of the dye-sensitized solar photovoltaic cell primarily covers a conductive base material (e.g., transparent conductive electrode), a nanoporous semiconductor film, a dye photosensitizer, an electrolyte, as well as a counter electrode. A nanoporous semiconductor film is prepared on a transparent conductive substrate. Subsequently, the dye molecules are adsorbed on the porous film surface; thus, a working electrode is formed, commonly termed as a photoanode. Since the photoanode outputs electrons, from the perspective of the power source, the photoanode refers to the negative electrode of the power source, and the counter electrode pertains to the anode electrode of the power source. On the whole, the counter electrode is a conductive film coated with a layer of platinum. Indeed, carbon or other metals can be exploited instead of platinum, whereas platinum exhibits the optimal conversion effect of the cell.

8.2.1 Conductive Substrate Material

Overall, transparent conductive glass has been adopted as the conductive electrode. TCO refers to a SnO_2 transparent conductive thin film doped with F (or Sb) on ordinary glass. The transparent conductive film may also be indium tin oxide (ITO). On the whole, the square resistance of the TCO should be 5 Ω/□, and the light transmittance is above 85%. TCO can be used to prepare a substrate for a photoanode and a photocathode, collect and transport electrons from the photoanode, and transmit electrons to the photocathode via a loop and then supply them to electron acceptors in the electrolyte.

8.2.2 Nanoporous Semiconductor Film

The photoanode formed by the nanocrystalline thin film, the nano mesoporous thin film, the nanocomposite thin film, and the nanowire structure takes up a major part of the DSSC. Besides TiO_2, suitable semiconductor materials for photoanode consist of ZnO, Nb_2O_5, WO_3, Ta_2O_5, CdS, Fe_2O_3, SnO_2, etc. To be specific, ZnO is employed in dye-sensitized solar cells for its abundant source, low-cost and simple preparation. Over the past few years, great progress has been achieved in the application of ZnO in flexible dye-sensitized solar cells.

Methods to prepare semiconductor thin films consist of chemical vapor deposition, powder sintering, hydrothermal reaction, RF sputtering, plasma spraying, screen printing, as well as colloidal coating. The most frequently used existing method of preparing nano-TiO_2 porous films is the sol–gel method. Overall, nano-semiconductor films to fabricate dye-sensitized solar cells should exhibit the following significant characteristics:

(1) It should have a large specific surface area, enabling it to effectively adsorb monolayer dyes and more effectively exploit sunlight;
(2) There should be close electrical contact between the nanoparticles and the conductive substrate and the nano-semiconductor particles, so carriers can be efficiently transported inside to ensure the conductivity of the large-area film.
(3) Redox couples in the electrolyte (generally I_3^-/I^-) can penetrate into the nano-semiconductor film, so the oxidation state dye can be effectively regenerated.

8.2.3 Dye Photosensitizer

Dye molecules are critical to light absorption of dye-sensitized solar cells. Overall, dye photosensitizers for dye-sensitized solar cells should satisfy the following conditions:

(1) It exhibits significant absorption characteristics for visible light, i.e., it is capable of absorbing most or all of the incident light, and its absorption spectrum should comply with the emission spectrum maximally.
(2) It can be closely adsorbed in the TiO_2 surface, i.e., it can efficiently reach the adsorption balance and is difficult to fall off. The dye molecules should generally contain the molecule groups (e.g., –COOH, $–SO_3H$ and $–PO_3H$) that are easy to binding on the surface of the nano-semiconductor. As indicated from the existing studies (taking carboxylic acid bipyridyl dye as an example), the carboxyl group on the dye is combined with the hydroxyl group on the TiO_2 film to form ester; as a result, the electron coupling between the 3d orbital of the TiO_2 conduction band and the n-orbital of the dye is enhanced to facilitate electron transfer.
(3) The oxidation state and excited state exhibit higher stability and activity, the excited state lifetime is sufficiently long, and it exhibits high charge transfer efficiency.
(4) It shows an adequately negative excited state redox potential to ensure that the dye-excited electrons are injected into TiO_2 conduction band, and a relatively low barrier is required in the redox process (e.g., the ground and the excited states), so the loss of free energy during the primary and secondary electron transfers is minimized.

After more than 20 years of research, it has been demonstrated that the polypyridyl complexes (e.g., porphyrin and Group VIII Os and Ru) can effectively satisfy the mentioned requirements, and the latter exhibits the optimal photosensitivity of the polypyridine complex.

8.2.4 Electrolytes

The electrolyte system primarily aims to restore dye and transport charge. By altering the energy levels of the TiO_2 photoanode, sensitizing dye and redox couple, the mechanical behavior of the carrier transport in DSSC is optimized to achieve the maximal photogenerated voltage. The redox pair extensively employed in liquid electrolytes has been I^-/I_3^-. Selecting a suitable electrolyte to enable I^-/I_3^- to efficiently transport electric charges is critical to enhance the performance of cells.

8.2.5 Counter Electrode

The counter electrode, also termed as the photocathode, consists of conductive glass coated with a platinum catalyst. It aims to collect electrons transferred from the photoanode via the outer loop and then transfer the electrons to the electron acceptor in the electrolyte to restore and complete the closed circuit. Besides collecting electrons for the counter electrode, the platinum catalyst can also promote charge transfer on the electrode/electrolyte interface to be efficient, down-regulate the probability of recombination of electrons in the I_3^- and TiO_2 conduction band, inhibit dark current, and up-regulate the open-circuit voltage of the cell. Besides platinum, gold, nickel, carbon, certain conductive polymers, and inorganic oxides can also act as catalysts for DSSC. Since the material, surface condition, preparation method, and substrate material of the catalyst significantly impact the catalytic performance of the counter electrode, the preparation of the high catalytic activity of the counter electrode has become a hotspot of subsequent DSSC research.

8.3 The Working Principle of DSSC Device

The dye-sensitized solar cell is primarily a sandwich structure composed of a photoanode, a liquid electrolyte and a photocathode (Fig. 8.1). Its working principle is significantly similar to the photosynthesis of photosynthetic membrane in the chloroplast. Figure 8.2 presents a schematic diagram of photosynthesis of the photo-

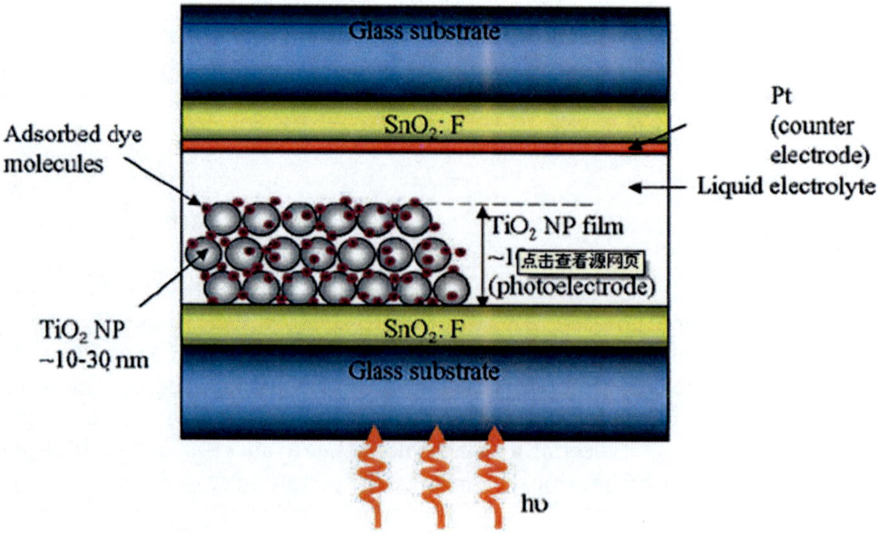

Fig. 8.1 Schematic diagram of the structure of the dye-sensitized solar cell

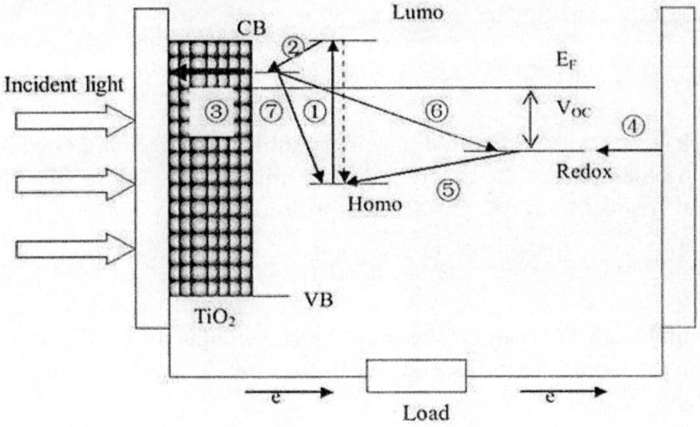

Fig. 8.2 Schematic diagram of the working principle of the dye-sensitized solar cells

synthetic membrane. As a matter of fact, the result of photons on the photosynthetic membrane is an electric field inside and outside the photosynthetic membrane. Electrons are transferred from the inside to the outside of the photosynthetic membrane, leaving holes in the membrane. As impacted by photons, the electrons move to form internal and external currents.

The processes of DSSC generating photocurrent are elucidated as follows:

(1) The dye molecules transition from the ground state (D) to the excited state (D*) after sunlight is absorbed, electrons transition from the highest occupied molecular orbital (HOMO) to the lowest unoccupied molecular orbital (LUMO) is expressed as

$$D + h^* \rightarrow D^*$$

(2) The dye molecules in the excited state inject electrons into the conduction band of the nano titanium oxide semiconductor (electron injection rate constant expressed as k_{inj}):

$$D^* \rightarrow D^+ + e^-(CB)$$

(3) The electrons in the conduction band (CB) are transmitted to the back contact (BC) in the nanocrystal network and subsequently flow into the external circuit

$$e^-(CB) \rightarrow e^-(BC)$$

(4) Electronics and I_3^- are combined to form I^-, which is equivalent to electrons entering the electrolyte

$$I_3^- + 2e^-(CE) \rightarrow 3I^-$$

(5) I^- reduction oxidation to regenerate dyes

$$3I^- + 2D^+ \rightarrow I_3^- + D.$$

(6) The electrons transported in the nanocrystalline film recombine with I_3^- ions entering the holes of the titanium dioxide film on the surface (the major dark current channel, the rate constant is expressed as k_{et})

$$I_3^- + 2e^-(CB) \rightarrow 3I^-.$$

(7) Recombination between electrons in the semiconductor conduction band and dyes in the oxidized state (secondary dark current channel, electron return rate constant is k_b)

$$D^+ + e^-(CB) \rightarrow D.$$

The longer the lifetime of the excited state of the dye, the more favorable the injection of electrons will be. Under the overly short electron lifetime of the excited state, the molecules in the excited state may not be capable of injecting electrons into the conduction band of the semiconductor; they have already transitioned to the ground state via non-radiative attenuation. Steps (2) and (7) are the vital steps to determine the efficiency of electron injection. The larger the ratio of the electron injection rate constant (k_{inj}) to the reverse reaction rate constant (k_b) (largely greater than 3 orders of magnitude), the smaller the probability of charge recombination and the higher the efficiency of electron injection will be. I^- ion reduced oxidation state dye regenerates the dye, so the dye is capable of continuously injecting electrons into the conduction band of the titanium dioxide. The greater the rate at which the I^- ions reduce the oxidation state, the greater the degree to which electron return is inhibited will be, which is equivalent to the I^- ions intercepting the return of electrons. Step (6) refers to the major cause of the current loss. With the increase in the transmission speed of electrons in the nanocrystalline network (step (3)), and the decrease in the recombination rate constant ket between electrons and I_3^- ions, the current loss will become smaller and the photogenerated current will become bigger. The I_3^- ion generated in step (5), diffuses to the counter electrode to obtain electrons and then becomes I^- ions (step (4)), so the I^- ions are regenerated and the current cycle is completed.

In a conventional semiconductor solar cell, the semiconductor should fulfill simultaneously two purposes: to capture incident light, as well as is to transmit photogenerated carriers. The capture of light is performed with a sensitizer (i.e., a dye) of light. After being excited by light, the dye molecules are transited from a ground state to an excited state and then the electrons and the holes are separated. For the excited state energy level of the dye molecule higher than the conduction band bottom energy level of the semiconductor, and under the consistent energy levels, the dye molecules in the excited state will inject electrons into the semiconductor conduction band. The electrons injected into the conduction band are transported rapidly in the

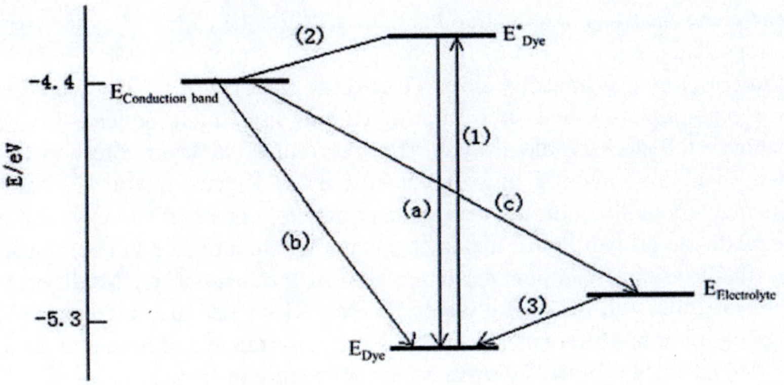

Fig. 8.3 Kinetics process of carrier transport in dye-sensitized cells

semiconductor film, which can reach the back contact of the film and conductive glass in an instant (<1 ms) and enter the external circuit. Besides load sensitizers, another major function of semiconductors refers to the collection and transport of electrons.

The principle of charge transfer is elucidated as follows:

Figure 8.3 shows the kinetics of carrier transport in DSSC. This refers to a kinetic equilibrium process that involves the excitation, transport, and recombination of carriers. It is primarily composed of the competition process between the electron loss reaction (a), (b), (c) and the desired reaction (1), (2), (3) in the solar cell. The specific analysis is presented below:

(1) The excited dye molecule is capable of directly relaxing to its ground state, causing a loss reaction (a). Compared with the electron injection reaction (2), the loss is negligible, primarily because of the extreme difference between the corresponding reaction rate constant k_2 and k_a ($k_2/k_a = 1000$). In such a process, the direct bonding of dye molecules to the surface of TiO_2 nanocrystals acts as the vital factor to satisfy $k_2/k_a = 1000$.
(2) The electrons in the conduction band may be trapped by the oxidized dye molecules (the reaction rate constant expressed as k_b), and such process leads to a loss reaction (b). Since such process competes with the I^- reducing dye molecules in the electrolyte, and the reaction rate constant of the latter is nearly 100 times that of the former, i.e., $k_3/k_b = 100$, the loss is small as well. In the process, the high-efficiency transmission of I^- and I_3^- in nanocrystalline porous membranes is critical to ensure that sufficient I^- participates in the competition, so the microstructure of the nanocrystalline porous membrane should be optimized. In some quasi-solid electrolyte systems, the diffusion coefficient of $I^-$and I_3^- is smaller than those of the liquid, so the recombination loss is significant.

(3) The electrons in the conduction band are likely to be captured by the oxidizing components (e.g., I_3^-) in the electrolyte. Such a process leads to a loss reaction (c), which primarily causes electron loss in DSSCs. To lower the electron recombination loss, an insulating coating layer is introduced between the nanocrystals and the electrolyte. The dye molecular layer refers to an insulating isolating layer. A single layer of dye molecules covering the surface of the nanocrystalline porous membrane completely is an effective way to down-regulate the probability of the electrons in the conduction band being captured by the oxidizing components (e.g., I_3^-) in the electrolyte. Modifying TiO_2 nanocrystals with insulating materials e.g., Al_2O_3 can also effectively reduce recombination. Moreover, the spectral response range and high quantum yield of dye molecules critically impact photon capture in DSSC.

To delve into the carrier transport in DSSC, Ni et al. adopted a Schottky barrier model, an electron diffusion differential model and a thermal ionization emission principle to theoretically analyze the effect of the TiO_2/TCO interface on the I-V characteristics of the DSSC. Figure 8.4 illustrates the band form of the DSSC. To be specific, Fig. 8.4a, gives the potential of TiO_2, TCO, and the counter electrode under dark conductivity, and such potential is equated with the redox potential of the electrolyte. Figure 8.4b presents the quasi-Fermi level of TiO_2 and TCO under illumination. As revealed from the figure, V_1 is the voltage loss at the TiO_2/TCO interface, V_2 is the voltage loss at the electrode/electrolyte interface, and V_0 denotes the difference between the TiO_2 Fermi level (E_f) and the redox potential of the electrolyte (E_{redox}). Obviously, the reference voltage V is expressed as

$$V = V_0 - V_1 \tag{8.1}$$

Since the TiO_2/electrolyte contact is in series with the TiO_2/TCO contact, the current density J flowing outside the porous TiO_2 film is equated with the current density flowing via the TiO_2/TCO contact. Thus, the relationships between V and

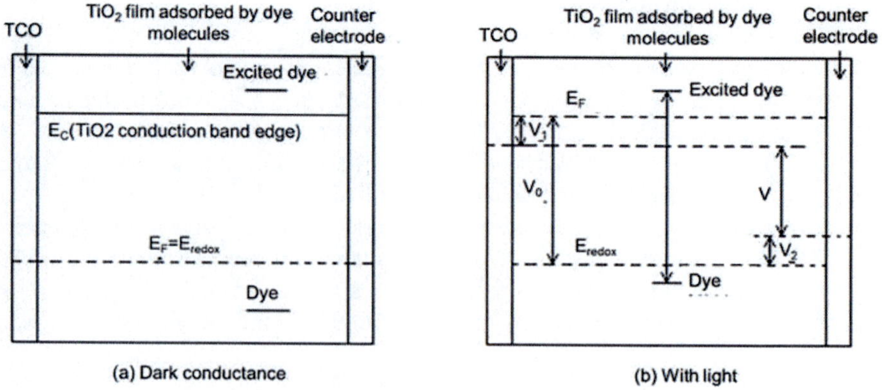

Fig. 8.4 Energy band of dye-sensitized cells under dark conductance and illumination

J can be characterized by solving V_0 and V_1. The potential difference V_0 can be obtained by solving a simple dynamic model presented by Sodergren et al. Under steady-state conditions, the electron generation, transport, and recombination in DSSC can be expressed with electron diffusion differential equations

$$D\frac{\partial^2 n(x)}{\partial x^2} - \frac{n(x) - n_0}{\tau} + \varphi\alpha e^{-\alpha x} = 0 \quad (8.2)$$

where x denotes the coordinate measured from the TiO_2/TCO interface to the inside of the electrode film; $n(x)$ represents the excess electron concentration at x; n_0 is the electron concentration at dark conductance; τ is the free electron lifetime of the conduction band; ϕ is the incident light flux; α is the light absorption coefficient of the nanoporous electrode; D represents the diffusion coefficient.

Under short-circuit conditions, electrons are easily extracted as photocurrents, and no electrons are directly transported to the counter electrode. Accordingly, the following boundary conditions are yielded

$$n(0) = n_0 \quad (8.3)$$

$$\left(\frac{dn}{dx}\right)_{x=d} = 0 \quad (8.4)$$

Thus, the short-circuit current density is written as

$$J_{sc} = \frac{q\varphi L\alpha[-L\alpha\cosh(d/L) + \sinh(d/L) + L\alpha\exp(-d\alpha)]}{(1 - L^2\alpha^2)\cosh(d/L)} \quad (8.5)$$

where q denotes the electron charge; L is the electron diffusion length; d is the film thickness.

Under illumination, the electron density at $x = 0$ will rise from n_0 to n, which is expressed as

$$n(0) = n \quad (8.6)$$

In such a way, the relationship between V_0 and J can be characterized by solving the differential equation

$$V_0 = \frac{kTm}{q}\ln[\frac{(J_{sc} - J)L\cosh(d/L)}{qDn_0\sinh(d/L)} + 1] \quad (8.7)$$

where k denotes the Boltzmann constant; m is the ideal factor; T is the thermodynamic temperature.

Since TCO is highly doped, it exhibits high electrical conductivity and can be considered metal. The TiO_2/TCO contact can be simulated by a Schottky barrier. Under light conditions, the electron at the TiO_2/TCO interface will cause a voltage loss V_1. When excessive electrons are injected from the excitation dye inside the TiO_2 valence band, electron recombination will not take place in TiO_2 and TCO; thus, only electrons are transferred from TiO_2 to TCO. In accordance with metal–semiconductor contact theory, thermal electron emission and electron tunneling will act as two transport mechanisms for electron transfer at the interface. Since TiO_2 in DSSC is significantly doped, and the cell operates at high-temperature (300 K), the electron tunneling process can be overlooked, and the electron transfer will be dominated by the thermionic emission process.

The current density J at the TCO/electrolyte interface is expressed as

$$J = A^{*}T^{2}\exp(-\frac{q\varphi_b}{kT})[\exp(\frac{qV_1}{kT}) - 1] \tag{8.8}$$

$$A^{*} = \frac{4\pi m^{*} q K^{2}}{h^{3}} \tag{8.9}$$

where φ_b denotes the Schottky barrier height; h is the Planck constant; m^* is the effective quality of the electron; A^* is Richardson constant of TiO_2. From the Eq. (8.8), V_1 is expressed as

$$V_1 = \frac{kT}{q}\ln[\frac{J}{A^{*}T^{2}\exp(-q\varphi_b/kT)}] \tag{8.10}$$

Substituting Eqs. (8.7) and (8.10), into Eq. (8.1), there is

$$V = \frac{kTm}{q}\ln[\frac{(J_{sc} - J)L\cosh(d/L)}{qDn_0\sinh(d/L)} + 1] - \frac{kT}{q}\ln[1 + \frac{J}{A^{*}T^{2}\exp(-q\varphi_b/kT)}] \tag{8.11}$$

As can be seen from Eq. (8.11), the effect of the TiO_2/TCO interface on DSSC photovoltaic performance is primarily reflected in the Schottky barrier height φ_b, temperature T, and the recombination current J at the TCO/electrolyte interface. Furthermore, the Schottky barrier height φ_b at the TiO_2/TCO interface will impact V_1; V_1 will increase with φ_b and directly affect the open-circuit voltage V_{OC}. Moreover, the value of φ_b also affects the size of J, i.e., with a decrease in φ_b, J increases. Figure 8.5 presents the effect of TiO_2/TCO Schottky barrier height $_b$ on the maximum output power P_{max} of DSSC. It is suggested that when $\varphi_b < 0.6$ eV, the value of P_{max} is a constant; when $\varphi_b > 0.6$ eV, the value of P_{max} decreases linearly with the increase in φ_b.

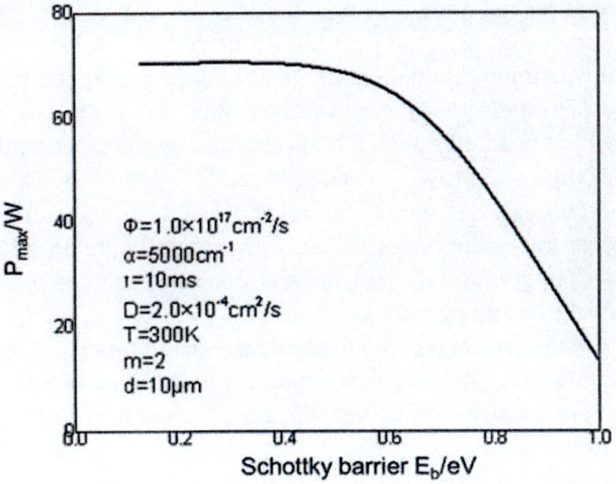

Fig. 8.5 Numerical simulation of the effect of the TiO_2/TCO Schottky barrier on the maximum output power of the DSSC [7]

8.4 Device Manufacturing Process

8.4.1 Preparation of Titanium Dioxide Nanocrystalline Film Electrode

The conventional method to prepare nano TiO_2 film is to adopt sol–gel method: TiO_2 sol is prepared by hydrolysis of titanate compound as the precursor, and TiO_2 slurry is prepared by high pressure heat treatment, evaporation to remove the solvent, and the surfactant is added to grind, or commercial grade nano-TiO_2 powder (P25, Degussa) is added with surfactant and appropriate solvent to produce TiO_2 slurry. Subsequently, TiO_2 is deposited on a conductive substrate by screen printing, direct coating or spin coating, and the nano-sized TiO_2 porous electrode is prepared by high-temperature sintering. Besides the mentioned methods, nano-TiO_2 film can be prepared with a variety of methods (e.g., chemical vapor deposition, electrodeposition, magnetron sputtering, and plasma spraying on a conductive glass or other conductive substrate material), and subsequently it undergoes high-temperature sintering at 450–500 °C to remove the surfactant [8–11].

The sol–gel method refers to preparing TiO_2 ultra-fine colloid solution by hydrolyzing butyl titanate (or inorganic salt titanium source, e.g., $TiCl_4$), and subsequently depositing TiO_2 on a conductive substrate based on dip-coating, screen printing, direct coating, spin coating, etc. However, the TiO_2 film at this time almost acts as an insulator, so a nano-TiO_2 electrode should be prepared by sintering at 450 °C to form a close electrical contact between the particles. The advantage of such method is that the sol is stable, uniform, small particle, and easy to be doped; it can make a variety of composites exhibit uniform and adjustable composition distribution, thus improving the performance of titanium dioxide. However, to achieve a suitable thickness of the titanium dioxide film, multiple sintering is required, which requires significant time and effort. Moreover, the process parameters are difficult to control (e.g., composition of the sol, pH, solvent, additives, sol aging time, film formation method, heat treatment temperature, and time of the film).

The hydrothermal synthesis method is an optimized method of the sol–gel method, which primarily adds a hydrothermal aging process, thereby regulating the crystallization and growth of the product, the particle size and distribution of the semiconductor oxide, as well as the void ratio of the film. Hydrothermal treatment temperature determines the crystal of TiO_2 and critically impacts the particle size. With the rise in the temperature of the hydrothermal treatment, the particle size increases, and some rutile will be prepared.

The powder coating method refers to adding a suitable surfactant and solvent to the nano-TiO_2 powder, ultrasonically dispersing and fully grinding the TiO_2 slurry, and subsequently depositing TiO_2 on the conductive substrate based on screen printing, direct coating, spin coating, etc. Lastly, nano-TiO_2 electrode is prepared by high-temperature sintering activation. The type, amount, grinding time, and heat treatment of the surfactant significantly impact the performance of TiO_2 film.

Electrochemical deposition methods consist of anodic deposition and cathodic deposition. Nano TiO_2 porous film obtained with the mentioned two methods has a robust adhesion. Anodic deposition method refers to cleaning conductive glass with acetone, anhydrous ethanol, and twice deionized water to generate a blank conductive glass electrode, and adopting freshly prepared $TiCl_3$ as the electrolyte for potentiostatic electrolysis to form tetravalent titanium hydration film on the conductive glass electrode. The hydrated thin film is dried in an infrared lamp or at ambient temperature and subsequently placed in a muffle furnace. After a regulated temperature heat treatment, a TiO_2 porous film is formed. The reaction process is explained as follows:

$$\begin{array}{ll} Ti^{3+}(aq) + H_2O \rightarrow TiOH^{2+} + H^+ & \text{Fast reaction} \\ TiOH^{2+} - e \rightarrow Ti(IV)_{polymers} & \text{Slow reaction} \\ Ti(IV)_{polymers} - H_2O \rightarrow TiO_2 & \text{Baking at 450 °C for 1 h} \end{array}$$

Cathode deposition method directly adopts $TiOSO_4$ as raw material and prepares nano-TiO_2 porous membrane based on cathode electrodeposition.

8.4.2 *Filling of Dyes in TiO_2 Nanofilms*

After the TiO_2 nanoporous film is prepared, the N3 dye photosensitizer is adsorbed on the surface of the TiO_2 film. The film is immersed in the dye solution and left at an ambient temperature for 12–18 h to ensure that the TiO_2 film fully adsorbs the dye. Before being utilized, the film should also be washed with alcohol or acetonitrile to remove incompletely adsorbed dyes in the pores of the TiO_2 film.

8.4.3 Preparation of Electrolyte

The electrolyte taken in DSSC is commonly an organic solution containing iodine redox ions, of which the most representative is a relatively viscous nitrile organic solution (e.g., acetonitrile solution, propionitrile solution, methoxy acetonitrile solution, and methoxy propionitrile solution), exhibiting relatively high ionic conductivity. As revealed from the existing studies, imide derivatives (e.g., 1,2-dimethyl-3-hexylimidazolium iodide (DMHImI) and 1,2-dimethyl-3-propylimidazolium iodide (DMPImI)) can lower the resistance of electrolyte solution and enhance the photovoltaic performance of the device. In 1997, the Graetzel team fabricated a composite photosensitizer prepared by dissolving 0.5 M DMHImI, 0.04 M LiI, 0.02 M I_2, and 0.5 M tert-butylpyridine (TBP) in acetonitrile solution, significantly enhancing the performance of the device; the conversion efficiency reached 10% [5]. Then, TBP can reduce the conduction band of the TiO_2 electrode, thereby down-regulating the dark current of the device and increasing the open-circuit voltage.

8.4.4 Preparation of the Counter Electrode

On the whole, a sputtering method is adopted to deposit 200 nm-thick metal platinum on a TCO substrate to form a counter electrode. When Pt is sputtered onto the substrate, a specular effect is exerted, reflecting the light multiple times to increase the photocurrent. Moreover, when platinum is sputtered onto the TCO, it will deposit on the surface as platinum colloid and then further promote the electrocatalytic effect for reducing I_3^- particles. If a small amount of H_2PtCl_6 is adsorbed on the TCO substrate after sputtering the alcohol solution, heat treated at 385 °C for 10 min will continuously contribute to the formation of Pt colloid. The performance of the Pt on the electrode will directly impact the fill factor of the solar cell.

8.5 Progress of DSSC Devices

8.5.1 Working Electrode

As the dye molecule adsorption carrier, electron acceptor and electron transport layer, the working electrode critically impacts the performance of the dye-sensitized solar cell, and thus has aroused the attention of numerous researchers. As impacted by the small number of dye molecules that can be adsorbed on the surface of a single-layer semiconductor, people cannot enhance quantum and light-trapping efficiency simultaneously, limiting the development of dye-sensitized solar cells. This problem was addressed in 1991, with the first sensitization of Swiss scientists Gratzel using high-surface-area semiconductor electrodes (e.g., titanium dioxide nanocrystal electrodes) [4]. On the one hand, the TiO_2 nanoporous membrane is capable of absorbing more dye molecules; on the other hand, the mutual reflection between the crystal grains inside the film enhances the absorption of sunlight. Accordingly, dye-sensitized TiO_2 nanoporous semiconductor electrode can ensure high photoelectric conversion quantum efficiency and high light capturing efficiency simultaneously. From this perspective, the application of nanoporous TiO_2 has made the study on solar cells usher into a brand-new era, which has significantly boosted the development of photoelectrochemical cells. Just as some people compare dye-sensitized solar cells to leaves of plants, nano-TiO_2 electrode refers to a solar-driven molecular electronic pump converting solar energy into electrical energy. The leaves of plants convert solar energy into chemical energy. From this perspective, dye-sensitized solar cells exhibit the significance of biomimetic research.

Nano-TiO_2 electrodes are critical to dye-sensitized solar cells, and their performance directly affects the efficiency of the cell. The microstructure of nano-TiO_2 (e.g., particle size and porosity) noticeably impacts the photoelectric conversion efficiency of solar cells. The particle size is overly large, and the amount of dye adsorption is small, which adversely affects photoelectric conversion. However, the particle size is overly small, the interface is overly much, the grain boundary barrier hinders the transport of carriers, and the carrier mobility is low, which also limits photoelectric conversion. Over the past few years, studies on nano-semiconductor thin films in dye-sensitized solar cells have primarily focused on three aspects, i.e., the methods to prepare films, the physical–chemical treatment of films, and other semiconductor materials.

1. **Preparation technology of nano-TiO_2 semiconductor film**

The conventional method to prepare nano-TiO_2 films is by sol–gel method, whereas this preparation method is challenged in the application of DSC based on flexible substrates. Flexible materials, particularly plastic substrates, are not suitable for preparing electrodes based on high-temperature sintering. As an alternative, sintering [12, 13] or cold pressing [14, 15], technologies at low temperatures around 130 °C are commonly used to fabricate low-temperature nano-semiconductor thin films. Pichot et al. [12] used the preparation technology of nanometer TiO_2 without surfactant; after the treatment at 100 °C, the photoelectric conversion efficiency of the cell reached 1.2%. Hagfildt et al. adopted a mechanical cold pressing method [15], to prepare a TiO_2 film on a flexible substrate and achieved a photoelectric conversion efficiency of 3%. Yum et al. [11] prepared porous network nano TiO_2 film based on electrophoretic deposition and cold-pressed activation for flexible DSC, and a flexible solar cell exhibiting photoelectric conversion efficiency of 1.66% on ITO/PET conductive substrate was obtained. Miyasaka et al. [16, 17] employed electrophoretic deposition followed by chemical vapor deposition of titanate and microwave heat treatment at 150 °C to produce a flexible solar cell exhibiting a fill factor of 61% and an efficiency of 4.1%. Durr et al. [18] exploited a high-temperature sintered porous TiO_2 layer to transfer from gold-plated glass to an adhesive-coated ITO/PET flexible conductive substrate; subsequently, the TiO_2 was prepared at a certain temperature and high pressure, achieving a high photoelectric conversion efficiency of 5.8% (AM1.5). However, such a method is difficult to apply on a large-scale as impacted by the complicated preparation. Miyasaka et al. [19] adopted TiO_2 sol acidic aqueous solution as a connecting agent, coated TiO_2 slurry on an ITO/PET conductive film (sheet resistance as 13/Ω/□) and sintered at 150 °C to produce nanoporous TiO_2 film; the solar cell efficiency using with it as a photoanode reached over 6%. In 2006, the Gratzel group [20], developed a platinum-plated counter electrode flexible solar cell based on a titanium foil flexible substrate, exhibiting an efficiency of 7.2%, which is the highest value of the conversion efficiency of flexible cells. The mentioned results have made flexible solar cell applications promising.

At present, another vital development direction of nano-TiO_2 applied to DSSC is to prepare nano-TiO_2 electrodes with a regular and ordered nanostructured TiO_2 film. Existing studies reported that ordered grains help enhance the photoelectric conversion efficiency and photocurrent of the cell. Meng et al. [21, 22] employed DSC cells assembled with nanocrystalline three-dimensional periodic holes of nanoparticles exhibiting different particle sizes; the open-circuit voltage reached 0.9 V. Kim et al. [23] adopted $TiCl_4$ hydrolysis to prepare TiO_2 nanorods and nanotubes on the anode Al_2O_3 thin film template, and doped 10 wt% TiO_2 nanorods into P25-based nano TiO_2 film; as a result, the efficiency increased by 42%. Adachi et al. [24] attempted to synthesize TiO_2 nanotubes and exploited them as DSC electrode materials; they achieved a conversion efficiency of 5%. Recently, the group also built a nanonetwork electrode composed of TiO_2 nanowires at a low-temperature (80 °C) exhibiting a photoelectric conversion efficiency of 9.33% [25]. Zukalova et al. [26] adopted a layer-by-layer deposition method to form a nanostructured TiO_2 film with a regular structure on a pluronic P123 template. The efficiency of solar cells based on a 1-m-thick three-layer superposition film reached 50% higher than that of based on ordinary randomly oriented anatase nano-TiO_2 films.

2. **Physical and chemical modification of TiO_2 film**

Nanostructured semiconductors absorb considerable monomolecular-layer dyes in solar cells through their huge surface area, enhancing the collection efficiency of sunlight. Meantime, nano-semiconductors can transport electrons of excited state dye molecules to the electrodes. The large surface area of the semiconductor electrode also promotes the charge recombination of the electrode surface, thereby downregulating the photoelectric conversion efficiency of the solar cell. To enhance the photoelectric performance of cells, various physical and chemical modification techniques have been adopted to optimize the characteristics of TiO_2 electrodes (i.e., surface treatment, surface coating, and doping).

(1) **Surface treatment**

Treatment with a solution containing HCl, $TiCl_4$, H_2O_2, and titanium isopropoxide can effectively enhance the photovoltaic performance of the cell, because:

(1) The chemical treatment alters the surface structure of the electrode and enhances the electron diffusion and transmission properties of the nanoporous network microstructure.
(2) The chemical treatment improves the surface state of the semiconductor film, making the surface energy band more suitable for electron injection and transport;
(3) Chemical treatment up-regulates the density of surface states, thereby making the bonding force between the nano-TiO_2 surface and the dye molecules increase, and the electron injection efficiency is enhanced;
(4) Chemical treatment is capable of activating the surface of TiO_2, enhancing the surface roughness, and significantly up-regulating the molecular weight of the adsorbed dye.
(5) Chemical treatment alters the concentration and state of titanium ions on the TiO_2 surface, which can be better combined with dyes and facilitate electron injection.

Gratezel [8] treated the nano-TiO_2 photoanode with $TiCl_4$ aqueous solution, which can coat a layer of high-purity TiO_2 on the outside of TiO_2 core exhibiting low purity, enhance the electron injection efficiency, and form a barrier layer at the semiconductor-electrolyte interface. Identical to electrodeposition, the formation of novel nano-TiO_2 particles between the nano-TiO_2 films enhances the electrical contact between the nano-TiO_2 particles. Cai Shengmin et al. [27, 28] confirmed that $TiCl_4$ treatment narrows the BET specific surface area and decreases average pore diameter of the TiO_2 film, which improves the contact between the nano-TiO_2 particles, and increases the photocurrent of the cell. After the nano-TiO_2 film is treated with a proper concentration of $TiCl_4$ solution, the open-circuit voltage and short-circuit current of the cell are up-regulated, thereby enhancing the photovoltaic performance of the cell. Graetzel research reported [29] that after $TiCl_4$ treatment, though the specific surface area of the nano-TiO_2 film decreases, the amount of TiO_2 per unit volume increases, thereby broadening the surface area of the TiO_2 film and increasing the photocurrent of the cell. Sommling [30] studied the mechanism of $TiCl_4$ treatment of nano-TiO_2 photoanode to increase photocurrent. It is considered that $TiCl_4$ treatment reduces the position of the conduction band edge of TiO_2, while enhancing the photoelectron injection efficiency. Methods similar to the surface treatment of $TiCl_4$ consist of acid treatment and surface electrodeposition. Huang chunhui et al. [31] administrated organic dye-sensitized TiO_2 film with hydrochloric acid, and the photocurrent, photovoltage, and efficiency of the cell were significantly enhanced. Wu jihuai et al. treated TiO_2 film with different acids [32]; they found that hydrochloric acid achieves better treating effect than other inorganic acids.

(2) **Surface coating**

Surface coating refers to another vital method for surface modification of nano-TiO_2 electrodes. Since TiO_2 exhibits a high specific surface area and a small particle size, the number of surface states of the porous film is relatively large than that of single crystal materials, so the TiO_2 conduction band electrons are seriously recombined with the electron acceptor in the oxidation state dye or electrolyte. Thus, a semiconductor or insulating layer with a higher conduction band position is coated on the nano-TiO_2 surface to form a core–shell structure barrier layer to down-regulate the recombination probability. TiO_2 surface coated with ZnO, Nb_2O_5, SrO, $SrTiO_3$, and other metal oxides have significantly enhanced cell efficiency [33–39]. After the surface is coated with Nb_2O_5, the photocurrent is promoted, and the charge recombination is hindered for the potential barrier between the core and the shell, and the electron density of the TiO_2 conduction band increases; as a result, the open-circuit voltage increases, and the cell efficiency is significantly enhanced. Diamant et al. [37] considered that after coating $SrTiO_3$, the TiO_2 surface is depolarized, elevating the Fermi level of TiO_2, and therefore, increasing the photovoltaic voltage of the cell. Wang [40] adopted 1wt% $CaCO_3$ to coat the TiO_2 electrode, and the photoelectric conversion efficiency of the cell reached 10.2%. As revealed from other studies, TiO_2 coated with metal hydroxide can enhance the adsorption of dyes and elevate the short-circuit current, open-circuit voltage, fill factor, and photoelectric conversion efficiency of the cell. The identical effect can be achieved by coating with ZnO, TiO_2, Al_2O_3, MgO, Y_2O_3, and other insulating materials [41]. Based on the mentioned research, Gratezel [42], proposed to introduce a dense TiO_2 layer at the interface between conductive glass and nano-TiO_2 to inhibit conductive glass to electrolyte from being recombined. Yanagida [43] introduced a Nb_2O_5 insulation layer at the FTO/TiO_2 interface, and the open-circuit voltage, fill factor, and photoelectric conversion efficiency of the cell are significantly elevated.

(3) **Doped**

Ion doping of TiO_2, the doped ions is capable of relatively affecting the band structure of TiO_2 electrode material. It moves in a direction that facilitates charge separation and transfer and enhances photoelectric conversion efficiency. At present, the doping ions primarily include transition metal ions or rare earth elements.

Composite Membranes are also a vital direction in the current research of semiconductor electrodes. Compounding TiO_2 with other semiconductor compounds to make composite semiconductor films [44, 45], can alter the energy level structure of TiO_2 films [46, 47], so it becomes more conducive to electron transfer, and electron–hole recombination is inhibited; as a result, cell performance is enhanced. Common semiconductor compounds include ZnO, CdS, PbS, etc. The composite methods consist of in-situ composite and layered composite. In-situ composite refers to a method in which a semiconductor compound is mixed with TiO_2 to form a thin film meantime. Layered composite indicates a method where a compound semiconductor is separately formed into a thin film; subsequently a TiO_2 is deposited or a semiconductor thin film is deposited on the TiO_2 film to achieve a multi-junction structure of the semiconductor film [48]. The formation of composite membranes can alter the distribution of electrons in the TiO_2 film, inhibits the recombination of carriers during conduction and, therefore, enhancing electron transport efficiency. Composite membranes are likely to be a hotspot of subsequent research.

8.5.2 Electrolyte

According to the research development and structure of dye-sensitized thin film solar cells, they can be split into four types, i.e., organic solvent electrolyte cells, ionic liquid electrolyte cells, sol–gel (quasi-solid) electrolyte cells, and solid electrolyte cells. The anodes of the mentioned cells are made of nanoporous TiO_2 semiconductor film. The dye photosensitizer is primarily a complex of ruthenium as center particles, and the counter electrode is largely a platinum electrode or a platinum electrode exhibiting a monomolecular layer. The major difference between the mentioned four types of cells is the electrolytes. Besides restoring dyes and transporting charges, the major function of the electrolyte system can also alter the energy levels of titanium dioxide, dyes and redox couples, thereby causing the thermodynamic and kinetic properties of the system to vary, which significantly impacts the photovoltage.

1. **Organic solvent electrolyte system**

Organic solvents are the critical components of liquid electrolytes, creating an environment for dissolution and diffusion of active ions. Their physical properties involve nucleophilicity (DN value), dielectric constant and viscosity. The DN value noticeably impacts the open-circuit voltage and short-circuit current of the DSSC. A donor–acceptor reaction occurs between an organic solvent and iodide to generate I_3^-. In accordance with the DN value of the solvent, the transition from iodide to I_3^- requires a specific solvent. Existing studies suggested that the higher the DN value of the solvent, the higher the open-circuit voltage of the cell device and the lower the short-circuit current will be.

1,2-dichloroethane (DCE), acetone (AC), acetonitrile (ACN), ethanol (EtOH), methanol (Meoh), tertiary-butanol (t-BuOH), dimethylformamide (DMF), propylenecarbonate (PC), 3-methoxypropionitrile (MePN), dimethylsulfoxide (DMSO), dioxane (DIO), pyridine (PY), and others are commonly used as organic solvents in liquid electrolytes. The mentioned organic solvents should exhibit the characteristics below: not participating in the electrode reaction, lower freezing point, wide temperature range and low viscosity, high good wettability and permeability, as well as the ability to dissolve numerous redox couples, additives, and other organic and inorganic substances.

On the whole, the organic solvent liquid electrolyte exhibits the advantages of large ion diffusion coefficient, easy design and variation of composition, and high permeability for TiO_2 porous membranes. Thus, DSSC adopting an organic solvent liquid electrolyte exhibits high photoelectric conversion efficiency. However, some unavoidable defects exist in the organic solvent liquid electrolyte: the presence of the liquid electrolyte easily causes the desorption of the sensitizing dye; the solvent will volatilize and may interact with the sensitizing dyes to cause dye degradation; the sealing process is sophisticated, and the sealant may also react with the electrolyte; the carrier mobility regulated by diffusion is significantly slow, and the photocurrent becomes unstable during high-intensity illumination; the irreversibility of ion migration cannot be completely ruled out, since other reactions besides the redox cycle cannot be completely avoided [49].

The redox couple in the liquid electrolyte is primarily I^-/I_3^-, and the I^-/I_3^- redox couple electrode potential complies with the energy level of the nano-semiconductor electrode and the LUMO level of the dye significantly better than other redox pairs (Br^-/Br_2, $SCN^-/(SCN)_2$ and $SeCN^-/(SeCN)_2$). The most common I^-/I_3^- redox couple counter cations include alkyl imidazole barrier cations and Li^+. Alkyl imidazole cations are adsorbed on the nano-TiO_2 surface to form a Helmholz layer, thus preventing the contact between I_3^- and nano-TiO_2 film, effectively inhibiting the recombination of conduction band electrons and I_3^- in the electrolyte solution on the surface of nano-TiO_2 film; as a result, fill factor, output power, and photoelectric conversion efficiency of the DSSC are significantly improved. On the other hand, alkyl imidazole cations achieve a large ionic radius and weaker I^- ion binding. Thus, the alkyl imidazole iodide salt exhibits a greater solubility in an organic solvent, and the iodide ion also shows a higher activity, elevating the rate of regenerating the oxidation state dye to the ground state dye and enhancing light utilization efficiency and photocurrent, as well as dye stability.

4-tert-butylpyridine (TBP) or N-methylbenzimidazole (NMBI) is a common additive in DSSC electrolyte solutions. The addition of the mentioned additives can inhibit dark current and enhance the photoelectric conversion efficiency of the cell. Since the organic solvent electrolyte exhibits high permeability to the nanoporous membrane, and the redox is fast, the maximum record of DSSC photoelectric conversion efficiency is obtained in a solar cell based on an organic solvent electrolyte, particularly a highly volatile organic solvent electrolyte. However, organic electrolytes are subject to volatile organic solvents, easy leakage of electrolytes, difficulty in sealing of cells, and degraded performance of cells during long-term operation, thus shortening the service lifetime of solar cells.

2. **Ionic liquid-based electrolyte**

Compared with conventional organic solvents, ambient temperature ionic liquid (RTILs) have a series of prominent advantages: ionic liquids are almost non-volatile; they have better chemical stability and wider electrochemical window; they are not flammable; they exhibit a high conductivity as impacted by ionic nature; low toxicity, and when the anion is I^-, the ionic liquid can be used as both a solvent and a source of I^- [50]. Accordingly, RTIL has become the new favorite of DSSC electrolytes over the past few years. Though the RTILs in DSSC are liquid, they eliminate many of the defects of conventional solvent liquid electrolytes. Anions constituting the ionic liquid include I^-, $N(CN)_2^-$, $B(CN)_4^-$, $(CF_3COO)_2N^-$, BF_4^-, PF_6^-, NCS^-, etc. Ionic liquids are liquid at ambient temperature and exhibit a significantly higher viscosity than organic solvent electrolytes. The rate of I_3^- diffusion to the counter electrode is slow and the mass transfer process occupies a dominant position. The ionic liquid employed in solar cells, the common redox couple is I^-/I_3^-. By adding I_2 to I^- to form I_3^-, the volume of anions increases, and the viscosity of the ionic liquid decreases. Thus, I_3^- in a solar cell based on an ionic liquid medium shows a higher concentration than that in a liquid electrolyte. The organic cation constituting the ionic liquids are commonly an alkyl imidazole cation (e.g., 1-methyl-3-propylimidazolium iodide and 1-methyl-3-hexylimidazole iodide (HMII)). Compared with the two ionic liquids, MPII exhibits low viscosity, good solubility in many organic and inorganic substances, and a large diffusion rate of working substances; however, the long aliphatic chain in HMII can effectively inhibit conduction band electrons on the TiO_2 film surface and I_3^- in solution, which also significantly impacts gel electrolytes based on ionic liquid media. Wataru et al. delved into the physical properties of 1-methyl-3-propylimidazolium iodide liquids of different length alkyl chains and prepared them as solvents to prepare ionic liquid electrolytes. As suggested from the results, the ionic liquid electrolyte based on HMII is better than the ionic liquid electrolyte based on MPII. Mazille et al. introduced a cyano functional group at the end of the 3-propyl chain of MPII, and the photovoltaic performance of the cell varies slightly.

In 1996, Papageorgiou [51], reported that RTILs are critical to extend the lifetime of electrochemical devices. However, as impacted by the high viscosity of RTILs, the cell current is an order of magnitude smaller than that of conventional liquid electrolytes. Accordingly, the synthesis and application of low viscosity RTILs is the major means to enhance the efficiency of cells. In 1999, Hagiwara et al. [52], prepared a low viscosity RTILs by reacting oxidized 1-ethyl-3-methylimidazole with hydrogen fluoride, and the conductivity at 25 °C reaches 1.2×10^{-4} S/cm. Matsumoto et al. [50] employed this ionic liquid to DSSC and achieved a battery efficiency of 2.1%. Over the past few years, polyiodotrialkyl phosphonium salts have suggested high electrical conductivity properties [53], its conductivity is 10^{-3}–10^{-4}S/cm, and increases with increasing iodine content. To be specific, DSSC using $(Bu_2MeS)I$ as electrolyte has exhibited an efficiency of 3.7% under AM1.5 simulated sunlight, which is equivalent to DSSC adopting imidazole salt RTILs.

3. **Sol–gel (quasi-solid) electrolyte system**

From a practical perspective, replacing solid electrolytes with liquid electrolytes will be a trend in dye-sensitized solar cells development. The solid electrolyte does not add any solvent, and uses organic solids, dyes, and inorganic materials (TiO_2) to form solid-state cells. However, the conversion efficiency of all-solid-state dye-sensitized solar cells remains not high. Quasi-solid electrolytes have always been a hotspot in this field. The so-called quasi-solid electrolytes refer to their mechanical properties between liquid and solid electrolytes, the appearance is gel-like, and the conductive mechanism is dependent on ion conduction, as well as liquid electrolytes. Quasi-solid electrolytes exhibit considerable advantages over liquid and all-solid electrolytes: Firstly, it is relatively stable compared with liquid electrolytes, and can basically overcome many problems of liquid electrolytes (e.g., difficulty in packaging, leakage, and degradation of dyes); secondly, some liquid electrolytes do not reduce the efficiency of the battery after they are converted into quasi-solid electrolytes, and the solar cells still exhibit high efficiency. A vital means of preparing a quasi-solid electrolyte is to add some other substances to the liquid electrolyte (e.g., small molecule gels, high molecular polymers, as well as nanoparticles). The mentioned substances can cause cross-linking in the electrolyte system, so the liquid electrolyte acts as a quasi-solid electrolyte, a polymer gel electrolyte, and a nanoparticle-added gel electrolyte.

(1) **Organic small molecule gel electrolyte**

Organic small molecule gelling agents employed in dye-sensitized solar cell primarily include sugar derivatives, amino acid compounds, amido compounds, biphenyl compounds, etc.

In 2001, Yanagida [54], group reported four types of organic gel small molecule compounds as gel liquid electrolyte. The composition of the liquid electrolyte adopted is 0.6 mol/L of 1,2-dimethyl-3-propylimidazolium iodide (DMPII), 0.1 mol/L of I_2, 0.1 mol/L of LiI, 1 mol/L of 4-tert-butylpyridine (TBP) and 3-methoxypropionitrile (MePN). Sol–gel quasi-solid electrolytes with gel temperatures of 47–49, 58–60, 61–63, and 85–87 °C were obtained by altering the length of the fatty chain and the amount of gelling agent added. A specific comparison was drawn of the effects of four different gel molecules on the photoelectric conversion performance and lifetime of the cell.

In 2004, researchers in Germany and Switzerland employed sorbitol to solidify liquid electrolytes and achieved effective results. The gelling agent they used was bis (3,4-dimethyl-dibenzylidene sorbitol), and the solidified electrolyte consisted of 0.6 mol/L of 1,2-dimethyl-3-propylimidazolium iodide, 0.1 mol/L of I_2, 0.5 mol/L of N-methylbenzimidazole, and the solvent is methoxypropionitrile. The liquid electrolyte can be solidified by adding 1.5% gel, and the prepared DSSC has significant photoelectric properties. The performance of the cell after solidification is not reduced, and it exhibits high thermal stability [55].

The molecules of small gelling agents are only dependent on weak intermolecular forces to form unstable physical crosslinks. Thus, such electrolytes tend to exhibit poor mechanical properties, and such quasi-solid electrolytes are thermally reversible and become liquid electrolytes at relatively high temperatures. As a result, the stability of the cell will decrease, and the lifetime will be shortened.

(2) **Polymer gel electrolyte**

The Gratzel group took PVDF-HFP (poly-vinylidenefluoride-co-hexafluoropropylene) as a gelling agent and successfully gelled a liquid electrolyte with an organic solvent and an ionic liquid as a medium. Though the viscosity of the gelled electrolyte is significantly larger than that of the liquid electrolyte on the macro scale, there is a channel for liquid transport in the system, which does not affect the diffusion I^-/I_3^-; thus, the photoelectric conversion properties are on the same order as the liquid electrolyte. An organic polymer compound (e.g., a copolymer of polyoxyethylene ether, polyacrylonitrile, epichlorohydrin, and ethylene oxide) can also form a gel network structure in a liquid electrolyte for generating a quasi-solid polymer electrolyte. Cao et al. [56] employed polyacrylonitrile (PAN) to introduce a liquid electrolyte (composed of I_2, NaI and acetonitrile, ethylene carbonate (EC) and propylene carbonate (PC)) into a gel state, and a photoelectric conversion efficiency of 4.4% was obtained.

In China, in 2004, Dai Songyuan, etc. of the Chinese Academy of Sciences [57], used a vinylidene fluoride and hexafluoropropylene copolymer gel 1-methyl-3-mercaptoimidazolium electrolyte to obtain a quasi-solid DSSC exhibiting a photoelectric conversion efficiency of 6.61%. Lin Yuan et al. used a long-chain polymer material to react with a polysiloxane containing a quaternary ammonium salt side chain to prepare a quasi-solid electrolyte. In 2003, the solar cell prepared with this quasi-solid electrolyte exhibits a photoelectric conversion efficiency of 1.39% at a light intensity of 60 mW/cm^2. The gelled electrolyte displays a stable chemical cross-linking structure and a high electrical conductivity, pertaining to a type of electrolyte that shows a great development potential.

4. **Solid electrolyte system**

Though the quasi-solid sol–gel electrolyte can prevent electrolyte leakage to a certain extent, reduce the vapor pressure of the organic solvent, and slow the volatilization of the organic solvent, problems with long-term stability remain. Accordingly, the development of all-solid-state solar cells continues to be the final aims. In DSSC applications, hole-conducting materials have the following requirements: after the dye molecules inject electrons into TiO_2, the p-type semiconductor material can transfer holes generated by the oxidation dye, i.e., the valence band top of the p-type semiconductor material is above the ground state level of the dye, and the p-type semiconductor material can be deposited in the porous nanoparticle layer. In the process of depositing p-type semiconductor materials with a suitable method, the single-layer dye adsorbed on the TiO_2 nanoparticles will not dissolve or degrade. Such p-type semiconductor material is required to transmit light in the visible spectral range, even if it leads to light absorption, it should exhibit the effective electron injecting ability of the dye.

(1) **p-type inorganic semiconductor**

Numerous existing inorganic p-type semiconductor materials can meet the mentioned conditions. However, common wide-bandgap p-type semiconductor materials (e.g., SiC and GaN) are not suitable for DSSC since they require high-temperature deposition which will inevitably degrade the dye. After extensive research, typical inorganic p-type semiconductor materials based on copper compounds e.g., CuI, CuBr, and CuSCN [58–61], can meet the mentioned requirements.

The mentioned copper-based inorganic semiconductor materials can form a relatively complete hole transport layer with the use of the solution or vacuum deposition technology, and the conductivity of CuI and CuSN exceeds 10^{-2} S/cm, which achieves hole conductivity energy.

In 1995, Tennakone et al. [62], initially reported the solid-state DSSC of CuI. When the sunlight intensity is 800 W/m^2, the device short-circuit current can reach 1.5–2.0 mA/cm^2. They replaced the ruthenium-pyridine complex dye [63, 64] with Huajing dye to generate a CuI solid-state solar cell exhibiting a conversion efficiency of 2.4%. However, the performance of this cell using CuI is not significantly stable. In 2003, Sirimanne et al. [65], found that the performance of CuI solid-state devices deteriorates rapidly, even faster than the degradation rate of liquid DSSC. One of the major reasons is that excess I_2 in the CuI film severely reduces the photocurrent of DSSCs. CuI is easy to oxidize after being continuously irradiated. Taguchi [66] and Kumara [67] reported DSSCs with porous films of TiO_2 core–shell structure covered by MgO and combined with CuI hole conductive materials to obtain higher photoelectric conversion efficiency and stability, which is because the shell layer blocks the transport of photo-generated holes in the CuI layer and inhibits the photo-oxidation ability of TiO_2.

Another critical factor affecting the instability of solar cells made with CuI is the loose contact between CuI grains and the surface of TiO_2. With acetonitrile as a solvent is easy to deposit to obtain larger-sized crystal grains, and cannot effectively penetrate into the voids inside the nanoparticles, so only weak contact can be formed. Tennakone et al. [68] reported that the addition of a trace amount of 1-methyl-3-ethylimidazolium thiocyanate (MEISCN) to the CuI acetonitrile solution enhanced the stability of the device. MEISCN is capable of inhibiting the growth of CuI crystals, while it exists at the interface between sensitized TiO_2 particles and CuI grains to promote hole transport. Meng et al. [69] also employed MEISCN to produce solar cells exhibiting a conversion efficiency of 3.8% and good stability. However, the purification of MEISCN requires chromatographic separation techniques, so it is expensive and not suitable for mass production. In subsequent studies, Tennakone et al. [55], found that a simple structure of triethylamine thiocyanate (THT) exerts the identical effect and can even hinder CuI crystals growth more effectively, and it is likely to replace MEISCN to fabricate solid-state solar cells.

The major problem in selecting CuSCN instead of CuI is suitable deposition methods. In 2001, Kumara et al. [70], successfully coated the ruthenium dye-sensitized TiO_2 porous film with a solution of dipropyl sulfide, enhancing the battery conversion efficiency by 1.25%. O'Regan et al. [71] reported in 2002, that the conversion efficiency of solid-state DSSC devices made of CuSCN reached 2%. It is noteworthy that the pore filling rate reached almost 100% when the film thickness was 2 μm. However, the efficiency of the cell prepared by CuSCN remains lower than that of CuI, which may be as impacted by the low hole conductivity of CuSCN.

(2) **p-type organic small molecule solid**

Compared with p-type inorganic semiconductor materials, p-type organic hole conductive materials are extensively employed in organic solar cells as impacted by their wide sources, easy film formation and low-cost [72–74]. The small organic molecules currently employed in DSSC are primarily 2′2′7′7-tetrakis (N, N-di-p-methoxyphenylhydrazineamino)-9, 9′-spirobiguanide (Spiro-OMeTAD), and triphenylamines (TPD). In 1998, Gratzel [75], initially reported high-efficiency solid-state DSSCs using amorphous organic hole transport materials (Spiro-OMeTAD). In 2001,

Kruger et al. [76, 77], adopted OMeTAD to regulate the recombination of heterogeneous interface charges, which significantly enhanced the DSSC conversion efficiency. The conversion efficiency of the battery achieved by adding a small amount of tBP and Li $(CF_3SO_2)_2$ N to this organic conductive material reaches 2.56%. In 2007, Gratzel [78], fabricated DSSC devices with diblock alkoxy and alkanyl pyridine ruthenium complexes as sensitizers, adopting Spiro-OMeTAD and silver counter electrodes to achieve the conversion efficiency of 5%. In 2006, Jessica et al. [79], synthesized six types of triphenylamine hole conductive organic small molecules and delved into the effects of substituted functional groups on the molecular conductivity and electrochemical physical properties.

(3) **p-type organic conductive polymer**

Conductive polymeric materials are typically synthetic materials exhibiting electrical, electronic, magnetic, and optical properties similar to those of metal or semiconductor materials and maintaining the mechanical properties of conventional polymers. Moreover, they can be fabricated by chemical or electrochemical methods with relatively low process requirements. Such materials often exhibit reversible transitions between metal and semiconductor properties, and they are commonly employed in a variety of electronic devices (e.g., photovoltaic and photoelectrochemical cells).

The gel polymer electrolyte is capable of enhancing the contact characteristics of the solid electrolyte and the titanium oxide nanoporous film. As a matter of fact, quasi-solid DSSC gels have been reported to exhibit almost the identical properties to their liquid precursor sols. The gel electrolyte should have the conditions that polymerization can proceed in the presence of I_2. The polymerization should be conducted at a temperature lower than the decomposition of the dye. In the presence of several impurities (e.g., oxygen, water, and ions), the polymerization can be initiated and completed. The polymerization process does not produce by-products detrimental to the performance of the cell, and it can also be conducted without an initiator.

8.5.3 Dye Sensitizer

Given whether the photosensitive dye contains metal, it can be classified into pure organic photosensitive dye and metal complex photosensitive dye. Pure organic photosensitivity dyes have become a hotspot in the field of photosensitivity dyes over the past few years as impacted by the advantages of no precious metals, low-cost, diverse structures, and high molar absorption coefficients. Metal complex photosensitive dyes cover metal ruthenium (Ru), osmium (Os), platinum (Pt), ruthenium (Re), copper (Cu), and dye iron (Fe), and the ligands are commonly various substituted bipyridines or poly bipyridine. The most extensively used are metal ruthenium complex dyes. Moreover, there have been many researches on the application of phthalocyanine and porphyrin series metal complexes in DSC. The representative

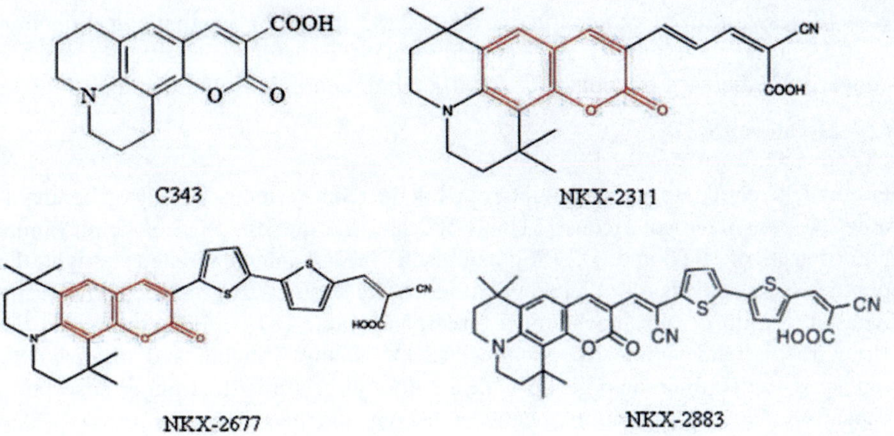

Fig. 8.6 Structure of coumarin dye

compounds with better performance in the two types of compounds are expressed below.

1. **Pure organic dye**

Organic dyes have many advantages, e.g., variety, low-cost, high-absorption coefficient, and easy structural design. Over the past few years, dye-sensitized solar cells based on organic dyes have developed rapidly. Photosensitive dyes currently applied to DSC consist of coumarin dyes, polymethine, carotenoids, cyanine dyes, hemicyanine dyes, rosewood dyes, chlorophyll, and derivatives thereof. To be specific, the organic dyes employed in DSC with high sensitization efficiency are as follows.

(1) **Coumarin dye**

C343 is a frequently used coumarin compound (Fig. 8.6), and is a good photosensitive dye. However, the overall energy conversion efficiency of DSC prepared with this dye is significantly low (less than 1%). To enhance the efficiency of the dye-sensitized cell, in 2001, the Arakawa research group in Japan, synthesized a series of excellent sensitizing dyes with high sensitization efficiency by modifying the structure of C343 [80–85]. Their design focuses on how to broaden the absorption spectrum of the dye and inhibit the aggregation of dye molecules. To be specific, NKX-231 exhibits the optimal performance, and the value of IPCE is more than 70% in the wavelength range of 400–600 nm. The absorption spectrum in ethanol is redshifted by 70 nm compared with C343, and the photoelectric conversion efficiency of 6% is obtained [81]. In 2003, they synthesized compounds NKX-2593 and NKX-2677 by introducing a thiophene ring into a π conjugated system; as a result, the absorption spectrum of NKX-2677 is broadened, the LUMO orbital energy level of the dye shifts negatively, and the power of the excited state dye to inject electrons into the TiO_2 conduction band increases, the cell performance is further enhanced, and a photoelectric conversion efficiency of 7.7% is achieved [82]. In 2007, Wang et al. [83], synthesized a coumarin dye NKX-2883 exhibiting a high molar extinction coefficient and good stability, and its sensitized DSC can still maintain a photoelectric conversion efficiency of about 6% after 1000 h of continuous illumination [84].

(2) **Hemicyanine dye**

Peking University Huang Chunhui Group conducted a systematic study on hemicyanine dye-sensitized solar cells [31]. At 500 nm, the highest IPCE of hemicyanine photosensitizers BTS and IDS [86], can reach 100%. Under simulated sunlight, the photoelectric conversion efficiency can reach 5.1% and 4.2%, respectively. Wang Xuesong group of physicochemical research institute [87], further optimized the structure of hemicyanine dye synthesized by Huang Chunhui and his team: ① replacing the sulfonic acid group with a carboxyl group with stronger adsorption capacity, enhancing the coupling between the dye and the semiconductor; ② shorten the length of the carbon chain between the adsorption group and the dye skeleton to make the distance between the dye and the semiconductor closer, which synthesizes a series of new benzothiophene hemicyanine dyes. The photoelectrochemical behavior of HC-1 with a carboxyl group and hydroxyl group is the best in DSC. The maximum IPCE is 73% and the overall energy conversion efficiency increases to 5.2%.

(3) **Polyene dye**

In 2003, Arakawa [88] research team discovered a class of polyene NKX-2569 (Fig. 8.7). N, N-dialkylaniline acts as the electron-donating group, carboxyl and cyano groups serve as the adsorption group and the electron-withdrawing group, and the methine unit acts as the conjugated bridge to link the donor and acceptor. With the absorption spectrum of such compound enhanced, the degree of conjugation of the olefin is redshifted, and the intramolecular charge transfer and charge separation of the compound are expedited. The overall energy conversion efficiency of NKX-2569 sensitized solar cells reaches 6.8%, demonstrating that polyene dyes also show broad applications in the DSC field (Fig. 8.8).

Fig. 8.7 Structure diagram of polyene dye

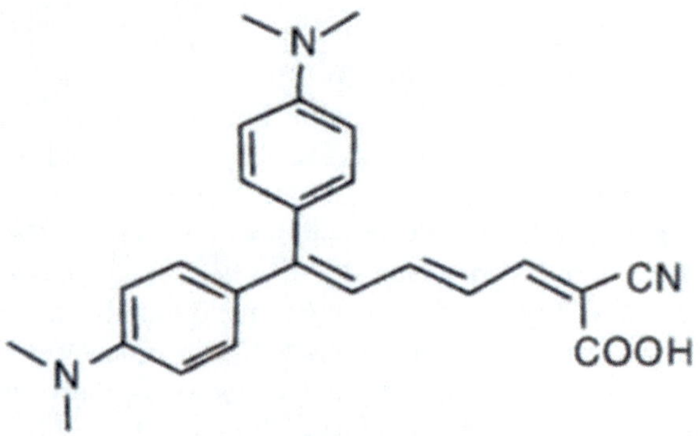

NKX-2569

Fig. 8.8 Structure diagram of indole dye

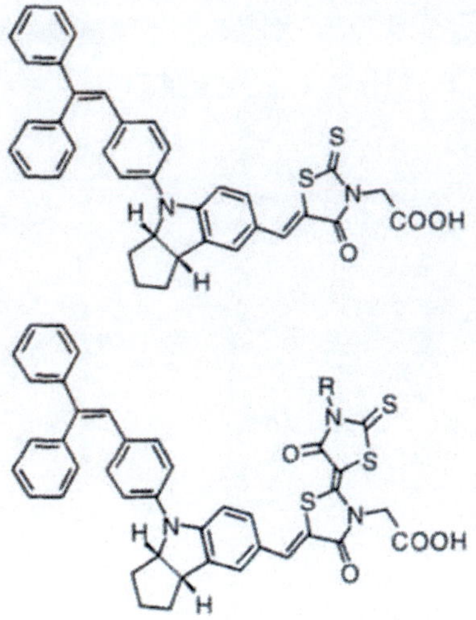

D102

D149 R=-C_2H_5
D205 R=-C_8H_{17}

(1) Indole dyes. In 2003, Horiuchi et al. [85], reported a class of indole photosensitizing dye D102 (Fig. 8.7), and the photoelectric conversion efficiency was 6.1%. Under the identical conditions, the photoelectric conversion efficiency of N3 dye reached 6.3%, and NKX-2311 sensitized cell exhibited the efficiency of only 3.3%. In 2004, Horiuchi et al., by adding rhodanine groups to red shift the absorption spectrum of D102, [89] obtained another indole dye D149; as a result, the photoelectric conversion efficiency of DSSC was enhanced to 8.0%. In 2006, Gratzel [90] research group enhanced the efficiency of D149 dye-sensitized solar cells to 9.0% by strengthening the structure of TiO_2 electrode. In 2008, with the use of D149 dye, they substituted the ethyl group of the electron acceptor with the n-octyl group to produce an anthraquinone dye D205. As impacted by the introduction of a long carbon chain, it effectively prevented the electrons and electrolytes in TiO_2 from being recombined. Furthermore, by adding deoxycholic acid to the dye solution, the dye aggregation was prevented, and the D205 dye-sensitized cell exhibited an efficiency of 9.5% [91].

(2) Other dyes. In 2006, Hagberg et al. [92], reported a class of simple triphenylamine dyes D5, adopting cyanoacrylic acid, triphenylamine, and thienyl ethylene as the electron acceptor, the electron donor and the bridge group, respectively. The overall efficiency of the prepared cell is 5.0%, and the overall efficiency of the N719 sensitized cell under the identical conditions is 6.0%. In 2007, Hwang et al. [93] used the benzene ring instead of the thiophene ring of dye D5 to fabricate the dye TA-St-CA, exhibiting the efficiency of 9.1%. Under the identical test conditions, the efficiency of N719 dye reaches 10.1%. In 2008, Hagberg et al. [94], introduced a methoxy group and a plurality of triphenylamine groups on triphenylamine to synthesize triphenylamine dyes D7, D9, and D11(Fig. 8.9); the introduction of methoxy group can remarkably enhance

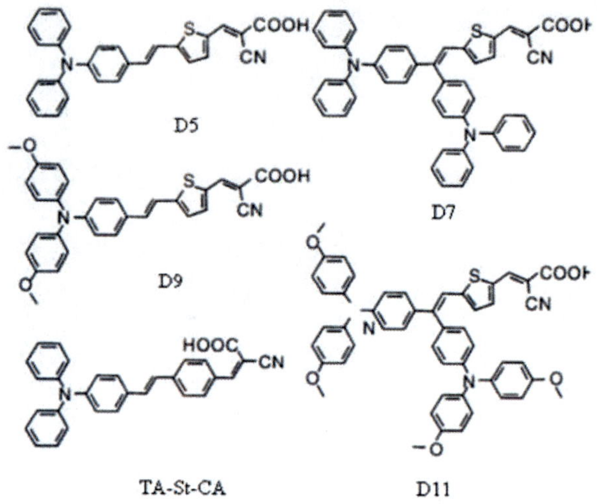

Fig. 8.9 Structure diagram of triphenylamine dyes

the efficiency of dyes. To be specific, D11 dye exhibits the maximal efficiency, which can reach 7.0%. D9 dye exhibits the maximal efficiency when employed in all-solid-state solar cells, reaching 3.3%.

In 2007, a series of phenothiazine organic light-sensitive dyes were reported [95] (Fig. 8.10). The phenothiazine dye T2-1 with cyanoacetic acid as the electron acceptor exhibits a photoelectric conversion efficiency of 5.5%. However, if Rhodanine acetic acid acts as the phenothiazine dye T2-3, DSSC exhibits the photoelectric conversion efficiency of only 1.9%. Under the identical test conditions, the solar cell using N3 dye as the photosensitive dye exhibits the photoelectric conversion efficiency of 6.2%.

In 2008, the C101 [96], fabricated by the Wang Peng research group of the Changchun Institute of Applied Chemistry of the Chinese Academy of Sciences exhibits a conversion efficiency of 11.9% in the international standard laboratory. In the same year, dye-sensitized solar energy exhibiting a conversion efficiency of 9.7% was prepared with a high-absorption coefficient of ruthenium dye C103 in combination with low-volatility electrolyte and ionic liquid electrolyte (Fig. 8.11); the cell exhibits long-term photothermal stability [97]. The DSSC developed by C203 dye exhibits the photoelectric conversion efficiency of 8.0%. If the eutectic ionic liquid

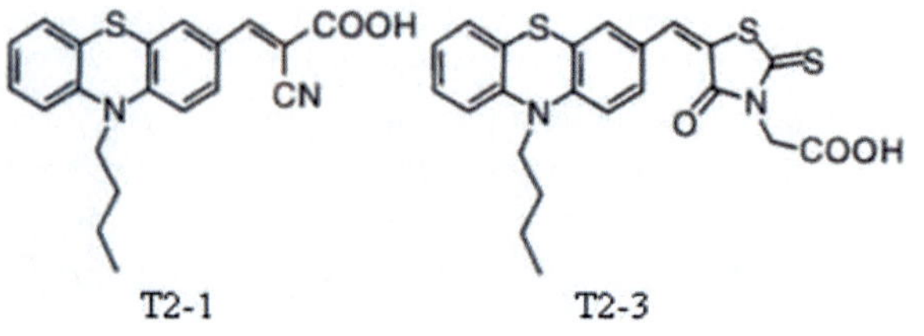

Fig. 8.10 Structure diagram of phenothiazine dyes

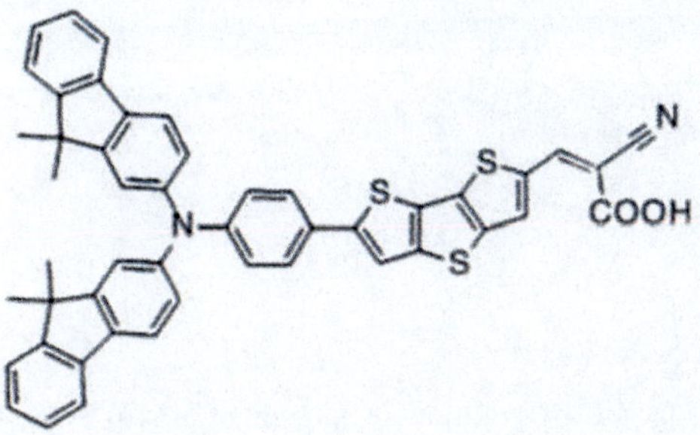

Fig. 8.11 Structure diagram of C203 dye

electrolyte is combined with C203 [98] dye, a practical dye-sensitized solar cell exhibiting long-term photothermal stability and conversion efficiency of 7% can be prepared.

2. **Metal complex photosensitizer dye**

Metal-organic complex dyes are the most extensively adopted dyes exhibiting special chemical stability, prominent redox properties, and high excited state reactivity. Moreover, their excited states exhibit lsong lifetime and good luminescence properties, thereby exerting robust photosensitization effects on energy transmission and electron transport. At present, besides complexes of ruthenium, cobalt, iron, platinum, and others, the most common organometallic complex dyes refer to ruthenium complexes. Most of the ligands exploit bipyridine, phthalocyanine, and porphyrin.

(1) **Bipyridine ruthenium dye**

The application of ruthenium bipyridine in photochemistry, electrochemical, and photoelectrochemical applications has aroused extensive attention for their high thermal and chemical stability. The bipyridinium ruthenium dye is capable of absorbing visible light to produce electron transitions from metal to ligand and inject electrons into the semiconductor conduction band. Accordingly, the bipyridine ruthenium complex has aroused rising attention as a photosensitizer for DSSC. The studies on it are conducted primarily in four aspects: ① increasing the degree of charge separation of the excited state energy level of the dye molecule; ② optimizing the bonding method of the dye and the TiO_2 film to make electron injection into the TiO_2 film more effective; ③ the dye exhibiting a wider absorption band and a larger molar extinction coefficient; ④ enhancing the stability of the dye. The metal complex sensitizer commonly covers an adsorption ligand and an auxiliary ligand, and the adsorption ligand acts as an emission group to adsorb the dye on the semiconductor surface. The role of the auxiliary ligand refers to regulating the overall performance of the complex, which is not directly adsorbed on the nano-semiconductor surface.

In 1979, the Wolfgang group first developed such dye. They introduced carboxyl groups into the parent of bipyridine, making dye sensitizers more effective in photolysis systems [99]. In 1985, Gratzel et al. [100], applied such dye to DSSCs, and later introduced electron dyes (e.g., Cl^-, H_2O, Br^-, I^-, CN^-, and NCS^-). Into the dye,

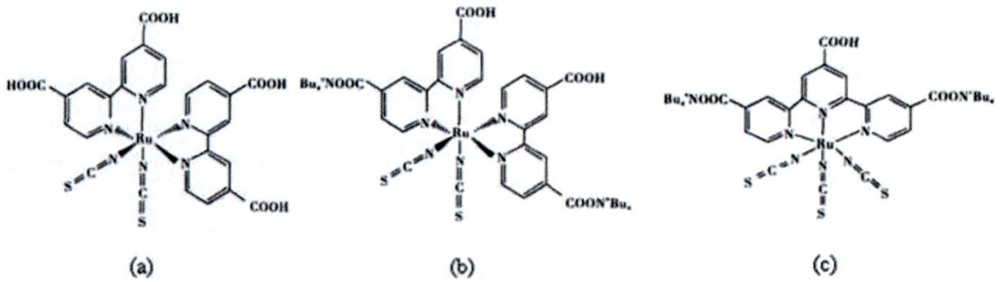

Fig. 8.12 Dye structure diagram of N3 (**a**), N719 (**b**), and N749 (**c**)

synthesized and systematically studied series of carboxylate polypyridine ruthenium dye sensitizers, including cis-Ru (L^1 = 2,2-bipyridyl-4,4′-dicarboxylic acid, X = Cl, Br, I, CN, NCS, etc.) [101–105]. Among the series of dyes, the "red dye" represented by N3 and N719 dyes exhibits the optimal performance (Fig. 8.12), and is the most extensively used dye sensitizer in DSC cells. In 2005, Gratzel et al. enhanced the photoelectric conversion efficiency of DSSC cells prepared with N719 as a sensitizer to 11.18% [106]. However, this dye lacks absorption in the long wavelength region of visible light. Thus, to further extend the range of the light effect of the dye, these researchers have improved the ligand and added an olefin bond between the carboxyl group and the bipyridine to promote the conjugated system and make the absorption spectrum red shift as to expand the light response range.

The black dye (N719) [9], reported in 2001, complies with the identical considerations: optimizing the conjugated system, expanding the bipyridyl ring to combine the carboxyl groups into one ligand and up-regulating the amount of thiocyanate, so the initial excitation wavelength generated by the photocurrent reaches 920 nm. With the decrease in the wavelength, the IPCE value tends to increase; after 700 nm, the IPCE reaches 80%. If the reflection and refraction of light by the conductive glass is overlooked, the DSC sensitized by the black dye achieves almost 100% IPCE value in the entire visible light region; thus, the dye is well absorbed in the entire visible light range. Under AM1.5 white light, the DSC sensitized with a black dye produced an open-circuit voltage of 721 mV, a short-circuit current of 20.53 mA/cm^2, a fill factor of 0.704, as well as a photoelectric conversion efficiency of 10.4%.

The critical part of DSSC refers to the dye, so it is required to achieve effective broadband absorption for sunlight, and also requires to exhibit high molar extinction coefficient and long-term stability to light, water, and high-temperature [107]. The Gratzel group first introduced the π conjugated chain to the bipyridine ligand and synthesized K8 [108], Z907 [109], Z910 [107], K-19 [110] (Fig. 8.13), and other compounds. In comparison with N3, the molar extinction coefficient of dye K8 increases by 30%, Z907 can work stably for 1000 h at 80 °C and exhibits high thermal stability. Mixed single molecule dye layer formed by Z907 and DPA can promote the stability of voltage output while enhancing the photoelectric conversion efficiency to 7.3%. Z910 substitutes the π conjugated system on the dye Z907 bipyridine with a non-carboxyl group to red shift the metal-to-ligand charge transfer (MLCT), which enhances the molar extinction coefficient and the ability to capture visible light. The

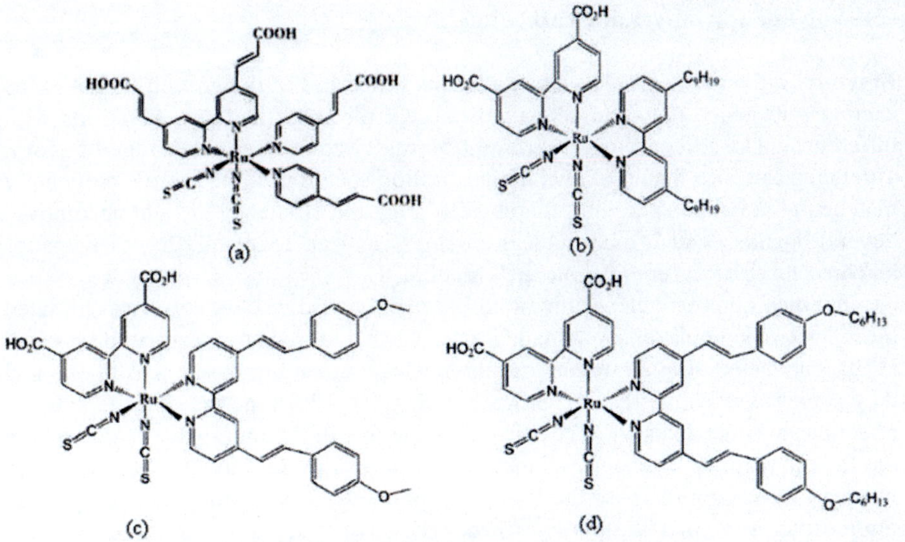

Fig. 8.13 Dye structure diagram of K8 (**a**), Z907 (**b**), Z910 (**c**), and K-19 (**d**)

total photoelectric conversion efficiency is 10.2%, and all parameters of the cell are also stable after 1000 h of light. The increase of the molar extinction coefficient reduces the thickness of the dye-sensitized semiconductor oxide film and enhances the electron injection efficiency.

Islam et al. [111–113] replaced the bipyridine ligand with a biquinoline ligand and a phenanthroline to promote the maximum absorption wavelength of the photosensitive dye to be redshifted to 700 nm. Taking $Ru(dcpq)_2(NCS)_2$ as an example, after being adsorbed on a nanocrystalline TiO_2 electrode, the excited state energy level is slightly higher than the conduction band potential of TiO_2, and the absorption spectrum can reach 900 nm, and the sensitized DSC monochromatic photoelectric conversion efficiency reaches 55%. Renounard et al. [114, 115] studied the tetrapyridine ruthenium photosensitizer dye. The electrons of the tetrapyridine were more delocalized than the bipyridine, and the thiocyanate ligand was transformed from homeopathic to trans. Under the mentioned factors, the maximum absorption wavelength of the tetrapyridine photosensitizer is redshifted by 44 nm compared with the bipyridine ruthenium photosensitizer (N3), and the maximum photoelectric conversion efficiency of the sensitized DSSC monochromatic light is 75%.

Ghanem et al. [116] simulated photosynthesis in Photosystem II (PSII) and introduced tyrosine (Tyr) to the bipyridine ruthenium ligand. Tyr is capable of effectively inhibiting the recombination of injected electrons, achieving long-life charge separation, and critically impacting electron transfer. Pan et al. [117] introduced pyridine on Tyr, which more effectively promotes the formation of phenolic hydroxyl radicals and elevates the electron transfer rate in the molecule.

(2) **Polynuclear bipyridine ruthenium dye**

Bipyridyl ruthenium complexes can alter the properties of the ground state and the excited state since they can selectively accept electrons and give electrons with different ligands. Accordingly, different bipyridyl complexes are bridged to form a Polynuclear. The ligand allows the absorption spectrum to be more significantly matched to the solar spectrum, thereby elevating the efficiency of light absorption. Several ligands of such multinuclear complex can transfer energy to other ligands and are said to have "energy antenna" functions.

The study on polynuclear ruthenium-based polypyridine complexes was initiated by reports of complexes by Amadeli et al. [118]. Moreover, Nazeerudding et al. [119], conducted similar studies on polynuclear ruthenium-based sensitizers, and they achieved effective results. Gratzel et al. [103, 120] reported that the antenna effect elevates the absorption coefficient of the dye in the long-wave region where the light absorption efficiency of the mononuclear bipyridyl dye is extremely low, whereas it does not increase the light absorption efficiency. Mononuclear dyes are more difficult to enter the pores of nano-TiO_2 as impacted by the larger volume, which limits the light absorption efficiency.

(3) **Phthalocyanine dyes**

Phthalocyanine refers to a sixteen-ring conjugate composed of four isoindoles. The metal atom is located in the middle of the ring and is connected to four isoindoles. After being introduced into the molecule, the groups (e.g., sulfonic acid group and carboxylic acid group) that can be combined with the surface of TiO_2 can act as photosensitive dyes. The metal atoms in the molecule are likely to be Zn, Cu, Fe, Ti, Co, etc. In 2007, Reddy et al. [121], reported that DSSC made of phthalocyanine photosensitizing dye PCH001 with ionic liquid electrolyte achieves a maximum IPCE of 75% as well as a photoelectric conversion efficiency of 3.1%. When this dye is employed in solid-state solar cells, IPCE can reach 43%. The phthalocyanines (Fig. 8.14) dyes have long been used as photosensitizers. It exhibits good semiconductor properties, high-absorption efficiency for sunlight, as well as chemical stability. By altering the metal species, dye molecules of different energy, levels can be obtained, which facilitates photoelectric conversion. However, phthalocyanine exhibits a poor solubility state in organic solvents, and it can easily form an aggregate state on the surface of TiO_2. Meantime, phthalocyanine in the solution easily generates optically active dimers, adversely affecting the photoelectric conversion efficiency of the device.

(4) **Porphyrin dyes**

Porphyrins could not compete with N3 and black dye sensitizer because of the lack of absorption in infrared and near-infrared regions, whereas after 2000, porphyrins have made great breakthroughs recently. They can combine with iron, magnesium and other metal ions to form a planar square structure containing four N atoms, which can be adjusted electronic properties by substituting the porphyrin peripheral ring. Moreover, porphyrin exhibits a strong absorption in the range of 400–450 nm and

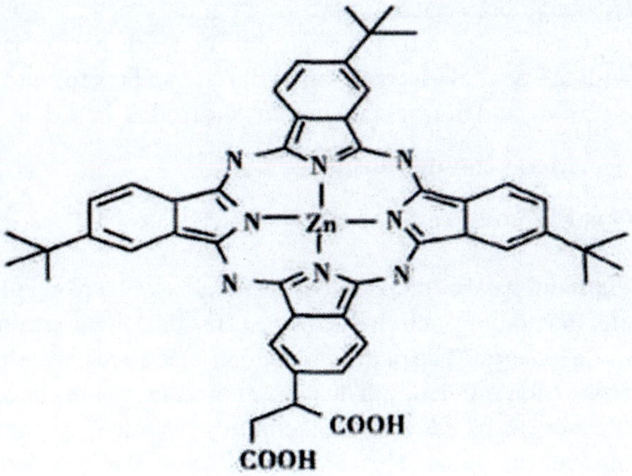

Fig. 8.14 Structure diagram of PCH001 dye

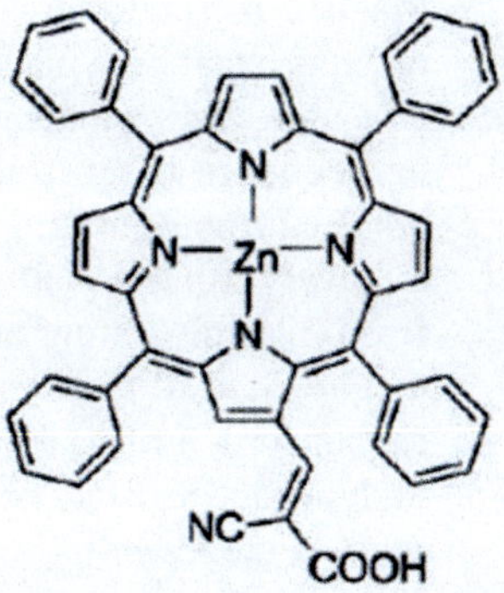

Fig. 8.15 Structure diagram of Zn-3 dye

500–700 nm, so it is expected to be employed in photovoltaic devices. The bonding state, the number of bonds, and the position of the substituents of the mentioned dyes with TiO_2 critically affect DSSC. In 2005, Wang et al. [122], reported a Zn-3 porphyrin compound, which introduced cyanoacrylic acid onto the porphyrin with an IPCE of 80% and a conversion efficiency of 5.6%. In 2007, Campbell et al. [123] based on Zn-3 dyes (Fig. 8.15), replaced polycyanuronic acid with cyanoacrylic acid to obtain another porphyrin dye with an efficiency of 7.1%. This dye is introduced in solid-state solar cells with spiro-MeOTAD as a hole transport material, and the conversion efficiency can reach 3.6%.

8.5.4 Counter Electrode

The counter electrode, also termed as photocathode, is primarily adopted to collect electrons [50]. Another major function is catalysis, increasing the rate of electron exchange between I_3^- and the counter electrode, which requires modification of the counter electrode to enhance its catalytic performance [124–126]. In accordance with the type of electrode material, it can be split into two types, i.e., metal counter electrode and non-metal counter electrode.

1. **Metal counter electrode**

(1) **Platinum counter electrode**

Platinum has become the earliest counter electrode material for DSSCs for its high electrocatalytic activity for I_3^-. Besides, it is currently the most common electrode material. The platinum counter electrode can exhibit the maximal photoelectric conversion efficiency. Thus far, numerous studies have been conducted on platinum counter electrodes, and the technology is relatively mature. The methods to prepare a platinum counter electrode primarily consist of magnetron sputtering coating method, thermal decomposition method, as well as chemical coating method [127, 128].

(1) Magnetron sputtering coating method. In the magnetron sputtering coating method, a platinum plate acts as an excitation source in vacuum 3×10^{-4} Pa, and a vacuum coating machine is adopted to sputter to form a platinum-modified film on a conductive substrate surface. The method exhibits the advantages of being non-toxic, non-polluting and no waste liquid, low-cost and energy consumption, relatively uniform platinum-modified film, robust adhesion on the substrate, and feasibility for coating on a flexible substrate. The defect of such method is that the prepared film has a dark color and poor reflectivity to transmitted light; it cannot form a regularly arranged platinum film and has numerous defects and high surface resistance (the higher the charge transfer resistance of the electrode surface, the lower the short-circuit current and fill factor of the cell will be).
(2) Thermal decomposition method. The thermal decomposition method refers to adding a mixed solution of chloroplatinic acid and water and an organic solvent to a conductive glass under heating; after drying, a platinum-modified electrode is achieved. The method exhibits the advantages that the preparation process is simple, the membrane is relatively uniform, and has a porous structure (large specific surface area). Since the counter electrode prepared with such method exhibits a large specific surface area and a prominent catalytic function for the formation of I_3^-, it produces a larger exchange current density when the electrode is working, causing less potential loss and being relatively stable, so it is expected to be applied to a glass-based DSSC on a large scale. The defect of the counter electrode prepared by the method is that thermal decomposition cannot completely reduce the high-valent platinum to the zero valence state, and the high-temperature tends to enhance the surface resistance of the conductive glass, and there are numerous defects on the surface, which cannot be employed for the flexible dye-sensitized solar cell.
(3) Chemical coating method. The chemical coating method adopts a platinum plate as an anode and a conductive substrate as a cathode to perform electroplating in a plating solution with a suitable concentration. The method exhibits the

advantages that the crystal grains grow neatly along the surface current direction of the conductive substrate during electroplating, and a bright platinum mirror is achieved on the conductive surface of the substrate, and the interference factor is small, and it is difficult to introduce impurities into the film.

The platinum film prepared by electroplating has robust adhesion, uniformity and compactness, few impurities and defects, bright surface of the film, high reflection performance, and the thickness of the film can be relatively thick. Moreover, the high-priced platinum can be fully reduced to the zero-valent state, the required temperature is low, the surface resistance is small, and the catalytic efficiency of the film is high. Thus, the electroplating method is more advantageous in the preparation of the counter electrode of the dye-sensitized solar cell. The defect is that the specific surface area of the electrode is small, and the catalytic ability of platinum is restricted.

In glass-based DSSC, platinum-modified FTO conductive glass is commonly used as a counter electrode (Pt/FTO). The modified electrode exhibits high catalytic activity for I_3^-, whereas its higher surface resistance and lower filling factor in the cell limit its applications (especially in large-area components). To further enhance the electrical properties of the platinum counter electrode, researchers have also attempted to construct platinum counter electrodes with other methods.

Fang et al. [129] studied the effect of the thickness of the prepared platinum layer on the performance of solar cells. It was reported that under the thickness of the platinum layer greater than 100 nm, the thickness slightly impacted the resistance and cell performance. Kim et al. [130] used pulse electrodeposition to deposit platinum on a conductive glass substrate, and obtained a solar cell exhibiting a photoelectric conversion efficiency of 5.0%. By thermally decomposing H_2PtCl_4, a counter electrode was prepared on a conductive glass substrate. Compared with a conventional Pt counter electrode, the obtained counter electrode has enhanced light reflection, the production cost is reduced, and the charge exchange resistance is reduced to 0.15 Ω/□, the sheet resistance is only 0.5 Ω/□, further increasing the light collection efficiency and the fill factor of the cell, and the photoelectric conversion efficiency increases from 5.6 to 8.3%, an increase of 33%. Hao Sancun prepared platinum counter electrode on ITO conductive glass using three methods: vacuum coating (magnetron sputtering), electrochemistry, and thermal decomposition. As revealed from the comparison, the platinum counter electrode prepared by the vacuum coating method declined the short-circuit current of the cell because the resistance was overly large, and the performance of the DSSC was not enhanced. The platinum counter electrode obtained by the electrochemical method increased the maximum output power of DSSC by nearly 7 times, and the open-circuit voltage and short-circuit current of the battery also elevated. The platinum content of the counter electrode prepared with the electrochemical method is relatively high, which is not compatible with the low price of DSSC. The thermal decomposition approach is simple and rapid and can achieve the control of platinum ion concentration and film thickness. Accordingly, it is a subsequent research direction to find a suitable coating solution formula, remove the residue of organic matter, and fabricate a platinum-modified counter electrode with a porous network structure, uniform thickness, and high specific surface area.

Wei et al. [131] exploited a coating method to prepare polyvinylpyrrolidine copper-coated platinum nanoclusters as a counter electrode and achieved a photoelectric conversion efficiency of 2.8%. The method exhibits a low-temperature, easy preparation and low platinum loading. Chen Jinmao et al. [132] developed a platinum counter electrode on FTO conductive glass by drawing upon the advantages of L-B film and self-assembly method, combined with the preparation of platinum colloid. The platinum particles are small in size, dense and uniform, and low in platinum loading. In the presence of approximate platinum loading, the catalytic performance of the self-assembled platinum counter electrode is 2–10 times higher than that of the thermal decomposition platinum counter electrode, and the photoelectric conversion efficiency is nearly 1 time higher. Khelashvili et al. [133] compared platinum counter electrodes prepared with different methods, e.g., hydrogen reduction, thermal decomposition, polyhydric compound reduction, and borohydride metal compound reduction. It was reported that the preparation temperature was low in hydrogen reduction, polyhydroxylation, and reduction of boron hydride metal compound. The H_2PTC_{16} precursor was thermally decomposed to prepare platinum above 580 °C, and the size of the platinum particles by XRD and TEM analysis falls in the nanometer range. The hydrogen reduction method follows a simple process, and the obtained platinum counter electrode exhibits the optimal electrochemical performance, which is a promising preparation method.

To keep abreast with the advancement of flexible DSSC, increasing flexible counter electrodes obtained by combining platinum with conductive polymers have been explored. Conductive polymers exhibit lower cost and are relatively simple to prepare, whereas they also exhibit relatively low catalytic activity and relatively low conductivity. Thus, while conducting in-depth research on platinum-modified electrodes, people have come to stress higher performance and more stable, simpler, and cheaper alternative materials [134–142].

By magnetron sputtering, Ma and Fang et al. [143, 144], prepared a platinum counter electrode on a flexible substrate of metal and plastic. To be specific, the platinum counter electrode prepared on three flexible substrates of stainless steel, nickel sheet, and ITO-encapsulated polyethylene terephthalate (ITO-PEN) achieves the photoelectric conversion efficiency similar to that of platinum counter electrode prepared on the conductive glass. The platinum counter electrode prepared on the flexible stainless steel substrate reduces the internal resistance of the cell, enhances the filling factor of the cell, and enhances the photoelectric conversion efficiency of the large-sized DSSC. Seigo et al. [145–147] electrochemically produced a platinum/ITO-PEN counter electrode on an ITO-PEN substrate. The assembled flexible dye-sensitized solar cell obtained photoelectric conversion efficiency of 7.2%, showing the application prospect of flexible substrate material in the preparation of the counter electrode. As revealed from existing studies, electrolytes containing I_3^- can corrode platinum to generate iodides (e.g., PtI_4), whereas long-term studies have not confirmed the degradation of platinum counter electrode performance, demonstrating that platinum counter electrodes have good stability. Though a small amount of Pt (approximately 50 mg/m^2) can be used for the counter electrode to obtain the required catalytic effect, the production cost of per watt electricity remains significantly high, especially when the electricity produced by the solar energy conversion system should be measured in megawatts, so the catalyst materials used are expected to be abundant and cheap.

(2) **Gold counter electrode**

Sapp et al. [148, 149] deposited 25 nm chromium and 150 nm gold on FTO conductive glass by thermal evaporation method to obtain a gold counter electrode (7.1×10^{-2} cm^2). A DSSC was prepared with I^-/I_3^- electrolyte supplemented by Co(II/III), and the relationships between electron transfer kinetics and electrode surface were obtained by cyclic voltammetry. Corresponding to the 4,4′-dmb and dtb-bpy electrolytes, the gold electrode achieved the optimal reversible complete voltammetry curve, the carbon counter electrode was quasi-reversible, and the platinum counter electrode was irreversible. The results of electrochemical tests showed that the performance of the platinum counter electrode is inferior to that of the gold counter electrode, and the gold counter electrode does not show corrosion during the test. 0.5 mol/L of LiI, 0.05 mol/L of I_2, and 0.2 mol/L of 4-tert-butylpyridine (TBP) is dissolved in methoxyacetonitrile as the electrolyte, and the maximal photoelectric conversion efficiency of the gold counter electrode is measured = 1.58%, J_{SC} = 5.32 mA/cm^2, V_{OC} = 507 mV, FF = 0.52, demonstrating that the gold electrode has a certain photoelectric conversion efficiency.

(3) **Nickel counter electrode**

Fan Leqing et al. [32] used 52 g of $NiSO_46H_2O$, 2.2 g of NaCl, 8 g of H_3BO_3, 0.24 g of $C1_2H_{25}NaO_4S$, and arranged into a 250 mL aqueous solution for nickel counter electrode deposition. In the preparation process, the conductive glass and platinum plating should be oiled and activated together and subsequently added to an electroplating solution at ambient temperature. The current density is regulated to 0.01 A/cm^2, the electroplating time is 1 min, and the thickness of nickel plating is about 0.2 μm. After testing, it is reported that the performance of the graphite electrode is inferior to that of nickel electrode. After modification by nickel, the photocurrent of the cell is enhanced. IPCE is the largest at 460 nm light wavelength, about 41.8%. However, since nickel can react with I_3^- in the electrolyte, V_{OC} declines from 478 mV before nickel modification to 468 mV after the modification, limiting the application of nickel counter electrode in DSSC.

Researchers of dye-sensitized solar cells have also attempted to use other metal materials [127, 128] (e.g., palladium, stainless steel, copper, as well as aluminum), as their counter electrodes, whereas their electrocatalytic performance is far inferior to that of platinum-modified counter electrodes. The new metal counter electrode materials and preparation methods need to be further developed.

2. **Non-metallic counter electrode**

(1) **Carbon counter electrode**

To improve the performance of the counter electrode of the DSSC and reduce the manufacturing cost of the cell, researchers have focused on finding alternative materials for the counter electrode to focus on the carbon-based material. As impacted by a good conductor of electricity, lightweight, readily available raw materials and non-toxic and pollution-free, carbon materials are significantly good electrode materials and have become a hotspot for electrode material research. At present, the research on carbon counter electrode primarily includes the following:

(1) Graphite counter electrode. The graphite counter electrode [150–154], is primarily prepared by physical coating method and heat treatment. Graphite is a layered structure and the conduction rate of electrons is limited at its intermediate interface. Moreover, the process of preparing the graphite electrode is unmatured, and the graphite and the conductive substrate are not bonded, which increases the surface resistance of the electrode and adversely affects the stability of the graphite electrode.
(2) Carbon nanotube counter electrode. As impacted by the large specific surface area, high catalytic activity, high longitudinal conductivity, and corrosion resistance to electrolytes, carbon nanotubes are a promising counter electrode material. Accordingly, the carbon nanotube counter electrode is expected to be introduced in a large-area dye-sensitized solar cell module group in the future. In 2003, the Suzuki group prepared single-walled carbon nanotube counter electrodes on conductive glass and Teflon films, respectively [142], exhibiting high catalytic activity for DSSC electrolytes. In 2008, the Lee group [23, 93], by spray coating, prepared a multi-layer carbon nanotube counter electrode on the FTO conductive film. The power conversion efficiency of the cell prepared by this electrode reached 7.59%, which is higher than that of the single-walled carbon nanotube electrode. Meantime, the relationship between the filling factor, the charge transport resistance and the spraying time was investigated. The counter electrode of the dye-sensitized solar cell fabricated by a research team at the New Jersey Institute of Technology exploits a carbon nanotube complex [152], which displays a cylindrical carbon molecular structure. They combine carbon nanotubes with a C_{60} regular icosidodecahedron (i.e., fullerene). Though the C_{60} regular icosidodecahedron cannot generate current, it can capture electrons. The sunlight will activate the polymer, and the C_{60} regular icosidodecahedron will trap electrons. Then, the nanotubes act like wires to generate electrons or current. Applying this carbon nanotube composite electrode to the organic solar cell can also enhance its performance. However, because the conventional preparation technology of carbon nanotubes is not perfect and the cost is high, the mentioned counter electrodes have not been extensively used.
(3) Carbon black counter electrode. Imoto et al. [141] prepared counter electrodes with activated carbon, graphite, carbon black, and glassy carbon, and its performance was studied: The improvement of the performance of the electrode was

determined by the roughness (specific surface area) of the carbon material. To be specific, carbon black exhibits the advantages of high specific surface area, strong catalytic activity, corrosion resistance to electrolyte, mature preparation process, and low-cost. Carbon black is a good alternative to the platinum counter electrode. In 1996, Kay et al. [155], introduced 20% carbon black as a carbon counter electrode (FTO glass substrate) to graphite. The high specific surface area of the carbon black was exploited to enhance the catalytic activity of the electrode, and the carbon black aggregate filled a part of the void between the graphite structures, enhancing the conductivity of the electrode. $I_{SC} = 4.425$ mA, $V_{OC} = 825.9$ mV, FF = 0.712, $\eta = 3.89\%$ of the cell, lower than the conversion efficiency of platinum counter electrode prepared based on magnetron sputtering of 4.3%. To reduce the charge exchange resistance of the I^-/I_3^- ion pair on the carbon counter electrode, to inhibit the back-reaction of photoelectrons and oxidation dyes (or I_3^-), and to improve the filling factor, people have increased the roughness of the carbon counter electrode maximally. Meantime, the V_{OC} of the carbon counter electrode DSSC is approximately 60 mV larger than that of the platinum counter electrode as impacted by the positive displacement of the I^-/I_3^- potential. In 2006, Murakami et al. [156], reported another high-performance carbon counter electrode prepared at a carbon black and TiO_2 mass ratio of 93: 7, respectively. The thickness of the carbon layer of the counter electrode impacts the fill factor and photoelectric conversion efficiency. Under the thick carbon layer, the charge exchange resistance of the carbon counter electrode is less than three times that of the platinum counter electrode. When the carbon layer thickness is 14.47 μm, this cell exhibits $J_{sc} = 16.8$ mA/cm^2, $V_{oc} = 789.8$ mV, $FF = 0.685$, = 9.1%. Huang et al. used hydrothermal and high-temperature treatment of sugar solution to obtain two types of amorphous carbon materials, HCS-1 and HCS-2, of which HCS-2 displays a high specific surface area (800 m^{2}g), the carbon counter electrode prepared by the screen printing technique on the FTO conductive glass exhibits a photoelectric conversion rate of 5.7%. Suzuki et al. employed a carbon nanotube to prepare a counter electrode, which achieves a conversion efficiency of 4.5% (0.25 cm^2).

Researchers have also reported carbon counter electrodes prepared on flexible plastic substrates. Lindstrom employed a mixture of graphite and carbon black to prepare a carbon counter electrode (0.39 cm^2) on ITO-PET flexible substrate by pressing. The flexible dye-sensitized solar cell assembled with this counter electrodes achieves a certain photoelectric efficiency, whereas the photoelectric conversion efficiency at a large current density remains to be studied.

The catalytic activity of the carbon material is located on the crystal edges. Since carbon black has considerable crystal edges, its catalytic activity is higher than that of highly oriented carbon materials (e.g., graphite and carbon nanotubes). When a carbon material acts as the counter electrode catalyst, its catalytic effect and resistance are dependent on the thickness of the carbon layer. The catalytic reaction rate of the carbon counter electrode is slower than that of the platinum counter electrode for the significantly large charge transition resistance R_{ct} of the redox reaction of I^-/I_3^- on the carbon electrode surface. Such a situation exists for all types of carbon counter electrodes. Thus, the photoelectric conversion efficiency of the carbon counter electrode DSSC is smaller than that of the platinum counter electrode.

(2) **Polymer counter electrode**

Saito et al. [157] employed poly (p-toluenesulfonic acid) (TsO) doped poly (3,4-dioxyethylthiophene) to prepare polymer PEDOT:TsO counter electrodes. The method is to add monomer 3,4-dioxoethylthiophene to the butanol solution of iron p-toluenesulfonate and imidazole, drip coating on the top surface of FTO glass and rotating (1000 r/min) to form a film, and heating at 110 °C for 5 min to polymerize the membrane, wash off the iron p-toluenesulfonate in the membrane with methanol, and obtain a PEDOT:TsO counter electrode after drying. The film thickness was regulated by the number of coatings and the solution concentration. Then, the PEDOT:TsO counter electrode was porous structure, the specific surface area of the film increases as the thickness increases, and R_{ct} tends to decrease. If an ionic liquid acts as the electrolyte, a 2 μm-thick PEDOT-TsO counter electrode has $J_{sc} = 10.2$ mA/cm^2, $V_{oc} = 603$ mV, $FF = 0.7$, $\eta = 3.93\%$, which is better than the performance of the porous structure platinum electrode prepared by magnetron sputtering.

Muto et al. [158] adopted poly (3,4-dioxyethylthiophene) doped polystyrene sulfonic acid (PSS) as the PEDOT-PSS counter electrode to prepare the cell on a flexible ITO-PEN conductive substrate, and obtained photoelectric conversion efficiency of 2.72%. To enhance the reduction catalytic performance of iodine, nano-TiO_2 particles were added to the dispersed phase of PEDOT-PSS in water–ethanol to produce a slurry, and a translucent counter electrode was prepared by the imprint coating method, which enhanced the efficiency of the flexible DSSC to 4.38%.

Hayase et al. [157] found that the performance of PEDOT counter electrode will vary noticeably with the type of electrolyte (i.e., organic liquid, ionic liquid, and ionic gel). Accordingly, by selecting a suitable electrolyte and optimizing the structure of PEDOT, e.g., porosity, thickness, doping ions, etc., the photoelectric performance of the polymer counter electrode can be further improved and the production cost can be reduced.

(3) **Copper oxide counter electrode**

CuO refers to a narrow bandgap (1.2 eV) p-type semiconductor material. As impacted by its good optoelectronic, photochemical, catalytic properties, and low price, it has attracted the attention of DSSC researchers.

Anandan et al. placed copper foil (1 cm^2) treated with hydrochloric acid in an aqueous solution of ammonia and NaOH with a certain concentration. Then, CuO nanorods arranged neatly are formed on the copper foil surface; Subsequently, they are washed with deionized water and dried to generate a CuO counter electrode. The surface morphology of CuO nanorods is dependent on the pH of the solution. The p-type CuO counter electrode is assembled with a photoelectrode prepared from n-type TiO_2 to fabricate a dye-sensitized heterojunction cell. The photoelectric generation mechanism is nearly identical to that expressed by Tennakone. The CuO prepared by the reaction in the solution of pH 12.3 for 14 days exhibits high catalytic performance, $I_{sc} = 0.45$ mA, $V_{oc} = 564$ mV, $FF = 0.17, = 0.29\%$. The fill factor of the cell can be promoted by further enhancing the performance of the CuO counter electrode, reducing interface defects, down-regulating the impedance of the CuO nanorod film, as well as introducing Al_2O_3 or the MgO barrier layer reduces the electron–hole recombination. However, the studies on CuO counter electrode has just begun, its catalytic effect remains relatively poor compared with platinum, and the photoelectric conversion efficiency requires further enhancement.

(4) **CoS counter electrode**

Wang et al., with the electrochemical method, deposited a layer of nanometer CoS particles on a flexible ITO/PEN conductive substrate, and fabricated the counter electrode with good catalytic activity on iodide (I^-/I_3^-); its catalytic capacity is close to that of the platinum-modified electrode. The photoelectric efficiency of the DSSC with this counter electrode can reach 6.5%. However, its long-term stability is not ideal. After one month, the open-circuit voltage and fill factor of the cell have dropped noticeably.

Though the platinum counter electrode exhibits high energy conversion efficiency, its high cost limits its application in the industrialization of dye-sensitized solar cells. The catalytic activity of other metal electrodes is significantly less than that of platinum, which requires further development. The future development direction of the electrode should be the electrode material exhibiting stable development performance, low-cost, low surface resistance, prominent catalytic activity, high stability, simple preparation process, and suitable for preparing large-area cells. Though the photoelectric conversion efficiency of a cell prepared with a carbon material and a high molecular polymer counter electrode remains lower than that of a platinum counter electrode cell, the performance of a cell prepared by a counter electrode material (e.g., a porous high specific surface area carbon material and a conductive high molecular polymer) is gradually enhanced. The existing research on the counter electrode of dye-sensitized solar cells remains dominated by platinum, and the research and exploration of other counter electrodes tend to show a diversified trend.

8.6 Outlook

As a high-efficiency and low-cost solar cell, dye-sensitized solar cells have attracted widespread attention from scientists all over the world. Given the current status of research and development, there are several aspects of work to be conducted subsequently:

(1) In the transmission of electrons, as impacted by the existence of a large number of grain boundaries between the nanocrystals, the transmission efficiency of photogenerated electrons is low, and reverse recombination is easy to occur, making

it difficult to further enhance the photoelectric conversion efficiency. If a semiconductor single crystal nanowire array structure with high spatial orientation can be constructed on a conductive substrate, it is a feasible way to significantly enhance the photoelectric conversion efficiency of the dye-sensitized cell.

(2) Since an electrolyte solution is exploited by the dye-sensitized cell, it is difficult to encapsulate and easy to leak; as a result, the stability of the photovoltaic cell is reduced, and the service life is shortened. Thus, new electrolyte materials should be found. For instance, alkaline components to the solution are added to increase the photovoltage, high-conductivity components (e.g., ionic liquids) are added to enhance the conductivity and filling factor, and polymer gels are added to synthesize quasi-solid gel polymers, etc. It is noteworthy that the use of a solid electrolyte can maintain the high conductivity and high conversion efficiency of the liquid electrolyte, and problems (e.g., easy leakage of the liquid electrolyte and difficulty in packaging) can be addressed.

(3) Looking for dye sensitizers with excellent performance (e.g., pure organic dyes or co-sensitized dyes) will also become a vital development direction for dye-sensitized cells. The development of new dyes aims to enable them to absorb photon energy and exhibit high quantum efficiency in the full visible spectral range and near-infrared spectral region. The major problem in the existing study on extended absorption spectrum is that after the absorption light is expanded, the energy difference between the dyes I^-/I_3^- decreases with the increase in HOMO, and the recovery rate of the dye decreases; with the decrease in the LUMO, the rate of electron injection into the TiO_2 conduction band decreases. Pure organic dyes exhibit better absorption in the long wavelength range. The most significant advantage is that it exhibits a low-cost and can exactly remedy the defects of N3 dye. However, the conversion efficiency of pure organic dyes remains far from N3. Though pure organic dye-sensitized solar cells exhibit conversion efficiencies of only 1–2%, their successful synthesis allows for assembling co-sensitized dye solar cells.

(4) Conducting more in-depth research on the internal mechanism of electron injection and transmission, which will more effectively optimize the cell and design dye-sensitized solar cells more conducive to light absorption, electron injection, and transmission. A solid foundation has been laid for the practical use of dye-sensitized solar cells.

(5) Increase the open-circuit voltage of the cell. Thus far, there have been rare reports on the nature of photovoltage and its major factors, and the underlying mechanism remains unknown. The dye-sensitized solar cells, currently prepared, exhibit lower open-circuit voltages, generally less than 1 V, limiting its application in practice. How to increase the open-circuit voltage will be a direction of subsequent research. Promoting the open-circuit voltage will significantly enhance the photoelectric conversion efficiency.

(6) Electrode materials and their modifications. At present, the common cathode modification material is platinum, whereas the price of platinum is overly high and the modification process is difficult to control, so it is also one of the

problems to be concerned within the search of an electrode material exhibiting a low price and prominent performance.

(7) Preparation of large-area cells. To make the dye-sensitized solar cell practical and industrialized, research on a large-area cell is required.

In brief, dye-sensitized solar cells are characterized by a lower cost and simple manufacturing process, which outperforms other types of solar cells. Though some problems remain at present, with the continuous advancement of technology, its broad application prospects will be highlighted, and it will indeed be more practical, which will more effectively cope with human energy needs.

References

1. C.A. Bequerel, Acad. Sci. **9**, 145 (1893)
2. J. Morser et al., Chem. **8**, 373 (1887)
3. H. Tsubomura et al., Nature **261**, 402 (1976)
4. B. O'Regan et al., Nature **353**, 737 (1991)
5. S. Ruile et al., Inorg. Chimica. Acta. **261**, 129 (1997)
6. T. Yoshida et al., in *The 16th International Conference of Photochemical Conversion and Solar Storage* (2006)
7. T. Miyasak et al., in *The 16th International Conference of Photochemical Conversion and Solar Storage* (2006)
8. M.K. Nazeeruddin et al., J. Am. Chem. Soc. **115**, 6382 (1993)
9. M.K. Nazeeruddin et al., J. Am. Chem. Soc. **123**, 1613 (2001)
10. M. Gratzel, Nature **6861**, 338 (2001)
11. J.H. Yum et al., J. Photochem. Photobiol. A **173**, 1 (2005)
12. F. Pichot et al., Langmuir **16**, 5626 (2000)
13. C. Longo et al., J. Phys. Chem. B **106**, 5925 (2002)
14. H. Lindström et al., Nano Lett. **1**, 97 (2001)
15. H. Lindström et al., J. Photochem. Photobiol. A **145**, 107 (2001)
16. T. Miyasaka, Chem. Lett. **31**, 1250 (2002)
17. T. Miyasaka et al., J. Electrochem. Soc. **151**, A1767 (2004)
18. M. Dürr et al., Nat. Mater. **4**, 607 (2005)
19. Miyasaka T., et al., in *The 16th International Conference of Photochemical Conversion and Solar Storage* (2006)
20. S. Ito et al., Chem. Commun. **38**, 4004 (2006)
21. Q.B. Meng et al., Appl. Phys. Lett. **77**, 4313 (2000)
22. Q.B. Meng et al., Chem. Mater. **14**, 83 (2002)
23. J.H. Yoon et al., J. Photochem. Photobiol. A Chem. **180**, 184 (2006)
24. M. Adachi et al., J. Electrochem. Soc. **150**, G48 (2003)
25. M. Adachi et al., J. Am. Chem. Soc. **126**, 14943 (2004)
26. M. Zukalova et al., Nano Lett. **5**, 1789 (2005)
27. Y.J. Ren et al., Electro Chem. **8**, 5 (2002)
28. L.Y. Zeng et al., Chin. Phys. Lett. **21**, 1835 (2004)
29. S. Ito et al., Chem. Commun. **41**, 4351 (2005)
30. P.M. Sommeling et al., J. Phys. Chem. B **110**, 19191 (2006)
31. Z.S. Wang et al., J. Phys. Chem. B **105**, 9210 (2001)
32. S.C. Hao et al., Sol Energy **76**, 745 (2004)
33. Z.S. Wang et al., Chem. Mater. **13**, 678 (2001)
34. S.S. Kim et al., J. Photochem. Photobiol. A Chem. **171**, 269 (2005)

35. F. Lenzmann et al., J. Phys. Chem. B **105**, 6347 (2001)
36. Y. Diamant et al., J. Phys. l Chem. B **107**, 1977 (2003)
37. S.G. Chen et al., Chem. Mater **13**, 4629 (2001)
38. S.M. Yang et al., Chem. Mater. **14**, 1500 (2002)
39. J. Bandara et al., Sol. Energy Mater. Sol. Cells **88**, 341 (2005)
40. Z.S. Wang et al., Chem. Mater. **18**, 2912 (2006)
41. A. Kay et al., Chem. Mater. **14**, 2930 (2002)
42. B. Peng et al., Chem. Rev. **248**, 1479 (2004)
43. J.B. Xia et al., Chem. Lett. **35**, 252 (2006)
44. Y.Q. Wang et al., J. Mater. Sci. **34**, 2773 (1999)
45. J. Ferber et al., Sol. Energy Mater. Sol. Cells **54**, 265 (1998)
46. K. Temnakone, Semiconductor. Sci. Technol. **10**, 1689 (1995)
47. C. Fenabdi et al., Sol. Energy Mater. Sol. Cells **58**, 337 (1999)
48. 孟庆波等, 物理 **33**, 177 (2004)
49. L.M. Peter et al., Electrochem. Commun. **1**, 576–580 (1999)
50. 郝三存等, 感光科学与光化学 **12**, 175–181 (2004)
51. 戴松元等, 高等学校化学学报 **26**, 1102–1105 (2005)
52. B. O'Regan et al., Nature **353**, 737–740 (1991)
53. P.V. Kamat, J. Phys. Chem. C **111**, 2834 (2007)
54. M. Gratzel, Prog. Photovolt Res. Appl. **8**, 171–185 (2000)
55. A.K. Ghosh et al., J. Appl. Phys. **45**, 230–236 (1974)
56. L. Gang et al., Nat. Mater. **4**, 864–868 (2005)
57. J. Desilvestro et al., J. Electrochem. Soc. **133**, 331–336 (1986)
58. G. Kunara et al., Sol. Energy Mater. Sol. Cell **69**, 195–199 (2001)
59. B. Oregan et al., Chem. Mate. **10**, 1501–1509 (1998)
60. K. Tennakone et al., App. Phys. Lett. **77**, 2367–2371 (2000)
61. J.T. Park et al., J. Nanosci. Nanotecho. **11**, 1718–1721 (2011)
62. D.B. Kuang et al., in *Ionic Liquids IV: Not Just Solvents Anymore*, vol. 975 (2007), pp. 212–219
63. K. Tennakoke et al., Semicond. Sci. Technol. **12**, 128–132 (1997)
64. K. Tennakone et al., J. Phys. D Appl. Phys. **31**, 1492–1495 (1998)
65. P. Simimanne et al., Sci. Technol. **18**, 708–712 (2003)
66. T. Taguchi et al., Chem. Commun. **19**, 2480–2481 (2003)
67. G. Kumara et al., J. Photochem. Photobiol. A. **164**, 183–185 (2004)
68. G. Kumara et al., Chem. Mater. **14**, 954–955 (2002)
69. Q. Meng et al., Langmuir **19**, 3572–3574 (2003)
70. G. Kumara et al., Langmuir **18**, 10493–10495 (2002)
71. B. O'regan et al., Chem. Mater. **14**, 5023–5029 (2002)
72. L. Sicot et al., Synth. Met. **102**, 991–992 (1999)
73. M. Catellani et al., Thin Solid Films **203**, 66–70 (2002)
74. J. Schn et al., Nature **403**, 408–410 (2000)
75. U. Bach et al., Nature **395**, 583–585 (1998)
76. J. Kruger et al., Appl. Phys. Lett. **79**, 2085–2087 (2001)
77. J. Kruger et al., Appl. Phys. Lett. **81**, 367–370 (2002)
78. H.J. Snaith et al., Nano. Lett. **7**, 3372–3376 (2007)
79. V. Jovanovski et al., Thin Solid Films **511**, 634–637 (2006)
80. K. Hara et al., Chem. Commum. **6**, 569–570 (2001)
81. K. Hara et al., J. Phys. Chem. B **107**, 597–606 (2003)
82. K. Hara et al., New J. Chem. **27**, 783–785 (2003)
83. Z.S. Wang et al., J. Phys. Chem. B **109**, 3904–3907 (2005)
84. Z.S. Wang et al., Adv. Mater. **19**, 1138–1141 (2007)
85. K. Hara et al., Sol. Energy Mater. Sol. Cells **77**, 89–103 (2003)
86. Q.H. Yao et al., New J. Chem. **27**, 1277–1283 (2003)
87. Y.S. Chen et al., J. Mater. Chem. **15**, 1654–1661 (2005)
88. K. Hara et al., J. Chem. Commun. **2**, 252–253 (2003)

89. S. Cherian et al., **104**, 3624—3629 (2000)
90. F. Fungo et al., J. Mater. Chem. **10**, 645–650 (2000)
91. S. Ito et al., J. Adv. Mater. **18**, 1202–1205 (2006)
92. D.P. Hagberg et al., J. Chem. Commun. **21**, 2245–2247 (2006)
93. S. Hwang et al., J. Chem. Commun. **46**, 4887–4889 (2007)
94. D.P. Hagberg et al., J. Am. Chem., Soc, 130, 6259–6266 (2008).
95. H. Tian et al., J. Chem. Commun. **36**, 3741–3743 (2007)
96. Y. Bai et al., Nat. Mater. **7**, 626–630 (2008)
97. D. Shi et al., J. Phys. Chem. C **112**, 17046–17050 (2008)
98. H. Qin et al., J. Am. Chem. Soc. **130**, 9202–9203 (2008)
99. J. Ferguson et al., J. Chem. Phys. Lett. **68**, 21–24 (1979)
100. J. Delilvestro et al., J. Am. Chem. Soc. **107**, 2988–2990 (1985)
101. P. Liska et al., J. Am. Chem. Soc. **110**, 3686–3687 (1988)
102. R. Argazzi et al., J. Inorg. Chem. **33**, 5741–5749 (1994)
103. O. Kohle et al., J. Inorg. Chem. **35**, 4779–4787 (1996)
104. M.K. Nazeeruddin et al., J. Inorg. Chem. **38**, 6298–6305 (1999)
105. A.M. Bond et al., J. Electrochem. Soc **146**, 648–656 (1999)
106. M.K. Nazeeruddin et al., J. Am. Chem. Soc. **127**, 16835–16847 (2005)
107. P. Wang et al., Adv. Mater. **16**, 1806–1811 (2004)
108. C. Klein et al., J. Inorg. Chem. **44**, 178–180 (2005)
109. P. Wang et al., Adv. Mater. **15**, 2101–2104 (2003)
110. P. Wang et al., J. Am. Chem. Soc. **127**, 808–809 (2005)
111. A. Islam et al., Inorg. Chim. Act **322**, 7–16 (2001)
112. A. Islam et al., Chem. Lett. **29**, 490–491 (2000)
113. A. Islam et al., New J. Chem. **26**, 966–968 (2002)
114. T. Renouard et al., Inorg. Chem. **41**, 367–378 (2002)
115. T. Renouard et al., Tetrahedron **57**, 8145–8150 (2001)
116. R. Ghanem et al., Inorg. Chem. **41**, 6258–6266 (2002)
117. J. Pan et al., J. Phys. Chem. B **108**, 12904–12910 (2004)
118. R. Amadelli et al., J. Am. Chem. Soc. **112**, 7099–7103 (1990)
119. M.K. Nazeeruddin et al., J. Helv. Chim. Acta. **73**, 1788–1803 (1990)
120. C.A. Bignozzi, J. Chem. Soc. Rev. **29**, 87–96 (2000)
121. P.Y. Reddy et al., J. Angew. Chem. Int. Ed **46**, 373–376 (2007)
122. Q. Wang et al., J. Phys. Chem. B **109**, 15397–15409 (2005)
123. W.M. Campbell et al., J. Phys. Chem. C **111**, 11761–11762 (2007)
124. S. Dai et al., Sol. Energy Mater. Sol. Cells **84**, 125–133 (2004)
125. R. Sastrawan et al., Prog. Photovolt Res. Appl. **15**, 612–616 (2006)
126. 曾隆月等, 中国科学院研究生院学报 **21**, 393–397 (2004)
127. 李国等, 材料导报 **12**, 16 (2007)
128. 尹艳红等, 材料导报 **5**, 110 (2009)
129. X.M. Fang et al., J. Photochem. Photobiol. A **164**, 179 (2004)
130. S.S. Kim et al., Electrochimca. Acta **51**, 3814 (2006)
131. T.C. Wei et al., Appl. Phys. Lett. **88**, 103122 (2006)
132. 陈今茂等, 科学通报 **50**, 28 (2005)
133. G. Khelashvili et al., Thin Solid Films **511**, 342 (2006)
134. M.A. Green et al., Prog. Photovolt Res. Appl. **12**, 55 (2004)
135. M. Gratzel, C.R. Chimie, **9**, 578 (2006)
136. J.J. Lagref et al., Synth. Met. **138**, 333 (2003)
137. J. Lagemaat et al., J. Phys. Chem. B **105**, 11194 (2001)
138. B.O. Regan et al., Chem. Mater **14**, 5023 (2002)
139. M. Shijni et al., J. Pbotochem. Photobiol. A **148**, 33 (2002)
140. K. Imoto et al., Sol. Energy Mater. Sol. Cells **79**, 459 (2003)
141. K. Suzuki et al., Chem. Lett **32**, 28 (2003)
142. T.L. Ma et al., J. Electroanal. Chem. **574**, 77 (2004)

143. X.M. Fang et al., Thin Solid Films **472**, 242 (2005)
144. L. Seigo et al., Chem. Commun. **38**, 4004–4006 (2006)
145. X.M. Fang et al., J. Electroanal. Chem. **570**, 257 (2004)
146. H. Ago et al., Adv. Mater **11**, 1281 (1999)
147. S.A. Sapp et al., J. Am. Chem. Soc. **124**, 11215 (2002)
148. H.J. Snaith et al., Adv. Mater. **19**, 387 (2007)
149. H.S. Wroblowa et al., J. Electroanaly Chem. **42**, 329 (1973)
150. M.R. Tarasevich et al., in *Modem Aspects of Electrochemistry* (Plenum Press, 1989)
151. H. Zhen et al., Electrochem. Commun. **9**, 596 (2007)
152. T.N. Murakami et al., J. Electrochem. Soc. **153**, 2255 (2006)
153. H. Tetsuo, Carbon **44**, 880 (2006)
154. A. Kay et al., Sol. Energy Mater. Sol. Cells **44**, 99 (1996)
155. S. Yasuteru et al., J. Photochem. Photobiol. A **164**, 153 (2004)
156. T. Muto et al., in *16th International Conference of Photochemical Conversion and Storage of Solar Energy* (2006)
157. Hayase S., et al., Organic photovoltaics IV, in *Proceedings of SPIE*, vol. 52 (2006), pp. 15–16
158. S. Anandan et al., Mater. Chem. Phys. **93**, 35 (2005)

Chapter 9
Organic Solar Cells

Solar energy is inexhaustible, clean and pollution-free, and it has become increasingly popular over the past few years. Solar cells represented by inorganic semiconductor materials (e.g., single crystalline silicon, polycrystalline silicon, and III–V compounds) are highly efficient, whereas considerable energy is required, and serious pollution is caused in the manufacturing process. It is noteworthy that the dominant solar cells based on silicon have a relatively high cost, which has significantly impeded the development and popularity of solar cells. Organic materials have aroused widespread attention for their low-cost, high light absorption coefficient, light texture, prominent flexibility, and simple manufacturing process [1].

A wide variety of organics are employed in solar cells (e.g., macromolecular polymers, micromolecular materials, and fullerene derivatives), whereas the most successful combination to date refers to a polymer-fullerene solar cell. In such a solar cell, the polymer and the fullerene act as an electron donor and an electron acceptor, respectively. In 1992, Sariciftci et al. [2], first identified the effective electron transfer from organic semiconductor materials to fullerene C_{60}. Earlier polymer-fullerene organic solar cells comply with the soluble polymer Poly(2-methoxy-5-(2-ethylhexyloxy)-1,4-phenylenevinylene) (MEH-PPV) and fullerene C_{60}. It displays a double-layer heterojunction structure and exhibits a low efficiency of only about 0.4% [3], whereas it has started to boom in the study on organic solar cells. The next milestone is considered the application of bulk heterojunction structures in the organic solar cell [4]. Mixing the donor material and the acceptor material together to form an interconnected network structure can effectively expand the contact area and enhance the efficiency. The bulk heterojunction solar cells based on MEH-PPV and phenyl-C_{61}-butyric acid methyl ester ($PC_{60}BM$) can exhibit an efficiency of 2.5%, noticeably higher than that of the double-layer heterojunction structure. By optimizing the morphology of the active layer and the interface characteristics of the various layers of the device, the most frequently used organic bulk heterojunction solar cells based on Poly(3-hexylthiophene) (P3HT) and $PC_{60}BM$ systems can exhibit the efficiency above 5% [5]. Many original researches have highlighted the physical and chemical characteristics of materials, as well as the working

C. Zhang et al., *Semiconductor Photovoltaic Cells*,
https://doi.org/10.1007/978-981-15-9480-9_9

mechanism and structure of devices. Besides with a novel material system, the critical physical methods to enhance efficiency are introducing an interface layer between the effective layer and the electrode to optimize the interface characteristics, as well as advancing various methods to prepare the solar cells. For instance, annealing methods are employed in almost all devices. For chemical methods, new organic compounds have been increasingly developed and introduced in devices. Overall, materials exhibiting narrow bandgaps can facilitate the absorption of photons and ultimately enhance the efficiency of devices. The most successful acceptor materials refer to the materials in the fullerene system. The most common are the fullerene derivatives $PC_{60}BM$, the new fullerene derivatives $PC_{70}BM$, and $BisPC_{60}BM$ exhibiting higher efficiency [6, 7]. The present chapter will brief organic solar cells.

9.1 Features of Organic Semiconductors

Organic semiconductors are carbon-based materials, exhibiting the basic properties of semiconductors as well. It is all known that inorganic semiconductor materials are all composed of covalent structures. In contrast, the atoms in organic semiconductor molecules are linked by conjugated π bonds, and molecules are connected via van der Waals forces. Such bonding structure endows organic semiconductors with unique flexibility, lightweight, and relatively low sublimation temperatures which facilitate the manufacturing.

From a macroscopic perspective, the band structure of organic semiconductors can be handled as inorganic semiconductors. In organic semiconductors, the Highest Occupied Molecular Orbital (HOMO) is similar to the top of the valence band, while the Lowest Unoccupied Molecular Orbital (LUMO) presents similarity to the bottom of the conduction band. HOMO and LUMO represent the bonding and anti-bonding states of conjugated π electrons, respectively. The valence band is almost filled by electrons, while the conduction band is nearly empty.

Organic semiconductors consist of organic molecules formed by π bonds. The carbon atoms and the surrounding atoms form three robust σ bonds, and the residual P orbitals of the carbon atoms form a shared electron cloud via forming π bonds. Such bonding structure makes the conjugated organic semiconductor molecules exhibit a quasi-one-dimensional structure. Depending on the overlap of the wave functions of the electrons on adjacent atoms, the π-bond system is capable of exhibiting different bonding states. For instance, Fig. 9.1, suggests two different states of the π bond, i.e., the bonding state and the anti-bonding state in accordance with different energy levels.

Different hybrid states of the π bond form a range of energy levels in organic semiconductors, and the HOMO and LUMO of organic semiconductors refer to the energy bands formed based on the mentioned different energy levels. After the electrons are excited from the HOMO to the LUMO of the organic semiconductor, the molecule itself is also excited to a higher energy state, which differs from the

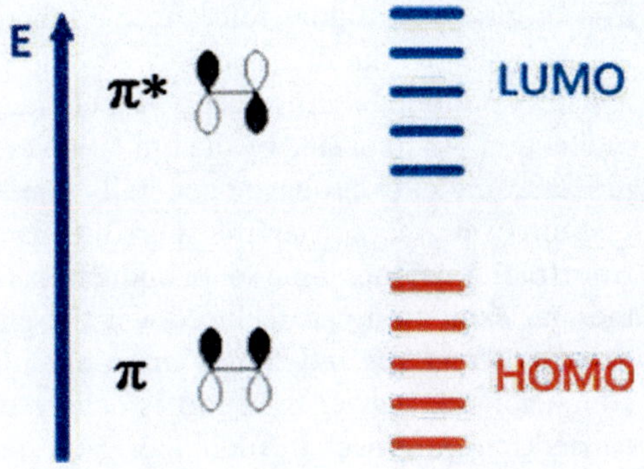

Fig. 9.1 Schematic of bonded and anti-bonded states in organic semiconductors [8]

case where the electron is excited from the valence band to the conduction band in the inorganic semiconductor.

When the energy of incident photons exceeds the optical absorption bandwidth of the organic semiconductor material, the photon is absorbed by the organic material. However, unlike the case of inorganic semiconductors, the absorption of photons in organic semiconductors is robustly bound electron-hole pairs—excitons on the whole, instead of being free electrons and holes. This is because the relatively low dielectric constant of organic materials leads to a robust Coulomb attraction between electrons and holes. Meantime, organic semiconductor molecules are connected via van der Waals force, causing weak interaction between them, as shown in Fig. 9.2.

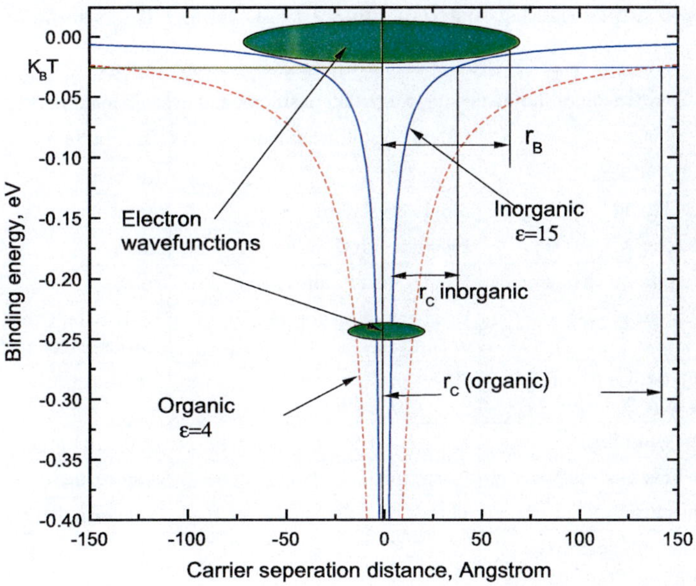

Fig. 9.2 Comparison of the binding energy between charges in an organic and inorganic matter [9]

The origin in the figure refers to the position of a positive charge. The Coulomb radius indicates the distance between the charges under the equal thermal energy and the Coulomb binding energy. V_B characterizes the strength of the intermolecular interaction. As impacted by the limitation of the electronic spin state by the transition process, singlet excitons are generally generated.

Moreover, the carrier transport mechanism in organic semiconductors differs from traditional inorganic semiconductors. In organic semiconductors, carriers under thermal excitation can tackle down the energy barriers in disordered conjugated polymer structures and "jump" from one place to another, i.e., in the organic semiconductor, the carrier transport is achieved by "jumping", which differs from the carrier transport mechanism in inorganic semiconductors. In inorganic semiconductors, free carriers can be transported in the conduction band or valence band for the presence of a robust shared motion. The presence of the "jump" transport mechanism makes the mobility of carriers in organic semiconductors noticeably lower than that in inorganic semiconductors. The mobility of holes in silicon can reach 450 $cm^2 V^{-1} s^{-1}$, whereas that of holes in micromolecular organic semiconductors can only reach 1.5 $cm^2 V^{-1} s^{-1}$. Besides, the mobility of electrons in silicon can exceed 1400 $cm^2 V^{-1} s^{-1}$, whereas that of electrons in micromolecular organic semiconductors can only reach about 0.1 $cm^2 V^{-1} s^{-1}$. In the application of solar cells, the mobility of electrons and holes in the most common P3HT: PCBM system is on the order of 0.001–0.0001 $cm^2 V^{-1} s^{-1}$. Compared with inorganic semiconductor materials, the lower mobility of organic semiconductors is obviously considered a defect, so highlight should be placed on device design and manufacturing. Single crystal silicon is taken as an example, and the differences between organic materials and inorganic materials are summarized, as listed in Table 9.1 [9]. Table 9.1 lists some of the prop-

Table 9.1 Comparison of inorganic single crystalline silicon and organic materials

Material	Single crystalline silicon	Organic semiconductor
Basic building blocks	Atom	Molecule
Crystal mechanism	Single crystalline	Amorphous
Mobility μ ($cm^2 V^{-1} s^{-1}$)	Electron: 1500 Hole: 450	$\ll$0.1
Carrier transport mechanism	Band transmission	Jump
Transport characteristics	Be able to transport two types of carriers	Usually only one kind of carrier can be transmitted
Relationship between carriers and temperature T	$T\uparrow \rightarrow \mu\downarrow$	$T\uparrow \rightarrow \mu\uparrow$
Optical bandwidth E_{go} (eV)	1.1	Generally greater than 2
Absorption near E_{go} (cm^{-1})	Approximately 10^3	Approximately 10^5
Relative permittivity	11.9	Approximately 2–4
Ratio of Coulomb potential well to Bohr radius ($\gamma = r_C/r_B$)	<1	>1
Binding energy of excitons, 300 K (meV)	<26	>100

erties of typical organic semiconductor materials, which may be offset for specific materials.

9.2 Materials for Organic Semiconductor Photovoltaic Devices

The general characteristics of organic materials are introduced previously. The present section will focus on materials employed in organic solar photovoltaic devices.

Polymer materials are generally used for p-type organics in organic solar cells. The bandgap and the design of LUMO in the mentioned materials are critical to the external quantum efficiency and fill factor of the device. Accordingly, the difference in energy levels between HOMO and LUMO is critical to organic polymers. To control the bandgap of organic semiconductors, the following six vital parameters should be rigorously considered: molecular weight, alteration in bond length, flatness, atomic resonance level, substitution energy, and intermolecular interaction. To obtain organic polymers with narrow band gaps, all six factors cannot be ignored, and they all impact device efficiency. To enhance the solubility of polymers, the alkyl chain should be lengthened; as a result, the substitution energy is affected, and eventually the twist angle increases. Thus, any novel material design is a process in which multiple factors compromise each other. Figure 9.3 presents several organic semiconductor materials employed in solar cells. Below, we briefly introduce the materials commonly employed in organic solar cells.

9.2.1 *Electron Donor Materials*

1. **Alternative derivatives of poly (p-phenylenevinylene)**

As shown in Fig. 9.4, derivative materials of two types of poly(p-phenylenevinylene) (PPV) are the most common: 1 is poly[2-methoxy-5-(2-ethylhexyloxy)-1,4-phenylene vinylene] (MEH-PPV); 2 is poly[2-methoxy-5-(3′,7′-dimethyloctyloxy)-1,4-phenylene vinylene] (MDMO-PPV). The cells with MEH-PPV and MDMO-PPV/$PC_{60}BM$ as organic layers can exhibit efficiency of 1.3% and 2.5%, respectively. After the fullerene derivative is introduced, the hole mobility of PPV will be enhanced, which facilitates the enhancement of device characteristics. However, the uneven carrier mobility between PPV and PCBM, and the broad bandgap of organic materials causes the relatively low efficiency of the cell.

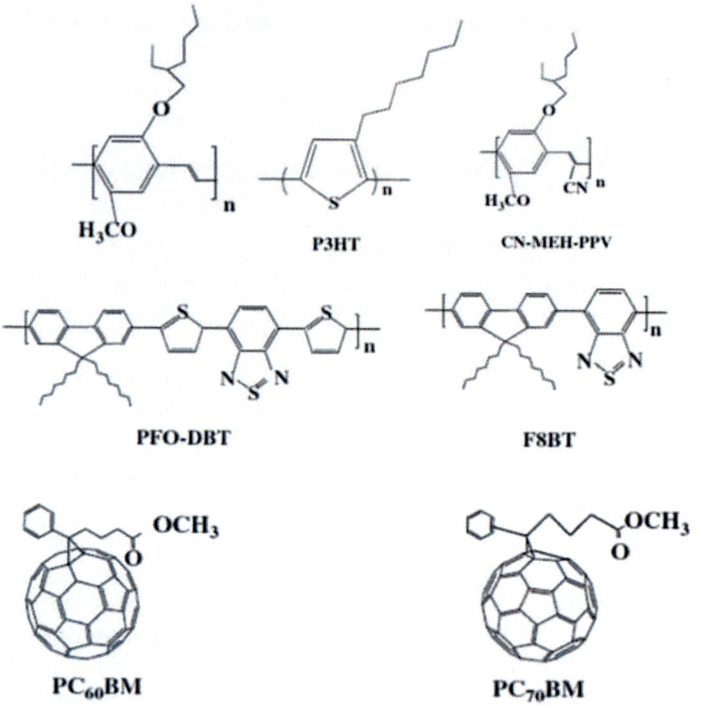

Fig. 9.3 Structures of several common organic materials

Fig. 9.4 MEH-PPV and MDMO-PPV [10]

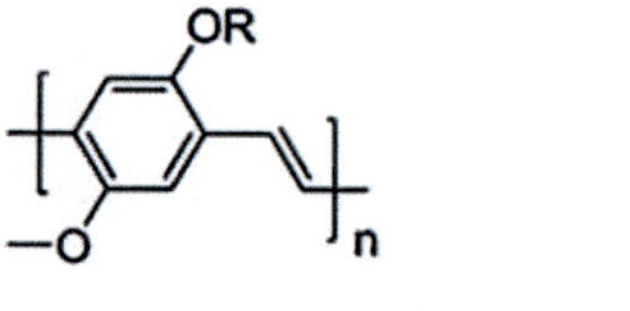

1 R: 2-ethylhexyl (MEH-PPV)
2 R: 3,7-dimethyloctyl (MDMO-PPV)

2. **Alternative derivatives of poly(3-alkyl)thiophene and poly(3-hexylsenophene)**

Such derivatives cover the most common rrP3HT (regular poly(3-hexylthiophene)), as shown in Fig. 9.5a. A cell that mixes P3HT and PC$_{60}$BM at a weight ratio of 1:1 can achieve a maximum efficiency of 5%. For the head-to-tail chain structure, such derivatives commonly exhibit spectral absorption capabilities and suitable hole mobility. The optical bandwidth of rrP3HT is nearly 2 eV, with LUMO of 2.8–3 eV and HOMO of about 5 eV.

Over the past few years, a series of rrP3ATs (poly (3-alkylthiophenes)) polymers with butyl or pentyl alkyl side chains have been successfully synthesized, with their

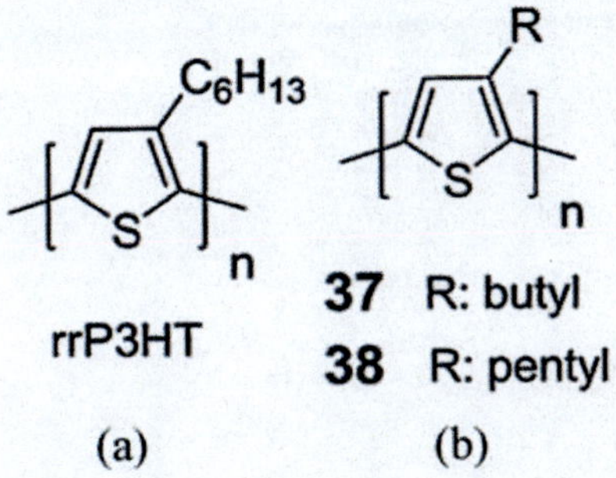

Fig. 9.5 Schematic diagram of P3HT with regular structure [10, 11]

characteristics comparable to P3HT, as shown in Fig. 9.5b. On the whole, both rrP3ATs can obtain high efficiency, whereas their unpredictable length of the side chain is considered their major problem. In the growth of the side chain, PCBM will be more inclined to aggregate into blocks. Under the overly obvious aggregation effect, the contact area of the two materials will be significantly narrowed, while the aggregation effect is extremely small, which hinders the continuous transport of carriers. Such contradiction limits the further enhancement of device performance. Table 9.2 presents the performance comparison of the mentioned devices.

3. **Acetylene-based polymers**

Figure 9.6 illustrates several acetylene-based polymers, numbered 27–30. Overall, they exhibit similar molecular structures, and the HOMO energy level is 0.3 eV lower

Table 9.2 Parameters of organic solar cell devices based on rrpoly (3-alkylthiophenes):$PC_{60}BM$ [10, 11]

Mixture	Ratio	J_{SC} (mA/cm^2)	V_{OC} (V)	*FF*	PCE(%)
37:$PC_{60}BM$	(1:0.8, w/w)	11.2	0.54	0.53	3.2
38:$PC_{60}BM$	(1:1, w/w)	12.5	0.55	0.62	4.3
rrP3HT:$PC_{60}BM$	(1:1, w/w)	12.0	0.57	0.68	4.6

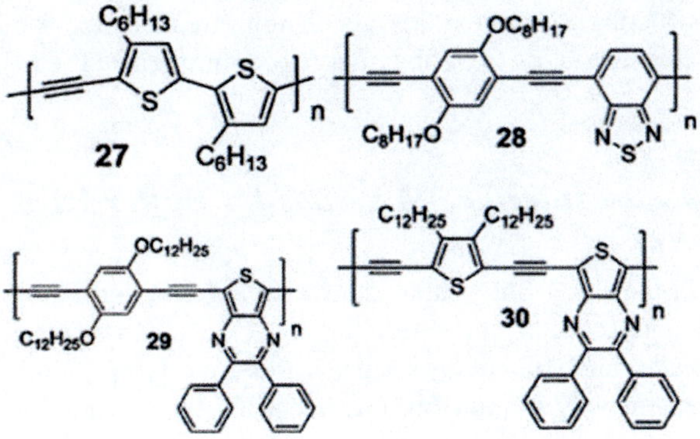

Fig. 9.6 Several acetylene-based polymers [10]

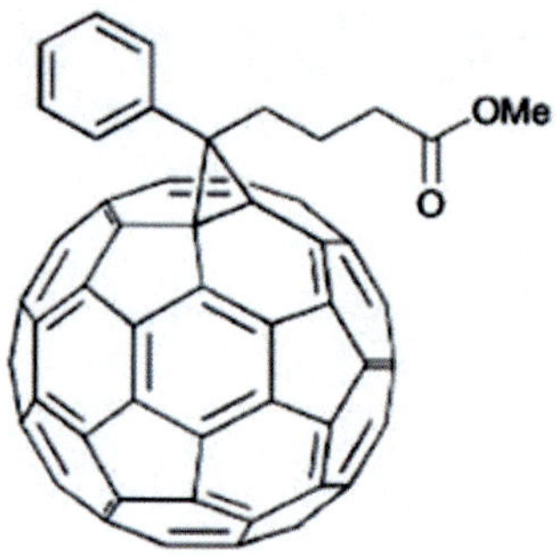

Fig. 9.7 Schematic diagram of the molecular structure of $PC_{60}BM$

than that of P3HT. For 27 and $PC_{60}BM$ mixed at a weight ratio of 1:4, an efficiency of 1.1% can be obtained. After unremitting efforts by researchers, the organic solar cells with acetylene derivatives can exhibit an efficiency of 3.8% [12].

9.2.2 Electron Acceptor Materials

Overall, organic polymer materials are p-type. All the mentioned materials are electron donor materials. Besides, since hole mobility is generally greater than electron mobility in organics, n-type polymer materials are not adopted to fabricate solar cells. Derivatives of fullerene C_{60}, including micromolecular materials (e.g., $PC_{60}BM$ and $PC_{70}BM$) are the major electron acceptors. Fullerene C_{60} and its derivatives have prominent electronic properties. On the whole, they exhibit high electron mobility, can be well dissolved in organic solvents, e.g., chlorobenzene, and can be well phase-separated from common organics (e.g., P3HT). The P3HT/PCBM system has become a classic polymer material employed in organic cells. The molecular structure of $PC_{60}BM$ is illustrated in Fig. 9.7. As mentioned earlier, the P3HT/$PC_{60}BM$ system is capable of exhibiting an efficiency of 5%, while the system based on $PC_{70}BM$ can exhibit higher efficiency. Additional derivatives of fullerene are presented in Fig. 9.8. The good electron capture rate of fullerene materials is achieved for their spherical structure, so these materials exhibit effective electronegativity, instead of prominent characteristics (e.g., the quadrupole moment) [13, 14].

9.2.3 Interface Materials for Buffer Layers

In the study on organic solar cells, prominent nano-sized surface morphology and good electrical characteristics are critical to improve device efficiency, so researchers have conducted many studies in this area. The polymer/fullerene active layer has been extensively studied thus far. Accordingly, the selection of suitable donor and acceptor

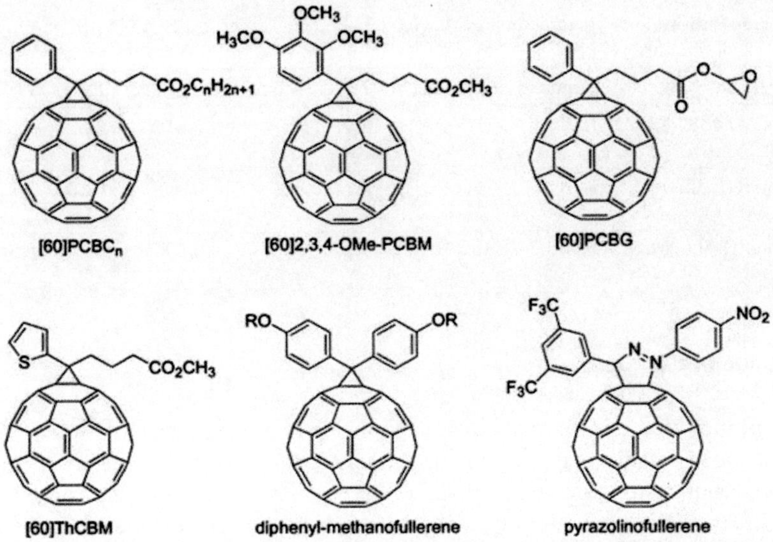

Fig. 9.8 Some derivatives of fullerene [15]

materials in such structure has become a research hotspot. The previous discussion has highlighted organic materials in the active layer. However, the efficiency of the device is dependent on the active layer and the choice of the buffer layer. The role of the metal electrode is primarily to collect carriers. Inserting a buffer layer between the active layer and the electrode makes it more conducive to collecting carriers, which is effective and critical to the device. The mentioned buffer layers are already a vital part of today's device structures. The buffer layer can be split into an anode buffer layer and a cathode buffer layer. After the buffer layer is introduced, the organic/electrode interface is separated, and the flow of carriers is affected by the energy level of the buffer layer. Overall, the cathode buffer layer creates a high barrier to holes and has an energy level promoting the flow of electrons. The role of the anode buffer layer is exactly the opposite. For such a buffer layer, the energy level structure of the device becomes asymmetric, making carriers more efficiently collected by the electrodes. Besides the benefits of collecting carriers, the buffer layer can protect the organic layer from the effect of substrate material or the top metal electrode, while acting as a trap for oxygen and water, which can significantly enhance the stability of devices. Moreover, in some well-designed structures, the distribution of the light field will vary as impacted by the buffer layer, so the efficiency can be enhanced. Thus, the buffer layer is also termed as an "optical spacer layer". For the inverted cell (the cell structure to be described later), the buffer layer is also an essential part, which can make the ITO layer as the cathode. For the role of the insertion level of the cathode buffer material (e.g., zinc oxide) and the anode buffer layer material (e.g., molybdenum trioxide), ITO is transformed from the usual anode to the cathode.

Table 9.3 Performance parameters of devices based on several PEDOT:PSS employed in conventional structure [17]

PEDOT:PSS	Conductivity (S cm^{-1})	J_{SC} (mA cm^{-2})	V_{OC} (V)	FF	PCE (%)
Baytron Al 4083 (H.C. Starck)	10^{-3}	8.12	0.617	0.526	2.64
Baytron P (H.C. Starck)	1	8.72	0.615	0.609	3.27
Baytron PH-500 (H.C. Starck)	100	9.27	0.596	0.641	3.54

1. **Anode buffer layer**

The work function of ITO is about 4.7 eV, which is not consistent with the HOMO level of most organic materials or the LUMO level of fullerene. If an ITO/donor or ITO/acceptor structure is employed directly, a non-ohmic contact is commonly formed. Thus, the open-circuit voltage of such a cell will decrease significantly, evidently lower than the difference between LUMO and HOMO. The energy levels of several common anode buffer layers can effectively comply with the energy levels of electrode and effective layer to form ohmic contacts and exhibit unilateral conductivity. Here are a few examples.

(1) Organic buffer layer

The most common is Poly (3, 4-ethylenedioxithiophene): poly (styrene sulfonic acid), which is PEDOT:PSS. This PEDOT organic compound doped with PSS was initially synthesized by a German laboratory in 1980. It exhibits prominent hole-collecting properties, non-toxicity, and effective dispersion in water after doped with PSS. Table 9.3 lists several device parameters with PEDOT:PSS materials. In the device structure of ITO/PEDOT:PSS/P3HT:PCBM/Al, the efficiency can reach 5% as mentioned above. The advantage of this material is that its light transmittance is sufficiently high, and it exhibits a high work function, about 4.8–5.2 eV [16], so it can effectively form an ohmic contact with most donor polymer materials. By annealing and UV treatment at about 120 °C, the work function, mobility, and other characteristics of PEDOT:PSS will be enhanced. Thus, annealing and UV treatment have also become standard processes in preparing the device. Moreover, polyalcohol, glycerol act as additives in PEDOT:PSS. After the mentioned additives are added, the series resistance of the device can decrease noticeably, and then the device efficiency will be elevated. The researchers incorporated Au nanoparticles into the aqueous solution of PEDOT:PSS at a ratio of nearly 20%. Thus, the maximum efficiency of the device is about 4.19%, and the efficiency of the device without gold is nearly 2.5%. Hydrophilic graphene oxide can also act as an impurity of PEDOT:PSS, and the efficiency can increase from 2.1 to 2.4%.

Fig. 9.9 Schematic diagram of several self-assembled layers after processing [17]

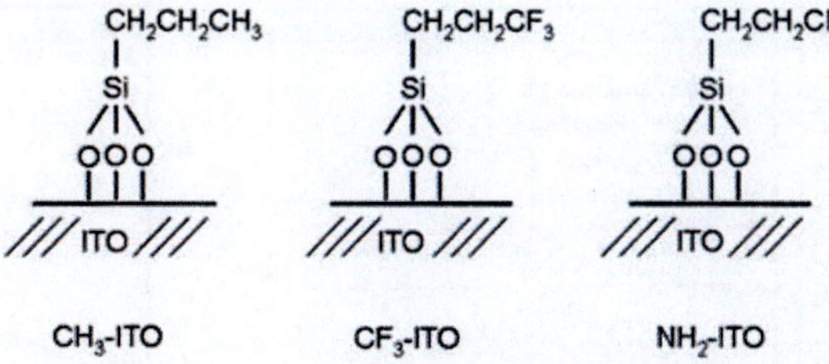

(2) Self-assembly layer (SAM layer)

The self-assembled layer can form a thin, highly regular dipole at the interface between the two materials, thereby minimizing the energy barrier during injecting or extracting carriers. Kim et al. [18] employed different silanizing agents (N-propyl trimethoxysilane, 3,3,3-trifluoropropyl trichlorosilane or 3-aminopropyltriethoxysilane) to treat the interface of P3HT:PCBM layer. Silane can form a 2-nm-thick self-assembled layer on the ITO surface. Thus, the work function of the anode will change. The work function of unprocessed ITO is 4.7 eV, while NH_2-ITO is 4.35 eV, and the work function of CF3-ITO decreases to 3.62 eV. The work function of the processed ITO is closer to the HOMO level of P3HT, so a better ohmic contact can be formed. Figure 9.9 presents a schematic of several self-assembled layers. Other self-assembled materials cover 4,4′-bis [(p-Trichlorosilylpropylphenyl)phenylamino]biphenyl ($TPDSi_2$)-ITO material, 3-ercaptopropyltrimethoxysilane (MPTMS)-Au material, etc. Self-assembling materials can effectively alter the interface characteristics and work function of the material, which facilitates the optimization of parameters (e.g., open-circuit voltage).

(3) Some inorganic compounds in the anode buffer layer

Some common inorganic materials for the buffer layer include V_2O_5, MoO_3, Au nanoparticles, NiO_x, WO_3, ReO_3, VO_x, etc. When the thickness of molybdenum oxide on the ITO surface reaches 3–5 nm, it is reported that it can replace the role of PEDOT:PSS, and the contact of molybdenum oxide with P3HT can effectively transport holes.

2. **Cathode Buffer Layer (CBL)**

Similar to the anode buffer layer, the cathode buffer layer functions to transport electrons and block holes. To be able to extract electrons from an electron acceptor (e.g., fullerene), a metal with a suitable work function is required, whereas a non-ohmic contact will be formed. Accordingly, a barrier layer that can form an ohmic contact with a metal, efficiently transport electrons, and block holes are required. Moreover, high light transmission and stability are also important.

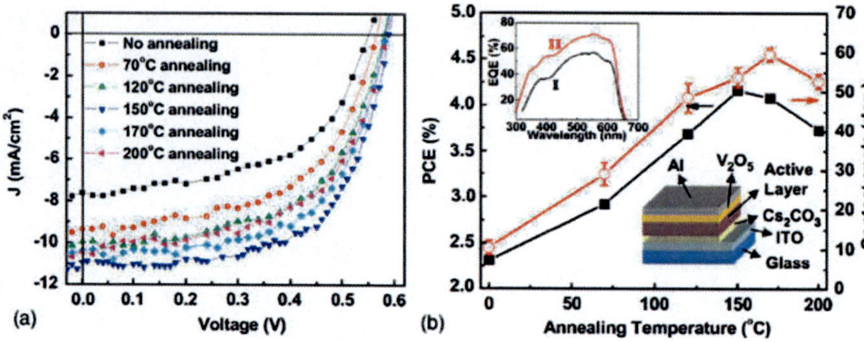

Fig. 9.10 Device performance and structure of $CsCO_3$ buffer layer [19]

(1) Alkali metal compounds

Lithium fluoride (LiF) refers to one of the most common CBL materials in conventional structures. It can form a dipole layer at the metal interface, thereby reducing the work function of the metal and promoting the formation of an ohmic contact. Caesium carbonate is also successfully inserted into the P3HT:PCBM layer, which can be thermally evaporated or spin-coated on the ITO surface. Figure 9.10 presents the performance of devices with $CsCO_3$ buffer layers at different annealing temperatures. Table 9.4 lists some common solar cells in which the alkali metal compound acts as a buffer layer. The brackets present the absence of a buffer layer.

(2) Metal compounds

Titanium oxide buffer layers are extensively employed in conventional and inverted structures. The bandgap of titanium oxide is about 3.7 eV, so it achieves good light transmission. The energy level of the valence band and the conduction band is −8.1 eV and −4.4 eV, respectively, so it can form an ohmic contact with the fullerene derivative and has a high hole barrier. The mobility of the electrons in amorphous material is about $1.7 \times 10^{-4}\ cm^2\ V^{-1}\ s^{-1}$, indicating that it can efficiently transport electrons. Since titanium oxide has the function of blocking oxygen and water, the reliability of the device can be significantly enhanced. As oxygen and water can pass through the gaps in the metal and affect the organic layer, and the diffusion of metal atoms to the effective layer, the efficiency of a cell with a normal structure and without a cathode buffer layer will rapidly decline, which is solved effectively by the insertion of titanium oxide. Similar to titanium oxide, zinc oxide is another effective n-type material exhibiting a bandgap of approximately 3.2 eV and a conduction band of −4.1 eV. Zinc oxide can be prepared with various methods (e.g., the sol–gel method and the sputtering method). In the inverted cell, the efficiency of P3HT/PCBM cell with zinc oxide and vanadium pentoxide intercalation layer can reach 3.09%, and the zinc oxide nanowires grown with the hydrothermal method can enhance the efficiency from 3 to 3.9% [20]. Calcium oxide acts as another effective material. The calcium oxide material exhibiting a thickness of 1 nm can make the P3HT:PCBM

Table 9.4 Solar cells with some common alkali metal compounds as buffer layers

Cathode buffer layer (CBL)	Thickness (nm)	Device structure	Inversed or not	J_{SC} (mA cm^{-2})	V_{OC} (V)	*FF*	PCE (%)
LiF	0.3	ITO/PEDOT:PSS/MDMO-PPV:PCBM/CBL/Al		5.25	0.82 (0.76)	0.61 (0.53)	~3.30
CsF	0.4	ITO/PEDOT:PSS/MEH-PPV:PCBM/CBL/Al		(5.26)	(0.72)	(0.37)	2.20 (1.40)
Cs_2CO_3	0.2	ITO/PEDOT:PSS/P3HT:PCBM/CBL/Al		10.20 (10.20)	0.61 (0.61)	0.66 (0.48)	4.10 (3.00)
Cs_2CO_3	1.0	ITO/PEDOT:PSS/P3HT:PCBM/CBL/Al		5.95 (7.44)	0.52 (0.42)	0.66 (0.52)	1.55 (1.25)
Cs_2CO_3	1.0	ITO/CBL/P3HT:PCBM/V_2O_5/Al	Yes	8.42 (6.97)	0.56 (0.30)	0.62 (0.41)	2.25 (0.66)
Cs_2CO_3	1.0	ITO/PEDOT:PSS/P3HT:PCBM/CBL/Al		9.50 (11.20)	0.56 (0.41)	0.60 (0.50)	3.10 (2.30)

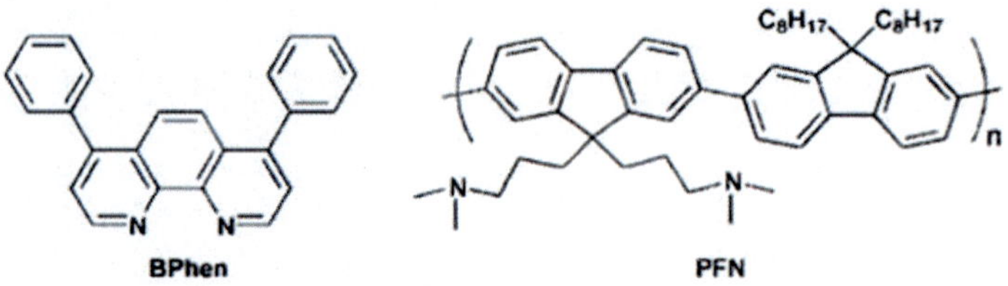

Fig. 9.11 Material of organic intercalation layer [17]

system exhibit an efficiency of 2.85%. If a Ca/Al electrode is used and Ca is oxidized to calcium oxide in the air, this system can exhibit the efficiency of 5% [21]. The thickness of the metal oxide can regulate the distribution of light in the device, and the thickness can be designed to maximize the efficiency.

(3) Organic insertion layer

Many organics and polymers exhibiting low relative molecular mass can be employed in the buffer layer. Bathocuproine (BCP) materials, N,N0-Tridecyl-3,4,9,10-perylenetetracarboxydiimide (PDI) and other materials, displaying a thickness smaller than 10 nm, can act as the buffer layer. p-type organic semiconductor materials (e.g., copper phthalocyanine (CuPC) and pentacene) can act as intercalation layers. Bphen and PFN in Fig. 9.11, are also common organic intercalation layers.

(4) Self-assembly layer (SAM)

The self-assembled layer in this part is primarily used to modify the cathode buffer layer, primarily used to modify zinc oxide and titanium oxide materials. Similar to the anode self-assembly layer, the cathode self-assembly layer will form a dipole at the interface to alter the work function, enhance ohmic contact, and elevate the open-circuit voltage.

9.2.4 Some Novel Materials Replacing ITO

ITO (Indium Tin Oxide) refers to a popular material, whereas some novel types of low-cost materials exhibiting high performances can effectively replace ITO in the future. PEDOT:PSS is one of the most promising substances in organics for its conductivity of 1×10^3 S/cm [23]. Besides, metal grids are another promising technology, and single-walled carbon nanotubes are also promising. The carbon nanotubes exhibit high transmittance, whereas the conductivity is not ideal. It is one of the noteworthy and important hotspots to use different materials together (e.g., spin coating a layer of PEDOT:PSS) on the surface of carbon nanotubes to enhance electrical conductivity. Moreover, many novel materials (e.g., metal nanoparticle layers and graphene layers) have aroused gradual attention, whereas their device

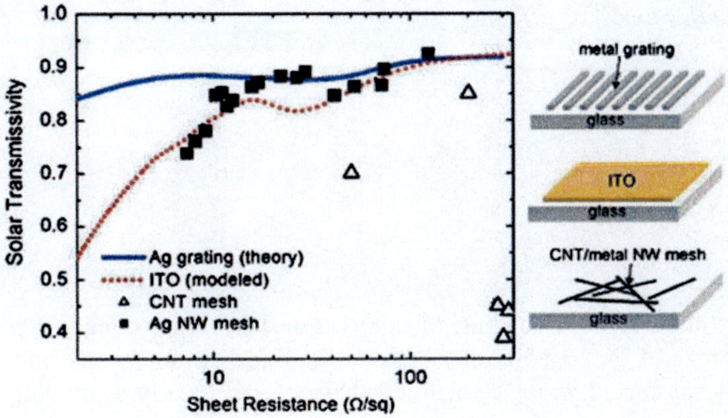

Fig. 9.12 Schematic diagram and characteristics of several transparent conductive materials [22]

efficiency is overall lower than ITO, and further research is required. Figure 9.12 presents the schematic and square resistance-transmittance characteristics of several materials (e.g., metal grid, ITO, and carbon nanotubes).

9.3 Basic Structure of Organic Solar Cells

The materials in organic solar cells are introduced early, and the basic structure of organic solar cells is introduced in the present section. The organic solar cells development has undergone several stages and can be split into the following types in line with the difference in the effective layer structure.

9.3.1 Single-Layer Organic Solar Cells

The first organic photoelectric conversion device was prepared by Kearns and Calvin in 1958 [24]. There is only one material in the effective layer, and its primary material is magnesium phthalocyanine (MgPc) dye. The dye layer is sandwiched between two electrodes exhibiting different work functions. Such structure can be termed as a single-layer organic solar cell. The principle of such device is illustrated in Fig. 9.13: The electrons in the organic semiconductor are excited from the HOMO level to the LUMO level under light, thus generating a pair of electrons and a hole-exciton. The electrons are extracted by the electrode exhibiting a low work function, and the holes are filled by the electrons from the electrode exhibiting a high work function; as a result, a photocurrent is formed under the light. In theory, when an organic semiconductor film is in contact with two electrodes with a range of work functions, different Schottky barriers will be formed, underpinning the directional

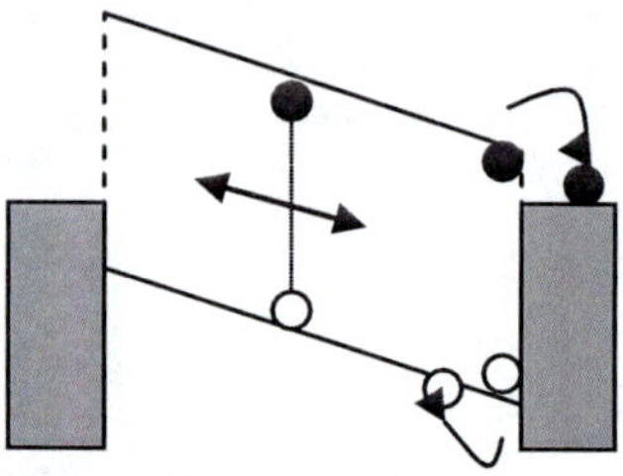

Fig. 9.13 Energy band diagram of single-layer organic solar cell—Energy band diagram of MIM model

transfer of photoinduced charges. Such structure takes up a dominant position in the early stage of the organic solar cells development. In the nearly three decades of initial development, organic solar cells have been basically complying with this structure, except that various organic semiconductor materials are replaced between two electrodes exhibiting different work functions.

In such a single-layer organic solar cell, the efficiency of exciton separation is problematic. Excitons formed by photoexcitation can be separated only at the interface between the organic material and the electrode or in the diffusion layer of the Schottky junction under the action of the electric field in the nodal region. Excitons formed at other positions should move to this area before contributing to the photocurrent. However, the migration distance of excitons in organic dyes is significantly limited, so most excitons recombine before separating into electrons and holes. For the lack of effective means of separating excitons in this structure, the efficiency is low on the whole.

9.3.2 Double-Layer Heterojunction Organic Solar Cell

In 1986, Dr. Qingyun Deng of Kodak Company initially prepared a novel organic solar cell structure, i.e., two-layer heterojunction organic solar cell [25], in which two layers of organic materials were employed, i.e., a double-layer film consisting of a derivative of tetrahydroxy perylene and copper phthalocyanine (CuPc). The double-layer structure of the film forms a heterojunction. This effort suggests that the concept of heterojunction has been initially in organic solar cells, and the photoelectric conversion efficiency has reached about 1%. The heterojunction structure of the double-layer membrane remains one of the focuses of the studies on the organic solar cell. The introduction of the concept of heterojunction refers to a landmark of the organic solar cells development. The future development of solar cells basically complies with this heterojunction structure.

The schematic diagram of the band structure of the double-layer heterojunction organic solar cell is given in Fig. 9.14. Overall, the working principle is: the donor organic semiconductor material generates excitons after absorbing photons (process (a)), and the excitons move to the interface between donor and acceptor (process (b)), where electrons are injected into the acceptor organic semiconductor material

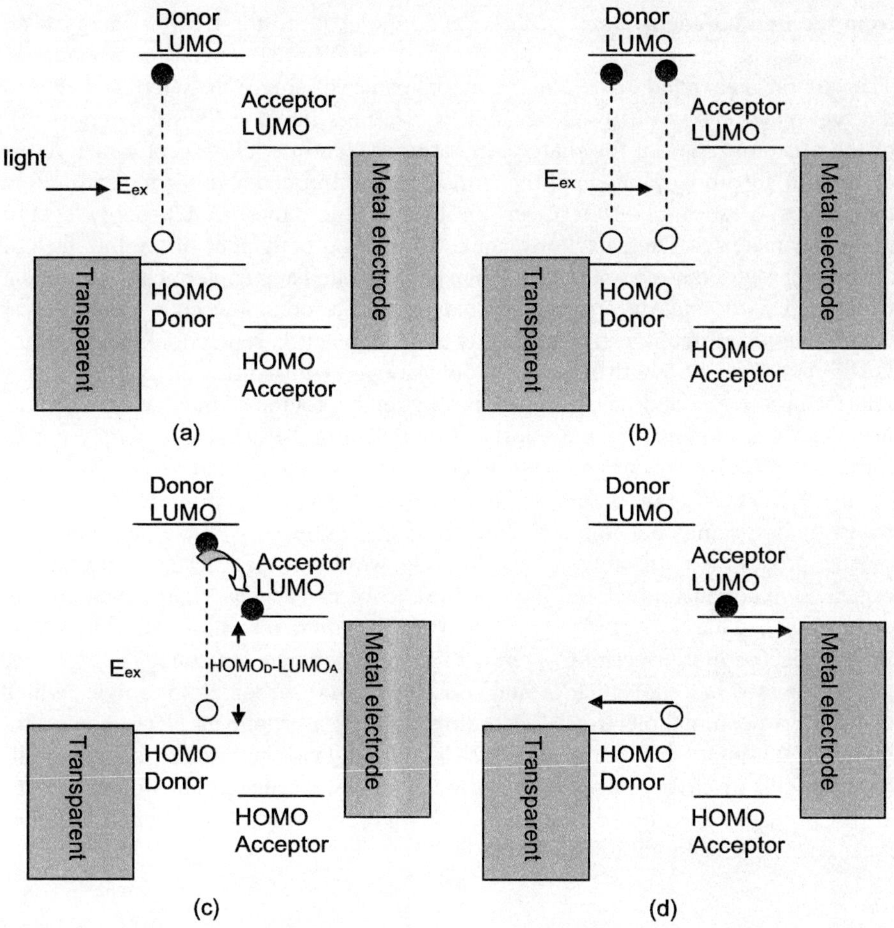

Fig. 9.14 Schematic diagram of the band structure of a working double-layer heterojunction organic solar cell: **a** exciton generation, **b** exciton migration, **c** exciton separation, **d** carrier collection

to separate holes and electrons (Process (c)). In such a system, the electron donor is p-type, and the electron acceptor is n-type. As a result, holes and electrons are transmitted to the two electrodes, thereby forming a photocurrent (process (d)). Such a process includes the critical photoelectric conversion in organic solar cells, which will be elucidated in the later organic solar cell mechanism.

Compared with single-layer organic solar energy, the double-layer heterojunction structure introduces a mechanism of charge separation, i.e., the interface between the donor and acceptor material. Excitons exist for a limited time, and electrons and holes having not been fully separated will be recombined, releasing the energy they absorb. Obviously, excitons that fail to separate into free electrons and holes do not contribute to the photocurrent. Introducing the interface between the donor and acceptor material significantly enhances the efficiency of exciton separation. Electrons are injected from the LUMO energy level of the excited molecule to the LUMO energy level of the electron acceptor. Such a process is essentially the separation of excitons. The excitons generated in the donor material can easily reach the interface between the two materials by diffusion, and inject electrons into the LUMO energy level of the acceptor material for charge separation. Meantime, the acceptor material is capable of absorbing photons of the corresponding frequency to form excitons and then inject holes from its HOMO energy level back into the HOMO energy level of the donor material. Thus, excitons can be formed on both sides of the interface of the bilayer membrane meantime, and then get separated at the interface by diffusion. Compared with single-layer organic solar cells, the donor–acceptor double-layer heterojunction structure can significantly expedite exciton separation. However, for the limited diffusion length of excitons, only a region with a diffusion length from the interface between donor and acceptor is an effective region of the device. Excitons generated outside this region are lost before reaching the interface, thereby down-regulating the photoelectric conversion efficiency of organic solar cells.

In 1992, Sariciftci of Turkey reported that excited state electrons can be injected from organic semiconductor molecules into C_{60} molecules efficiently, while the reverse process is noticeably slower. In other words, at the interface between the organic semiconductor material and C_{60}, the exciton can achieve charge separation at a high rate, and the charges after separation are difficult to recombine at the interface. This is because the surface of C_{60} refers to a large conjugate structure. The electrons are delocalized in a molecular orbital consisting of 60 carbon atom orbitals, which can stabilize external electrons. Accordingly, C_{60} is a prominent electron acceptor material. In 1993, Sariciftci made PPV/C_{60} double-layer heterojunction solar cells based on this discovery. Since then, various double-layer heterojunction solar cells with C_{60} as the electron acceptor have emerged, and studies on organic-fullerene solar cells have been initiated.

9.3.3 Bulk Heterojunction Organic Solar Cells

With the organic solar cells development, a novel structure—a bulk heterojunction organic solar structure has emerged, as shown in Fig. 9.15. The proposal of "body heterojunction" primarily aims at the limitations of exciton separation and carrier transport in the photoelectric conversion. The exciton can be effectively separated only at the interface between the donor and the acceptor material. Excitons generated far away from the interface are often recombined before reaching the interface. Under the bulk heterojunction structure, a donor material and an acceptor material are mixed; besides, a mixed film is formed based on co-evaporation or spin coating. The donor and acceptor form a single component region in the mixed film and finally form a conductive network where the donor and acceptor materials penetrate each other. Excitons generated at any position can reach the interface between donor and acceptor via a short path, so the efficiency of charge separation is enhanced. Meantime, the positive and negative carriers formed on the interface

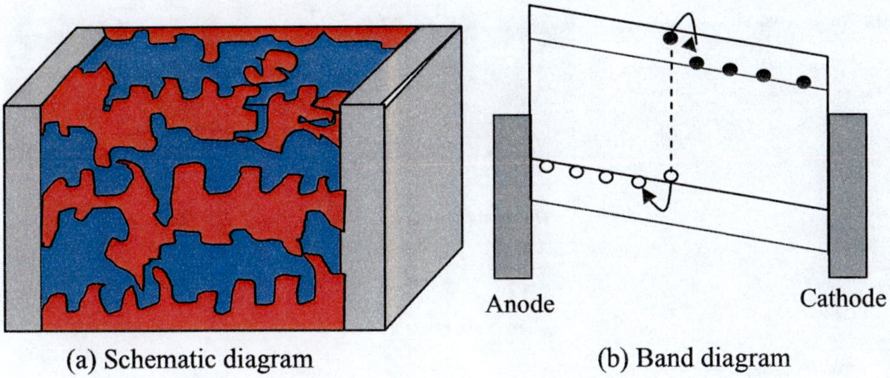

Fig. 9.15 Bulk heterojunction organic solar cell. **a** Schematic diagram. **b** Band diagram

can also reach the electrode via a relatively short path, thereby remedying the insufficient carrier mobility. Compared with the double-layer heterojunction structure, the bulk heterojunction structure significantly enhances the efficiency of exciton separation, while effectively ensuring the transmission and collection of carriers, and remarkably enhancing the efficiency of organic solar cells. Such structure has become another milestone in the organic solar cells development. At present, almost all high-efficiency organic solar cells comply with bulk heterojunction structures, and they are still under development. The content of the present chapter is basically based on this structure.

9.3.4 Conventional and Inverted Structure of Organic Solar Cells

Bulk heterojunction organic solar cells can be divided into conventional solar cells and inverted solar cells in accordance with the polarity of their electrodes, as shown in Fig. 9.16.

Figure 9.16a is a schematic diagram of an organic solar cell with a conventional structure, and Fig. 9.16b is a schematic diagram of an inverted organic solar cell, in which ABL (Anode Buffer Layer) is the anode buffer layer, and CBL (Cathode Buffer Layer) acts as the cathode buffer layer. It is suggested that the difference between the two structures is whether the transparent electrode acts as an anode (i.e., conventional structure) or a cathode (i.e., inverted structure), i.e., electrons or holes flow to the transparent electrode (the direction of carrier flow presented in Fig. 9.16). In the most common organic solar cells of the P3HT:PCBM system, the materials of the CBL layer consist of zinc oxide, titanium dioxide, caesium carbonate and other materials, and the most common materials of the ABL layer include PEDOT:PSS and molybdenum trioxide (MoO_3).

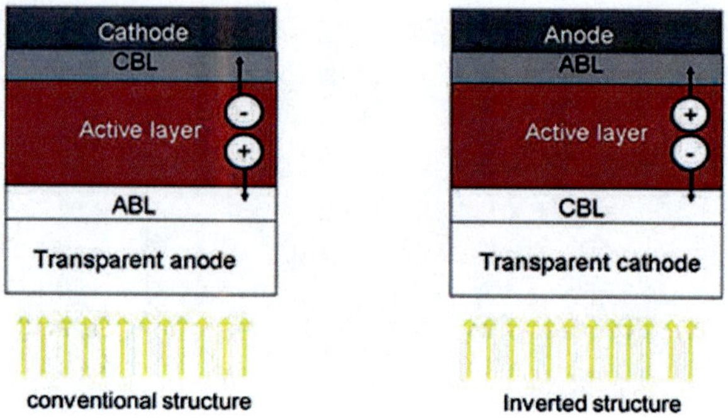

Fig. 9.16 Two structures of bulk heterojunction organic solar cells

In conventional cells, light is radiated from the anode, and ITO is generally adopted as a transparent electrode. The bandwidth of ITO is nearly 3.5–4.2 eV, thus pertaining belongs to a wide bandgap material that can effectively pass light. In the organic absorption layer, the light transmittance can reach 90% in the wavelength range of 300–800 nm, and the conductivity is high, and the square resistance can reach 10 Ω/□ or less. Though some alternative materials (e.g., FTO (fluorine-doped tin oxide) and AZO (aluminum-doped zinc oxide)) can replace ITO, the most efficient devices are achieved on ITO. ITO exhibits a relatively high work function, and it was mostly used as an anode in the early production of organic solar cells. Accordingly, the device produced also displays a conventional device structure. The upper cathode is made of an opaque metal with low work function. The most typical example is Al, and others include Ag, etc.

Inverted cells exhibit the opposite structure, with ITO as the cathode, which requires the transparent electrode to be modified with interface materials. Compared with conventional structures, inverted structures have two main merits: 1. they have better stability than conventional structures; 2. they exhibit better flexibility, facilitating the design of cascaded cells or multi-junction cells. According to related reports, an inverted cell covering a zinc oxide CBL layer manufactured with ALD (Atom Layer Deposit) technology is capable of stably maintaining an efficiency over 90% in the air for 500 days, while the efficiency of normal cells will decay rapidly. Note that attenuating to 90% typically takes a few hours or days.

9.4 Working Principle of Organic Solar Cells

Figure 9.17 presents the basic process of photoelectric conversion in organic solar cells, and Fig. 9.18 also presents such a process on the band diagram. When the energy of the incident photon exceeds the optical absorption gap of the donor material, the photon is absorbed by the donor material (Fig. 9.18 (1)). For the relatively low dielectric constant of organic materials and the relatively weak interactions between molecules (Fig. 9.2), the absorption of photons does not generate free charges but

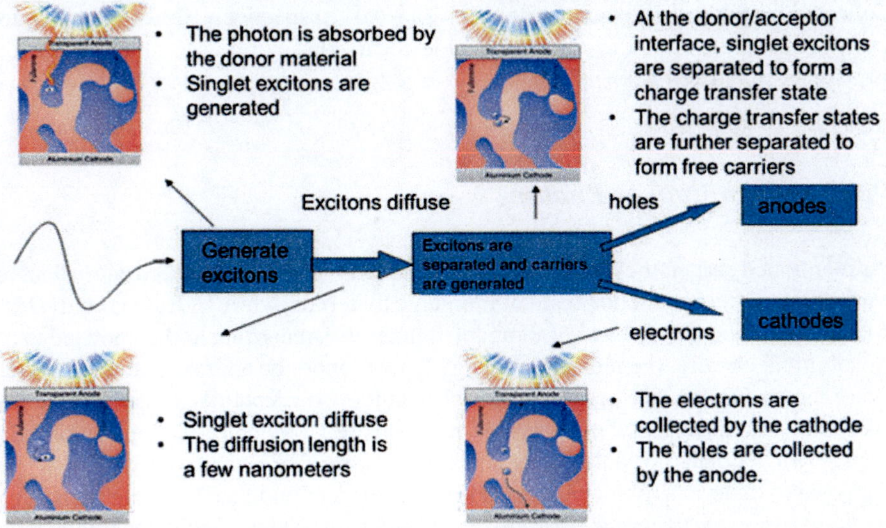

Fig. 9.17 Basic process of photovoltaic generation in organic solar cells

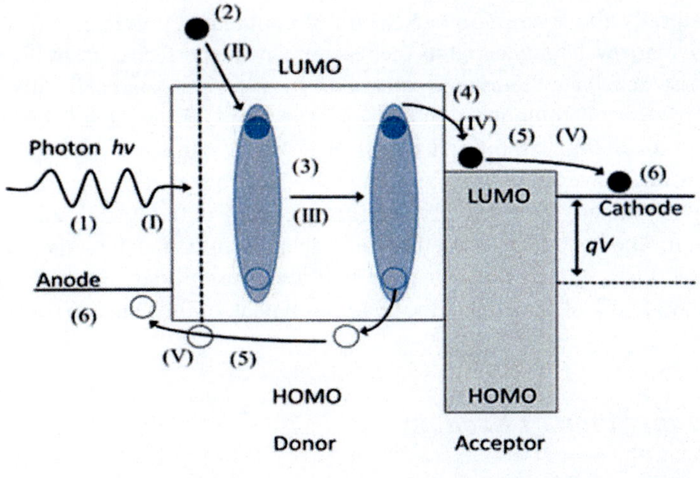

Fig. 9.18 Mechanism of photoelectric conversion in bulk heterojunction solar cells

robustly bound electron-hole pairs, i.e., singlet excitons (Fig. 9.18 (2). To efficiently convert light energy into electrical energy, singlet excitons should diffuse to the heterojunction interface between the donor and acceptor material before being recombined (Fig. 9.18 (3)), in which electrons will transfer efficiently from the donor material to the acceptor material to achieve the separation of the singlet exciton (Fig. 9.18 (4)). Free electrons and holes are transported to the electrodes (Fig. 9.18 (5)), and collected by the cathode and anode (Fig. 9.18 (6)), respectively, thus the photocurrent was generated and the effective photoelectric conversion was achieved. Such a process is elucidated below.

9.4.1 Absorption of Photons

As mentioned earlier, the LUMO of organic materials can be considered equivalent to the bottom of the conduction band of inorganic materials, while HOMO is equivalent to the top of the valence band of inorganic materials. According to the knowledge of quantum mechanics, the molecular energy levels cannot be approximated as continuous energy bands since organic materials cannot form repetitive crystal structures like inorganic materials. For inorganic materials (e.g., silicon), photons exhibiting energy greater than the bandgap are absorbed to form free electron-hole pairs, while for organic materials, photons are absorbed to form bound electron-hole pairs (i.e., excitons). Organic materials generally have a light absorption coefficient greater than $10^7\ m^{-1}$, so materials exhibiting a thickness of 100–300 nm can absorb photons well. In contrast, inorganic semiconductor materials (e.g., silicon) require hundreds of microns to achieve sufficient absorption. However, compared to inorganic semiconductor materials, the absorption bandwidth of conjugated polymer semiconductors is relatively narrow, largely explains the reason why the performance of organic solar cells is worse than that of inorganic solar cells. In an organic solar cell, only part of its absorption spectrum complies with the solar spectrum. For instance, the absorption spectrum of a material exhibiting a bandgap of 1.1 eV can cover 77% of the AM1.5 solar spectrum. However, in organic solar cells, the bandgap of semiconductors, e.g., P3HT is greater than 1.9 eV, which can only cover 50% of the AM1.5 solar spectrum. Moreover, for the lower carrier mobility of organic semiconductors, the thickness of the effective layer in organic solar cells is limited, usually around 100 nm. Under thin thickness, 40% of the light at the absorption peak of the material is wasted [26].

9.4.2 Generation of Excitons

When the energy of the incident photon is greater than the absorption edge of the organic material, the electron transitions from HOMO to LUMO, generating an exciton. Since excitons are critical to solar cells, the excitons are elucidated in this study. The actual reason why inorganic materials forming free carriers and organic materials forming bound excitons is that the attractive forces between the electrons and the holes of the exciton differ. There will be a Coulomb force between the electron and hole pairs. When the electrons and holes are separated by 1 nm and the relative permittivity of the material is 3, the Coulomb binding energy reaches approximately 0.5 eV. If the attractive force between electrons and holes is at this level, it has exceeded the thermal energy of the material at ambient temperature, at which point a so-called stable exciton is formed. Since the dielectric constant of the inorganic semiconductor is relatively large, the energy excitons are insufficient to reach a stable state; thus, stable excitons cannot be formed. The situation of organic materials is just the opposite. The typical binding energy of excitons in organic materials is 0.4 eV, which makes it easy for organic materials to form stable excitons. Moreover, one reason that excitons are relatively easy to form in organic materials is that the interaction forces between organic molecules are relatively weak. Carriers are transported through jumps, instead of being transported through conduction bands or valence bands as in inorganic semiconductors. Thus, for the slower movement of carriers, the chance of separating electron-hole pairs is smaller, and the locality of photogenerated electrons and holes is stronger. In organic materials, excitons usually stay in one molecule or are distributed along an extended polymer chain. In rare cases, excitons are formed between molecules. According to the knowledge of quantum mechanics, excitons are split into singlet excitons with a spin of 0 and triplet excitons with a spin of 1. If the exciton is generated by the interaction of the carriers injected though the electrode, instead of being formed under the action of light, the ratio of the singlet state and triplet state is theoretically about 3:1, which is a common situation in OLED. For the limitation of selection rules, the spins of photogenerated electrons and holes are opposite, so generally only singlet excitons are generated. Thus, singlet excitons play an important role in solar cells.

9.4.3 Diffusion of Excitons

After the exciton has a limited lifetime, radiation recombination occurs. For the singlet exciton, the lifetime is at the nanosecond level, and the lifetime of triplet exciton can reach milliseconds. This is because the triplet exciton is difficult to recombine for the restriction of the selection rule, which significantly increases its lifetime. The energy of triplet excitons is lower than that of singlet excitons. In photovoltaic cells, as in the solar cells we discussed, triplet excitons generally do not have an important position unless they take advantage of their merits with novel device structures.

For the presence of high-energy excitons (typical binding energy is 0.4 eV), the thermal energy is not enough to separate them at ambient temperature. To effectively realize photoelectric energy conversion, excitons should migrate to the interface between the donor material and the acceptor material. Since the exciton is electrically neutral, it cannot move under the action of an electric field, so its migration is primarily accomplished by diffusion. For the limitation of the lifetime of the exciton, the diffusion length L_D of the exciton is small. For instance, in PPV polymer materials, L_D is generally 5–8 nm [27]. This means that excitons that exceed L_D from the interface between the donor material and the acceptor material will recombine without contributing to effective photoelectric conversion. The value of L_D is noticeably smaller than the thickness of the effective layer in the organic solar cell (about 100 nm). Accordingly, in a two-layer heterojunction structure, only the light absorbed by the absorbing layer in the distance L_D from the heterojunction interface promotes the final efficiency, so the device with this structure exhibits a relatively low performance. Ideally, all excitons generated can reach the donor–acceptor interface before being recombined. The body heterojunction structure is formally proposed based on this idea. In the bulk heterojunction structure, the donor material and the acceptor material are mixed together to build a conductive network that penetrates each other, which significantly broadens the effective area of the interface and enables almost all excitons to reach the heterojunction interface before they are recombined. For the long-distance between the exciton and the heterojunction interface in the double-layer solar cell, the excitons are largely recombined. Moreover, the material interface of the bulk heterojunction material is far more than that of the double-layer material, the exciton will be effectively separated at a higher probability, so the efficiency will be significantly improved. As mentioned earlier, currently effective solar cells are basically based on bulk heterojunction structures. A larger diffusion length of excitons allows for the presence of larger donor material regions and acceptor material regions in the bulk heterojunction, facilitating the absorption of light and the transport of carriers; as a result, higher device performance is achieved. In various different polymer materials, the diffusion length of excitons varies extensively, and its value is primarily distributed between 5 and 20 nm [28, 29].

9.4.4 Separation of Excitons

Since the binding energy of the exciton reaches over the thermal energy in the organic semiconductor, an additional force is required to separate the exciton. As shown in Fig. 9.18, when a difference is identified in the energy band between the two materials, there will be a force. When moving to the donor–acceptor interface, the exciton will be separated by this force [30], i.e., the process (4). However, not any material interface will generate an effective exciton separation. Set E_{ex} as the energy difference between electrons and holes in an exciton. As shown in Fig. 9.19a (X denotes HOMOD-LUMOA in the Fig. 9.19), when

$$E_{ex} > HOMO_D - LUMO_A$$

effective separation of excitons can occur by charge transfer

$$D^* + A \rightarrow D^+ + A^-$$
$$D + A^* \rightarrow D^+ + A^-$$

where D denotes the donor material; A is the acceptor material; D^* and A^* are the excited states in the donor material and the acceptor material, respectively; D^+ and A^- suggest the state of the donor material and the acceptor material after the charge transfer takes place, respectively. However, when the interface between donor

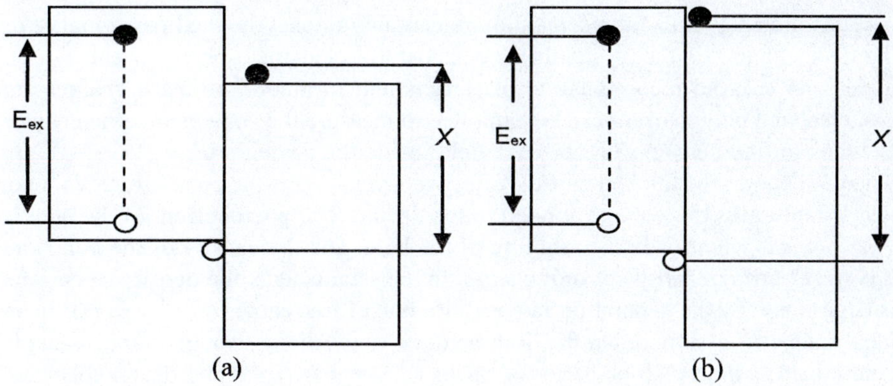

Fig. 9.19 Schematic diagram of the band structure required for exciton separation at the donor and acceptor interface. **a** Interfaces suitable for exciton separation. **b** Interfaces not suitable for exciton separation

material and acceptor material is

$$E_{ex} < HOMO_D - LUMO_A$$

(Fig. 9.19b), it hinders the transfer of charges in energy, which is not an effective exciton separation interface.

When the electron transfer takes place at the interface and the electron enters the acceptor material from the donor material, the electron may act as a free electron, or it may be bound to the holes left in the donor material to form charge pairs (dipole) as impacted by the Coulomb force. The mechanism of exciton separation has long been concerned with by researchers. According to the extinction test of the PL spectrum of the exciton and the study of the photocurrent, the formation of the exciton to free charge in the trapezoid polymer refers to a two-step process. To be specific, after the charge transfer takes place at the interface, the holes and electrons in the original exciton first form electron-hole pairs bound by Coulomb forces; subsequently, the mentioned electron-hole pairs are separated into free charges. Is there such a two-step process in solar cells? If there are interface-bound electron-hole pairs, it is in contrast to the high internal quantum efficiency of solar cells (100% internal quantum efficiency has been observed). Some people have proposed that after the exciton is separated, the electrons and holes exhibit excess kinetic energy. The excess kinetic energy can help the exciton separate. As a result, the exciton is fully separated into free electrons and holes without forming an interface dipole. It has also been suggested that the excess kinetic energy of electrons and holes after charge transfer causes them to be separated by a relatively large distance at the beginning, which remarkably facilitates the formation of free charges [31]. Moreover, the presence of interfacial dipoles and the relatively high mobility of carriers are also beneficial for exciton separation.

The efficiency of exciton separation remarkably impact the final performance of the device, and this factor should be considered to calculate photocurrent. For an ideal solar cell, no recombination of carriers and formation of space charges are assumed, and photocurrent can be exploited to measure the conversion efficiency of excitons to free charges. The electric field inside the device is $E = (V_{OC} - V)/L$, where L denotes the thickness of the effective layer, V_{OC} is the open-circuit voltage, and V represents the voltage applied to the device. The photocurrent of the hole is $J_{ph} = qp\mu E$, where μ is the mobility of the hole, p is the density of the hole, and q is the electric quantity per unit charge. In the steady state, the density of carriers is determined by the generation rate and lifetime of free carriers, i.e., $p = \tau G$. In an ideal state, it is assumed that the lifetime of carriers is long enough that its value is determined by the transition period of carriers, i.e., $\tau = L^2/\mu J$. Thus, the photocurrent is [32]

$$J_{ph} = qGL \tag{9.1}$$

The same will be true for electrons. Thus, when the generation rate of free carriers G is constant, J_{ph} is independent of the voltage V. J_{ph} and G have a linear relationship. G can be directly measured with J_{ph}, so it can also be obtained whether there are bound electron-hole pairs (also termed as dipoles) at the interface after exciton separation. The mentioned relationship is based on the assumption that the carrier lifetime is unlimited. In practical devices, the recombination time of carriers is about milliseconds, and the transition period of carriers is also in about this order of magnitude. Accordingly, when the device is operated at a relatively high negative voltage, the transition period of carriers is reduced, and the above formula holds. For low bias voltages, the above formula does not hold. If the diffusion effect of carriers is included, with the identical assumptions above, the value of the photocurrent is revised to [33]

$$J_{ph} = eGL\left[\frac{\exp\left(\frac{qV}{kT}\right) + 1}{\exp\left(\frac{qV}{kT}\right) - 1}\right] - \frac{2kT}{eV} \tag{9.2}$$

where k is the Boltzmann constant and T is the temperature. The fitting of the test results of the MDMO-PPV:PCBM solar cell with the above formula is shown in Fig. 9.20. The solid line in the Fig. 9.20 represents the calculated value, $G = 1.46 \times 10^{27}$ m^{-3} s^{-1}. This formula can fit the experimental data well when $V_0 - V < 0.1$, indicating that diffusion plays an important role in this region. From the fitting results in Fig. 9.20, it is suggested that when the voltage exceeds 1 V, the photocurrent increases with the voltage, instead of being saturated in eGL. This is because the above formula does not consider the presence of electron-hole pairs bound by the interface. Their re-separation also contributes to the photocurrent.

We assume that after the charge transfer takes place at the interface, the number of bound electron-hole pairs at the interface is G_{MAX}, and the relationship between

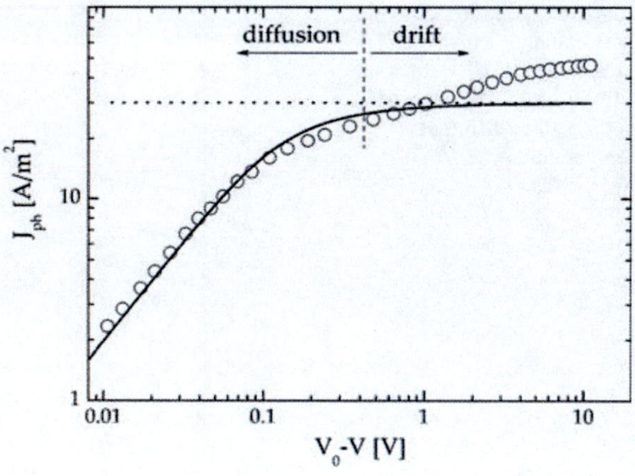

Fig. 9.20 Current and voltage characteristics of MDMO-PPV:PCBM (20:80 wt%) organic solar cells at ambient temperature

the generation rate of free carriers G and G_{MAX} is [33, 34]

$$G(T, E) = G_{\mathrm{MAX}} P(T, E) \tag{9.3}$$

$P(T, E)$ is the probability of the separation of the electron-hole pair bound at the interface, which can be calculated with the Onsager-Braun model, where T is the temperature and E is the electric field. In the calculation of $P(T, E)$, there are only two tunable parameters: the initial distance a of the electron-hole pair formed by the exciton separation at the interface and their decay rate kF to the ground state. The fitting of MDMO-PPV:PCBM cell with variable temperature with the above formula is shown in Fig. 9.21. The solid line in the figure represents the fitting result. At a low effective bias, $G(T, E)$ starts to saturate, fitting the photocurrent well with the previous formula (9.2); At a high effective bias (above 10 V), the photocurrent is saturated, independent of the field and temperature, and all electron-hole pairs at the interface have been separated at this time, with the mentioned assumptions and formulas, the photocurrent of the MDMO-PPV:PCBM cell is well fitted. According to this result, it can be said that at least for MDMO-PPV:PCBM cells, interfacial electron-hole pairs do exist at the interface. Whether such conclusion is applicable to other organic material systems remains a major controversy. The study of exciton dynamics at the interface of organic solar cells remains an important research direction.

9.4.5 Transport of Charges

The free electrons and holes finally formed after exciton separation are left in the acceptor material and the donor material, respectively. The mentioned free electrons and holes then move towards the cathode and anode in the acceptor and donor

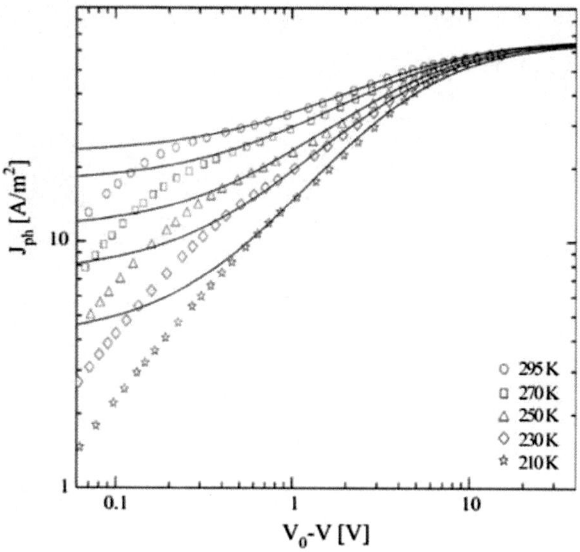

Fig. 9.21 Relationship between photocurrent and effective bias of MDMO-PPV:PCBM organic solar cells at different temperatures

networks, respectively, thereby generating a photocurrent. Here, if the two materials are well separated, i.e., there are continuous channels for carriers to flow, the efficiency of carrier recombination is noticeably lower; If the materials are not well separated and there are no continuous channels for carriers to flow, the current to the electrodes will be significantly reduced. The study of the process of electrons in the acceptor material and holes in the donor material is important to understand the photoelectric characteristics of bulk heterojunction cells. Noticeably research has been done on the mobility of carriers in single organic material employed in solar cells. However, for practical solar cell devices, the effective layer is formed by mixing the donor material and the acceptor material. In such a structure, the mobility of carriers differs from that in a single material. For instance, the mobility of electrons in a single PCBM material is higher than the mobility of holes in a single MDMO-PPV by more than three orders of magnitude. However, in many experiments, it has been found that the mobility of holes in MDMO-PPV:PCBM organic solar cells is significantly improved, so the difference between the mobility of electron and hole is only about one order of magnitude. It is worth studying why the mobility of carriers increases in blended material systems. Numerous tests now reported that the mobility of the identical organic material measured in solar cells, transistors, and light-emitting diodes is different. Thus, this expounds the necessity to fully understand why the mobility of holes in polymer increases when it is blended with PCBM in solar cells. The mobility of holes and electrons in practical devices should be directly measured, and using devices with different structures (e.g., transistors) is considered inappropriate.

To directly measure the space charge limiting current (SCLC) of one type of charge in a blended organic, the transport of another type of carrier should be suppressed. In other words, a device with only electron transport or hole transport should be

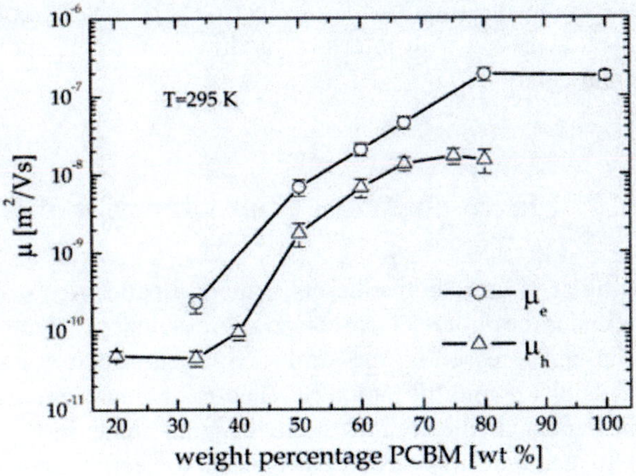

Fig. 9.22 Mobility of carriers in MDMO-PPV:PCBM as a function of the composition of PCBM at ambient temperature [35]

developed. This method has been employed in experiments. As shown in Fig. 9.22, the mobility of electrons and holes in the MDMO-PPV:PCBM mixture measured with this method varies with the composition of PCBM. With the composition of PCBM rising from 33 to 88%, the mobility of the electrons increases first and then saturates. Likewise, the mobility of holes also presents an identical trend, which differs from our intuition. Intuitively, with the increase in the composition of the PCBM material, the organic material is diluted, so the mobility of the holes should be lower. In practical measurement, however, the mobility of holes is enhanced with the increase in the composition of PCBM, and it eventually saturates. The mechanism of this phenomenon is not well understood. However, recent studies reported that the addition of PCBM components alters the morphology of the effective layer, which more significantly enhances the interaction among polymer molecules and then facilitates the transfer of charges among polymers. If such a conclusion is correct, the increase in holes in the blend with the increase in the PCBM component can be explained.

9.4.6 Collection of Electrodes

As the optimization of the composition and appearance of the active layer, the modification of the electrode can significantly enhance the performance of the organic solar cell device. For instance, when with Al as the cathode of a cell, inserting a thin layer of LiF between the active layer and metal Al, not only can increase V_{OC}, but also increase J_{SC} and FF, and the final efficiency can be improved by about 20%. It is not fully understood why the addition of LiF can significantly increase the efficiency of the device. One possible explanation is that the addition of LiF significantly reduces the series resistance of the device, thereby improving the performance of the device. The effect of interface modification on improving device performance fully illustrates the importance of electrodes on charge collection. Various results have been achieved with interface modification to enhance the performance of organic solar cells.

9.5 Macro-electrical Characteristics of Organic Solar Cells

Similar to inorganic solar cells, the critical device parameter of organic solar cells remains the photoelectric conversion efficiency ?. Photoelectric conversion efficiency can be expressed by the product of open-circuit voltage V_{OC}, short-circuit current density J_{SC} and fill factor *FF*. The *FF* is represented by the ratio of the maximum power point (the area of the rectangular shade in Fig. 9.23a) to the product of the V_{OC} and the short-circuit current. Photoelectric conversion efficiency is the ratio of the maximum output power to the intensity of the initial incident light P_{in}. Similar to inorganic solar cells, it is expressed by the following equation.

$$\eta = \frac{J_{SC} V_{OC} FF}{P_{in}} \tag{9.4}$$

The relationship between the current and voltage of organic solar cells can also be expressed by the Shockley diode equation. To be consistent with the actual device, the resistance R_S in series with the diode describing the contact resistance, and R_{sh} in parallel with the diode describing the parallel effect between the two electrodes, are introduced. The equivalent circuit is shown in Fig. 9.24.

For organic solar cells, the relationship between current and voltage can be expressed by the formula [36]

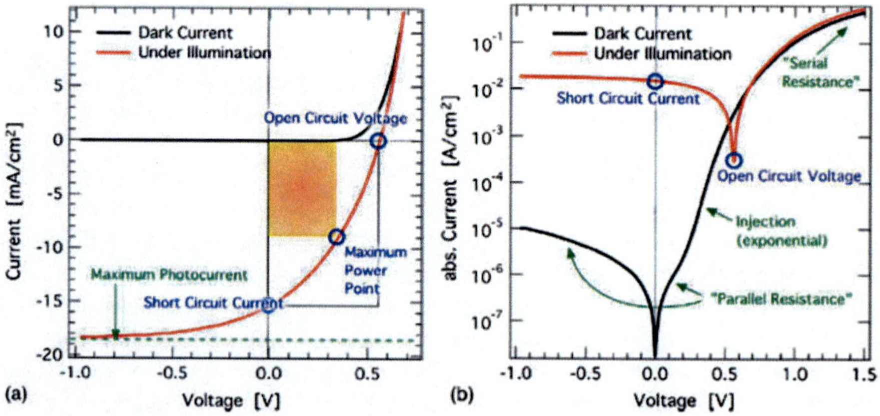

Fig. 9.23 **a** Schematic diagram of current curves and parameters of organic solar cells. **b** Photocurrent and dark current curves of organic solar cells in logarithmic coordinates

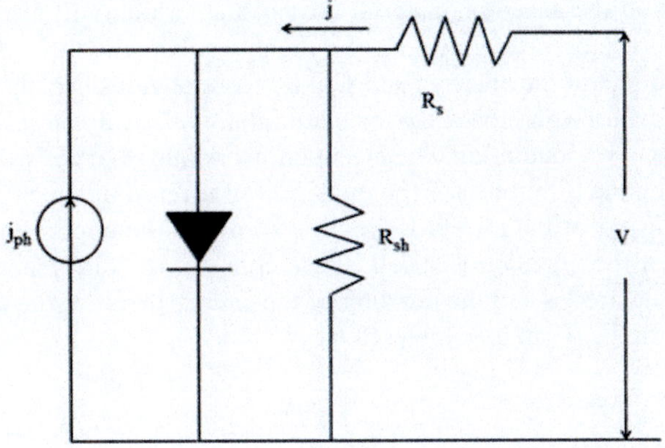

Fig. 9.24 Equivalent circuit diagram of an organic solar cell

$$J(V) = J_0\left(\exp(\frac{q(V - R_s J)}{nk_B T}) - 1\right) + \frac{V - R_s J}{R_{sh}} - J_{ph} \tag{9.5}$$

where J_0 is the reverse saturation current density, q is the basic charge amount, $k_B T$ is the thermal energy, and n is the ideal factor for the diode. The photogenerated current is expressed as J_{ph}.

As shown in Fig. 9.23b, similar to inorganic solar cells, the dark current of organic solar cells can be split into three areas: the area where the parallel resistance dominates when the voltage is low, the injection area (the exponential area), and the series resistance area. Organic solar cells have some differences from inorganic solar cells. In organic solar cells, currents related to electric field are generated for the separation of photoinduced excitons and the extraction of electron-hole pairs by external circuit pairs. It results in an apparent parallel resistance R_{sh} dependent on the electric field; Meantime, compared with inorganic semiconductors, the series resistance of organic semiconductors is noticeably larger. This is because the carrier transport in organic semiconductors is primarily accomplished by transitions, and its transport process is dependent on the electric field and carrier density. This results in the space carrier limiting current no longer strictly obeying the relationship proportional to the square of the voltage in Mott-Gurney's law.

9.5.1 Open-Circuit Voltage

Open-circuit voltage is critical to solar devices. The magnitude of the open-circuit voltage is first associated with the energy of the dipole ($E_{g, DA}$, which is a quantity associated with $HOMO_D$-$LUMO_A$) after the exciton is separated. The magnitude of this energy value is dependent on the donor and acceptor materials. The V_{OC} is also associated with the difference in work function between the metal electrodes. Moreover, in the case of an open circuit, the energy bands of the donor material and the acceptor material are not flat, which will also affect the magnitude of the open-circuit voltage. For an organic solar cell with a double-layer structure, the tilt effect of the energy band will be more obvious because the dimensions of the donor material and the acceptor material are relatively large.

For double-layer heterojunction organic solar cells, the V_{OC} is basically dependent on the difference of the quasi-Fermi level in the open-circuit condition. The energy $E_{g,DA}$ of the interface dipole determines the upper limit of the open-circuit voltage of the organic solar cell. If we consider the difference in the work function of the electrodes and the bending of the energy bands of the donor and acceptor materials, the V_{OC} can be expressed as

$$V_{OC} = \frac{E_{g,DA}}{q} + BB_D + BB_A - \Delta\varnothing_D - \Delta\varnothing_A \quad (9.6)$$

BB_D and BB_A are determined by the degree of band bending of the donor and acceptor material layers; $\Delta\varnothing_D$ denotes the height of the potential barrier between the anode and the donor material; $\Delta\varnothing_A$ represents the height of the potential barrier between the cathode and the acceptor material. They indicate the difference in work function between the two electrodes. In the bulk heterojunction structure, the donor and acceptor materials are mixed, and the V_{OC} is noticeably more sophisticated than the double-layer heterojunction structure. Given the differences between bulk heterojunction cells, Scharber et al. [37], developed an empirical formula for the V_{OC} of solar cells exhibiting bulk heterojunction structure, which exploits a linear relationship

$$V_{OC} = \frac{HOMO_D - LUMO_A}{q} - 0.3 \quad (9.7)$$

The above formula, as an approximate formula, points the relationships between the difference between the HOMO of the donor and the LUMO of the acceptor and the open-circuit voltage. Subsequently, it was reported that a dipole can be formed at the interface between the donor and the acceptor material. The energy of the dipole impacts the V_{OC}. Vandeval et al. [38] adopted

$$V_{OC} = \frac{E_{g,DA}}{q} - 0.43 \quad (9.8)$$

to approximate theV_{OC} of a bulk heterojunction organic solar cell. Such formula considers the relationships between the molecular orbital, the energy of the dipole pair, as well as the V_{OC}. It has been recently found that different directions of the dipoles also affect the V_{OC} of the device. Thus, if a dipole with a certain energy and direction can be artificially introduced, the V_{OC} can vary significantly. For simplicity, the impact of the dipole is not considered on the whole, and $\frac{HOMO_D - LUMO_A}{q}$ acts as the upper limit of the V_{OC}, indicating that the HOMO and LUMO of the material should be matched to obtain good characteristics.

9.5.2 *Short-Circuit Current and Fill Factor*

Short-circuit current and *FF* are also vital parameters in solar cells. It is directly associated with the ability of the material to absorb light, the specific structure of the device, the efficiency of exciton separation, the transport and recombination of carriers, as well as the ability of the electrode to extract charge. For a given material system, the optimization of the thickness of material and other methods to regulate the distribution of light field can significantly enhance the overall light absorption capacity of the device. The specific optimization of the morphology can enhance the efficiency of exciton separation, hinder the recombination of carriers, and shorten the transition period of carriers. The modification of the device interface is conducive to the extraction of carriers. The mentioned measures can significantly increase the short-circuit current and *FF* of the device. The mentioned factors are discussed in the photoelectric conversion mechanism of organic solar cells, so they will not be elucidated in this study.

9.6 Tandem Organic Solar Cells

Single-junction organic solar cells are introduced earlier. Here we review the working mechanism of single-junction organic solar cells. During the photoelectric conversion of organic solar cells, the absorption of light first generates excitons. The excitons generated should first diffuse to the interface between the donor material and the acceptor material, where the exciton dissociates to form free electrons and holes. Subsequently, the electrons and holes are transported in the acceptor material and the donor material, respectively. Lastly, they are harvested by electrodes. Accordingly, the photoelectric conversion of organic solar cells can be split into six steps, as shown in Fig. 9.18: (1) absorption of photons; (2) generation of exciton; (3) diffusion of exciton; (4) separation of exciton; (5) transfer of charges; (6) collection of charges. With the mentioned six processes, there are five major efficiency limiting factors and loss mechanisms, as shown in Fig. 9.18: (I) loss of absorption; (II) loss of heat; (III) loss of excitons; (IV) loss of energy for the separation of exciton; (V) recombination of charges. The loss of absorption is primarily for the mismatch between the absorption spectrum of the material and the solar spectrum, and the incident photons cannot be absorbed by the material. Under the energy greater than the optical absorption bandwidth of material, the absorbed photons will first excite the hot carriers. The mentioned hot carriers will relax efficiently to form more stable excitons with lower energy, which will release heat in the process and cause the loss of heat (II). The exciton exhibits a large binding energy, as well as a diffusion length of nearly 5–10 nm. If it cannot be ionized into free charge timely, it will be lost without contributing to the efficiency of device (III). The separation of excitons requires the difference in energy between the minimum unoccupied orbital (LUMO) energy of the donor material and the acceptor material to overcome the binding energy of the exciton (IV). However, if this energy difference is noticeably larger than the binding energy of the exciton, loss of energy (IV) will also be caused. The recombination of charges is accompanied by the transfer and collection of charges (V).

The presence of the mentioned efficiency loss mechanisms has significantly limited the performance of organic solar cells, and each breakthrough in enhancing the performance of organic cells is for tackling down or inhibiting one or several loss mechanisms in the design of materials and structures. For instance, the introduction of the donor/acceptor heterojunction interface addresses the problem of how to effectively separate excitons and has made a landmark in organic solar cells. In the subsequent development of the bulk heterojunction structure, the donor material and the acceptor material form an interpenetrating network structure in the effective layer, facilitating the diffusion of excitons to the donor/acceptor interface and the transfer of charges. The design of this structure has revolutionized the organic solar cells development and become the mainstream structure of current devices. Recently, low-band-gap materials have boomed as to remedy the loss of light absorption maximally. According to the prediction of the Scharber model (Fig. 9.25), when the position of the LUMO and the bandgap of the polymer are adjusted, the maximum efficiency of a single-junction organic solar cell is about 11% for a polymer material exhibiting a bandgap of 1.5 eV. The experimental results tend to approach this value. However, further enhancements in the efficiency of single-junction organic solar cells will be challenging in the future. Fortunately, the tandem structure (Fig. 9.26), can

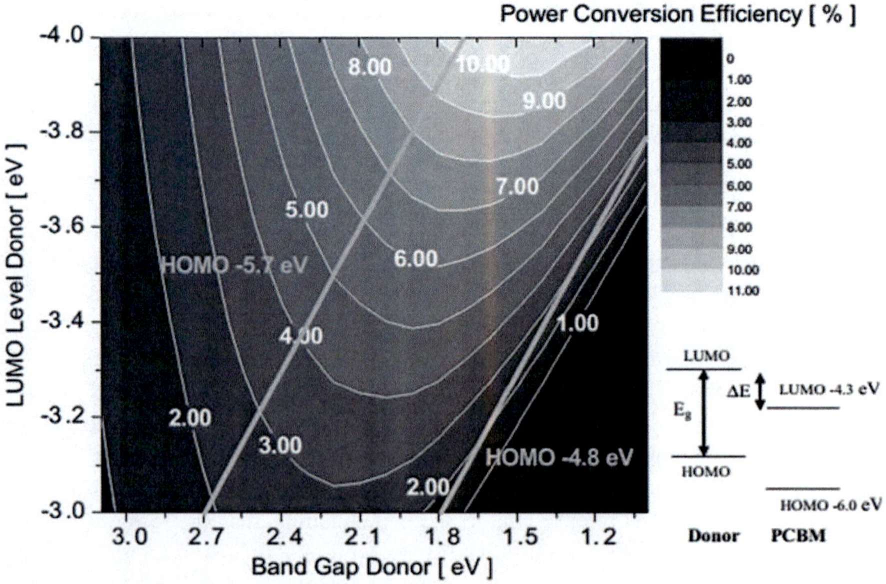

Fig. 9.25 The efficiency of the single-junction organic solar cell predicted by the Scharber model [37]

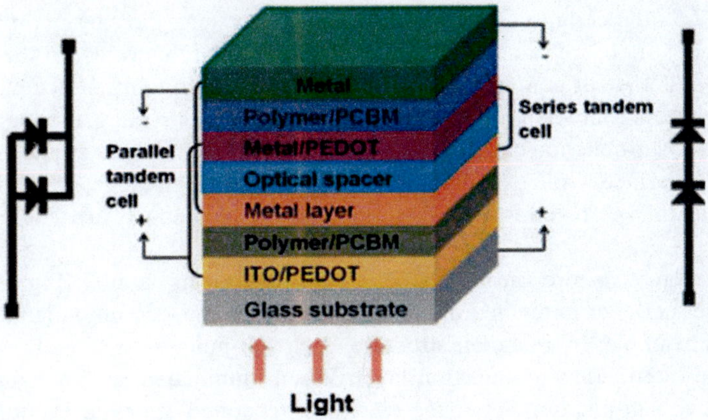

Fig. 9.26 Structure of a conventional tandem organic solar cell [39]. Two connection methods of the tandem structure, one of which is connected in series with a two-terminal structure, and the other is connected in parallel with a four-terminal structure

further suppress the loss mechanism of efficiency and enhance the performance of organic solar cells.

9.6.1 Mechanism of Tandem Organic Solar Cells

In the previous chapter, the tandem structure is covered. In this chapter, organic solar cells are briefed. The tandem structure has obvious merits in enhancing the overall performance of the device, which can weaken the loss mechanisms, e.g., (I), (II), and (IV). Firstly, the loss of absorption can be significantly reduced by the tandem structure. In a laminated structure, organic materials exhibiting different bandgap widths are stacked one after the other, and the absorption spectra of the mentioned materials form a complementarity, thereby maximizing the width of the absorption spectrum of the material and down-regulating the mismatch degree between the loss of the absorption spectrum of the material and the solar spectrum (loss mechanism I). Moreover, the loss of heat can be significantly reduced by the laminated structure. In a tandem structure, materials exhibiting narrower bandgaps are capable of absorbing lower-energy photons and converting them into lower-energy excitons, while materials exhibiting wider bandgaps absorb higher-energy photons and convert them into higher-energy photons. Thus, after the material absorbs photons, the energy of the hot carriers formed is down-regulated, thereby significantly reducing the loss of the heat in the relaxation process from the hot carriers to the exciton (loss mechanism II). Moreover, the tandem structure can also reduce the energy loss attributed to exciton separation (loss mechanism IV). As mentioned in the previous chapters, the energy difference at the interface should be greater than the binding energy of the exciton to effectively separate the exciton. The binding energy of excitons is typically

0.1–0.4 eV. In the common P3HT:PCBM material system, the energy difference at the interface for charge transfer is noticeably larger than this value. In the previous description of V_{OC}, it is mentioned that such a large energy difference will reduce V_{OC} and reduce the efficiency of the device. In a laminated structure, the energy bands of respective sub-junction can be individually designed to down-regulate the energy loss at the interface while expanding the spectral absorption range, which is noticeably simpler than performing the mentioned tasks simultaneously in a single-junction device.

The tandem structure can be connected in series or in parallel (Fig. 9.26). We discuss the merits of laminated organic cells with the most common tandem structure as an example. In a tandem structure, each sub-junction is directly connected through an intermediate connection layer. When illuminated, excitons are formed in respective sub-junction. The free electrons generated after the exciton separation in one sub-junction and the free holes formed in the other sub-junction reach the intermediate connection layer in which they are recombined, as illustrated in Fig. 9.27.

In a series tandem structure, the total current is determined by the minimum current generated in the sub-junction. Thus, to maximize the efficiency, the current of each sub-junction should be optimized so that the current of each sub-junction can be matched. Moreover, the voltage of the series tandem structure is a simple algebraic sum of the voltages of the sub-nodes. It is assumed that the sub-junctions are well connected in series, the LUMO of the acceptor material in each sub-junction is adjustable, each sub-junction is well optimized to generate a matching current, the electrode is in good contact with the effective layer, and there is no contact resistance. If the mentioned assumptions hold, the *FF* and quantum efficiency of the device can be used to predict the efficiency of the tandem cell, as shown in Fig. 9.28. The results show that the highest efficiency is 13% when the number of layers is 1, 19% when it is 2, and 24% when it is 3. It is noteworthy that the tandem structure can effectively tackle down the loss mechanism of the single-layer structure and enhance the performance of the device.

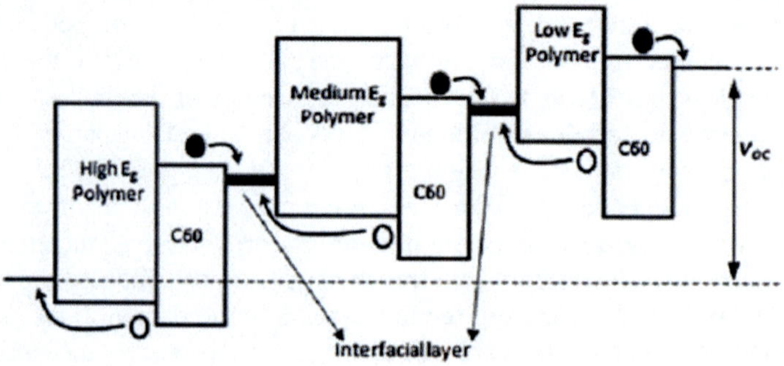

Fig. 9.27 Band diagram of a three-layer organic solar cell with a tandem structure when the circuit is open [30, 40]

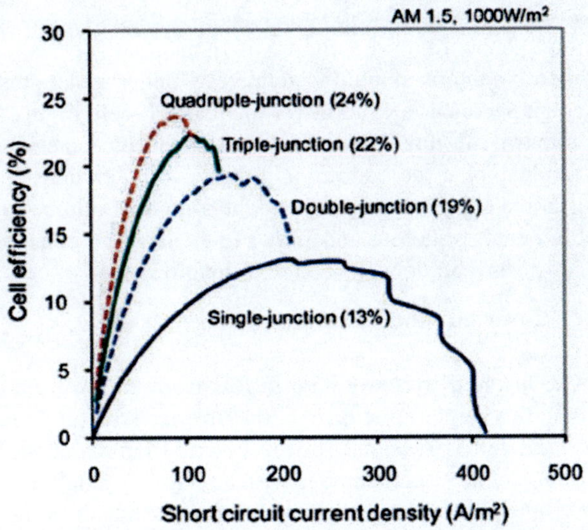

Fig. 9.28 Effect of J_{SC} on efficiency in tandem organic solar cells, assuming $FF =$ 65% and external quantum efficiency is 60% [39]

In the tandem structure, besides the identical requirements for materials as the single-junction cell structure (e.g., good light absorption capacity, suitable band structure, and high mobility), there are some special requirements. The absorption spectrum of each sub-junction should be complementary, and the total absorption spectrum of each sub-junction should be able to cover the solar spectrum as extensively as possible. In the laminated structure, proper selection and control of the material system in each sub-junction is critical to achieve relatively high efficiency. Extending the total absorption spectrum to near-infrared or even mid-infrared is an important direction to improve the overall efficiency. To achieve relatively effective absorption, materials exhibiting a bandgap less than 1.8 eV are required. Important progress has recently been made in such materials exhibiting low bandgaps.

9.6.2 Introduction to the Structure of Tandem Organic Solar Cells

Over the past few years, tandem organic solar cells have achieved rapid development. Various tandem organic structures have appeared, e.g., tandem structures in series, tandem structures in parallel, mechanical tandem structures, folded reflective structures, etc. We introduce the mentioned structures below.

1. **Normally tandem structure in series**

In a tandem solar cell with normal structure, the sub-cells are composed of organic polymers exhibiting different bandgaps. In this structure, a material exhibiting a relatively broad bandgap generally acts as a front sub-cell, and a material exhibiting a relatively small bandgap is generally used as a rear sub-cell. The light first passes through the front sub-cell for which the relatively broad bandgap makes it difficult to absorb photons exhibiting relatively long wavelengths. Accordingly, the front sub-cell is served as a window for the rear sub-cell. Figure 9.29 presents a laminated cell with two sub-junctions. It is suggested that the material exhibiting the relatively broad bandgap is at the bottom, and the material exhibiting the relatively small bandgap is at the top. Photons are incident from the bottom, so the bottom cell only absorbs high-energy photons and allows low-energy photons to pass through, and the top cell can absorb photons in such wavelength range.

2. **Inverted tandem structure in series**

The inverted structure here differs from the inverted structure in a single-junction cell. In a single-junction cell, the inverted structure means that the electrode polarity of the cell is reversed. In the inverted laminated structure, the inverted structure refers to the organic matter with a relatively small bandgap in the front cell, and the organic matter with a relatively broad bandgap in the back cell. It is in opposition to the placement of the material in the normal tandem laminated structure above. The reason for this structure is that the thickness of the organic cell is thin. For a single-junction cell, the thickness of the active layer is 100 nm on the whole. Even in a two-layer structure, the overall thickness of the active layer plus the intermediate connection layer is only about 200 nm. With such a thin thickness, the optical interference effect is strong. Accordingly, the distribution of the light field in the tandem structure does not decrease with the incident depth. Under such optical redistribution coupled with the light absorption characteristics of the material itself, the front junction in some cases contain an organic material exhibiting a relatively small bandgap, and the rear junction covers an organic material exhibiting a relatively large bandgap. Thus, this structure is more conducive to the absorption of light. Figure 9.30 presents the inverted tandem cell structure with light absorbing layers of P3HT and PCPDTBT. The absorption spectrum of the material is also presented in the figure. The PCPDTBT:[70]PCBM cell exhibiting a relatively small bandgap was placed

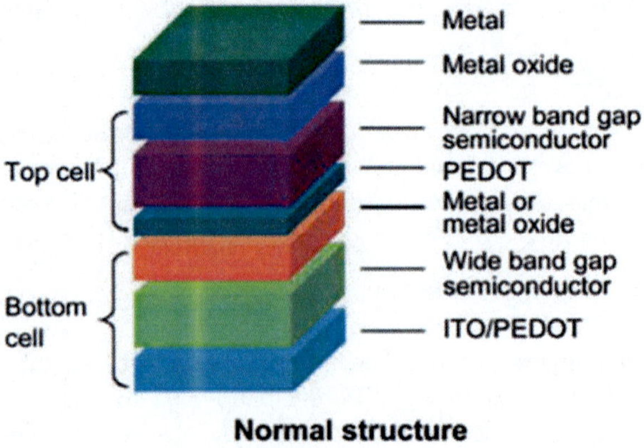

Fig. 9.29 Solar cells with normal structure in series

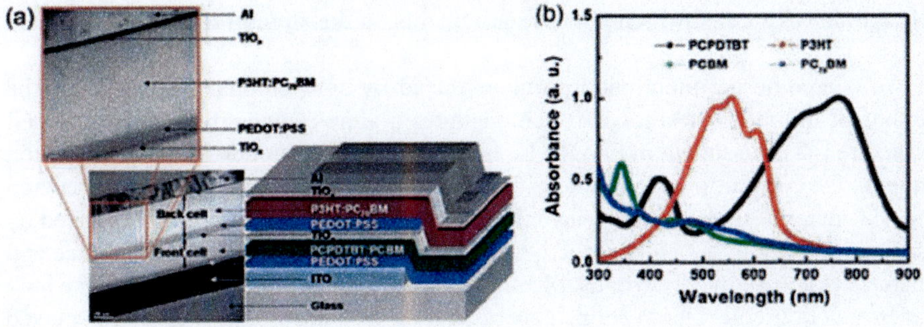

Fig. 9.30 Inverted organic tandem structure [41]

in the front junction, while the P3HT:PCBM with a relatively broad bandgap was placed in the rear junction. Experiments reveal that for the material system employed here, this inverted tandem structure can be optimized for efficiency.

3. **Simple tandem structure in parallel**

The structures of the cells introduced above are all the structure of conventional tandem cells, and the mentioned structures also have obvious defects. As shown in Fig. 9.26, in the conventional tandem structure, the sub-cells are superimposed on each other, and they are physically separated. The sub-cells are connected through an intermediate layer to form a whole. The defects of this structural design are also obvious. Firstly, the incident light should pass through the intermediate connection layer from one sub-junction to another sub-junction. Thus, the absorption of light by the intermediate connection layer is unavoidable, which will reduce the performance of the device. To increase the effective absorption of light, the intermediate connection layer should be as transparent as possible, which places a strict requirement on the optical properties of the intermediate connection layer. Moreover, the intermediate connection layer should ensure that the two sub-nodes can be effectively connected electrically. Accordingly, the resistance of the intermediate layer should be as low as possible, which also places a strict requirement on the electrical performance of the intermediate layer. The optimization of the optical and electrical properties of the intermediate connection layer is often contradictory in actual production. Moreover, the simple superposition of individual sub-junctions in the conventional tandem structure causes the number of material layers contained in the device to increase dramatically. The total number of layers of the device covers the number of layers inherent to each sub-junction itself and the intermediate connection layer. Furthermore, the practical intermediate connection layer itself often consists of two or more layers. Such considerable material layers complicate the device structure and significantly increases the difficulty of the manufacturing process. The mentioned shortcomings will eventually be reflected in increasing the process difficulty and cost of device manufacturing, and eventually reducing or even losing the cost advantage of organic solar cells, which raises requirements on the structural design of tandem cells.

To overcome the mentioned problems faced by conventional tandem cells, the author of this book designs and demonstrates a simple tandem organic solar cell structure [42]. As shown in Fig. 9.31a, in this structure, the intermediate connection layer in the conventional laminated structure is removed, and the CuPc material and PCBM material form a double-layer heterojunction cell. The excitons generated by CuPc absorbing light can be effectively separated at the CuPC/PCBM interface and then harvested by the electrodes. Meantime, PCBM and P3HT materials form a bulk heterojunction cell. The excitons generated by P3HT absorbing light are dispersed at the PCBM:P3HT interface of the mixed layer, separated and then harvested by the electrode. The choice of material also promotes the charge transfer (Fig. 9.31b). From the perspective of the equivalent circuit, the two sub-cells can be considered connected in parallel (Fig. 9.31c). This structural design makes full use of the fact that the photoelectric conversion of the organic solar cell is dependent of the interface between the donor material and the acceptor material, so the two donor materials share the identical acceptor material, thereby neatly removing the intermediate connection layer. The traditional solar cell represented by silicon relies on the PN junction rather than the interface between the donor material and the acceptor material. Thus, the design of this structure cannot be achieved in traditional solar cells. It is a novel laminated structure, termed as a simple tandem organic solar cell structure in parallel.

Compared with the conventional tandem structure, the structure design of simple tandem organic solar cells in parallel has obvious merits. Firstly, for the disappearance of the intermediate connection layer, the incidental absorption of light through this layer is avoided, facilitating the effective absorption of light by the device; Moreover, the disappearance of the intermediate connection layer avoids the resistance introduced by the presence of it, which is conducive to the optimization of the electrical performance of the device; what's more, the disappearance of the intermediate connection layer significantly reduces the total number of material layers of the tandem cell and evidently simplifies the structure. This makes the process simpler, the flow of process easier to control, and the device cost is remarkably reduced.

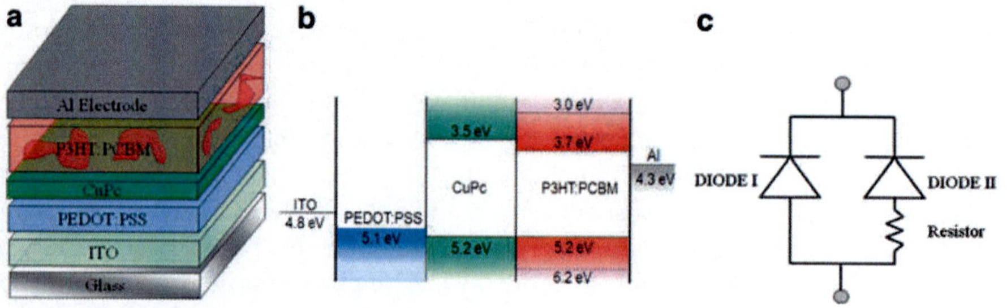

Fig. 9.31 Structural design of a simple tandem cell in parallel: **a** Schematic diagram. **b** Band diagram. **c** Equivalent circuit diagram

Simple laminated organic cells in parallel can continue to maintain cost merits while improving device efficiency, which has important research significance.

Research progress over the past few years has proven that the structural design of simple tandem organic solar cells in parallel is feasible. In the initial work of the authors, a simple laminated device in parallel was obtained with a J_{SC} of 8.63 mA cm^{-2} and a total photoelectric conversion efficiency of 2.79%, approaching to the sum of short-circuit current and photoelectric conversion efficiency of the corresponding single-junction CuPc/PCBM solar cell (J_{SC} = 2.09 mA·cm^{-2} and PCE = 0.43%) and the single-junction P3HT:PCBM solar cell (J_{SC} = 6.87 mA·cm^{-2} and PCE = 2.5%), demonstrating that the two sub-junction cells are indeed effectively connected in parallel on the circuit. In the subsequent work, the device was optimized to increase the efficiency of this simple laminated solar cell in parallel to 4.1%. Such work report has aroused widespread attention from research colleagues in the field, and its design ideas and methods have been extensively adopted. Note that several review articles on tandem organic solar cells over the past few years have introduced this structure. However, this structure also has its own weaknesses. It raises high requirements on the energy band of the material, and the process realization is relatively difficult. Nevertheless, this simple laminated structure in parallel has been employed in various researches on organic solar cells and has become one of the important ways to improve device performance.

4. **Mechanical tandem structure**

When the effective layer of a sub-cell is realized by a solution process, the fabrication of an organic tandem solar cell becomes more difficult. This is because, for the presence of the solution process, it may destroy the already prepared intermediate connection layer or the effective layer of the previous cell in the manufacturing process of the latter cell. To address the difficulty of making organic stacked cells by solution process, two single-junction tandem organic solar cells can be developed separately, and then one of the cells is mechanically placed on top of the other. Such a cell can be termed as a mechanically tandem organic cell. Given practical requirements, the two cells can be connected in series or in parallel, as shown in Fig. 9.32.

5. **Folded reflective tandem organic solar cell**

The structure of the folded reflective tandem solar cell is also proposed to reduce the transparent intermediate connection layer and tackle down the process difficulty of the solution method in the tandem structure. To reduce the loss of photons in the middle layer in solar cells in parallel or in series fabricated on the identical substrate, the cells on two substrates were fabricated and folded, as shown in Fig. 9.33a. In such structure, both sub-junctions have reflective electrodes, so light that cannot be absorbed by one sub-junction will be reflected on the other sub-junction, where it may be absorbed by the second sub-junction. In such folded structure, the incident light is reflected multiple times by the reflective electrodes of the two sub-junction

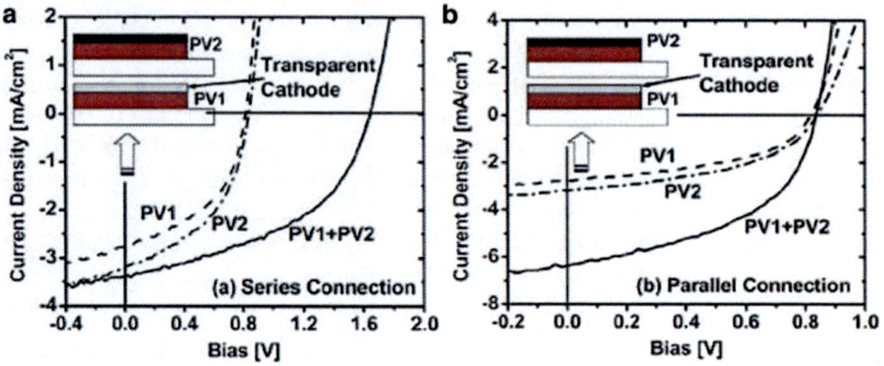

Fig. 9.32 Schematic diagram of organic solar cells with mechanical tandem structure **a** in series, **b** in parallel [43]

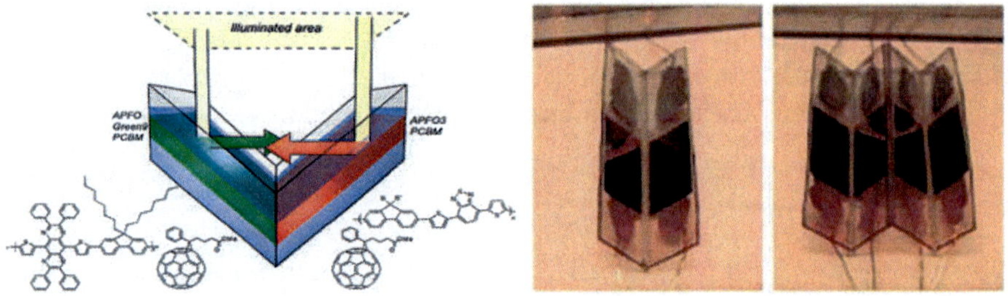

(a) (b)

Fig. 9.33 Laminated organic solar cell with a folded structure [44, 45]

cells to elevate the probability of light being absorbed by the device, i.e., a light trapping structure. As revealed from simulation results, for CuPc-based solar cells, this structure can triple the light absorption capacity; for materials systems exhibiting higher Bunsen efficiency (e.g., P3HT:PCBM material system), however, the optimization of light absorption capacity is relatively limited. Moreover, the degree of enhancement of light absorption is dependent on the angle at which the two sub-cells are placed. This structure also makes it possible to arbitrarily connect two sub-nodes in parallel or in series. As suggested from the results, when the folding structure solar cell based on the AFPO material system is altered from 180° to 40°, the efficiency of the device can be elevated from 2.0 to 3.7%. With a similar design, multiple devices can also be placed on an identical substrate, as shown in Fig. 9.33b.

9.6.3 Intermediate Connection Layer Engineering in Tandem Structures

In tandem solar cells, the effective layers of many efficient sub-cells are made based on the solution process. As we mentioned in the introduction of the tandem structure, this solution-based effective layer fabrication process faces many difficulties in the lamination. For instance, in the production of the effective layer of the rear cell, the previously produced cell may be destroyed. This possibility is even greater when the solvents employed in the active layers of the current and subsequent sub-cells are identical. Several new tandem structures we introduced earlier are proposed to overcome such process difficulty. Besides utilizing new tandem structures, to overcome the difficulties of such a process, thermally cracked organic materials have recently been employed in tandem cells. The thermal cracking material is first deposited on the front cell, and then the thermal cracking material is changed from soluble to insoluble material through heat treatment, thereby effectively protecting the front cell in the production of the rear cell. However, the efficiency of laminated cells with this approach is relatively low, and most of the effective tandem cells comply with a tandem structure in series. Accordingly, the intermediate connection layer engineering of laminated structures has an important role in achieving effective tandem cells, so we should introduce it.

The intermediate connection layer in a tandem cell should meet the following points:

(1) It can effectively protect the front cell in the manufacture of the rear cell;
(2) It will not be affected in the production of the subsequent cell, and the performance will not change;
(3) The light transmittance should be good enough to not affect the absorption of light by the front and rear cells;
(4) The electrical performance is good enough to form good contact between the front and back cells.

Over the past few years, various materials that can act as interlayers have been developed, e.g., metal interlayer materials, metal oxide interlayer materials, conductive organic interlayer materials, and their combinations. The mentioned materials can also act as interface materials in a single cell. For the mentioned specific materials, we will introduce them later in the section of organic solar cell materials. The present section does not involve much. Here the primary methods to achieve the mentioned intermediate layer materials and some special effects are discussed.

(1) Manufacturing method of sputtering of the intermediate connection layer material

A typical material for applying this method is ITO. Because ITO materials have high light transmission and conductivity, it is important to sputter a layer of ITO by sputtering after manufacturing a cell. Thus, the characteristics of the two sub-cells can be as high as possible, whereas because sputtering may damage the device and the process is complicated, the practicality is not high.

(2) Manufacturing method of thermal vapor deposition of the intermediate connection layer material

Many metals, metal oxides, and organic materials can be formed by thermal evaporation. Thus, the thermal evaporation process also plays an important role in the production of the intermediate layer. Figure 9.34 presents a tandem solar cell with an intermediate transition layer made by thermal evaporation. The intermediate transition layer is composed of thermally evaporated LiF, Al, Au/WO_3 layers. The thermal evaporation method is a common coating method and is also employed in industrial production. It is an important method for manufacturing thin films.

(3) Solution manufacturing method of the intermediate connecting layer material

Many organic interlayer materials and metal oxide materials can be made with the solution method, e.g., typical conductive organic compounds PEDOT:PSS and metal oxides ZnO, TiO_2 and other materials. Figure 9.35 presents a laminated solar cell with an intermediate transition layer made by the solution method. The intermediate transition layer consists of a ZnO layer and a PEDOT:PSS layer made by the solution method. The production process of the intermediate layer based on the solution method is technologically similar to the production process of the effective layer. Its potential cost is low, and it applies to the large-scale manufacturing process of roll-to-roll. It will be critical to the subsequent industrialization.

The mentioned three methods are commonly employed in the production of the intermediate connection layer of tandem organic solar cells. Since the intermediate connection layer of a tandem solar cell usually includes several layers of different materials, film formation of the mentioned different materials can be achieved with different processes. Accordingly, sometimes the middle connection layer of the identical tandem device may use two or even three of the mentioned three processes simultaneously. Moreover, the intermediate connection layer can use not only a film material but also a nanomaterial. As shown in Fig. 9.36, the middle Ag nanopar-

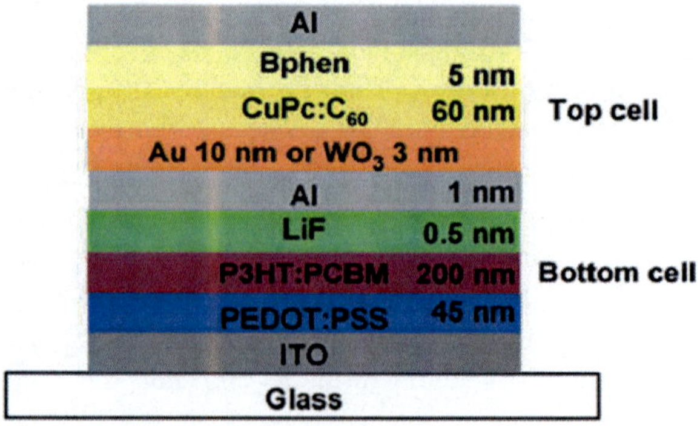

Fig. 9.34 A tandem solar cell with an intermediate transition layer made by thermal evaporation. The intermediate transition layer consists of thermally evaporated LiF, Al, Au/WO_3 layers [46]

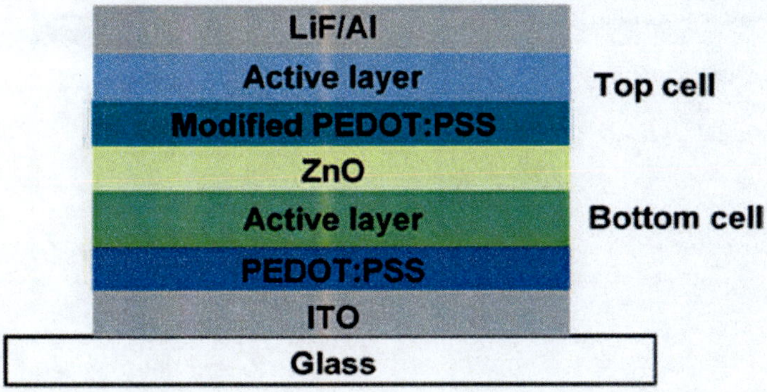

Fig. 9.35 A tandem solar cell with an intermediate transition layer made by the solution method. The intermediate transition layer consists of a ZnO layer and a PEDOT:PSS layer made by the solution method

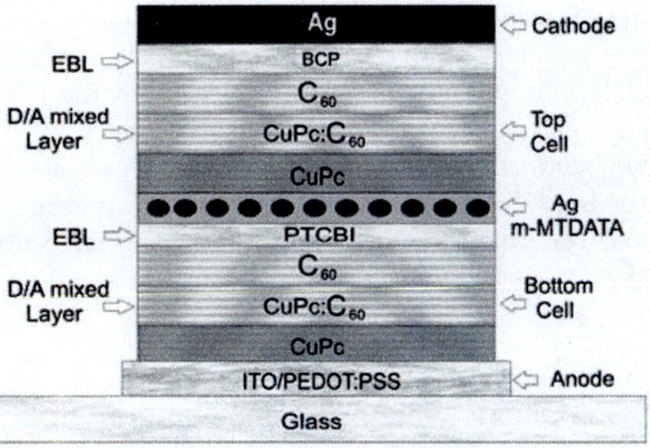

Fig. 9.36 Cascade solar cells containing metal nanoparticles [47]

ticle layer acts as a recombination center, and electrons and holes will recombine in the nanoparticle layer. The Ag nanolayer has a thickness of about 0.5 nm and is surrounded by a m-MTDATA layer exhibiting a thickness of 5 nm. The cladding layer is used to reduce the series resistance and increase the current.

Moreover, the intermediate connection layer in the tandem solar cell can simply connect the front and rear cells, while serving as an optical adjustment layer for the device, as shown in Fig. 9.37. Overall, the electrical characteristics of two solar cells are first matched, and a series or parallel structure is selected, and then the distribution of light in the device can be further optimized by adjusting the thickness of the optical isolation layer to obtain an optimal result. It has merits over simple devices with a combination center as the isolation layer. The design is relatively simple and efficient.

The tandem structure is an important structure for achieving high-efficiency organic solar cells. According to calculations, the efficiency of the two-junction, three-junction, and four-junction tandem structures can reach 19, 22, and 24%. To

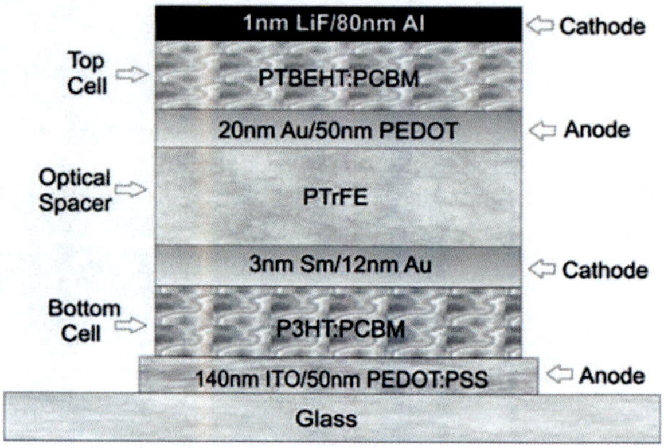

Fig. 9.37 A structure of the tandem organic solar cell. The intermediate connection layer also serves as an optical adjustment layer

obtain theoretical efficiency, it is necessary to increase the absorption of the active layer material to cover the solar spectrum maximally, especially to make full use of visible light and near-infrared light. Moreover, the intermediate connection layer plays an important role in the tandem structure. In the series structure, the intermediate connection layer serves as the combination center of the carriers in the front and rear cells and does not need too high conductivity. However, for a parallel structure, the intermediate connection layer should extract carriers from the sub-devices, so it needs a sufficiently high conductivity. The present section gives a basic introduction to the concept, structure, and intermediate connection layers of tandem organic solar cells.

9.7 Decay Mechanism of Organic Solar Cells

In the past decade, great progress has been achieved in organic solar cells. However, its commercialization should meet the requirements of price (process), efficiency, and lifetime, as shown in Fig. 9.38. The organic solar cells exhibit relatively low photoelectric conversion efficiency and relatively short lifetime. The mentioned shortcomings can be partially compensated by their relatively low prices. As a matter of fact, it is precisely for the high-cost performance of organic solar cells that people have enthusiastically researched and invested it. Over the past few years, the efficiency of organic solar cells has been elevated rapidly, already around 10%. The further optimization of its efficiency can be solved by improving organic semiconductor materials and transparent electrode materials, regulating incident light, and gaining insights into device physics. The reliability of organic devices is another factor limiting the commercialization of organic solar cells. It is known that silicon-based solar cells can work for over 25 years, and organic solar cells can only be more attractive if they have significantly enhanced reliability. Organic materials inherently exhibit higher susceptibility to the effects of oxygen and water vapor than inorganic

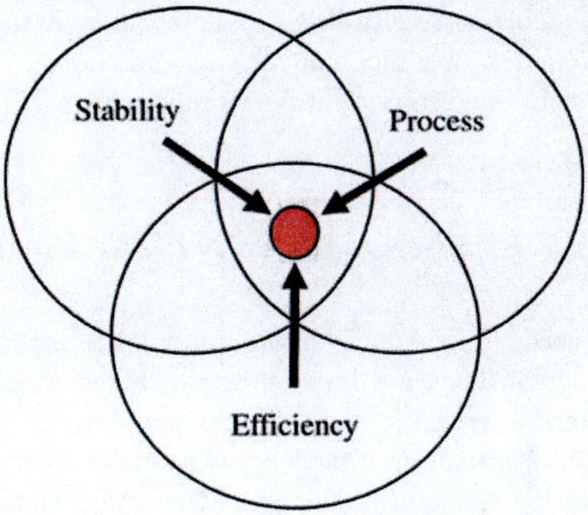

Fig. 9.38 Three important factors for the success of organic solar cells

materials. As impacted by light, this effect will be more obvious. Moreover, in practical work, organic solar cells exhibit relatively high internal temperature, which will to a certain extent accelerate the decline of organic materials and electrode materials. Since organic materials are sensitive to oxygen and water vapor, the device should be encapsulated to isolate the active layer from the air.

Figure 9.39 presents the multiple decay mechanisms that exist in bulk heterojunction organic solar cells. The mentioned decay mechanisms can be roughly split into external decay mechanisms and internal decay mechanisms according to the source of the factors that cause the decay. The external decay mechanism is attributed to factors, e.g., the diffusion of oxygen and water vapor, mechanical stress, and temperature fluctuations; The internal decay mechanism is attributed to internal factors, e.g., internal movable particles and changes in the morphology of the active layer. It can be split into chemical degradation mechanism and mechanical degradation

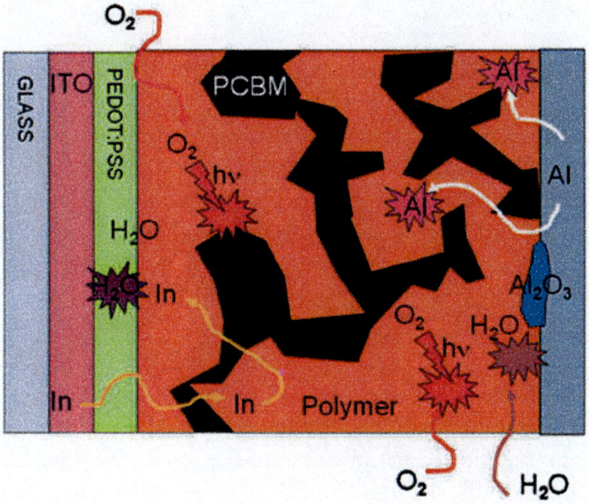

Fig. 9.39 Multiple decay mechanisms in bulk heterojunction organic solar cells [48]

mechanism according to whether the device degradation is attributed to a chemical reaction or a mechanical physical change. In the present section, we introduce the mentioned important decay mechanisms.

9.7.1 Chemical Decay of the Device

The chemical decay of organic solar cells primarily refers to the degradation of device performance attributed to chemical reactions associated with oxygen and water vapor in the device or the reaction between the electrode and the effective layer. During the fabrication of the device, a small amount of oxygen and water vapor will remain in the layers of the device. More importantly, they can enter the inside of the device through diffusion.

Under the irradiation of ultraviolet rays, oxygen can be activated to react with organic materials in organic solar cells. Some organic materials are easy to react with oxygen, while other organic materials are relatively difficult to react with oxygen. Thus, in the solar device, it is necessary to select materials that have good optoelectronic properties and are not prone to chemical or photochemical degradation. Common PPV materials in solar cells (e.g., MEH-PPV polymers) are more susceptible to chemical attack. Under the irradiation of the light with an intensity of 1000 Wm^{-2} in the air, the solar device produced by it can have serious performance degradation in a few minutes to several hours. The other common organic material, P3HT, is relatively stable. But this does not mean that the manufactured solar cells will not be affected by the chemical decline.

When the performance of a solar cell is measured as a function of time, the performance eventually declines over time. On the whole, the regression curve can be fitted with Eq. 9.9 [49]

$$\eta = ae^{-\alpha t} + be^{-\beta t} \tag{9.9}$$

The decay curve essentially reflects the total impact of the various factors that cause the device to decline. Sometimes, the regression curve also requires more complex equations to fit. For instance, in the solar cell of MEH-PPV, the decay is rapid at first, and then the decay rate is slowed down. It can be fitted with a second-order exponential equation. Existing studies reported that the initial rapid decline of solar cells is not related to whether they are exposed to air or nitrogen. It is most likely attributed to the reaction of the metal electrode (Al) with the polymer. A thin barrier layer e.g., C_{60} can be deposited between the polymer and the Al electrode to prevent rapid degradation of the device. In the presence of air, the subsequent slower device degradation is significantly accelerated, and it is accordingly associated with the reaction with oxygen. When measuring the degradation curve, the open-circuit voltage, short-circuit current, and fill factor are generally tested as a function of time. The mentioned parameters are the macro parameters of the device, and the micro details of the decline of the device are difficult to reflect. Among the mentioned three macro parameters, the open-circuit voltage is relatively difficult to be the object of studies on recession.

1. **Diffusion of oxygen and water into organic photovoltaic devices**

When organic photovoltaic devices are placed in the air, the degradation of the device is rapid. This is because water and oxygen in the air can diffuse into the device and then react with the effective layer of the organic solar cell, which accelerates the decay process in the device. Water and oxygen primarily enter the inside of the device through electrodes outside the device. Numerous microscopic holes exist in the electrode, which are the primary channels for water to enter the interior of the device. Moreover, there are many metal grain boundaries in the metal electrode, which may also be a channel for water and oxygen to enter the interior of the device. Whether under light or in the dark, oxygen and water can enter the device through the electrode. However, light is capable of accelerating the reaction of incoming oxygen with organics near the metal electrode. Oxygen and water can reach the interface of the effective layer via the electrode, and they can even reach the ITO electrode on the other side via the effective layer of the organic solar device. Thus, to enhance the device performance, micropores on the electrodes should be reduced or even eliminated.

2. **Photooxidation of organic matter**

PPV materials have been extensively employed in organic solar cells. Because it is easily oxidized under the light, which results in a short lifetime of the device. It is relatively easy to understand this chemical decay process. The light-excited polymer material transfers energy to the ground state oxygen adsorbed on it, thereby producing singlet oxygen, as shown in Fig. 9.40. Accordingly, the energy of the triplet state T1 of the polymer should be higher than that of the singlet state of oxygen to facilitate the transfer of energy. Meantime, the conversion from S1 to T1 state in the polymer should also be able to take place, and the lifetime of T1 should also be sufficiently long that allows energy to be transferred to oxygen. The formed singlet oxygen then

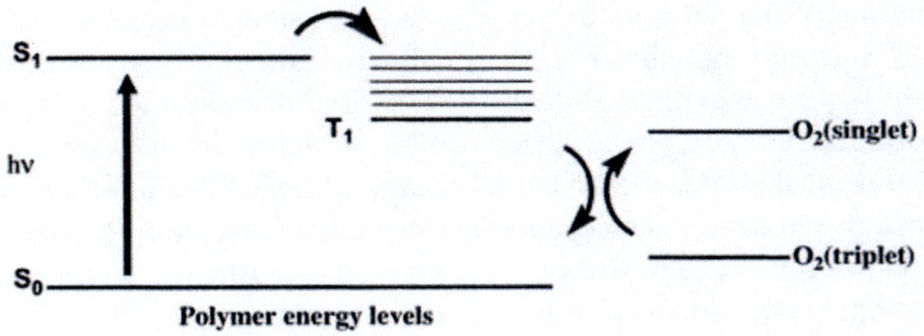

Fig. 9.40 Band diagram of polymer and oxygen. The low-energy excited state (S1) of the polymer can be transformed into the triplet state (T1), and then this excited state is transferred to the adsorbed O_2 to form a singlet oxygen [48]

reacts with vinylene groups in the PPV material; as a result, the molecular chain of the polymer is broken.

The reaction between oxygen and P3HT remains unclear. Existing studies reported that the formation of charge transfer complexes between oxygen and P3HT is reversible, and it is unlikely to cause the polymer to degrade. Since the triplet state in P3HT exhibits low formation efficiency, the singlet oxygen formed by the energy transfer of singlet P3HT exhibits a relatively low probability. It is likely that singlet oxygen is generated during the separation of the charge transfer complexes, which causes device performance to degrade.

3. **Chemical decay of the electrode**

Al is an electrode commonly employed in organic solar cells, whereas it is prone to chemical reactions. For instance, the theoretical study of the interface between PPV materials and Al presents that they can directly react to form Al–C bonds. PCBM and other fullerene materials exhibit high electron affinity and are relatively easy to react with metal electrodes. C_{60} can react with alkali metals to form K_3C_{60}. It is found experimentally that when the interface between Al and C_{60} is exploited to transfer charge, the reaction between them is critical. Inserting an interface layer on the interface or doping C_{60} can significantly improve the transmission capacity of the interface. Moreover, PEDOT:PSS materials are common to modify the ITO interface, whereas PEDOT:PSS itself is acidic, and it will react with ITO to affect device performance.

9.7.2 Physical and Mechanical Decay Mechanisms

In bulk heterojunction organic solar cell devices, efficiency is significantly dependent on the phase separation between the donor material and the acceptor material in the organic layer. Ideally, an interpenetrating network is formed between the donor and acceptor materials. Such a network facilitates the separation of excitons and the transport of carriers. The optimization of the morphology of the active layer has turned into one of the hotspots of the studies on organic solar cells. Micromolecular PCBM and even polymer P3HT will slowly diffuse or crystallize over time at relatively high temperatures. For the device, its internal structure is preferably thermodynamically stable, so the device exhibits higher reliability. Variations in the microstructure of the effective layers of organic solar cells will cause the device performance to degrade. It has been experimentally found that PCBM will grow over time in MDMO-PPV/PCBM devices, thereby affecting its stability. The variation in morphology is associated with the glass transition temperature of the material. P3HT exhibits a higher glass transition temperature than MDMO-PPV. As a result, P3HT-based organic solar devices exhibit higher stability.

9.7.3 Device Packaging

Organic solar cells are sensitive to both oxygen and air, and packaging can protect organic solar cells from them. At present, glass acts as a transparent substrate for organic solar devices, thus insulating the device. However, the metal electrode on the other side exhibits extreme susceptibility to attack by water and oxygen. If flexible substrates (e.g., PET and PEN) are employed instead of glass, the problem will be exacerbated since the mentioned materials are more permeable to water and oxygen than glass.

The transport of oxygen and water in the film is characterized by oxygen transmission rate (OTR) and water vapor transmission rate (WVTR). For materials employed in organic solar cells, the upper limit of OTR refers to 10^{-3} cm^3 m^{-2} day^{-1} atm^{-1}, and the upper limit of EVTR is 10^{-4} gm^{-2} day^{-1} atm^{-1} [48]. Commercial PET and PEN films exhibit a high water and oxygen transmission rate. Introducing an inorganic coating on the film can effectively enhance its ability to block water and oxygen. It is shown experimentally that the multilayer structure of SiO_x and PEN can effectively enhance the stability of MDMO-PPV/PCBM cells. Another reported packaging method refers to exploiting a thick Al backplane on one side, glass on one side, as well as epoxy seal in the middle. Such structure can effectively isolate water and oxygen and extend the lifetime of a large-area P3HT/PCBM cell to more than one year. Other packaging methods consist of device packaging with parylene and alumina.

9.8 Manufacturing Process of Organic Solar Cell

To manufacture inexpensive organic solar cells, common processes include spin coating, annealing, and vacuum evaporation are used. Moreover, printing processes, roll-to-roll processes, spray coatings, and other new processes have been continuously researched. For the foreseeable future, when organic solar cells are marketed, how to combine the cheapness, stability, and large-scale production of the process is the direction that researchers should work hard. If a material can provide high efficiency, whereas it is easily degraded or extremely unstable, or the preparation process is extremely difficult, it will become an obstacle to practicality.

Organic solar cells usually have a series of layered structures, and each layer requires independent technology to form a film. There are many technologies that can form films, whereas few are applicable to organic solar cells. There are three primary reasons: Firstly, many technologies require considerable material; Secondly, repeatability is sometimes difficult; Finally, many technologies are not suitable for small-scale use in the laboratory, so there is no way to fully verify their reliability. However, the spin coating method employed in the laboratory is difficult to apply on a large-scale application. Despite the mentioned difficulties, some technologies are useful for considerable papers, plastic, and plastic materials, which are usually continuous roll materials. Accordingly, people often call it roll-to-roll or rail-to-rail overlay technology (R2R for short). The processing steps usually include: flattening the material, covering, and re-rolling. More detailed steps include cleaning, pretreatment (heating, drying, etc.). Because such a process is suitable for large-scale high-speed production, it can significantly reduce costs. Though the merits of such technology are obvious, identifying a cladding material technology suitable for large-scale production is a major challenge. Since the one-dimensional scale of the material layer should be controlled, and the layout pattern should be prepared on the two-dimensional plane, a process that can cover multiple layers of materials while forming a three-dimensional structure is required. Second, the ink in the traditional printing process will change for many factors, whereas the characteristics of the organic cell require the device to be stable and it is best that there is no difference between the devices, which is a major challenge for the traditional method.

The ideal preparation process of the organic solar cell includes dissolving all materials in a solvent and then preparing a device with a three-dimensional structure on the flexible substrate through steps, e.g., covering and printing. It is best that the mentioned steps do not produce environmentally harmful substances and the device is stable, and the efficiency will not decay in a short time.

There is an essential difference between cladding and printing technology. Overall, printing refers to the transfer of a layer of ink to another substrate through a seal; Coating means that the ink is transferred to the substrate by spraying, staining and other methods. Printing technology usually means the formation of complex patterns (e.g., text), whereas cladding simply means forming a layer of material in a specific area. Printing technology primarily includes screen printing, transfer printing, letter-press printing, flexographic printing, and so on. The coating technology primarily includes spin coating, doctor blading, drawing, spraying, and so on. What can be classified as both printing and coating is inkjet printing. In essence, inkjet printing is a cladding technology, and it can also form a pattern, so it can also be considered a printing technology. In the present section, we will introduce some technologies employed in the manufacture of organic solar cells, some of which are mentioned earlier but raise unique requirements and exhibit characteristics for solar cells.

1. **Casting**

The casting method is considered the simplest film-forming process. It does not require any equipment other than a horizontal workbench, and its steps are to simply cover the solution on the substrate and then set it into a film by drying. For the lack of effective means to control the thickness of film, phenomena (e.g., frame effects) will occur after the film is dried. Since surface tension usually exists in the solution, it will become more uneven and even cracks appear after the film is formed. To avoid precipitation and other phenomena in the drying process, good dissolution characteristics of the coated material in the solution are required.

2. **Spinning coating**

The critical process for preparing organic solar cells is the spin coating process. Despite the complexity of film formation, such a process can produce thin films in large quantities and the repeatability of the films is high. It has some superiorities over other coating processes. It can form a large-area and uniform film (substrate size up to 30 cm). In the manufacturing of microelectronics, such process is extensively employed in the coating of photoresist during photolithography. The typical operation process is to drop the liquid to be coated on the substrate and then rotate the substrate at a certain angular velocity. For the effects of angular velocity and centrifugal force during substrate rotation, the excess solution will be thrown away, leaving only a thin film. Sometimes the substrate is rotated first, and then the liquid to be coated is dropped on the rotated substrate. Figure 9.41 illustrates the operation of the spin coating. The thickness and surface morphology of the film is highly repetitive. It is primarily dependent of the speed, the viscosity and volatility of the solution, the molecular weight and concentration of the material, etc., whereas it has little to do with the amount of solution introduced and the time of rotation. The thickness of the obtained film can be given by

$$d = k\omega^{\alpha} \tag{9.10}$$

ω is the rotational speed, and k and α are empirical parameters. The typical value of α is -0.5, and k is associated with many parameters, e.g., the viscosity of the solution.

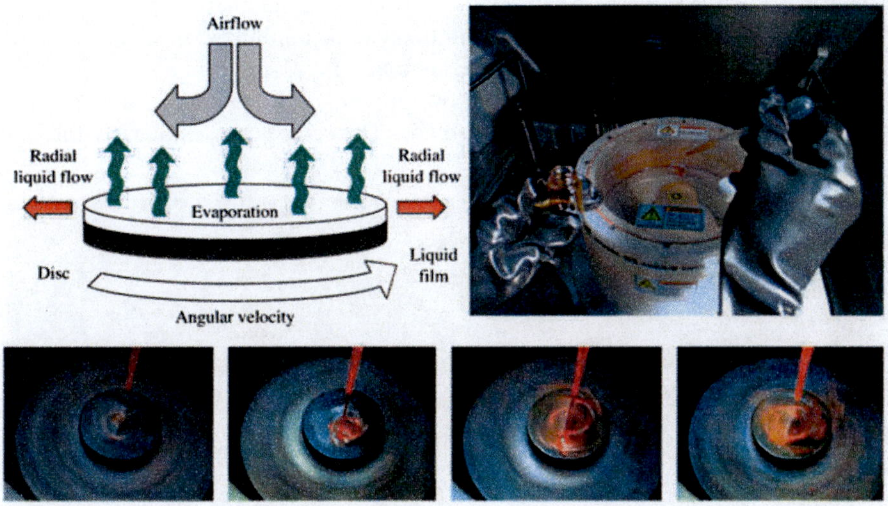

Fig. 9.41 A schematic diagram of the spin coating process in the upper left, and photos of the spin coating process in the glove box in the upper right. Below are photos of different periods during the spin coating process [50]

In the production process of the organic photovoltaic device with the spin coating process, most of the material is wasted during the rotation of the substrate, and this is not a serious problem, because even if only a small amount of material (e.g., 0.1 ml) is used, the required film can be obtained. In the process of spin coating to form a film, there are several changes, e.g., evaporation of the solvent, change of viscosity, radial flow of the solution, self-assembly of molecules on the solid–liquid and gas–liquid surfaces, and phase separation in the mixture. This is why the final performance of the device is closely associated with the process in many scientific reports. In the spin coating process, besides the thickness of the final film, the uniformity of the film, the density of defects (pinholes, etc.), and the interface characteristics of phase separation of the mixed materials are also critical to the final device performance. The spin coating process can better control the mentioned parameters, which is why it is extensively employed in the production of organic photovoltaic devices. One successful system for applying the spin coating method is the P3HT:PCBM system [5]. They are usually dissolved in 1,2-chlorobenzene. After they are spin-coated, the wet film can be dried slowly, and finally, an effective device layer is formed. Almost all related work in the laboratory is based on such a process. Though this method is effective in small-scale applications in the laboratory, when we apply it to large-scale production, production speed becomes the primary constraint. Moreover, as only a part of the materials can be employed in such a process, the applicability of this method is also tested.

3. **Scraper covering process**

In comparison with the spin coating, a scraper covering has been less implemented in the production of organic solar cells in the laboratory. The scraper covering process is shown in Fig. 9.42. This method can also well control the thickness of the film. Unlike the spin coating process, the waste of solution can be minimized in such a process. The lost solution is about 5%. Such a technique first places a scraper at a certain distance from the substrate (overall 10–500 microns). Subsequently, the covering solution is placed in front of the doctor blade. The scraper passes linearly through

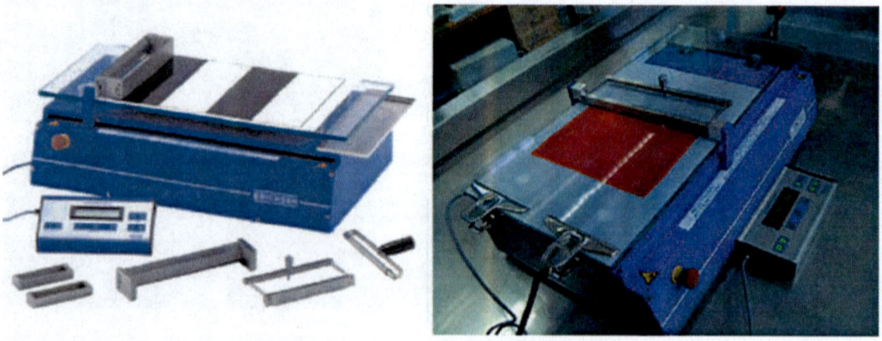

Fig. 9.42 The picture on the left is ErichsenCoatmaster 509 MC-I scraper equipment, and the photo on the right is MEH-PPV scraper process [50]

the substrate, leaving a thin film behind. The thickness of the film is associated with the distance between the scraper and the substrate, and it may also vary with variations in the surface topography of the substrate, surface energy, volatility of the solution and surface tension. The thickness d of the final film can be empirically expressed as

$$d = \frac{1}{2}\left(g\frac{c}{\rho}\right) \tag{9.11}$$

where g denotes the distance between the scraper and the substrate; c is the concentration of solid matter in the ink; ρ represents the density of the final matter. Compared with the spin coating process, the scraper process forms a film efficiently, suitable for large-scale rapid application. The scraper process and spin coating process exhibit the identical instrument cost and operation complexity, whereas the scraper process is compatible with the R2R process, and the spin coating process does not have this. The scraper process outperforms the spin coating process in large-scale production. Researchers have exploited the scraper process to form thin films in devices based on MDMO-PPV:PCBM system, whereas compared with the spin coating process, the related studies on the scraper process have been rarely conducted. According to reports, PCBM prepared by the scraper process has better crystal properties. This prominent characteristic can be attributed to the slow evaporation and drying process of the solvent, so the film will be more thermodynamically balanced in the scraper process.

4. **Screen printing**

Screen printing technology dates back to the early twentieth century. It differs from other printing and coating technologies since it requires a coating solution with high viscosity and low volatility. Such technology is illustrated in Fig. 9.43. It should be fixed to a braided screen (e.g., a metal grid), the part to be printed is hollowed out, and the screen is filled with a coating solution and then approaches the substrate. Squeegee is pressed on the screen to make it contact the substrate; subsequently, the squeegee moves in a straight line to make the solution reach the substrate via the screen to realize the pattern transfer. The thickness d of the finally formed film is

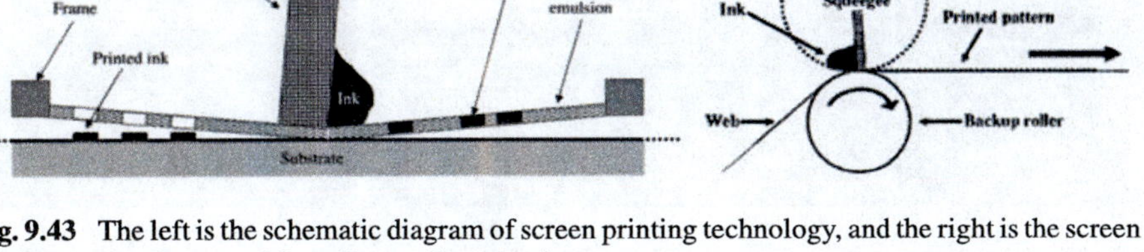

Fig. 9.43 The left is the schematic diagram of screen printing technology, and the right is the screen printing technology used for roll-to-roll production

associated with the volume of the solution V_{screen} complying with the substrate via the screen, and also with the concentration c of the coating solution and the density ρ of the dried material

$$d = V_{screen} k_p \frac{c}{\rho} \tag{9.12}$$

k_p denotes a process-related constant.

Screen printing technology requires a coating solution with high viscosity and low volatility, and the thickness of the film produced by such a process is relatively large. As a result, its application in organic solar cells has been limited, whereas recent developments have gradually addressed the mentioned problems. Screen printing is ideal for mass production and can be employed in roll-to-roll production processes. Screen printing technology is likely to become the critical technology for large-scale organic solar production. However, some difficulties remain. The solution of the active layer material in the device may not yet satisfy the requirements of such a process. However, conductive materials (e.g., PEDOT:PSS), conductive colloids (e.g., silver), and silver-aluminum have been employed for screen printing, and the mentioned materials are all required in the manufacture of solar cells.

5. **Inkjet printing**

Inkjet printing is illustrated in Fig. 9.44. From the perspective of the printing industry, inkjet printing is a relatively new printing technology, which is primarily developed by low-cost inkjet printer technology for office use. Its print head is ceramic or other

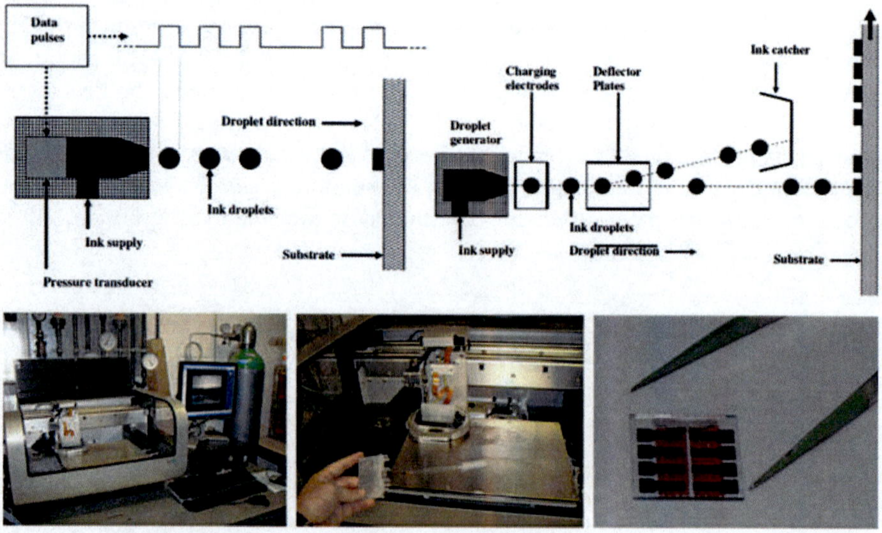

Fig. 9.44 Schematic of inkjet printing technology [51]

materials that are resistant to organic solvents, so a wide variety of solvents can be used to make the device. Inkjet printing technology can exhibit high resolution, and the pixels can be smoothly altered from 300 to 1200 dpi. Compared with other printing technologies for manufacturing organic solar cells, it requires no graphic sources (e.g., masks, screens, etc.), and its defect is limited printing speed. The thickness of the inkjet printing film after drying is determined by the number of ink droplets per unit area N_d, the volume of each droplet V_d, as well as the concentration of the material

$$d = N_d V_d \frac{c}{\rho} \tag{9.13}$$

Inkjet printing technology is significantly complex and it is based on the formation of small droplets. The formation of liquid droplets can be obtained by the nozzles compressing the ink machinery or heating the ink. Subsequently, the droplets should be charged. When an electric field exists between the substrate and the showerhead, the droplet can accelerate toward the substrate, which also makes the ink more difficult to prepare. Inks need lower viscosity, and electrostatic charges should be introduced. On the whole, inks are a mixture of solvents, at least one of which is highly volatile. Moreover, in order for the ink to form the ink flow well, the ink should have a certain surface stress effect. This requires adding additives to the solution, with a concentration of about 1%, which adversely affects the manufacture of organic solar cells. Whether such technology can become the mainstream technology of organic solar cells is dependent on its technological breakthrough as well.

6. **Roll-to-roll technology (R2R)**

The mentioned technologies are some important technologies that can be employed in a single device. Here we briefly introduce the concept of roll-to-roll for mass production. In the roll-to-roll production, the substrate material is long, and it can be bent into a roll, which requires a certain mechanical flexibility of the substrate material. In the printing and coating process, the material passes through the roller and gets flattened; then it passes through the printing machine or coating machine, and subsequently, the material is rolled into a roll by the roller. Besides printing or coating processes, other processes (e.g., heating, drying and UV treatment) will also be included. Ideally, the substrate material enters from one end of the machine, and after a series of standard process steps, it is output from the other end to be a complete device. No manual operation is required in the middle. A series of R2R steps for manufacturing organic solar cells are illustrated in Fig. 9.45.

R2R has been employed in the actual manufacturing of organic solar cells. To enable integrated production, some practical situations should be considered. In many printing or coating technologies, the emphasis is placed only on the formation of a wet film, and the drying of the film is overlooked. The drying step is relatively complicated. In brief, for systems exhibiting different speeds, the thickness of the material after drying is different. For the various operating speeds of different systems, the operating speed of the identical system may change as well, increasing the difficulty

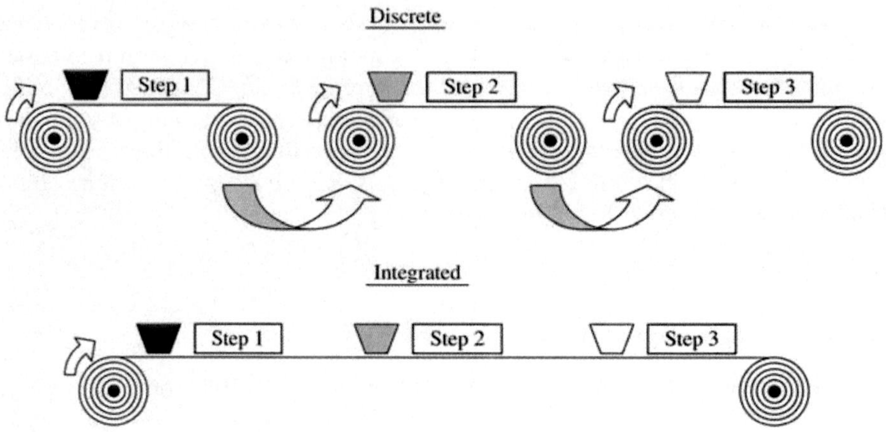

Fig. 9.45 Schematic of the R2R process of an organic solar cell with 3 layers of material. The upper figure presents the separated process, and the lower figure is the integrated process

of preparation. On the whole, a small R2R system covers a de-curling, i.e., drying and curling unit. Moreover, the system should integrate the control of the stress and speed of substrate material, substrate cleaning, charge removal, surface treatment, IR-heating, hot air drying, UV processing, and substrate cooling.

9.9 Development of Organic Solar Cells

Organic solar cells comply with organic materials. It is known that organic materials are easily chemically modified, which makes it relatively easy to design materials as required. By improving the spectral absorption capacity of the material and expanding the spectral absorption range, the degree of matching between the organic solar cell and the solar spectrum can be significantly improved. Moreover, the processing process is simple, and a large-area film can be formed, which is conducive to mass production. Low production costs, coupled with the low price of organic materials, make organic solar cells have a huge cost advantage. Moreover, the organic material has good flexibility and can be mechanically bent with large radians, which provides a guarantee for the flexible application of solar cells. Moreover, organic materials can degrade and have a small impact on the environment. Under the premise that people pay more attention to environmental protection today, it has also become one of its great merits.

Besides conventional applications, organic solar cells have some special uses. Organic solar cells are light in weight. Under the identical weight, the light receiving area will be significantly increased after deployment. If the reliability problem is solved, it will play a huge advantage in space applications, because for space launch, weight is one of the factors that should be considered. Organic solar cells are thin

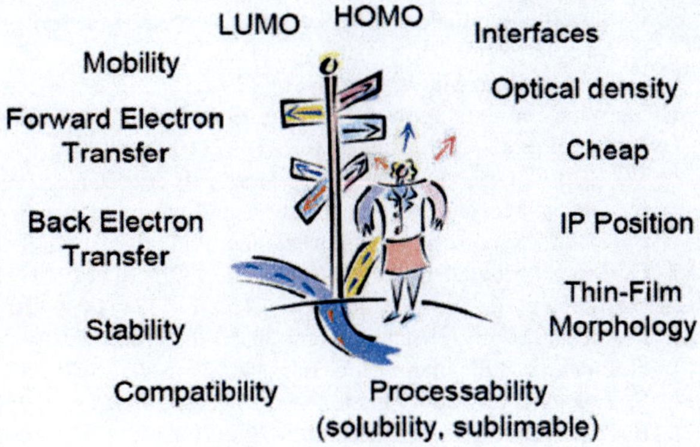

Fig. 9.46 Problems to be solved to further develop organic solar cells

and easy to carry, and can be combined with clothes, handbags, etc., and can also be applied to consumer electronics. Moreover, organic solar cells can be made into different types to meet various environments, e.g. transparent, translucent, and opaque solar cells. It can be made into various colors and patterns, can be combined with buildings for exterior walls, etc., and can also be employed in areas, e.g., automotive film. Organic solar cells have a wide range of uses and bright application prospects.

The organic solar cells development has experienced a single-layer organic cell structure, a double-layer heterojunction cell structure, and a bulk heterojunction cell structure, especially the introduction of the bulk heterojunction structure significantly promoted the organic solar cells development. Based on the bulk heterojunction structure, the efficiency of organic solar cells has increased rapidly in the past decade. Around 2000, the efficiency of organic solar cells was generally below 1%, whereas in the subsequent period, the efficiency continued to increase, and the highest efficiency has reached more than 10%, which is close to the threshold of marketization. However, the organic solar cells development still faces many problems, and scientists from the fields of chemistry, physics, and electronics should work together to find a corresponding solution as soon as possible. Figure 9.46 presents the problems that should be solved to further develop organic solar cells. The mentioned issues include both material issues and process issues. Indeed, in-depth studies on the mechanism of solar cells are also indispensable. The mentioned problems are more or less touched on in the description of organic solar cells in the present chapter, and further solutions to the mentioned problems will significantly promote the organic solar cells development. Given the organic solar cells development in the past decade and the existing experimental results, organic solar cells development is up-and-coming.

References

1. S.B. Darling, Energy Environ. Sci. **2**, 1266–1273 (2009)
2. N.S. Sariciftci et al., Science **258**, 1474–1476 (1992)

3. N. Sariciftci et al., Appl. Phys. Lett. **62**, 585–587 (1993)
4. G. Yu et al., Appl. Phys. Lett. **64**, 3422–3424 (1994)
5. G. Li et al., Nat. Mater. **4**, 864–868 (2005)
6. M. Wienk et al., Angew. Chem.-Int. Edit. **115**, 3493–3497 (2003)
7. M. Lenes et al., Adv. Mater. **20**, 2116–2119 (2008)
8. W.C. Choy, *Organic Solar Cells: Materials and Device Physics* (Springer, 2013)
9. C. Zhang, Mechanism investigation and structure design of organic photovoltaic cells for improved energy conversion efficiency. Ph.D thesis, NUS, 2009
10. C. Chochos et al., Prog. Polym. Sci. **36**, 1326–1414 (2011)
11. W. Zhang et al., J. Am. Chem. Soc. **132**, 11437–11439 (2010)
12. D.A. Egbe et al., J. Mater. Chem. **20**, 9726–9734 (2010)
13. H. Imahori, Bull. Chem. Soc. Jpn. **80**, 621–636 (2007)
14. S. Verlaak et al., Adv. Funct. Mater. **19**, 3809–3814 (2009)
15. R. Po, J. Phys. Chem. C **114**, 695–706 (2009)
16. S.A. Ponomarenko et al., Chem. Mater. **18**, 579–586 (2006)
17. R. Po et al., Energy Environ. Sci. **4**, 285 (2011)
18. J.S. Kim et al., Appl. Phys. Lett. **91**, 112111–112111–3 (2007)
19. H. Liao et al., Appl. Phys. Lett. **92**, 173303 (2008)
20. A.W. Hains et al. ACS Appl. Mater. Interfaces **2**, 175–185 (2009)
21. Y. Wang et al., Sol. Energy Mater. Sol. Cells **95**, 1243–1247 (2011)
22. J.-Y. Lee et al., Nano Lett. **8**, 689–692 (2008)
23. A. Gadisa et al., Synth. Met. **156**, 1102–1107 (2006)
24. D. Kearns et al., J. Chem. Phys. **29**, 950–951 (1958)
25. C.W. Tang, Appl. Phys. Lett. **48**, 183 (1986)
26. P.W. Blom et al., Adv. Mater. **19**, 1551–1566 (2007)
27. D.E. Markov, J. Phys. Chem. A **109**, 5266–5274 (2005)
28. M. Theander et al., Phys. Rev. B **61**, 12957 (2000)
29. T. Stubinger et al., J. Appl. Phys. **90**, 3632–3641 (2001)
30. P. Peumans, J. Appl. Phys. **93**, 3693 (2003)
31. P. Peumans et al., Chem. Phys. Lett. **398**, 27–31 (2004)
32. H. You, Chin. Phys. B **18**, 349 (2009)
33. V. Mihailetchi et al., Phys.Rev. lett. **93**, 216601 (2004)
34. L. Koster et al., Phys. Rev. B **72**, 085205 (2005)
35. V.D. Mihailetchi et al., Adv. Funct. Mater. **15**, 795–801 (2005)
36. C. Zhang, J. Appl. Phys. **110**, 064504–064504–7 (2011)
37. M.C. Scharber et al., Adv. Mater. **18**, 789–794 (2006)
38. K. Vandewal et al., Adv. Funct. Mater. **18**, 2064–2070 (2008)
39. M.K. Siddiki et al., Energy Environ. Sci. **3**, 867–883 (2010)
40. A. Hadipour et al., Adv. Funct. Mater. **16**, 1897–1903 (2006)
41. J.Y. Kim et al., Science **317**, 222–225 (2007)
42. C. Zhang et al., Appl. Phys. Lett. **92**, 083310 (2008)
43. V. Shrotriya et al., Appl. Phys. Lett. **88**, 064104–064104–3 (2006)
44. K. Tvingstedt et al., Appl. Phys. Lett. **91**, 123514–123514–3 (2007)
45. Y. Zhou et al., Appl. Phys. Lett. **93**, 033302–033302–3 (2008)
46. L. Hung et al., Mater. Sci. Eng. R **39**, 143–222 (2002)
47. J. Xue et al., Appl. Phys. Lett. **85**, 5757–5759 (2004)
48. M. Jørgensen et al., Sol. Energy Mater. Sol. Cells **92**, 686–714 (2008)
49. C. Féry, Appl. Phys. Lett. **87**, 213502–213502–3 (2005)
50. F.C. Krebs, Sol. Energy Mater. Sol. Cells **93**, 394–412 (2009)
51. T. Aernouts et al., Appl. Phys. Lett. **92**, 033306 (2008)

Chapter 10
High-Efficiency Semiconductor Photovoltaic Devices

10.1 Solar Cells Efficiency

Solar cells are devices capable of converting light energy into electrical energy. The incident photons (light energy) are absorbed by the solar cell to generate electrons and holes, which ultimately act as electrical energy in an external circuit to achieve energy conversion. Such a photoelectric conversion process can be measured by the amount of work of each incident photon in an external circuit, so it refers to a process of quantum energy conversion. Besides, the photoelectric conversion efficiency of a solar cell is a measure of the amount of work of total photons. In fact, enhancing the efficiency of a solar cell is an increase in the amount of work that each photon can ultimately do via photoelectric conversion, i.e., increasing the rate of energy extraction from each photon. The extraction ratio of photon energy can be elevated in many ways, both theoretically and practically.

Figure 10.1 presents a solar cell energy band diagram with only one bandgap. In this scenario, the incident photon $h\nu$ with energy less than the Bandgap width E_g is not able to be absorbed. The incident photon $h\nu$ with energy more than the forbidden band width E_g will be absorbed by the material. The electrons are excited to the conduction band to generate electron–hole pairs. The electrons and holes in the excited state are efficiently relaxed to the bottom of the conduction band and the top of the valence band, and the energy is lost as heat. When the energy of the incident photon is close to the forbidden band width of the material, the conversion of light energy to electrical energy is the maximum. In solar cells, the recombination process of electrons and holes via the forbidden band width is sufficiently slow, allowing them to be efficiently extracted before recombination.

As early as 1961, Shockley and Queisser discussed the ultimate efficiency of solar cells based on the principle of detailed balance [1], which is also the most common method for efficiency analysis of solar cells [2–4]. Based on the analysis of the detailed balance principle, it is known that a simple single-cell solar cell under the standard AM 1.5 spectrum exhibits the final efficiency of nearly 33%. However, this ultimate efficiency is achieved under a range of assumptions. Among the mentioned

C. Zhang et al., *Semiconductor Photovoltaic Cells*,
https://doi.org/10.1007/978-981-15-9480-9_10

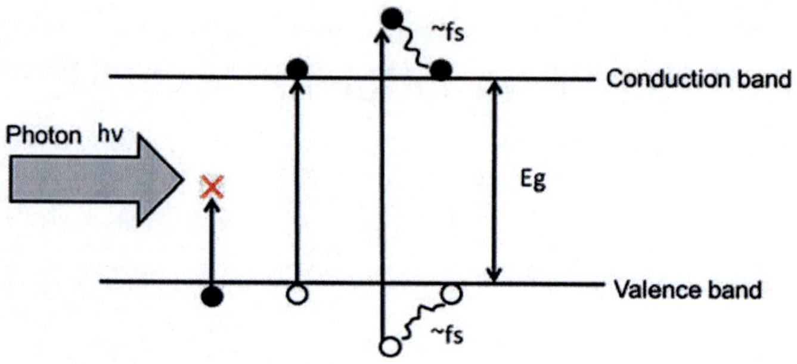

Fig. 10.1 Band diagram of a solar cell containing only one bandgap

assumptions, the more crucial assumption is that only one electron–hole pair can be generated for respective photon absorbed. After the formation of electrons and holes, the excess energy will be relaxed and finally transferred to the crystal lattice, thereby forming a thermal equilibrium with the crystal lattice, and all light is absorbed by the PN junction with only one bandgap width. Under the mentioned limited conditions, the main loss processes of sunlight energy include [5] the follows:

(1) When the incident photon energy is less than the bandgap of the material, it will not be absorbed by the material and energy loss will occur.
(2) When the energy of the incident photon is more than the bandgap, the excess energy will be dissipated as thermal energy.

The loss process of solar energy is shown in Fig. 10.1. With the optimized forbidden band width, the energy lost in process (1) and process (2), respectively, took up approximately 23 and 33% of the total solar energy. When the energy of the incident photon is less than the bandgap, the energy conversion rate is zero because the photon cannot be absorbed by the material. Under the energy of the incident photon more than the bandgap, the energy conversion efficiency will be relatively low for the existence of process (2). When the energy of the incident photon is close to the bandgap of the material, the light energy is maximally converted to the electrical energy; thus, the losses of process (1) and process (2) are avoided maximally. Such variation is clearly presented in the comparison between the output energy of the solar cell and the incident spectral energy (Fig. 10.2). It is suggested that the energy loss primarily originates from the mismatch between the bandgap of the solar cell material and the solar spectrum.

The principle of detailed balance elucidates the efficiency limit of solar cells under a range of assumptions. It is the most and mature theory we have applied in discussing the limit of solar cell efficiency. The previous chapters that clarify various solar cells, when it comes to material and device efficiency limitations, all comply with the detailed balance principle. However, too many assumptions are proposed in the discussion of the solar cell efficiency limit in accordance with the principle of detailed balance [6], and the mentioned assumptions are strict. Under the mentioned strict assumptions, we actually underestimate the limits efficiency of the

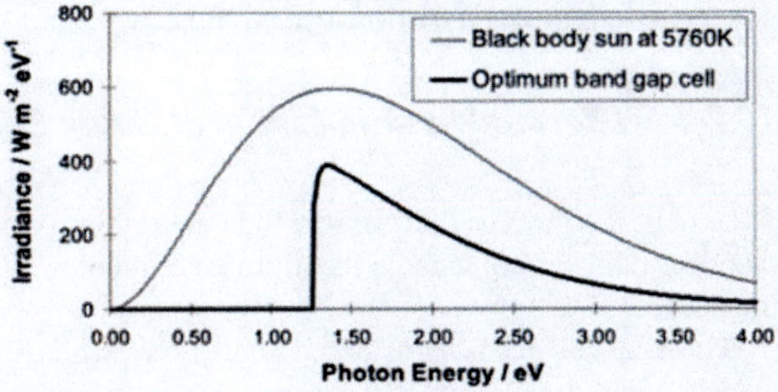

Fig. 10.2 Comparison of calculated solar cell output energy with incident spectral energy [5]

device. The present chapter discusses the theoretical efficiency of solar cells beyond the assumption of the detailed balance principle and presents some concepts and practical ways of high-efficiency solar cell development.

In the following sections, we will first review the efficiency limits based on the principle of detailed balance and the assumptions on which it is based. Next, we will discuss measures to enhance the energy conversion efficiency of each photon in the case where the hypothesis of the principle of detailed balance is relaxed. Simply put, the mentioned measures include the following:

(1) Increase the number of bandgap and use different bandgap widths to absorb photons of different energies (laminated and multiband solar cells);
(2) Increase the number of electron–hole pairs generated by each photon (collision ionized solar cell);
(3) Reduce thermal energy loss of photogenerated carriers ("hot-carrier" solar cells).

Among the above measures, the laminated structure has been mentioned many times in the introduction of solar cells in the previous chapters, and it is a practical and mature method. In addition to the laminated structure, the remaining measures to improve the efficiency of the solar cell are still some distance from the actual application. Although the mentioned measures are theoretically feasible and the corresponding phenomena are observed in the experiment, their real implementation relies on materials exhibiting special properties and the maturity of the mentioned materials needs further research. In any case, theoretical studies and experimental results have shown that the mentioned ways to improve the efficiency of solar cells are still some distance from practical applications, but they are feasible in future applications, so it is necessary to introduce them in this book. Among the wide variety of ways to improve solar cells presented below, the laminated structure is more introduced in the previous section, and some of the remaining concepts have also been mentioned in the previous chapters. In the present chapter, the emphasis is not placed on the specific device performance, whereas based on the discussion of J. Nelson [5], the concept introduction and theoretical prediction of various solar cell improvement approaches are highlighted. Thus, a complete insight into the extent to which the mentioned measures can ultimately enhance cell efficiency is gained.

10.2 Limits on Solar Cell Efficiency

10.2.1 Thermodynamic Limits of Solar Cell Efficiency

Solar cells are devices that convert light energy into electrical energy. If only thermal theory is considered, what is the ultimate efficiency of solar cells? Now let's discuss this issue.

The solar cell can be considered a cold body, which is coupled with the surrounding environment by radiation. By absorbing radiation with a higher temperature T_s as the heat source, the absorbed energy can be extracted for operation as working outside, so energy is converted. In accordance with the principle of detailed balance, the solar cell should also radiate at its own characteristic temperature T_c to achieve equilibrium. For simplicity, both the sun and the solar cells can be considered blackbodies, which radiate energy at their respective characteristic temperatures (Fig. 10.3). Such a process can be expressed by the model in the figure. It is assumed that all incident light can reach the solar converter. In the figure, T_c denotes the characteristic temperature of the converter, T_s represents the characteristic temperature of the heat source, and T_a indicates the characteristic temperature of the environment.

Given thermodynamics knowledge, the overall amount of energy emitted by a blackbody with a characteristic temperature T at unit time and unit area is $\sigma_s T^4$, where σ_s denotes the Stefan constant. If all incident energy can reach the solar cell, and the only way to lose energy is the spontaneous emission of the cell device itself, the net energy flow density received by the solar cell is defined as

$$\sigma_s T_s^4 - \sigma_s T_c^4 \tag{10.1}$$

The cell device can be considered a heat engine, and the energy conversion will be realized through this heat engine doing external work. If the solar cell operates in a Carnot cycle, the energy loss will be minimal. Thus, the work achieved by the heat engine is expressed as

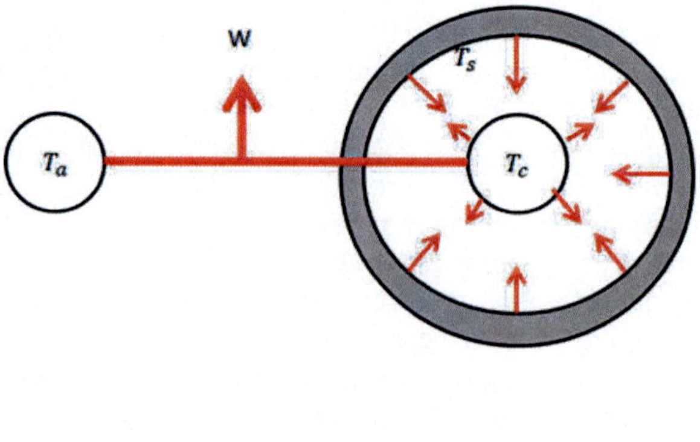

Fig. 10.3 Schematic diagram of an ideal solar converter

$$W = (\sigma_s T_s^4 - \sigma_s T_c^4)\left(1 - \frac{T_a}{T_c}\right) \tag{10.2}$$

The last term in the above equation represents the Carnot factor of the work done by the solar cell under constant entropy. The energy conversion efficiency is the ratio of the work done by the heat engine to the energy incident

$$\eta = \frac{W}{\sigma_s T_s^4} = \left(1 - \left(\frac{T_c}{T_a}\right)^4\right)\left(1 - \frac{T_a}{T_c}\right) \tag{10.3}$$

The sun has the blackbody characteristic temperature of 5760 K, and the blackbody characteristic temperature of the solar cell working environment is 300 K. According to the mentioned formula, the efficiency is maximum at the solar cell temperature of 2470 K, which is about 85%. It is noteworthy that in this calculation, some assumptions are also proposed: photons can be absorbed and the energy of each photon can be transformed to the maximum extent, and there is no thermal energy dissipation in energy conversion. Some aspects of the cell itself were not considered during the calculation. The efficiency of the solar cell from the energy band of the semiconductor material is analyzed in the following.

10.2.2 The Principle of Detailed Balance Limits of Solar Cell Efficiency

This section discusses the limiting efficiency of a solar cell that covers only a single bandgap E_g (only two energy levels of the conduction band and the valence band). The energy of the incident photon less than the forbidden band width will not be absorbed. Only photons exhibiting energy higher than the bandgap can be absorbed to generate electron–hole pairs, and the generated electron–hole pairs form a heat balance with the crystal lattice.

The solar cell device is assumed as a planar structure, and it can receive and emit energy in a hemispherical space. As shown in Fig. 10.4, for non-concentrated sunlight, the device is capable of receiving solar radiation in the azimuth of $0 < \theta < \theta_{sun}$ and environmental radiant energy in the azimuth angle of $\theta_{sun} < \theta < \frac{\pi}{2}$. Only the energy loss process is assumed as radiation recombination, so the output current is equated with the unit charge q multiplied by the difference between the photons absorbed by the device and the photons of the radiation.

$$I = q\left\{\begin{array}{l} \iint \alpha(E,s,\theta,\varphi)\beta_s(E,s,\theta,\varphi)dEdSd\Omega \\ + \iint \alpha(E,s,\theta,\varphi)\beta_a(E,s,\theta,\varphi)dEdSd\Omega \\ - \iint \alpha(E,s,\theta,\varphi)\beta_e(E,s,\theta,\varphi)dEdSd\Omega \end{array}\right\} \tag{10.4}$$

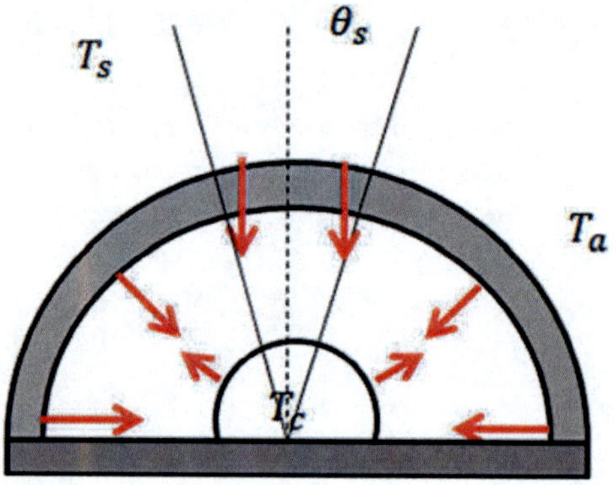

Fig. 10.4 Model of solar cell absorbing solar and environmental radiation and self-emission radiation

where $\alpha(E, s, \theta, \varphi)$ denotes the absorptance of the point s on the device surface to the light of the energy E which is incident in the (θ, φ) direction, i.e., the probability that the photon is absorbed. β_s and β_a are photon flows of energy E of point s incident on the device surface by the sun and the surrounding environment in the (θ, φ) direction, respectively, and β_e is a photon flow of energy E emitted by a point s on the device surface along the (θ, φ) direction (photon flow represents the number of photons passing through a unit area in a fixed angle per unit time). The integrals for β_s and β_a should be integrated at the corresponding azimuth, as shown in Fig. 10.4.

For simplicity, the sun and the surrounding environment are treated as blackbodies, producing an isotropic photon flow, as described by the Planck blackbody radiation formula. For the sun, its characteristic temperature is T_s, and it yields

$$\beta_s = \frac{2}{h^3c^2}\frac{E^2}{e^{E/kT_s} - 1} \tag{10.5}$$

It is incident in the angle range of $0 < \theta < \theta_{sun}$. For the surrounding environment, the characteristic temperature is T_a, and it yields

$$\beta_a = \frac{2}{h^3c^2}\frac{E^2}{e^{E/kT_a} - 1} \tag{10.6}$$

It is incident in the angle range of $\theta_{sun} < \theta < \frac{\pi}{2}$. Assume that the solar cell has a uniform chemical potential $\Delta\mu$ when irradiated with light, and it is balanced with the surrounding environment at the characteristic temperature T_a. Solar cells emit photons spontaneously, and the isotropic photon flow emitted by it is expressed as

$$\beta_e = \frac{2}{h^3c^2}\frac{E^2}{e^{(E-\Delta\mu)/kT_a} - 1} \tag{10.7}$$

It is incident in the angle range of $0 < \theta < \frac{\pi}{2}$.

To maximize the efficiency of a solar cell, it is assumed to be able to absorb all incident photons exhibiting energy greater than the forbidden band width, that is

$$\alpha(E) = \begin{Bmatrix} 1 \; E \geq E_g \\ 0 \; E < E_g \end{Bmatrix} \tag{10.8}$$

Moreover, assuming that the chemical potential is constant, it is equated with q times the bias voltage

$$\Delta\mu = qV \tag{10.9}$$

Such an assumption means that there is no loss of potential during the absorption of light and the transport of charge, i.e., the transport of charge is lossless, and the mobility of carriers is infinite. Moreover, the results of the general detailed balance principle are adopted

$$\alpha(E) = \varepsilon(E) \tag{10.10}$$

where ε(E) denotes the emissivity.

Based on the above assumptions, for a planar solar cell with a perfect back reflection of area A, the current generated under the offset of V is calculated by Eq. (10.4) integral.

$$\mathrm{I}(V) = qA\left\{\frac{2F_s}{h^3c^2}\int\frac{E^2}{e^{E/kT_s}-1}dE + \frac{2(F_a - F_s)}{h^3c^2}\int\int\frac{E^2}{e^{E/kT_a}-1}dE - \frac{2F_a}{h^3c^2}\int\int\frac{E^2}{e^{(E-qV)/kT_a}-1}dE\right\} \tag{10.11}$$

where $F_a = \pi$, $F_s = \pi f_s$, and $f_s = \sin^2\theta_s = 2.16 \times 10^{-5}$ are the angular range of the sun. If the light is concentrated X times, it yields

$$F_s = \pi X f_s \tag{10.12}$$

When $X = \frac{1}{f_s}$, it yields

$$F_s = \pi$$

It is in a fully concentrated state.

It is assumed that

$$N(E_{\min}, E_{\max}, T, \Delta\mu) = \frac{2\pi}{h^3c^2}\int_{E_{\min}}^{E_{\max}}\frac{E^2}{e^{(E-qV)/kT_a}-1}dE \tag{10.13}$$

Subsequently, $N(E_{min}, E_{max}, T, \Delta\mu)$ denotes the maximum photon flux density that can be absorbed or emitted between the energy intervals E_{min} and E_{max}.

Make

$$L(E_{\min}, E_{\max}, T, \Delta\mu) = \frac{2\pi}{h^3c^2}\int_{E_{\min}}^{E_{\max}} \frac{E^2}{e^{(E-qV)/kT_a} - 1}dE \tag{10.14}$$

Next, $L(E_{min}, E_{max}, T, \Delta\mu)$ represents the maximum energy flow density that can be absorbed or emitted between the energy intervals E_{min} and E_{max}.

Using $N(E_{\min}, E_{\max}, T, \Delta\mu)$, the current density is expressed as

$$J(V) = \frac{I(V)}{A} = q\{Xf_sN(E_g, \infty, T_s, 0) + (1 - Xf_s)N(E_g, \infty, T_a, 0) - N(E_g, \infty, T_a, qV)\} \tag{10.15}$$

The output energy power is

$$P(V) = V \times J(V) \tag{10.16}$$

It is suggested that the output energy is a function of E_g and V. For each fixed value of E_g, there is a V (defined as V_m) determined between 0 and E_g to maximize the output power.

$$P_{\max} = P(V_m) \tag{10.17}$$

V_m can be obtained from the differential derivative of P-to-V.

$$\frac{\partial P}{\partial V}|_{V=V_m} = 0 \tag{10.18}$$

Lastly, the maximum efficiency is a function of E_g and X,

$$\eta_{\max} = \frac{P(V_m)}{P_s} \tag{10.19}$$

where

$$P_s = Xf_sL(0, \infty, T_s, 0) \tag{10.20}$$

This is the energy received by the solar cell from the sun. For non-concentrated sunlight ($X = 1$), the η_{max} of the material exhibiting a bandgap of 1.3 eV is 31% [2].

As revealed from the above formula, concentrating can improve the balance between absorption and emission photons of the solar cell, i.e., the conversion efficiency of the device can be enhanced. At the maximum concentration of light, i.e., when $X = \frac{1}{f_s}$, the current density becomes

$$J(V) = q\{N(E_g, \infty, T_s, 0) - N(E_g, \infty, T_a, qV)\} \tag{10.21}$$

Under full concentration, the maximum energy conversion efficiency of solar cells can reach about 41% for materials exhibiting a bandgap of 1.1 eV. The relationships of the calculated solar cell efficiency and the material forbidden band width are illustrated in Fig. 10.5. In the figure, (a) is a non-condensing case of 6000 K blackbody radiation, (b) represents a full-condensing case of 6000 K blackbody radiation, (c) indicates a non-condensing case of AM 1.5D, and (d) is a non-condensing case of AM 1.5G [7].

The principle of detailed balance was reviewed above, and the limits efficiency of solar cells under this principle was discussed. It is suggested that this discussion process is based on a number of assumptions, and the conclusions of the mentioned assumptions are drawn below:

(1) Only photons exhibiting incident photon energy greater than the forbidden band width can be absorbed, one photon can only generate one electron–hole pair, and photons exhibiting incident photon energy less than the forbidden band width cannot be absorbed by the material.
(2) The generated electron–hole pairs form a thermal equilibrium with the lattice through relaxation, and the lattice characteristic temperature is T_a.
(3) The quasi-Fermi level values of electron and hole separation in the device are constant, with a magnitude of qV, where V denotes the potential difference in the device.
(4) The mobility of carriers tends to infinity, and the only mechanism of energy loss is spontaneous emission.

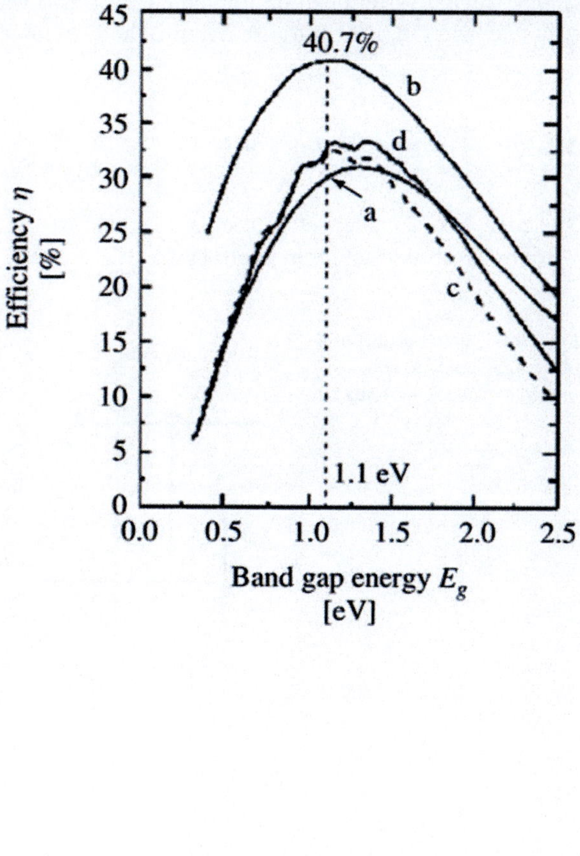

Fig. 10.5 Ideal solar cell limits efficiency when the forbidden band width changes

10.3 Cell with Multi-Bandgap Light-Absorbing Structure

In a solar cell, if photons of different energies can be absorbed by units of different bandgaps, the overall amount of energy that each photon can convert can be significantly increased. A solar cell with only one bandgap material has the highest photoelectric conversion efficiency when the incident photon energy is exactly equivalent to the forbidden band width. In other words, a solar cell with only one bandgap material has the highest photoelectric conversion efficiency under the illumination of monochromatic light having energy equated with the forbidden band width. Then, since the energy of the incident light is only equated with the forbidden band width, the incident photons are just absorbed by the solar cell, the generated electrons or holes exhibit no excessive kinetic energy, and kinetic energy is not lost. However, the energy of photons in the solar spectrum shows a significantly wide distribution, solar cells cannot work in monochromatic light in practice, and the loss of energy is inevitable. Figure 10.6 suggests that if the solar cell covers multiple bandgap units, and the solar spectrum is separated, each part of the photons are, respectively, incident on different forbidden band width units in the solar device. Accordingly, more solar energy is exploited, and the lost energy decreases noticeably, and then higher solar energy conversion efficiency is achieved [8]. The concept of this solar energy system is illustrated in Fig. 10.7, in which sunlight is split by a spectroscopic system and then absorbed by different bandgap units in the solar cell to maximize the efficiency. Such a system can be implemented in two ways, one is to use a laminated structure, and the other is to use a multi-bandgap material system.

10.3.1 Tandem Solar Cells

The concept of a tandem structure has been covered in the previous chapters, and we will discuss this structure in depth here.

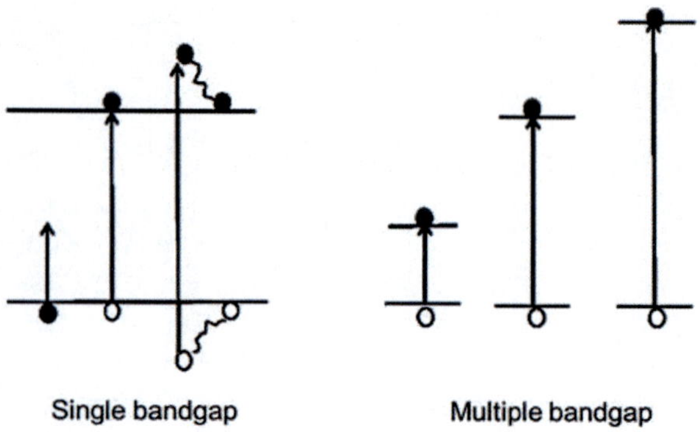

Fig. 10.6 Photon absorption in a single bandgap and multiple bandgap systems

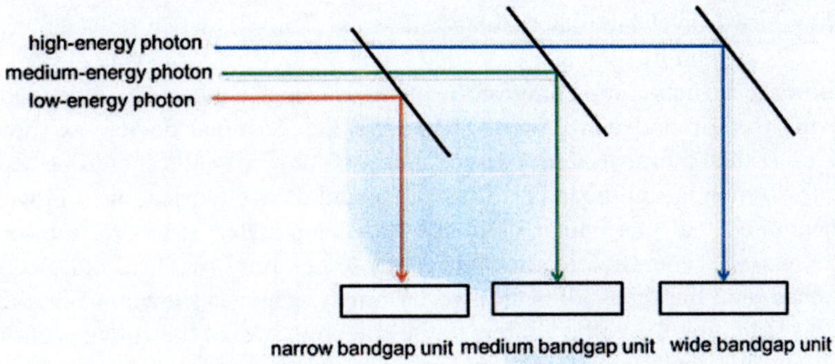

Fig. 10.7 Schematic diagram of multi-bandgap system

In the concept of the multi-energy band system illustrated in Fig. 10.7, incident photons should be split according to energy to reach different bandgap units in the solar cell. However, the solar spectrum is significantly difficult to be effectively split by optical lenses. A more practical approach is to optically connect PN junctions with different bandgap, so the material exhibiting a larger bandgap first absorbs more energy, and photons with less energy can be absorbed via the material region exhibiting a smaller bandgap, so different photons are absorbed by different bandgap materials (Fig. 10.8). As suggested from the figure, light is incident from the top, and the width of the bandgap from top to bottom decreases progressively.

Fig. 10.8 Schematic diagram of the tandem solar cell

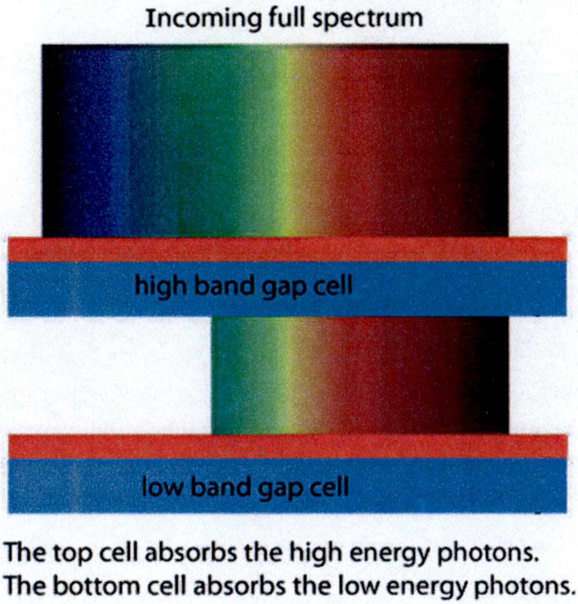

Taking a two-layer laminated solar cell as an example, when only the top and bottom cells are simply optically cascaded while retaining the two lead electrodes of each sub-cell, the maximum conversion efficiency can be achieved by independently optimizing the top and bottom cells. This is a four-terminal device, as shown in Fig. 9.26. In such a four-terminal device, the two sub-cells cells should be separate electrodes, which is significantly difficult in actual device fabrication, and we have also mentioned that in Chapter 4. In the actual laminated structure, the top and bottom cells are generally electrically cascaded, i.e., only two lead-out electrodes are retained, and the final cell is the two-terminal device, as shown in Fig. 9.26. In this series structure, the voltage refers to the algebraic sum of the voltages of the top and bottom cells, and the current should pass through the top and bottom cells, so the currents of the top and bottom cells can be matched to achieve the maximum solar cell efficiency. Because the current generated by the top and bottom cells changes under different sunlight, i.e., the matching current under a kind of illumination may become mismatched under the changed illumination, the design of the laminated structure must be considered in the actual working conditions.

Now let's calculate the efficiency of the two-junction laminated cell. It is assumed that the bandgap of the top cell is E_{g2}, the bandgap of the bottom cell is E_{g1}, and $E_{g2} > E_{g1}$.

In a four-terminal laminated structure, the maximum output efficiency refers to the algebraic sum of the two independent sub-cells. The spectrometry of the solar spectrum is assumed to be perfect, i.e., all photons exhibiting energy greater than E_{g2} are absorbed by the top cell, and all photons exhibiting energy between E_{g1} and E_{g2} are absorbed by the bottom cell. Under full concentrating conditions, the maximum output power is

$$\begin{aligned} \mathrm{P_{max}} = & \mathrm{q}V_{m1}\{N(E_{g1}, E_{g2}, T_s, 0) - N(E_{g1}, E_{g2}, T_a, q\mathrm{V}_{m1})\} \\ & + \mathrm{q}V_{m2}\{N(E_{g2}, \infty, T_s, 0) - N(E_{g2}, \infty, T_a, q\mathrm{V}_{m2})\} \end{aligned} \quad (10.22)$$

where V_{m2} denotes the voltage of the top cell, and V_{m1} is the voltage of the bottom cell. It is assumed that the two sub-cells cells can be optimized independently, and P_{max} only refers to a function of the forbidden band width of the two sub-cells.

In the two-terminal structure, the maximum efficiency is given by

$$\mathrm{P_{max}} = \mathrm{q}(V_1 + \mathrm{V}_2)\{N(E_{g1}, E_{g2}, T_s, 0) - N(E_{g1}, E_{g2}, T_a, q\mathrm{V}_{m1})\} \quad (10.23)$$

where V_1 and V_2 are not the maximum output voltages for each sub-cell, since each sub-cell can no longer be independently optimized for current matching requirements. According to current matching conditions, it yields

$$\begin{aligned} & \{N(E_{g1}, E_{g2}, T_s, 0) - N(E_{g1}, E_{g2}, T_a, \mathrm{qV}_{m1})\} \\ & = \{N(E_{g2}, \infty, T_s, 0) - N(E_{g2}, \infty, T_a, \mathrm{qV}_{m2})\} \end{aligned} \quad (10.24)$$

It is suggested that for a two-terminal device structure, the limiting efficiency is relatively low and more sensitive to the value of the forbidden band width. Figure 10.9 shows the limiting efficiency of a laminated cell containing two ideal sub-cells with AM 1.5D illumination as a function of the two sub-bandgap widths.

Regardless of the series (two-terminal) laminated structure or the parallel (four-terminal) laminated structure, increasing the number of bandgap can increase the limiting efficiency. Under an infinite number of junctions, the minimum bandgap is 0 eV, and the limiting efficiency under sunlight is achieved as 69%. Under full concentrating conditions, the limiting efficiency can be close to the thermodynamic limiting efficiency of 86%.

As fueled by the continuous development of solar cells, a range of material combination systems has been practically applied to laminated structures. Since it is relatively easy to technically integrate various different junctions into the same multilayer structure, the laminated device structure in series is more used than the parallel laminated device structure. In an actual series cell configuration, a tunnel junction is physically connected between the various PN junctions. The so-called tunneling junction is a heavily doped PN junction in which one terminal is connected to the p-type end of one sub-cell to form an ohmic contact, and the other end is connected to the n-side of the other sub-cell to form an ohmic contact. Since the III-V material has a relatively high absorption coefficient and its forbidden band width can be adjusted by changing the components, it is significantly suitable for fabricating a laminated device in practice. As mentioned in Chap. 4, GaAs single-element devices have been studied extensively and are often taken as one of the sub-cells in the laminated structure. The bandgap of GaAs is 1.42 eV, which is not ideal for the ideal top or bottom cell in a two-junction laminated structure; however, for its good material quality

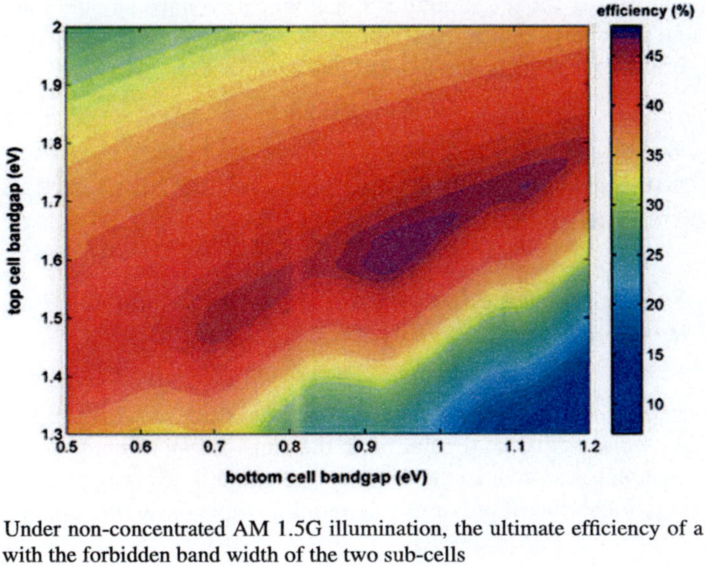

Fig. 10.9 Under non-concentrated AM 1.5G illumination, the ultimate efficiency of a laminated cell varies with the forbidden band width of the two sub-cells

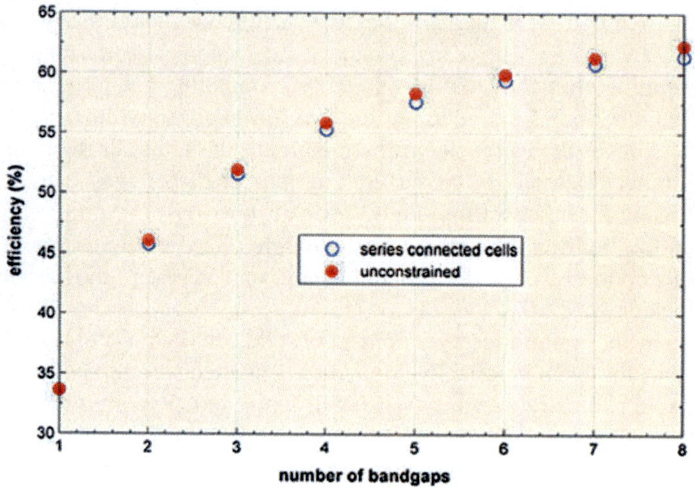

Fig. 10.10 The ideal laminated cell varies with the number of forbidden bands under AM 1.5D illumination [9]

and high carrier mobility, the actual laminated cell devices commonly outperform ternary alloy materials exhibiting ideal bandgaps. Moreover, in the actual stacked structure, the lattice and thermal conductivity matching problem between the layers of materials should be considered. The mismatched lattice constant material exhibits higher interface defects and increases the recombination loss of carriers. Different thermal conductivity factors introduce stresses when the device temperature varies. The manufacture of laminated cells is significantly expensive, and the main purpose of its development is for space applications, which we have already introduced in GaAs system solar cells. Figure 10.10 shows the variation of the ideal laminated cell with the number of bandgaps under AM 1.5D illumination.

10.4 Solar Cell Containing Light-Absorbing Structure with Multiple Bandgaps

10.4.1 Solar Cells with Intermediate Bandgap and Multiple Bandgaps

The laminated structure contains several sub-cells, each of which has a different bandgap. In each sub-cell, the separation of the quasi-Fermi level of photogenerated electrons and holes is similar to the chemical potential of exciting photons, thereby reducing energy loss and improving the overall energy conversion efficiency of the device. That is, if a system supports the separation of several quasi-Fermi levels at the same time, each photon in this system can convert more energy. If the separation of multiple quasi-Fermi levels is not achieved with several different materials, but with only one material, this system is undoubtedly very attractive.

If the active layer material in a solar cell contains only a single bandgap, separation of different quasi-Fermi levels is impossible. This is because the states in the conduction band are coupled to each other through phonons. As a result, all photogenerated electrons will relax and reach thermodynamic equilibrium. In the end, the electron has only a unified chemical potential μ_n. All photogenerated holes will have the same situation and have a uniform chemical potential μ_p. Although a PN junction can be designed, the width of the forbidden band in the PN junction can be changed, so that photons with different energies can be absorbed in different positions of the PN junction. But the coupling between phonons will eventually relax the photogenerated electrons and holes quickly and eventually have a unified chemical potential μ_n and μ_p. Ideally, the chemical potentials μ_n and μ_p are determined by the smallest bandgap in the device.

The ideal solution is that a certain material contains two or more bandgaps [10]. Figure 10.11 illustrates such a system. In this system, a certain material contains two separate conduction bands. When a photon is incident, the electrons are excited from the valence band into the middle conduction band and the uppermost conduction band, or they can be excited from the middle conduction band to the uppermost conduction band. Assuming that the middle conduction band and the uppermost conduction band are independent of each other, there is no coupling effect between them; then the electrons in the middle conduction band and the uppermost conduction band will each reach equilibrium, and eventually have their own quasi-Fermi energy level $\mu_{n,i}$. The intermediate conduction band can be introduced by impurity doping, or by a heterojunction with quantum characteristics, or more ideally, the material itself has this level structure. This type of solar cells with intermediate bandgaps has become an important direction for solar cell research. For example, the core mechanism of "quantum well" or "quantum dot" solar cells are based on the concepts we discussed above [11, 12].

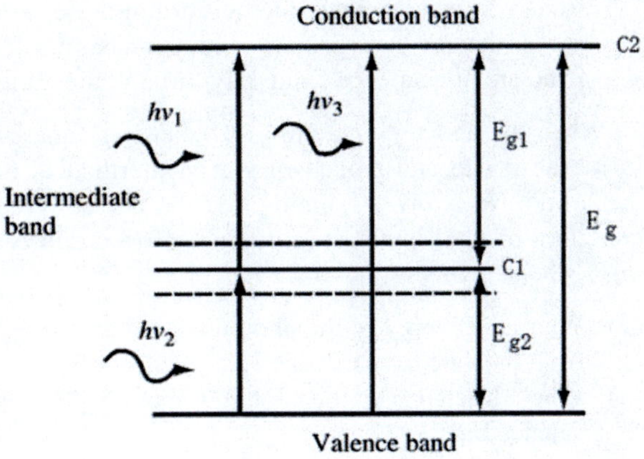

Fig. 10.11 Schematic diagram of the energy band structure of a material with an intermediate bandgap. There is one valence band and two conduction bands C1 and C2

In order for a solar cell with an intermediate bandgap or multiple bandgaps to work effectively, the following points need to be met:

(1) It is possible to have different quasi-Fermi levels in the same material. Carriers must satisfy collisions and scattering inside each sub-band more frequently than collisions and scattering between sub-bands. This requires the energy intervals between the sub-bands to be greater than the maximum energy of phonons, otherwise, the carriers are easily scattered by the phonons to a lower energy state.
(2) Structures that provide intermediate energy bands should be periodic. Periodicity means that commonization can occur, which is conducive to the transmission of carriers. An imperfect periodic arrangement will localize the carriers, which is not conducive to the transmission of carriers. In addition, the position of the intermediate energy level in the energy band should be able to suppress the scattering of phonons. For the intermediate energy band far from the conduction band or valence band within k_BT, due to the existence of thermal energy, scattering between the energy bands is always allowed from an energy perspective, but it is also limited by the conservation of momentum. For discrete impurity energy levels, since all momentum states are included in the wave function of the stationary defect, the momentum limit for the transitions between energy levels no longer exists. Therefore, it is necessary to provide a periodic arrangement of the intermediate energy band structure to limit the scattering between the energy bands. In addition, for deep energy levels far from the conduction band or the edge of the valence band exceeding multiple k_BT, the existence of the multi-phonon process will also cause scattering between the energy bands. In a periodic structure, this sequential relaxation process of multi-phonon is forbidden due to the requirement of symmetry, so it is also required that the structure of the intermediate band is arranged periodically.
(3) In order to thermally isolate the middle conduction band from the upper conduction band, only electrons can be extracted from one conduction band. Therefore, the electrode must be in selective contact with the upper conduction band and not in contact with the middle conduction band, otherwise, the generated electrons will achieve thermal equilibrium through contact. This requirement also means that the intermediate conduction band is coupled to the upper conduction band and the valence band only through the optical transmission.

When the above points are met, a separate quasi-Fermi level can be generated inside the device, and a solar cell with an intermediate bandgap or a multiple bandgaps can work effectively. Assuming that the device is a planar structure and is in the condition of total light concentration, there are three photon absorption processes inside the device.

(1) When the energy of the photon is greater than E_{g2}, the electron can be excited from the valence band to C1;
(2) When the energy of the photon is about E_g, the electron can be excited from the valence band to C2;

(3) When the energy of the photon is about E_{g1}, the electron can be excited from C1 to C2.

Since C1 and C2 are both conduction bands, that is, their states are basically empty, the excitation from C1 to C2 should be from the lower edge of C1 to the lower edge of C2, that is, $E_{g1} + E_{g2} = E_g$. In order to maximize the energy conversion rate of each photon, the electrons excited after photon absorption should transition to the highest conduction band allowed by the energy. In other words, the absorption coefficient of the V-C1 transition should be much larger than that of the C1-C2 transition, and the absorption coefficient of the V-C2 transition should be much larger than that of the V-C1 transition. Only in this way can we ensure that the three electron transition processes mentioned above all experience the largest range of energy changes: E_{g1} to E_{g2}, E_{g2} to E_g, and E_g to infinity. In this way, the output current density is expressed as

$$\begin{aligned} J(V) =& q\{N(E_g, \infty, T_s, 0) - N(E_g, \infty, T_a, \mu_{c2} - \mu_V)\} \\ &+ q\{N(E_{g2}, E_g, T_s, 0) - N(E_{g2}, E_g, T_a, \mu_{c1} - \mu_V)\} \end{aligned} \tag{10.25}$$

Among them, μ_V, μ_{c1}, and μ_{c2} are quasi-Fermi levels of the valence band, the middle conduction band, and the uppermost conduction band. Under steady-state conditions, the currents from V to C1 and C1 to C2 must be matched, so there is

$$\begin{aligned} & q\{N(E_{g1}, E_{g2}, T_s, 0) - N(E_{g1}, E_{g2}, T_a, \mu_{c2} - \mu_{c1})\} \\ & = q\{N(E_{g2}, E_g, T_s, 0) - N(E_{g2}, E_g, T_a, \mu_{c1} - \mu_V)\} \end{aligned} \tag{10.26}$$

The quasi-Fermi level should satisfy

$$(\mu_{c1} - \mu_V) + (\mu_{c2} - \mu_{c1}) = \mu_{c2} - \mu_V = qV \tag{10.27}$$

It can be seen that the current density is a function of E_g and E_{g1}. Optimized combinations can be found for varying E_g and E_{g1} to maximize conversion efficiency. If the sun is equivalent to a blackbody with 6000 K, when $E_g = 1.93$ eV and $E_{g1} = 0.7$ eV, the maximum output efficiency is 63%.

We now compare the solar cell with a middle bandgap to the laminated solar cell. For a two-cell solar cell with a laminated structure, it includes two different values of the bandgap, which can be equivalent to the case of including an intermediate bandgap. Compared with ideal solar cells with intermediate bandgaps, the limiting efficiency of laminated solar cells should be lower than that of solar cells with an intermediate bandgap. This is because a solar cell with an intermediate bandgap contains a total of three possible energy-level transition processes. Strictly speaking, it should be compared with a three-cell laminated solar cell, so it is more efficient than a two-cell laminated solar cell. In addition, for cascaded laminated solar cells, due to the requirements of current matching, in order for the electrons to reach the external circuit, there must be a transition process in both sub-cells at the same time. For solar cells with an intermediate bandgap, the process of electrons reaching the external circuit can require either one transition or two transitions, so its efficiency is higher. However, at the current stage, due to the limitations of technology and materials, the actual realization of the solar cells with intermediate bandgaps is much more difficult than the laminated solar cells.

Various schemes for realizing solar cells with an intermediate bandgap or multiple bandgaps have been proposed. Among them, the simplest solution is to use the in-band state inside the material, and the in-band state can be introduced by the doping process in the semiconductor. These in-band states can help absorb photons with energy below the forbidden band width to cause electrons to transition from the conduction band to the impurity state, or from the impurity state to the conduction band. The photocurrent increase effect caused by the absorption of photons below the bandgap has been observed in In-doped Si. However, since separated impurity atoms can also be used as non-radiative recombination centers through multi-phonon processes, there is insufficient evidence to show that separated quasi-Fermi levels can be generated in this structure.

Another approach is to use materials with narrow multi-bandgap structures, such as polar semiconductors, including some II–VI materials, oxide semiconductors, and molecular semiconductors.

Another more promising approach is to use the heterojunction of low-dimensional materials to achieve the desired band structure, for example, using a quantum dot structure, as shown in Fig. 10.12. In these nano-sized semiconductor particles, the dimensions of the material are limited, resulting in a discrete band structure with an atom-like structure. If quantum dots are grown in a semiconductor material with a large bandgap, these quantum dots can introduce discrete energy levels in the semiconductor material. The regular periodic arrangement of quantum dots will introduce sub-bands in the host material. The conservation of momentum required for band-to-band conversion reduces the possibility of inter-band scattering of phonons, which makes it possible to produce separate quasi-Fermi levels. If the size and spacing of the quantum dots can be controlled, then the energy size and width of the sub-bands can be adjusted.

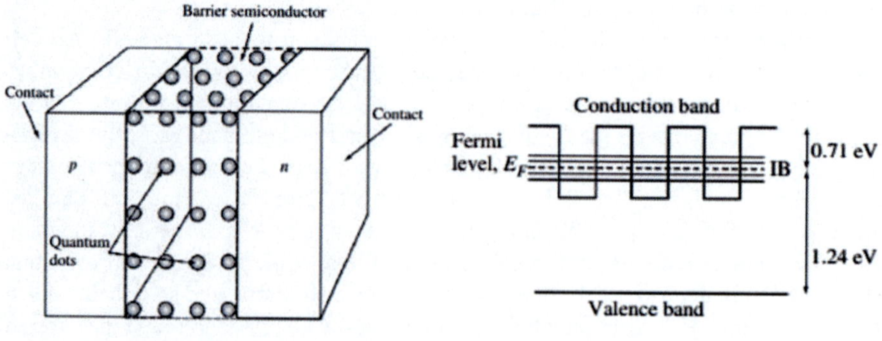

Fig. 10.12 Multiband mechanism in a quantum dot array [3]

10.5 Hot-Carrier Solar Cell

Solar cells with multiple bandgaps use units with different bandgaps to absorb photons with different energies, thereby reducing the loss of optical energy. Another way to improve the energy conversion efficiency of a single photon is to make full use of the excess kinetic energy before the relaxation of photogenerated carriers. There are two main ways to use the excess kinetic energy of carriers. One is to reduce the interaction speed between electrons and phonons so that the photogenerated carriers still have excess kinetic energy, even though they are still "hot" when they are collected. In addition, the excess kinetic energy of hot carriers can be used to generate more carriers in a "collision current" manner, thereby improving the energy conversion efficiency of a single photon [13–16]. Among them, the former method is to increase the voltage, and the latter method is to increase the current. Both of these methods use the excess kinetic energy of photogenerated carriers, but because their physical behavior is different (one method is to increase the voltage and the latter method is to increase the current), in this section and the next section we will separately discuss them.

If the energy of the incident photon is greater than the bandgap of the material, when the photon is absorbed by the semiconductor material and the electrons are just excited from the valence band to the conduction band, the photogenerated carriers have excess kinetic energy. The magnitude of this kinetic energy is distributed according to the incident solar spectrum and the absorption spectrum of the material. In the next few hundred femtoseconds, the scattering between carriers and carriers quickly makes these photogenerated carriers reach a self-equilibrium state. This equilibrium state can be represented by a characteristic temperature T_H, and the resulting chemical potential is μ_H. T_H is higher than the ambient temperature T_a, which can be equal to T_s under certain circumstances. In the next few picoseconds, hot carriers lose energy through collisions with phonons until the characteristic temperature of their statistical distribution is the same as the lattice temperature T_a. As a result, the carriers are "cooled" and are no longer hot carriers. This process is shown in Fig. 10.13.

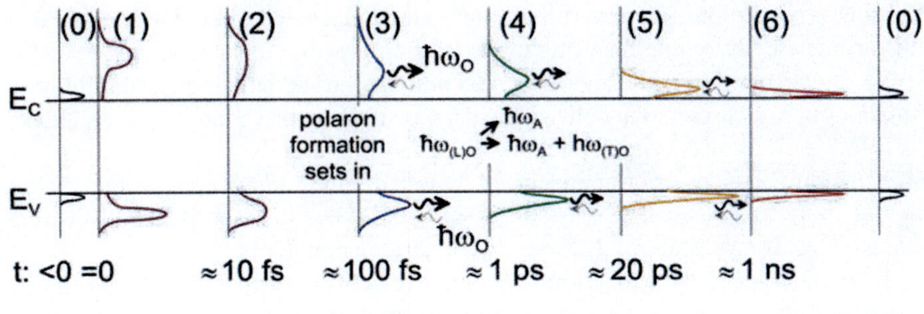

Fig. 10.13 Cooling mechanism of carriers in semiconductors [17]

At this time, the carrier and the lattice are in a quasi-thermal equilibrium state in the concept of statistical distribution, and the absorption and emission of phonons are equal. At the same time, there is no longer a thermal equilibrium between the number of electrons and holes. The number of electrons in the conduction band and the number of holes in the valence band have exceeded the equilibrium number in the absence of light, so the quasi-Fermi level of the electron moves to the bottom of the band, while the quasi-Fermi level of the hole moves to the top of the valence band. The separation of Fermi levels of electrons and holes increases with increasing intensity of light. In the "cooling" state, the carrier transfers kinetic energy to the crystal lattice in the form of heat, and the entropy of the system increases.

Over a longer period of time, carriers will recombine by radiative and non-radiative means. Radiation recombination depends on the absorption coefficient of the material, and its time is in the interval of nanoseconds to milliseconds. The function of recombination is to reduce the number of electrons and holes, so that the quasi-Fermi level of the carriers moves to the position at equilibrium. In the case of constant intensity of light, when the generation rate of photogenerated carriers is equal to the sum of the recombination rate and the extraction rate (current) of carriers, the number of carriers will no longer change. This equilibrium state determines the power output of the solar cell. For an effective photoelectric conversion process, the extraction rate of carriers should be greater than the recombination rate of carriers. The self-equilibrium and "cooling" processes among carriers are relatively rapid. In the previous chapters, our discussion of solar cells generally assumed that it was completed before the carriers were extracted. When analyzing the energy output of the device, only the recombination process needs to be considered. In this section, we will discuss how to use the energy before the hot carrier cools.

The central concept of hot-carrier solar cells is to extract the carriers before they are "cooled". If the speed of the extraction process can be accelerated or the speed of the "cooling" process can be reduced, the carriers may be extracted in a "hot" state. The excess kinetic energy of carriers will also be output by the solar cell in the form of electricity, which can be achieved in special electrical and optical structures. We know that the "cooling" process is related to the number of phonons with suitable energy and momentum; in addition, the "cooling" process is related to the number of states with lower energy that can be scattered, which can be used to control the "cooling" speed of hot carriers. In addition, electrodes need to be designed so that hot carriers are not "cooled" during the extraction process of carriers.

Now, ignoring the scattering of carriers and the lattice, let's look at the limiting efficiency of hot-carrier solar cells. Consider a material with a bandgap E_g, and make the following assumptions: All photons with energy greater than E_g can be absorbed, each photon produces only one electron–hole pair, and radiative recombination is the only way to lose energy. It is still assumed that the sun is a blackbody with a characteristic temperature of T_s, and its structural parameter is F_s.

After photogenerated carriers are generated, electrons reach equilibrium through collisions between themselves, as do holes. In equilibrium, Gibbs free energy is the smallest, so its value

$$\sum_i \eta_i dn_i \quad (10.28)$$

is conserved in any scattering process, where η_i is the chemical potential of the carrier i, and dn_i is the corresponding number of carriers. For electron scattering events

$$e_1 + e_2 = e_3 + e_4 \quad (10.29)$$

because $dn_i = 1$ for a carrier, there is

$$\eta_{1e} + \eta_{2e} = \eta_{3e} + \eta_{4e} \quad (10.30)$$

Since the scattering of carriers is elastic, the kinetic energy obeys

$$E_{e_1} + E_{e_2} = E_{e_3} + E_{e_4} \quad (10.31)$$

When the relationship between η_i and the kinetic energy is linear

$$\eta_i = \eta_0 + \gamma E_i \quad (10.32)$$

Then the above equation is true for all pairs of carriers. If $\gamma = 0$, this is the case where all carriers have the same chemical potential at equilibrium. Assuming the same is true for holes, the chemical potential of the electron–hole pair is

$$\Delta\mu = \eta_{e0} + \eta_{h0} + \gamma(E_e + E_h) \quad (10.33)$$

Therefore, the chemical potential of an electron–hole pair excited by a photon with energy E is

$$\Delta\mu = \mu_0 + \gamma E \quad (10.34)$$

among them

$$\mu_0 = \eta_{e0} + \eta_{h0} - rE_g \quad (10.35)$$

The constants η_{e0} and η_{h0} depend on the extraction conditions.

Assuming the energy of the photon is E, the distribution of the carriers obeys

$$f = \frac{1}{e^{(E-\Delta\mu)/k_B T_a} + 1} \quad (10.36)$$

Here, T_a is the ambient temperature, and $\Delta\mu$ is as described above. Therefore, f can be written as

$$\mathrm{f} = \frac{1}{e^{\frac{E(1-\gamma)-\mu_0}{k_B T_a}} + 1} \tag{10.37}$$

Define

$$T_H = \frac{T_a}{1-\gamma} \tag{10.38}$$

then the distribution function can be described by a single chemical potential μ_H. Then

$$\Delta\mu = \mu_H \frac{T_a}{T_H} + E\left(1 - \frac{T_a}{T_H}\right) \tag{10.39}$$

It can be seen from the above analysis that the carriers can be represented by the temperature of the hot carriers.

Now let's look at the current output of hot-carrier solar cells. J is determined by the difference between the number of absorbed and emitted photons

$$\mathrm{J}(V) = \mathrm{q}\left\{X\mathrm{f}_s N\left(E_g, \infty, T_s, 0\right) + (1 - X\mathrm{f}_s)N\left(E_g, \infty, T_a, 0\right) - N\left(E_g, \infty, T_H, \mu_H\right)\right\} \tag{10.40}$$

Under the condition of full condensing, $\mathrm{Xf_s}$ is 1.

For hot-carrier solar cells, V depends in particular on the chemical potential μ_H. During extraction, the carriers must be "cooled" to the ambient temperature T_a. In this process, the chemical potential of the extracted carriers increases. In the state of maximum energy output, the entropy should remain unchanged during the cooling process. When electrons and holes are extracted by an electrode with a very narrow bandgap ($\Delta E \ll k_B T$), it is possible that there is no loss of kinetic energy during the cooling of the carriers, and the entropy remains unchanged.

When the entropy is constant, the expression of $\Delta\mu$ remains unchanged, so the extracted chemical potential of the carriers is

$$\mu_{\text{out}} = \mu_H \frac{T_a}{T_H} + E_{\text{out}}\left(1 - \frac{T_a}{T_H}\right) \tag{10.41}$$

Here, E_{out} is the energy difference between the electron extraction electrode and the hole extraction electrode. Therefore, the output voltage is $V = \mu_{\text{out}}/q$, and the extracted energy is

$$\mathrm{P}(V) = \frac{V_{\text{out}}}{q} \times J \tag{10.42}$$

The above discussion holds when the extraction speed of the carriers is slow compared to the scattering between the carriers and the carriers.

Now let's calculate the power output of a hot-carrier solar cell. Because energy is brought in by the absorption of photons, the loss of energy is caused by the extraction of energy and the emission of hot carriers, so there are

$$JE_{out} = q\{Xf_sL(E_g, \infty, T_s, 0) + (1 - Xf_s)L(E_g, \infty, T_a, 0) - L(E_g, \infty, T_H, \mu_H)\} \quad (10.43)$$

It can be seen that the performance of hot-carrier solar cells depends on three parameters: E_{out} and E_g and V. E_{out} greatly affects the performance of the device. For a given E_{out} and E_g, the value of V can vary from 0 to V_{OC}, and the varying output voltage controls the temperature of the carrier. In the case of a short circuit, all carriers can be extracted, $\mu_{out} = 0$ and T = Ta, that is, there is no hot-carrier effect. When V increases to V_{OC}, due to the increased radiation recombination, J gradually decreases, and T_H gradually increases, and its value is close to T_s until the open-circuit state.

This can be understood from the perspective of energy balance. The excess kinetic energy of photogenerated carriers is conserved during elastic collisions, and is not lost through cooling. In the short circuit, all the carriers generated by light are extracted, so the excess kinetic energy does not contribute to the distribution of the carriers, and $T_H = T_a$.When the output voltage increases, the density of the carriers in the solar cell increases, and the kinetic energy of the carriers increases, so the characteristic temperature of the carrier distribution also increases. In the open circuit, no carrier is extracted, and T_H has a maximum value.

The efficiency of the device can be determined by the parameters E_{out} and E_g. For full concentration, the efficiency is up to 85%, and for non-concentrating, the highest efficiency is about 65%.

For hot-carrier solar cells, two material systems are necessary: one is the material of the effective layer, in which the cooling rate of the hot carriers is lower than the rate at which the carriers are transferred to the contacts. The second is the contact material, which allows electrons and holes to pass selectively through a very narrow energy band, as shown in Fig. 10.14.

In a typical bulk semiconductor device, the cooling process is generally less than 10 picoseconds, and the transmission time of carriers generally requires nanoseconds or even longer. The rate of cooling depends on the number of phonons with the appropriate energy and momentum required to scatter the excited electrons into a lower energy state. There are two types of phonons: acoustic phonons, whose energy is at a few millielectron volts at room temperature, and longitudinal (LO) optical phonons, whose energy is relatively high. In III-V materials, its energy is about 30–40 millielectron volts. LO phonons are most likely involved in the cooling process of hot carriers. In bulk materials, the momentum distribution of phonons is approximately isotropic, which depends on the band structure of the crystal. Since the density of electronic states in the bulk material is continuous and isotropic, carriers in the excited state are more likely to be scattered by phonons into the empty states with lower energy. This process can be attenuated by proper phonon deletion and absence of vacant electron states. The first case can be satisfied when the optical intensity

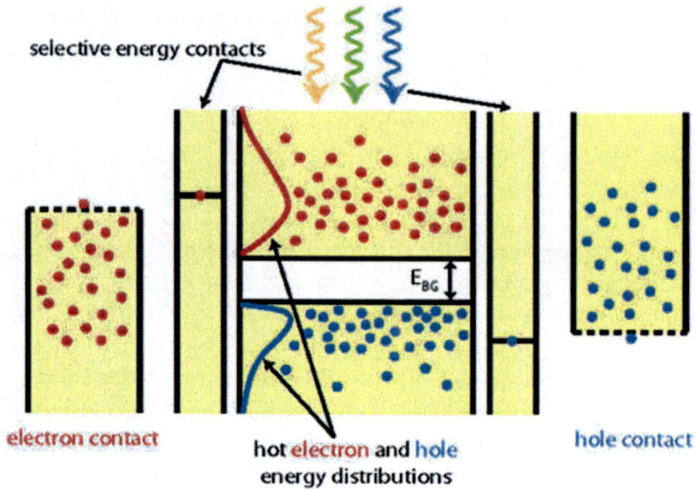

Fig. 10.14 Structure of a hot-carrier solar cell

is large. Under the condition of sufficiently high light injection, there may not be enough phonons to relax the photogenerated carriers, so the energy distribution of the carriers is still "hot" at steady state, that is, there are hot carriers. In addition, the strong reaction between high-energy phonons and hot carriers affects the equilibrium state of the distribution of phonons, so that the phonons themselves become hot, which eventually slows down the cooling process. However, this requires very high optical density, and the intensity requirements are orders of magnitude higher than the intensity of sunlight itself.

Reduced cooling rates can also be achieved in low-dimensional semiconductor structures. In the structure of a quantum well, carriers are confined in a two-dimensional space, and the quantization of the electronic structure generates sub-bands. Although the density of states is still continuous, it is indeed anisotropic, so that an excited electron can be scattered to a state with lower energy only when the momentum of the phonon is appropriate. In a quantum well with a GaAs/AlGaAs structure, the cooling time of hot carriers can be increased from 10 picoseconds in the GaAs bulk structure to 1,000 picoseconds in the quantum well structure when the optical intensity is 10^4 times higher than the sunlight.

Quantum dot structures also have great potential for use in hot-carrier solar cells. The electrons in a quantum dot are quantized into discrete energy levels due to the reduction of the dimensions in all directions. If the distance between these energy levels is higher than the maximum energy of the phonon, then the situation where the carrier is scattered from the excited state to a state with low energy by only one phonon cannot occur. The relaxation of carriers can only occur through a multi-phonon scattering process, which is relatively slow. In the quantum dot structure, a reduced cooling rate of hot carriers has been observed, which also occurs under the irradiation of the light with higher intensity, but its value is much lower than expected. This may be due to the existence of other relaxation processes such as Auger processes.

Another way to achieve hot-carrier solar cells is to use a structure that makes the separation of charges extremely fast in it. The charge transfer process occurring at the interface of the semiconductor electrolyte junction and the organic semiconductor heterojunction is very rapid. As in dye-sensitized cells, the charge transfer process can be less than 1 picosecond in some cases. After the charge transfer, the injected carriers are inevitably hot. Hot carriers can be used if they are extracted by the electrodes before they cool down. This requires a very thin charge transport layer to make the transit time of charges very fast. The difficulty lies in ensuring both high optical absorption and very fast extraction speed of charges.

In addition, the realization of hot-carrier solar cells also requires the use of selective electrodes with a very narrow band. Common contact materials such as metals and heavily doped semiconductors are not suitable, while low-dimensional semiconductor materials may achieve this function. One possible approach is to use a superlattice structure, in which sub-bands are formed. By selecting materials and designing the structural period, the sub-bands can be narrowed and be separated to achieve the required contact.

10.6 Collision Ionization Solar Cell

Now we discuss collision ionization solar cells. In the above discussion, we only discussed the relaxation phenomenon of hot carriers through interaction with phonons. In fact, hot carriers can also relax by generating electron–hole pairs through collision ionization.

The collision ionization process, or Auger process, is the opposite of Auger recombination. Auger recombination is a three-body process. In this process, one electron collides with another electron or impurity, thereby recombining with the hole to transfer the chemical potential energy to the second electron in the form of kinetic energy. In this opposite process, an electron collides with a crystal lattice, releasing its own kinetic energy to excite an electron–hole pair, as shown in Fig. 10.15. In a solar cell, this means that the quantum efficiency of the device will be greater than 1 when the energy of light is more than twice the E_g. High-energy photons can generate multiple electron–hole pairs. This effect has been observed when the energy of photons is greater than 2.5 eV ($E_g = 0.67$ eV) in Ge photodiodes and the energy of photons is greater than 3.3 eV ($E_g = 1.12$ eV) in Si diodes.

Energy and momentum must be conserved during the Auger process. Assume that the kinetic energy of electron e1 is E_{e1}, the momentum is k_{e1}, and the electron–hole pair generated by Auger is e2, h2. There are

$$k_{e1} = k'_{e1} + k_{e2} - k_{h2} \tag{10.44}$$

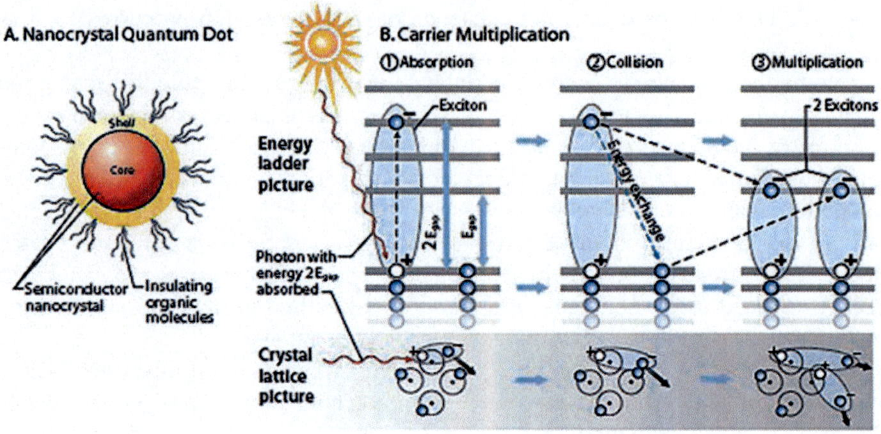

Fig. 10.15 The hot carriers in a quantum dot structure with precise controllable size can be multiplied by collision ionization

$$\mathrm{E}_{e1} = E_{e1}^{'} + \mathrm{E}_{e2} - \mathrm{E}_{h2} \tag{10.45}$$

Auger generation depends on the number of electron and hole states with appropriate k and E, so the band structure is very important to determine the probability of Auger generation.

Let's take a look at the ultimate efficiency of collision ionization solar cells. In order to meet the meticulous equilibrium conditions, the Auger recombination, the inverse process of collision ionization, needs to be considered. We assume that the relaxation time of carriers determined by the action of the phonon is slow enough that we ignore this factor. Therefore, although the photogenerated carriers can be in equilibrium, they will not be cooled, and the recombination of carriers can only be achieved by radiative recombination or Auger recombination. As can be seen from the previous section, the chemical potential of an equilibrium carrier pair depends on the energy of the carrier

$$\Delta\mu = \eta_{e0} + \eta_{h0} + \gamma(\mathrm{E}_e + \mathrm{E}_h) \tag{10.46}$$

When Auger recombination exists, a balance can be reached between Auger recombination and Auger generation. At equilibrium, the number of electron–hole pairs dn_1 with energy E_{eh1} and chemical potential $\Delta\mu_{eh1}$ is transformed into the number of electron–hole pairs dn_2 with energy E_{eh2} and chemical potential $\Delta\mu_{eh2}$. Since the energy conservation and free energy $\sum_i \mu_i dn_i$ need to be minimum, this means that the chemical potential must be proportional to the energy of carriers, i.e.,

$$\Delta\mu = \gamma(\mathrm{E}_e + \mathrm{E}_h) \tag{10.47}$$

Assuming the energy of the photon is E, the distribution of the carriers obeys

$$f = \frac{1}{e^{(E-\Delta\mu)/k_B T_a} + 1} \tag{10.48}$$

Here, T_a is the ambient temperature, and $\Delta\mu$ is as described above. Therefore, f can be written as

$$f = \frac{1}{e^{\frac{E(1-\gamma)}{k_B T_a}} + 1} \tag{10.49}$$

Define

$$T_H = \frac{T_a}{1-\gamma} \tag{10.50}$$

The distribution can be represented by the characteristic temperature T_H, and the chemical energy $\mu_H = 0$.

When calculating the energy conversion efficiency of a solar cell, since new electron–hole pairs are generated during the collision ionization process, it should be noted that the current at this time is no longer the difference between the current generated by incident sunlight and the current generated by radiative combination. Although the number of carriers is not conserved during the Auger process, the energy is still conserved. At this time, the current should be calculated by the principle of energy balance

$$JE_{out} = q\{Xf_s L(E_g, \infty, T_s, 0) + (1 - Xf_s)N(E_g, \infty, T_a, 0) - L(E_g, \infty, T_H, 0)\} \tag{10.51}$$

Output power is

$$P = JV = q\left(1 - \frac{T_a}{T_H}\right)\{Xf_s L(E_g, \infty, T_s, 0) + (1 - Xf_s)N(E_g, \infty, T_a, 0) - L(E_g, \infty, T_H, 0)\} \tag{10.52}$$

The output voltage V is related to the extracted energy E_{out}, which is calculated as

$$V = \frac{1}{q}\mu_{out} = E_{out}\left(1 - \frac{T_a}{T_H}\right) \tag{10.53}$$

In the above formula, we can see that although there are three parameters in the J-V relationship of the device, E_g, E_{out}, and V, in fact only two quantities change independently. The output power P depends on two parameters, E_g and T_H, and T_H is related to changing V or E_{out}. In other words, P depends on the same two variables. The determinants of ultimate efficiency in collision ionization solar cells are the same as in hot-carrier solar cells, so they should have the same ultimate efficiency. As in hot-carrier solar cells, as V changes from zero to V_{out}, that is, T_H gradually increases from zero during the process from short circuit to open circuit. T_H at the maximum power point depends on the intensity of the incident light.

For the non-light-condensing case, when the bandgap is 1 eV, efficiency is the largest, and the value is 55%. In the case of full condensing, when the bandgap approaches zero, the efficiency is the largest, with a value of 85%. In the case of full condensing, the characteristics are the same as those of hot-carrier solar cells. The ultimate efficiency of the solar cell at this time is determined by the thermodynamic limits.

It should be noted that, except for the case of full condensing, compared with hot-carrier solar cells, the limiting efficiency of collision ionization solar cells is lower. This means that if the energy band structure of the material facilitates collision ionization, it may not be desirable for efficient solar cells. However, since collision ionization solar cells do not need to optimize E_{out} to obtain maximum efficiency, that is, the selectivity requirements for the electrodes are not so high, it is easier to design and implement.

The collision ionization process and the phonon scattering process are two competing processes. In collision ionization solar cells, we need to suppress phonon scattering, so the methods discussed in hot-carrier solar cells to suppress phonon scattering and extend the "cooling" time of hot carriers are still applicable.

10.7 Summary

When considering the fine balance between light absorption and radiation recombination, the ultimate efficiency of single-junction solar cells is around 33%. This limit efficiency can be breached under the premise of relaxing the assumptions of one or more detailed balance principles. Considering the principles of thermodynamics, the ways to increase solar cells are as follows:

(1) Photons with different energies are absorbed by materials with different bandgaps, that is, laminated cell structures and solar cell structures with multiple bandgaps can be used;
(2) Quickly collect the photogenerated hot carriers before relaxation, using the excess kinetic energy of the carriers;
(3) Use the large energy of hot carriers to generate collision current to increase the number of electron-hole pairs.

The realization of all these solutions depends on effective semiconductor material systems. The ideal quality of the material is high enough to overcome non-radiative recombination. These materials should also have suitable electrical and optical properties. Nanomaterials, such as the low-dimensional structures and quantum dots in semiconductor, are important material structures for achieving high-performance solar cells [18, 19].

References

1. W. Shockley et al., J. Appl. Phys. **32**, 510–519 (1961)
2. G.L. Araújo et al., Sol. Energy Mater. Sol. Cells **33**, 213–240 (1994)
3. A. Luque, S. Hegedus, *Handbook of Photovoltaic Science and Engineering* (Wiley, Com, 2011)
4. C. Mathers, J. Appl. Phys. **48**, 3181 (1977)
5. J. Nelson, The physics of solar cells (Properties of semiconductor materials), Imperial College (2003)
6. P. Würfel et al., *Physics of Solar Cells: From Basic Principles to Advanced Concepts* (Wiley, Com, 2009)
7. R. Hulstrom et al., Sol. Cells **15**, 365–391 (1985)
8. M.A. Green, Prog. Photovoltaics Res. Appl. **9**, 137–144 (2001)
9. S. Bremner et al., Prog. Photovoltaics Res. Appl. **16**, 225–233 (2008)
10. A. Martí et al., Sol. Energy Mater. Sol. Cells **43**, 203–222 (1996)
11. K. Barnham et al., Appl. Phys. Lett. **59**, 135–137 (1991)
12. I. Kramer et al., Chem. Rev. **114**, 863–882 (2013)
13. P. Würfel, Sol. Energy Mater. Sol. Cells **46**, 43–52 (1997)
14. R.T. Ross et al., J. Appl. Phys. **53**, 3813–3818 (1982)
15. J.H. Werner et al., Appl. Phys. Lett. **67**, 1028–1030 (1995)
16. P. Landsberg et al., J. Phys. D: Appl. Phys. **35**, 1236 (2002)
17. D. König, et al., Phys. E, **42**, 2862–2866 (2010)
18. M.A. Green, Nanotechnology **11**, 401 (2000)
19. R.D. Schaller et al., Phys. Rev. Lett. **92**, 186601 (2004)